The movement of oceanic water has important consequences for a variety of applications, such as climate change, sealevel change, biological productivity, weather forecasting, and many others. This book addresses the problem of inferring the state of the ocean circulation, understanding it dynamically, and even forecasting it through a quantitative combination of theory and observation. It focuses on so-called inverse methods and related methods of statistical inference. Both time-independent and time-dependent problems are considered, including Gauss-Markov estimation, sequential estimators, and adjoint/Pontryagin principle methods.

This book is intended for use as a graduate-level text for students of oceanography and other related fields. It will also be of interest to working physical oceanographers.

THE OCEAN CIRCULATION INVERSE PROBLEM

THE OCEAN CIRCULATION INVERSE PROBLEM

CARL WUNSCH
Massachusetts Institute of Technology

CAMBRIDGE
UNIVERSITY PRESS

Published by the Press Syndicate of the University of Cambridge
The Pitt Building, Trumpington Street, Cambridge CB2 1RP
40 West 20th Street, New York, NY 10011-4211, USA
10 Stamford Road, Oakleigh, Melbourne 3166, Australia

First published 1996

Printed in the United States of America

Library of Congress Cataloging-in-Publication Data

Wunsch, Carl.
The ocean circulation inverse problem / Carl Wunsch.
p. cm.
Includes bibliographical references and index.
ISBN 0–521–48090–6 (hardcover)
1. Ocean circulation–Mathematical models. 2. Inverse problem (Differential equations). I. Title.
GC288.5.W86 1996 96–10921
551.47–dc20 CIP

A catalog record for this book is available from the British Library

ISBN 0 521 48090 6 Hardback

To Marjory, Jared, and Hannah

Contents

Preface		*page* ix
Notation		xiii
1	Introduction	1
1.1	Background	1
1.2	What Is an Inverse Problem?	7
1.3	What's Here	14
2	Physics of the Ocean Circulation	17
2.1	Basic Physical Elements	17
2.2	Observations	45
2.3	The Classical Problem	73
2.4	Hidaka's Problem and the Algebraic Formulation	82
2.5	The Absolute Velocity Problem in Retrospect	88
3	Basic Machinery	92
3.1	Matrix and Vector Algebra	92
3.2	Simple Statistics; Regression	101
3.3	Least Squares	113
3.4	The Singular Vector Expansion	133
3.5	Using a Steady Model–Combined Least Squares and Adjoints	170
3.6	Gauss-Markov Estimation, Mapmaking, and More Simultaneous Equations	179
3.7	Improving Solutions Recursively	203
3.8	Estimation from Linear Constraints–A Summary	210
4	The Steady Ocean Circulation Inverse Problem	212
4.1	Choosing a Model	215
4.2	The Initial Reference Level	217
4.3	Simple Examples	221
4.4	Property Fluxes	242
4.5	Application to Real Data Sets	249

4.6	Climatologies and Box Models	284
4.7	The $\boldsymbol{\beta}$–Spiral and Variant Methods	288
5	Additional Useful Methods	297
5.1	Inequality Constraints; Nonnegative Least Squares	297
5.2	Linear Programming and Eclectic Models	301
5.3	Quantifying Water Mass; Empirical Orthogonal Functions	306
5.4	Kriging and Other Variants of Gauss-Markov Estimation	311
5.5	Nonlinear Problems	312
6	The Time-Dependent Inverse Problem	324
6.1	Some Basic Ideas and Notation	327
6.2	Estimation	338
6.3	Control Problems: Pontryagin Principle and Adjoint Methods	362
6.4	Duality and Simplification: Steady-State Filter and Adjoint	380
6.5	Controllability and Observability	383
6.6	Nonlinear Models	385
6.7	Assimilation	391
6.8	Other Minimization Methods and the Search for Practicality	397
6.9	Forward Models	398
6.10	A Last Word	403
References		405
Author Index		427
Subject Index		435

Preface

Physical oceanography is a branch of fluid dynamics and is a part of classical physics. As such, the oceanographer's job is to produce quantitative descriptions and explanations of the behavior of the fluid ocean. The movement of oceanic water has consequences for a bewildering variety of applications–climate change; biological productivity; sealevel change; weather forecasting; fisheries prosperity; the chemical history of the earth; the dynamics of the earth–moon system; the movement of pollutants; and so forth. Understanding of the fluid circulation and the properties it carries is of great and growing importance.

Fluid flows are diverse and often very complicated. For this reason, most understanding of particular situations has resulted from an intimate partnership of theory with observation and with laboratory experimentation. But as compared to fluid dynamics as practiced in its innumerable applications–meteorology, aerodynamics, hydraulics, heat transfer, etc.–the problem of observing the ocean is particularly difficult, rivaled perhaps only by the observational problems of inferring the fluid properties of the earth's interior, or of other planets and of stellar interiors. The ocean is very large, turbulent, and inaccessible to electromagnetic radiation. Armed mainly with slow-moving, expensive ships, and instruments which have to work in a corrosive, high-pressure environment, oceanographers have over the years built up in somewhat painful fashion a picture of how the ocean operates. But the picture is known to be badly distorted by the very limited observational base, leading for example, to the need to assume that the large-scale fluid flow is steady with time, so that measurements obtained over many years could be combined in the inferential process. Much of the description available is only qualitative.

But physical oceanography is changing rapidly. New, global-scale, high-resolution general circulation models have been appearing which demand myriad more observations to initialize and to test. New technologies, which are capable of producing those observations, are now at hand. But observations will nonetheless remain expensive and scarce relative to the need. The question of what is best measured, and how to best exploit what data there are, will become even more urgent. Furthermore, as the question of the influence of the oceanic circulation on climate has come to the fore, governments have begun to insist upon the need for quantitative estimates of the future state of the ocean–its fluxes of heat and nutrients, sealevel rise, etc. The stakes are much higher than in the past.

This book is directed at a discussion of the problems of data analysis in the presence of dynamical models with a focus on the general circulation of the ocean. It is an amalgam of ocean dynamics, rudimentary statistics, and bits of linear algebra. The result should be practical for students, working oceanographers, and others faced with the problem of extracting quantitatively useful results from their hard-won observations. As oceanography has emerged as an important component of global change problems, it has attracted specialists from outside the field, for example, those entering from backgrounds in satellite altimeters, or acoustics, or meteorology. I have tried to provide some rough guidance between the Scylla of pure geophysical fluid dynamics and the Charybdis of the overwhelming observational detail already known, in an effort to help newcomers steer a safe and useful course.

But some of the theory of data analysis is interesting in its own right, for example, the Hamiltonian structure of the Riccati equation used for computing the error in sequential estimation. This theory is generally unknown to theoretically inclined oceanographers, and I hope that some of them will find it sufficiently interesting to push the frontiers of our understanding. Following the well-known book by Whittaker and Robinson (1944), we might call the subject "the calculus of observations."

In contrast to the recent work of Bennett (1992), the treatment is almost wholly devoted to finite-dimensional problems and methods. Experience suggests that finite-dimensional techniques are more readily accessible to students and working scientists. These techniques are sufficiently powerful and flexible in their own right that there are few, if any, practical problems which cannot be adequately handled in this way.

I have tried to be careful about terminology, although not claiming to be wholly consistent. The subject is difficult and confusing enough without the added burden of the use of terms whose meaning is unclear or ambiguous. So, for example, the term *barotropic*, which derives originally from the expression *autobarotropic* fluid (meaning one whose pressure and density surfaces coincide), has been so debased by sloppy usage that it can no longer be used without specific definition in context. Similarly, *Sverdrup-relation* is in grave danger of losing its precise meaning, to describe the relationship between the windstress-curl and the integrated meridional flow, and is slipping into the vague implication of validity of the linear vorticity balance. Such loss of precision diminishes the scientist's ability to express ideas in an efficient, clear manner.

Acknowledgments

I have received highly useful comments and corrections from numerous students who put up with early mistake- and typographical error–ridden versions of this book, used as class notes over several years. A number of colleagues read all or part of the book in great detail, and I am especially indebted to Detlef Stammer, Jochem Marotzke, Joseph Pedlosky, Jürgen Willebrand, Harry Bryden, and Andrew Bennett for saving me from many blunders and numerous obscurities. They are not to blame for those that remain. Particular thanks are due to Dimitris Menemenlis for a virtual line-by-line check of the whole book. Gordon (Bud) Brown patiently worked over the endless diagrams, and the book would not exist without the tireless word-processing skills of William McMechan, who made the initial conversion to TEX and then dealt with endless revisions–all without ever seeing me face-to-face–a working relationship possible only in the era of overnight mail service, fax machines, and e-mail. The sabbatical hospitality of the Geophysical Fluid Dynamics Laboratory, Princeton, and of the Group de Recherche en Géodésie Spatiale, Toulouse, France, contributed considerably to the completion of the manuscript.

Over many years I have had financial support from several government agencies, but it is a pleasure to acknowledge particularly the National Science Foundation and its many program managers who provided much of the funding that permitted me to learn the material described here. Specific funding to help produce this book came from the Massachusetts Institute of Technology through the Cecil and Ida Green Professorship.

Notation

$\mathbf{A}$	$N \times N$ model state transition matrix, or $p \times N$ model constraint matrix.
$A, A_v \ldots$	Coefficients of eddy diffusion (eddy coefficients).
$\mathbf{B}$	$N \times p$ distribution matrix for known model controls.
$\mathbf{C} \equiv < (\tilde{\mathbf{x}} - < \tilde{\mathbf{x}} >)(\tilde{\mathbf{x}} - < \tilde{\mathbf{x}} >)^T >$	$N \times N$ solution covariance matrix. Sometimes subscripted.
$\mathbf{E}$	$M \times N$ observation or design matrix.
$\mathbf{G}$	Green's function; also matrix of orthogonal vectors (Ch. 3).
J, J'	Objective functions.
M	Number of observations.
N	Size of the state vector.
$\mathbf{P} \equiv < (\tilde{\mathbf{x}} - \mathbf{x})(\tilde{\mathbf{x}} - \mathbf{x})^T >$	$N \times N$ solution uncertainty matrix.
$\mathbf{P}_{nn}$	$M \times M$ residual (noise) uncertainty matrix.
$\mathbf{Q} \equiv < (\tilde{\mathbf{u}} - \mathbf{u})(\tilde{\mathbf{u}} - \mathbf{u})^T >$	Model error second moment matrix.
$\mathbf{R}_{xy} \equiv < (\tilde{\mathbf{x}} - \mathbf{x})(\tilde{\mathbf{y}} - \mathbf{y})^T >$	Second moment matrix.
$\mathbf{S}$	Salinity distribution in a hydrographic section.
T	As superscript, the matrix transpose.
$\mathbf{T}$	Mass transport matrix in a hydrographic section (Ch. 4).
$\mathbf{T}_u$	$M \times M$ data resolution matrix.
$\mathbf{T}_v$	$N \times N$ solution resolution matrix.

$\mathbf{U}$	$M \times M$ matrix of $\mathbf{u}_i$ singular vectors.
$\mathbf{U}_K$	$M \times K$ matrix of first K $\mathbf{u}_i$ singular vectors.
$\mathbf{V}$	$N \times N$ matrix of $\mathbf{v}_i$ singular vectors.
$\mathbf{V}_K$	$N \times K$ matrix of first K $\mathbf{v}_i$ singular vectors.
c, b	Reference-level velocities (flows at a reference level used to computer u_R, v_R) so absolute geostrophic velocity is $u_R + c$, $v_R + b$.
$\mathbf{n}$	$M \times 1$ noise or residual vector.
$\mathbf{n}(t)$	$M \times 1$ noise or residual vector at time t, in which case $\mathbf{n}$ is $M \cdot$ (number of time steps) $\times 1$.
p	Pressure field.
t	Time variable, either continuous or discrete.
$\mathbf{v} = (u, v, w)$	Components of fluid velocity in (x, y, z) directions in Cartesian coordinates.
u_R, v_R	Geostrophic relative velocities (computed relative to a reference level).
$\mathbf{u}(t)$	Control vector (Ch. 6) in time-evolving models.
$vec(\ )$	Vector operator, converting a matrix to a vector by column stacking.
$(x,\ y,\ z)$	Cartesian coordinates.
$\mathbf{x}$	$N \times 1$ state vector.
$\mathbf{x}(t)$	$N \times 1$ state vector at time t, in which case $\mathbf{x}$ is $N \cdot$ (number of time steps) $\times 1$.
$\boldsymbol{\Gamma}$	Control matrix in model evolution equation.
$\boldsymbol{\Lambda}$	$M \times N$ diagonal matrix of singular values λ_i.
$\boldsymbol{\Lambda}_K$	$K \times K$ reduced version of $\boldsymbol{\Lambda}$.
α^2	Trade-off parameter.
θ	Temperature, or potential temperature.
λ	Longitude.
λ_i	Singular values.
$\boldsymbol{\mu}$	Vector of Lagrangian multipliers.
ρ	Fluid density.
σ	Standard deviation, sometimes with subscripts.
$\boldsymbol{\tau} = (\tau_x, \tau_y)$	Wind-stress vector.
ϕ	Latitude.
χ_v^2	"Chi-square" probability variate, with v-degrees of freedom.

1
Introduction

1.1 Background

Ocean modelers, in the formal sense of the word, attempting to describe the ocean circulation have paid comparatively little attention in the past to the problems of working with real data. Thus, for example, one drives models, theoretical or numerical, with analytically prescribed wind or buoyancy forcing without worrying overly much about how realistic such assumed forms might be. The reasons for approaching the problem this way are good ones–there has been much to learn about how the models themselves behave, without troubling initially about the question of whether they describe the real ocean. Furthermore, there has been extremely little in the way of data available, even had one wished to use, say, realistic wind and buoyancy flux fields.

This situation is changing rapidly; the advent of wind measurements from satellite-borne instruments and other improvements in the ability of meteorologists to estimate the windfields over the open ocean, and the development of novel technologies for observing the ocean circulation, have made it possible to seriously consider estimating the global circulation in ways that were visionary only a decade ago. Technologies of neutrally buoyant floats, long-lived current meters, chemical tracer observations, satellite altimeters, acoustical methods, etc., are all either here or imminent.

The models themselves have also become so complex (e.g., Figure 1–1a) that special tools are required to understand them, to determine whether they are actually more complex than required to describe what we see (Figure 1–1b), or if less so, to what externally prescribed parameters or missing internal physics they are likely to be most sensitive.

The ocean is so difficult to observe that theoreticians intent upon explaining known phenomena have made plausible assumptions about the behavior

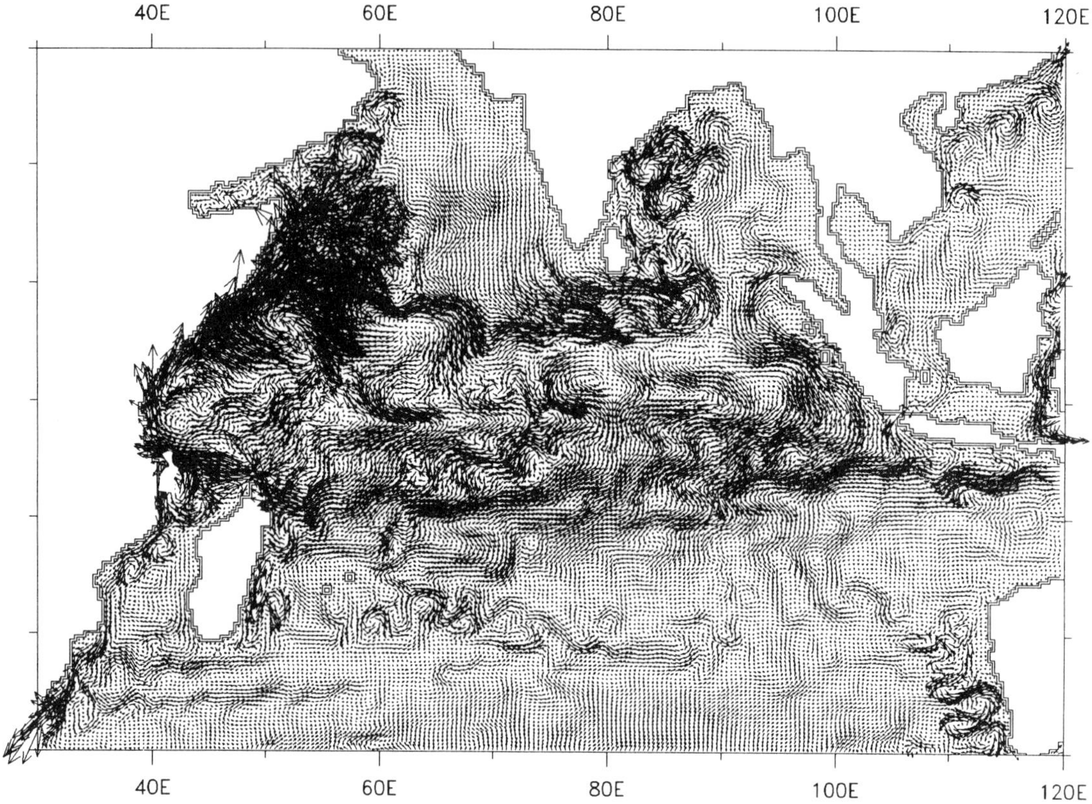

Figure 1–1a. The seasurface velocity on 15 September 1988 as estimated from a high-resolution global, oceanic general circulation model. Only a fraction of the global domain is depicted. The model is a nominal $^1/_4°$ latitude-longitude resolution version of the earlier computation described by Semtner and Chervin (1992). Much of the visual structure is time-dependent and raises very serious issues of observational sampling and of mathematical representation. (Courtesy of R. Tokmakian and A. Semtner.)

of the system and proceeded to construct systems of equations that are then solved. Some of these assumptions are indeed so plausible, and the resulting calculations so interesting, that it has become forgotten that they were assumed rather than demonstrated. The assumptions become elevated to the level of textbook dogma as known. One thinks immediately of the use of eddy coefficients in Laplacian diffusion/mixing terms, the assumption that Ekman layer divergences drive the large-scale interior circulation, that this circulation is in Sverdrup balance, that steady models are adequate descriptors of the circulation, etc. Consequently, the subject is rife with myths. Myths are important elements in human views of the world and often contain major components of historical truth. They have much that

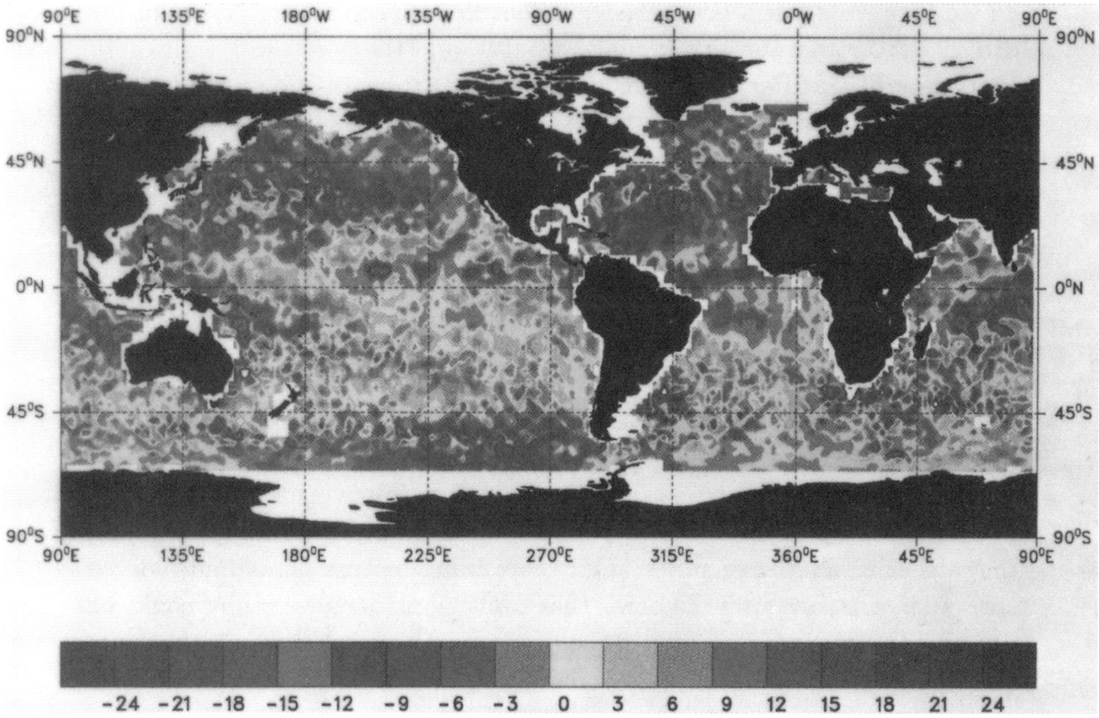

Figure 1–1b. Surface elevation of the ocean in cms during 10–20 March 1993 as seen by the TOPEX/POSEIDON altimeter satellite (after Stammer & Wunsch, 1994). The elevation is relative to a 2–year mean, has been averaged over 2° squares, and the 10–day "window" blurs the most rapidly changing features, therefore rendering the result somewhat simpler than a true instantaneous picture. Nonetheless, both data and model confirm the essentially turbulent nature of the circulation. (A color reproduction of Figure 1–1b is shown in the color insert.)

same character in science, but unless recognized for what they are, they ultimately inhibit progress because it is not always clear to the student where the critical questions lie, what is really known, and what is merely assumed for convenience.

One of the great achievements of oceanography and fluid dynamics, in partnership with meteorology and other branches of geophysics, was the creation in the period following about 1950 of what is now known as *geophysical fluid dynamics*. This field created a dynamical and mathematical framework for discussion of the circulation, one that hardly existed prior to that time (see the remarks by Stommel, 1982). The elegance and rigor of this branch of fluid mechanics has permitted the growth of discussions of theories of the circulation which are interesting, useful, and even beautiful. But a side effect of this "applied mathematics of oceanography" has been to create two, sometimes nearly independent and contradictory, views of the ocean: the ocean as observers understand it, and the ocean as the theoreticians describe it. Part of my motivation in writing this book has been to try to bring these two parts of the subject back into a closer relationship. I believe that we will progress most rapidly by continually calling attention to the real or apparent discrepancies between the theoretical picture and what we think observations imply.

Much of what we think we know about the ocean at any time is a direct consequence of the technology available at that point in history. One can usefully recall that in the Nansen bottle era, everyone knew that oceanic profiles of temperature and salinity were smooth functions of depth. The demonstration beginning in the middle 1960s with salinity/conductivity-temperature-depth profiling devices (STDs and CTDs) of what is now called *fine* and *microstructure* was initially greeted in many quarters with firm disbelief (the observations being attributed to faulty instruments). Until the advent of long time series of currents from drifting floats and moored current meters, everyone "knew" that nonsynoptic hydrography could be combined to produce a picture of the ocean circulation that was both steady and a true climatological average.

In general, the message is that as in all science, our understanding of the ocean comes through a distorting prism of our observational and theoretical technology; the student should maintain a very substantial degree of skepticism about almost anything said in textbooks about the ocean (beyond its being wet, salty, describable by the Navier-Stokes equations, and interesting), including this one.

Finally, oceanography is now struggling to become a true global science. The field is moving rapidly from what many now view with nostalgic regret as the romantic period of exploration–when "real" oceanographers were those who made all their own observations at sea and whose interpretations were limited mainly to drawing pictures. We are entering an era when the subject will necessarily become more like meteorology–with global data bases, obtained from observational networks run by large, impenetrable governmental agencies. The science will become less romantic, regrettably so (although like much of nostalgia, the romance of oceanography is often more apparent in the retelling than it was at the time). The excitement will still be there (it will be intellectual excitement), we will learn much more about what is going on, cherished myths will go on the trash heap, and new views of the ocean will surely emerge.

Some scientific fields, where there is little or no supporting theoretical framework, develop world views or paradigms through a process of scenario development: Observations are made and a plausible story is told about how those data might be explained.[1] Subsequent observations may lead to elaborations of the original story, but a new or conflicting story is not told unless there is overwhelming evidence that the old one cannot be sustained. Fields that develop in this mode include geology and much of descriptive physical oceanography.

Until the development of geophysical fluid dynamics, physical oceanography did not have an adequate theoretical framework, and the scenario approach was the only one possible. Despite the existence of detailed understanding of fluid flows, much of physical oceanography has continued in the story-telling mode, as perusal of most any issue of a modern journal will show. The development of inverse and related methods–the focus of this book–is an attempt to shift the emphasis from the scenario-based methodology to one whose emphasis is more upon the quantitative (that is, numerical) testing of data against the known equations of fluid dynamics (and of chemistry and biology where available).[2] The mere use of numbers does not, however, embody the required shift: One often sees scenario-based discussions of the ocean circulation based on use of many millions of numbers acquired from computer models. It is the quantitative testing, and potential rejection, of models and scenarios using estimation methods that is required. It is only in this way that one can expect to see a convergence upon a consensus view of the ocean circulation, and an escape from the present circumstance in which dozens of apparently conflicting estimates of the circulation exist. As more data and understanding accumulate, uncertainty estimates should shrink, and prior estimates would be identified as either in contradiction with the new picture or consistent within the uncertainties.

Over the years, oceanography has benefited from being a junior partner of meteorology. Meteorology is much more mature, as a result of the technically easier problem of observing the atmosphere and the drive by governments to forecast the weather. There are many analogies between the two fluid systems (e.g., Charney & Flierl, 1981) and for which meteorological insight has proven most helpful to understanding the much more opaque,

[1] Some of these scenarios are reminiscent of Rudyard Kipling's *Just-So Stories*: "How the Leopard Got His Spots," "How the Camel Got His Hump," etc. We have "How the Ocean Circulates...."

[2] Recall the words of William Thomson, Lord Kelvin: "When you can measure what you are speaking about, and express it in numbers, you know something about it; but when you cannot measure it, when you cannot express it in numbers, your knowledge is of a meager and unsatisfactory kind: it may be the beginning of knowledge, but you have scarcely in your thoughts, advanced to the stage of *science*."

literally and figuratively, ocean. As one looks to the future, however, some divergences are apparent.

Much meteorological effort is directed at operational forecasting problems. It is not pejorative to characterize this work as engineering in part. If a forecasting technique works well, one should use it, whether or not it is fully understood. If the forecasts are going astray, the forecaster normally gets fast and vehement feedback from the public. Thus, what the meteorologist refers to as *assimilation*, the combination of model analysis with data for the purpose of making the best forecast, has tended to pay comparatively little attention to the uncertainty estimates of the process–if the forecasts work well, one knows it and that is enough. But the oceanographer has no such clientele (at the moment), the goals are directed more at understanding the system than forecasting it, and when I deal with assimilation methods, much more emphasis is placed upon sensitivity and uncertainty than a meteorologist might regard as necessary or even sensible. Here, the view is taken that because the error or uncertainty estimates describe what we don't know, they may be more important than the formal solution to a set of model equations. So some of this book addresses techniques that differ from normal meteorological practice.

The book is intended to be at a level accessible to first- or second-year graduate students with only the beginnings of a knowledge of ocean dynamics. I hope, too, that it will prove a useful guide to workers on the edge of physical oceanography, including biologists and chemists, as well as those entering from fields such as satellite altimetry and scatterometry, who seek some guidelines as to what is important and what is peripheral to this subject.

Beginning in the middle 1970s, it began to be realized that the classical *dynamic method* could be readily extended to produce absolute estimates of the ocean circulation. The thermal wind balance, coupled with such simple statements that mass and salt are conserved, largely sets the strength of the ocean circulation. The gross tilt of thermocline throughout the world ocean is one of the best known of all oceanic phenomena. It sets the clock that determines how fast the system is moving and what the ocean transports and exchanges with the atmosphere.

To some extent, this book is a tribute to the power of geostrophy–in the form of the thermal wind balance. The simple assertion that the vertical derivatives of the horizontal velocity are proportional to the horizontal derivatives of the density field is so familiar that the enormous quantitative power of the relationship tends to be submerged in discussions of measurements that are technically more interesting (e.g., of transient tracers).

But temperature and salinity are tracers, too; they are the easiest of all to measure; we have much better coverage of the world ocean for them than anything else; they are an intimate element of the global climate system; and their immediate relation to the density field means that they must be the central focus of any effort to understand the general circulation.

1.2 What Is an Inverse Problem?

What I mean by the title of this book, *The Ocean Circulation Inverse Problem*, is the problem of inferring the state of the ocean circulation, understanding it dynamically, and even perhaps forecasting it, through a quantitative combination of theory and observations. It may help the reader in what follows to understand why it is called an inverse problem and why the label is partially a misnomer, and to connect the ocean circulation problem to the many other such problems both in oceanography and in science as a whole. In particular, I wish to emphasize the difference between an inverse problem and what we are really discussing in this book–what are called inverse *methods*.

We digress slightly and explore some conventions of mathematics as applied to familiar differential systems, relying heavily on the reader's experience. Consider a very familiar problem:

> Solve
>
> $$\nabla^2 \phi = \rho \tag{1.2.1}$$
>
> for ϕ, given ρ, in the domain $\mathbf{r} \in D$, subject to the boundary conditions $\phi = \phi_0$ on the boundary ∂D, where $\mathbf{r}$ is a spatial coordinate.

This statement is the Dirichlet problem for the Laplace-Poisson equation, whose solution is well-behaved, unique, and stable to perturbations in the boundary data, ϕ_0, and the source or forcing, ρ. Because it is a familiar boundary value problem, it is labeled a *forward* (or *direct*) problem.

Now consider a different version of the above: Solve (1.2.1) for ρ given ϕ in the domain D.

This latter problem is even easier to solve than the forward problem: merely differentiate ϕ twice to obtain the Laplacian, and ρ is obtained directly from (1.2.1). Because the problem as stated is inverse to the conventional forward one, it is labeled an *inverse problem.* It is inverse to a more familiar boundary value problem in the sense that the usual unknowns ϕ have been inverted or interchanged with (some of) the usual knowns ρ. Notice

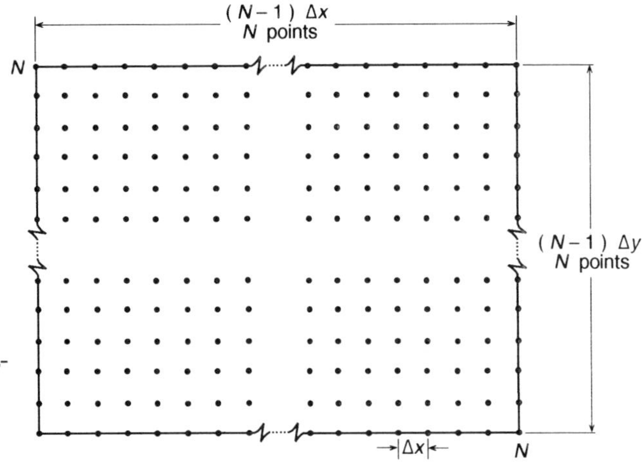

Figure 1–2. Simple square, homogeneous grid used for discretizing the Laplacian, reducing the differential equation to a set of simultaneous equations.

that both problems, as posed, are well-behaved and produce uniquely determined answers (ruling out mathematical pathologies in any of ρ, ϕ_0, ∂D, or ϕ). This is not the only inverse problem that could be set; we could, for example, demand computation of the boundary conditions, ϕ_0, from given information about some or all of ϕ, ρ.

Because this book is based upon discrete methods, write the Laplace-Poisson equation in finite difference form for two Cartesian dimensions:

$$\phi_{i+1,j} - 2\phi_{i,j} + \phi_{i-1,j} + \phi_{i,j+1} - 2\phi_{i,j} + \phi_{i,j-1} = \rho_{ij}, \qquad i, j \in D . \quad (1.2.2)$$

To make the bookkeeping as simple as possible, suppose the domain D is the square $N \times N$ grid displayed in Figure 1–2, so that ∂D is the four line segments shown. There are $(N-2) \times (N-2)$ interior grid points, and Equations (1.2.2) are then $(N-2) \times (N-2)$ equations in N^2 of the ϕ_{ij}. If this is the forward problem with ρ_{ij} specified, there are fewer equations than unknowns. But if we append to (1.2.2) the set of boundary conditions:

$$\phi_{ij} = \phi^0_{ij} , \qquad i, j \in \partial D , \quad (1.2.3)$$

there are precisely $4N - 4$ of these conditions, and thus the combined set (1.2.2) plus (1.2.3), which we write as

$$\mathbf{A}_1 \boldsymbol{\phi} = \mathbf{d}_1 , \quad (1.2.4)$$

is a set of $M = N^2$ equations in $M = N^2$ unknowns, with

$$\boldsymbol{\phi} = \begin{bmatrix} \phi_{11} \\ \phi_{12} \\ \cdot \\ \cdot \\ \phi_{NN} \end{bmatrix}, \qquad \mathbf{d}_1 = \begin{bmatrix} \rho_{11} \\ \rho_{12} \\ \cdot \\ \cdot \\ \rho_{N-2,N-2} \\ \phi^0_{11} \\ \cdot \\ \phi^0_{ij} \end{bmatrix}.$$

The nice properties of the Dirichlet problem can be deduced from the well-behaved character of the matrix $\mathbf{A}_1$. Thus the forward problem corresponds directly with the solution of an ordinary set of simultaneous algebraic equations (Lanczos, 1961, has a much fuller discussion of this correspondence).

A complementary inverse problem says: "Using (1.2.4), compute ρ_{ij} and the boundary conditions, given ϕ_{ij}," an even simpler computation–it involves just multiplying the known $\boldsymbol{\phi}$ by the known matrix $\mathbf{A}_1$. The problem can be written formally as

$$\mathbf{A}_2\mathbf{d}_1 = \mathbf{d}_2\,,$$

where $\mathbf{A}_2$ is the identity, and $\mathbf{d}_2 = \mathbf{A}_1\boldsymbol{\phi}$. Presumably all this is obvious.

But now let us make one small change in the forward problem, changing it to the Neumann problem:

Solve

$$\nabla^2\phi = \rho \tag{1.2.5}$$

for ϕ, given ρ, in the domain $\mathbf{r} \in D$ subject to the boundary conditions $\partial\phi/\partial\mathbf{m} = \phi_0'$ on the boundary ∂D, where $\mathbf{r}$ is a spatial coordinate and $\mathbf{m}$ is the normal to the boundary.

This new problem is another classical, much analyzed forward problem. It is, however, well-known that the solution to (1.2.5) with these new boundary conditions is indeterminate up to an additive constant. This indeterminacy is clear in the discrete form: Equations (1.2.3) are now replaced by

$$\phi_{i+1,j} - \phi_{i,j} = \phi^{0'}_{ij}\,, \qquad i,\, j \in \partial D' \tag{1.2.6}$$

etc., where $\partial D'$ represents the set of boundary indices necessary to compute the local normal derivative. There is a new combined set:

$$\mathbf{A}_3\boldsymbol{\phi} = \mathbf{d}_3\,. \tag{1.2.7}$$

Because only *differences* of the ϕ_{ij} are specified, there is no information concerning the mean value of ϕ. When we obtain some proper machinery in Chapter 3, we will be able to demonstrate that even though (1.2.7) appears to be M equations in M unknowns, in fact only $M-1$ of the equations are independent, and thus the Neumann problem is an underdetermined one. This property of the Neumann problem is well-known, and there are many ways of handling it, either in the continuous or discrete forms. In the discrete form, a simple way is to add one equation setting the value at any point to zero (or anything else).

Notice however, that the inverse problem remains unchanged, well-posed, and unique. Although somewhat trivial in form, we can write it as a set of simultaneous linear equations, by mere rearrangement of (1.2.7),

$$\mathbf{A}_4\mathbf{d}_3 = \mathbf{d}_4 \tag{1.2.8}$$

where $\mathbf{A}_4$ is the identity matrix and $\mathbf{d}_4 = \mathbf{A}_3\phi$.

These are examples in which a forward problem is badly posed in the sense of missing a piece of information necessary to determine the solution uniquely, while the inverse problem is fully posed. This point is labored a bit, because later we will encounter some inverse problems that are also missing some of the relevant information. Among the techniques for solving problems which are underdetermined are a class sometimes known as *inverse methods*. But one must carefully distinguish between inverse *problems* and the *methods* available for solving undetermined systems of equations, whatever the origin of the problem–whether forward or inverse.

Many examples of inverse problems are mathematically and computationally well behaved. One elegant example is Abel's problem (Aki & Richards, 1980): An observer stands on the floor of a symmetric valley and wishes to determine the shape of the valley. For reasons we need not inquire into, his method is to set a ball in motion, with initial speed s_0; he then measures the time it takes for the ball to return to him, yielding a set of travel times $t(s_0)$. Abel (1826) showed that the shape of the valley, $f(x)$, could be obtained as the solution to the integral equation (nondimensionalized):

$$t(s_0) = \int_0^{s_0} \frac{f(\xi)}{\sqrt{s_0 - \xi}} d\xi \tag{1.2.9}$$

(an Abel integral equation), with solution

$$f(\xi) = -\frac{1}{\pi}\frac{d}{d\xi}\int_\xi^a \frac{t(x')}{\sqrt{x' - \xi}} dx' , \tag{1.2.10}$$

a stable, well-behaved solution. So the Abel problem is an example of an inverse one that is well-behaved [but it involves a differentiation, and (1.2.10) may be undesirable if real, and not mathematical, data are being used].

Another familiar example of a *forward* problem with missing information is found in Maxwell's equations. It is well known that the complete solution for the magnetic field B can be written

$$B = \nabla \times \mathbf{A} + \frac{\partial q}{\partial t} \tag{1.2.11}$$

where $\mathbf{A}$ is the vector potential and the scalar potential q is indeterminate (gauge transformations make differing choices of q to suit the convenience of the investigator; see Jackson, 1975).

One might call indeterminacies of this type *structural*–some piece of information that could render the solution unique is missing. Another closely related indeterminacy might be called *statistical.* As an example, consider again the Dirichlet problem; as stated above, it has a unique, stable solution. But suppose the boundary data are given as

$$\phi = \phi_0 + n \text{ on } \partial D \tag{1.2.12}$$

where n is a noise element, representing an uncertainty of the boundary conditions. To the extent that n is commonly describable only through certain statistical properties, for example, its mean and variance, there is no longer a unique solution to (1.2.1) subject to (1.2.12). Rather, one could find a smoothest solution, or one of minimum mean square, or even a most probable one, or a maximum or minimum one. But no solution is then definitive; that is, it is not uniquely determined without some auxiliary requirement (like smoothest). For such reasons, all problems that mix observations with models are inevitably ill-posed.

Although ill-posedness owing to a lack of sufficient information is prominent in oceanography, other forms of ill-posedness arise and are very important. The problem of too much information–overspecification–is one type. Consider the simple example of the well-posed Dirichlet problem (1.2.4). With the extra information that $\phi_{i_0,j_0} = \bar{\phi}$ at some interior grid point, i_0, j_0, perhaps obtained from an observation, we have too much information and the problem is ill posed. In Chapter 6, we will encounter an initial value problem in the sense of Cauchy in which the existence of information about the solution at a later time also overspecifies the problem. Then there is ill-posedness in the sense that solutions are unstable to slight perturbations in some problem parameter, be it initial conditions, boundary conditions, or some coefficient. In some problems, more than one of these conditions of ill-posedness exists simultaneously, and the ability to solve them is of great practical importance.

Many of the problems encountered in this book will be at least approximately representable by a set of linear simultaneous equations, written commonly as

$$\mathbf{Ex} + \mathbf{n} = \mathbf{y}$$

where $\mathbf{n}$ is used to show explicitly that $\mathbf{y}$ is typically imperfectly known, containing noise elements, n_i, and the problem is to understand the solutions $\mathbf{x}$. We will see cases in which there are precisely as many equations, fewer equations, and more equations as there are unknown elements of $\mathbf{x}$. Whether the equations arose from a forward problem, like the Laplace-Poisson equation, or through an inverse problem, is essentially irrelevant. Depending upon the conditioning of the equations, the extent of our statistical knowledge of $\mathbf{n}$ or of $\mathbf{x}$ and our needs, we may or may not need an inverse *method* to solve the problem. Notice especially that $\mathbf{n}$ is a set of unknowns, too. Adding the number of elements of $\mathbf{x}$ to the number of elements of $\mathbf{n}$ shows immediately that there are *always* fewer equations than unknowns in such problems.

Much fuss is made here about the problem of determining the error, or better, uncertainty, of solutions. "Mathematical inverse problems" deal with perfect data–which are unachievable in the real oceanographic world. Useful inverse methods *always* recognize and deal with the problem of imperfect data.

The importance of making uncertainty estimates can hardly be overestimated. Consider a simple, familiar problem. We have five equations in three unknowns:

$$\begin{aligned}
1.2550x_1 + 1.6731x_2 - 1.3927x_3 &= .3511\\
0.4891x_1 + 0.0943x_2 - 0.7829x_3 &= -1.6710\\
-0.1755x_1 + 1.8612x_2 + 1.0972x_3 &= 6.838\\
0.4189x_1 + 0.2469x_2 - 0.5990x_3 &= -.8843\\
-0.2900x_1 + 0.7677x_2 + 0.8188x_3 &= 3.7018\,,
\end{aligned} \tag{1.2.13}$$

which can be solved by ordinary least squares, and the solution so obtained is $x_1 = 1.0000$, $x_2 = 2.0000$, $x_3 = 3.0000$. Such a solution is very neat, and the investigator will surely be tempted to say "this is a nice solution and so it must be the right one." Unfortunately, if the value on the right-hand side of the first equation is shifted from 0.3511 to 0.3600, the least-squares solution is now $x_1 = 1.9843$, $x_2 = 1.7482$, $x_3 = 3.5851$, and unless one is certain that the right-hand value of the first equation is really exactly 0.3511, and that all the other values are perfectly known, the first solution, tidy and attractive as it is, and despite the fact that it was obtained by a familiar, simple methodology, should be regarded as misleading and irrelevant. If one published the first solution (or more interesting analogues in more complicated problems), one runs the risk of appearing very foolish. But the remedy for guarding against such foolishness is simple. Suppose the right-hand side of (1.2.13) were believed accurate to ± 0.1. Then a straightforward estimate of the uncertainty (Section 3.3) shows that the first solution should really be written

$$x_1 = 1.0 \pm 26,\ x_2 = 2.0 \pm 6.8,\ x_3 = 3.0 \pm 16\,,$$

which gives one a very different impression. One often sees published calculations of fields made using models combined with observations and whose appearance pleases the author, who takes that fact as an indicator that the solution should be seriously discussed. It is worth recalling here the comment of Thomas H. Huxley in 1869:

> I do not presume to throw the slightest doubt upon the accuracy of any of these calculations made by such distinguished mathematicians as those who have made the suggestions I have cited.[3] On the contrary, it is necessary to my arguments to assume that they are all correct. But I desire to point out that this seems to be one of the many cases in which the admitted accuracy of mathematical processes is allowed to throw a wholly inadmissible appearance of authority over the results obtained from them. Mathematics may be compared to a mill of exquisite workmanship, which grinds you stuff of any degree of fineness; but nevertheless, what you get out depends on what you put in; and as the grandest mill in the world will not extract wheat-flour from [peapods], so pages of formulae will not get a definite result out of loose data.

Returning now to the question posed above, what makes an inverse problem? An inverse problem is one that is inverse to a corresponding forward or direct one, interchanging the roles of at least some of the knowns and unknowns. A forward problem might be thought of as one of the familiar well-posed problems of mathematical physics. But as has already been indicated, any such problem when discussed in the context of real data, is automatically and irretrievably ill posed and thus a candidate for solution by an inverse method.

Inverse methods in a solid-earth geophysics context are associated with

[3] Huxley is discussing 19th century estimates of the age of the earth, which turned out to be gross underestimates of the true age. See the interesting discussion in Körner (1988), who quotes this passage. Ironically, most conspicuous among the "mathematicians" referred to was Lord Kelvin, who was quoted above. Science isn't easy.

the pioneering work of Backus and Gilbert (1967, 1968, 1970). These interesting papers are not discussed here; their method has been discussed at length by Parker (1977, 1994), Aki and Richards (1980), and others. The method as they laid it out is conceptually more difficult than those described here, being basically one of functional analysis (see Backus, 1970a,b and Bennett, 1992). Menke (1989) gives a brief elementary account. Experience suggests that finite-dimensional vector spaces are more intuitively accessible and adequate for most everyday scientific use. Tarantola's (1987) book might be thought of as a hybrid, focused on Bayesian inference. Backus (1988a) has an interesting and accessible discussion contrasting Bayesian methods with those that are used here. Most classical geophysical inverse problems are *static*–that is, the medium under study does not change with time. But compared to solid-earth geophysics, oceanography differs profoundly because it deals with a fluid medium undergoing constant change. The challenge of such problems makes the oceanographic inverse problem considerably more difficult than for the solid earth, partially mitigated by the availability of dynamical models, based on the equations of fluid mechanics.

1.3 What's Here

Chapter 2 is devoted to a discussion of the most basic elements of the large-scale ocean circulation, adequate to appreciate the observational issues. The assumption is that the reader has a background in dynamics at about the level discussed in Pond and Pickard (1983)–that is, the reader knows about geostrophy, Ekman layers, Sverdrup balance, etc. The chapter takes some pains to discuss the extent to which these simple dynamical elements are supported by observations. The stage is then set for discussion of the simplest ocean circulation inverse problems, by showing their structure as sets of simultaneous equations.

Chapter 3 addresses some of the basic elements of linear algebra and statistics required to understand estimates made from sets of simultaneous equations derived from noisy, incomplete observations. But the discussion provided cannot be a substitute for one of the many excellent textbooks available in these areas, some of which are mentioned in the appropriate chapters. I hope that enough information has been provided to make the book intelligible and perhaps to stimulate the interested reader to master the basics by following them up in the references.

The approach is to provide examples that are nearly trivial in scale compared to the real problems one wishes to study. In general, it is possible to find small problems, "toy models" for example, involving matrices that are 5×5 and that exhibit the conceptual issues raised by problems in which the matrices are $10^6 \times 10^6$, but which are obviously much more easily studied on a small computer. Among the problems we address here, many have a form analogous to the set of simultaneous linear equations $\mathbf{Ex} = \mathbf{y}$ and for which we can write trivially $\mathbf{x} = \mathbf{E}^{-1}\mathbf{y}$. The existence of this solution in a mathematical sense is sometimes not in doubt. But if the matrix really does have 10^{12} elements, its construction, much less its inversion, may be far removed from anything one could do practically. There are two distinguishable issues; the first, which this book is mainly about, concerns the interpretation of the mathematical operation $\mathbf{E}^{-1}$ under different circumstances. As will be seen, there are many interesting problems here that need to be understood. Only then can one sensibly grapple with the question of whether sense can be made of problems in which matrices with 10^{12} elements are desirably inverted. One might also notice that the maximum size for which operations like matrix inversion can be employed is constantly increasing. Until the advent of computers, the inversion of even a 5×5 matrix would make any analyst hesitate, knowing he is in for hours of work. Today, the inversion of 100×100 matrices is routine on personal computers, supercomputers can easily handle matrices larger than 5000×5000, and we can anticipate these dimensions to grow continually in the future. This size is still a far cry from the super-large values required for oceanic general circulation models, but we are getting there.

Chapter 4 then returns to the problem of determining the quasi-steady, nearly geostrophic, ocean circulation, employing the available mathematical machinery. The emphasis is on the many practical issues of solution and interpretation that arise. Treatment of the ocean circulation as steady is ultimately indefensible–insistence upon it leads to distortion of the physics, kinematics, chemistry, and biology. Nonetheless, the time-dependent problems taken up in Chapter 6 cannot be understood without the foundation of the static problems. Chapter 5 discusses various additional topics related to the problem of determining the ocean circulation, including empirical orthogonal functions, linear programming, and other methods. The remainder of the book turns then to the very interesting problems of determining the time-varying circulation, discussing in turn sequential estimation (the filtering/smoothing problem), and its close cousin, the solution through the Pontryagin principle (adjoint modeling).

It is equally important to state what is not here. This book is not a monograph on the general circulation of the ocean. Although there is much discussion of the circulation, the intention is mainly to discuss *methods* of estimation for the circulation problem. A full discussion of what is known about the oceanic circulation warrants a volume of its own, which might someday exist.

The book is also not a textbook of linear algebra or statistics. Many good books exist on these subjects, and although a heuristic description is given of tools derived from these areas, there is no attempt to be either rigorous or complete. Among the wholesale omissions here, made to keep the volume from getting completely out of hand, are the extremely important problems of hypothesis testing, the statistical tests of adequacy of regression models, and formal Bayesian methodologies.

2

Physics of the Ocean Circulation

2.1 Basic Physical Elements

2.1.1 Equations of Motion

The ocean circulation is governed by Newton's laws of motion plus those of the thermodynamics of a heat and salt stratified fluid. That these physics govern the system is a concise statement that a great deal is known about it. Our problem is to exploit this knowledge to the fullest extent possible in the context of anything we can observe about the ocean. Elaborate theoretical studies of the ocean circulation exist; here we assume that the reader has a knowledge of this theory at a basic level, with a working knowledge of Ekman layers, geostrophy, the simplest theories of western boundary currents, and the existence of internal waves and similar phenomena. Extended treatments are provided by Fofonoff (1962), Phillips (1963), Veronis (1981), Gill (1982), Pedlosky (1987a), and others.

The full equations of motion describing the ocean are the Navier-Stokes equations for a thin shell of temperature and salinity–stratified fluid on a bumpy near-spheroidal body undergoing rapid rotation. Appropriate boundary conditions are those of no flow of fluid into the bottom and sides, statements about the stress exerted on these boundaries, and those appropriate to exchange of momentum, heat, and moisture with the atmosphere at the surface. It is not really possible to separate the study of the ocean and atmosphere; a rigorous treatment must describe the movement of both fluids together, but both meteorology and oceanography remain at a stage of understanding where a great deal is to still to be learned by discussing them separately.

A useful approximate set of equations (Newton's laws of motion) on a spherical earth rotating at rate Ω is of form (e.g., Phillips, 1966; Veronis, 1981; Gill, 1982; Pedlosky, 1987a)

$$\frac{du}{dt} - \frac{uv\tan\phi}{a} + \frac{uw}{a} - 2\Omega\sin\phi v = \frac{1}{a\cos\phi}\frac{\partial p}{\rho\partial\lambda} + F_u \qquad (2.1.1)$$

$$\frac{dv}{dt} + \frac{u^2\tan\phi}{a} + \frac{vw}{a} + 2\Omega\sin\phi u = -\frac{1}{a}\frac{\partial p}{\rho\partial\phi} + F_v \qquad (2.1.2)$$

$$\frac{dw}{dt} - \frac{u^2+v^2}{a} = -\frac{\partial p}{\rho\partial z} - g \qquad (2.1.3)$$

$$\frac{1}{a\cos\phi}\left(\frac{\partial u}{\partial\lambda} + \frac{\partial(v\cos\phi)}{\partial\phi}\right) + \frac{\partial w}{\partial z} = 0 \qquad (2.1.4)$$

$$\frac{d\rho}{dt} = \frac{\partial\rho}{\partial t} + \frac{u}{a\cos\phi}\frac{\partial\rho}{\partial\lambda} + \frac{v}{a}\frac{\partial\rho}{\partial\phi} + \frac{w\partial\rho}{\partial z} = 0\,, \qquad (2.1.5)$$

where

$$\frac{d}{dt} \equiv \frac{\partial}{\partial t} + \frac{u}{a\cos\phi}\frac{\partial}{\partial\lambda} + \frac{v}{a}\frac{\partial}{\partial\phi} + w\frac{\partial}{\partial z}\,.$$

The latitude is ϕ, longitude is λ, the radius is a, the radial position is $r = a+z$ (see Figure 2–1), and u, v, w are the velocities in the λ, ϕ, r directions, respectively, ρ is density, and p is the pressure. These equations have been simplified by neglecting molecular viscosity and the nonsphericity. The so-called Boussinesq approximation[1] has been made only in part, in (2.1.4), primarily because in the dynamic method, described below, it has been customary to retain the full density field in the computations. The flow field is supposed to represent large-scale movement of fluid in the presence of a smaller-scale turbulent motion represented in F_u, F_v and whose details are regarded as inaccessible. These terms are intended to represent mechanisms by which the large scales are broken down into smaller ones in a process normally considered to be primarily dissipative.

Equation (2.1.4) is often called the incompressibility condition, but more precisely it expresses approximate mass conservation [the thermodynamic equations can still include compressibility effects without destroying the validity of (2.1.4)]. Equation (2.1.5) represents density conservation derived from the first law of thermodynamics. For a general discussion, see Batchelor (1967) or Kamenkovich (1977).

If we are to discuss processes that represent changes in density owing to diffusivelike behavior of the temperature and salinity controlling the density distribution, Equation (2.1.5) may have to be modified so that the right-hand side no longer vanishes. Determining the most useful modification is

[1] The Boussinesq approximation treats the density field as constant, except where it directly multiplies gravity, g. It effectively ignores the inertial mass variations in the momentum equations and is generally an excellent oceanographic approximation.

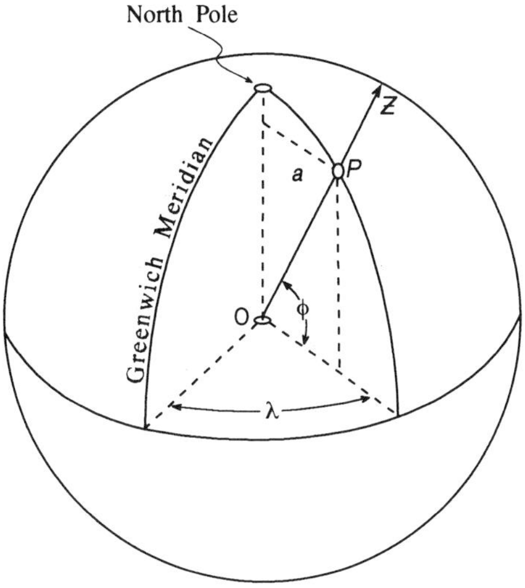

Figure 2–1. Spherical geometry used to define the elements of the dynamical equations: a is the radius of the earth, λ is the east longitude, and ϕ is the latitude of point P; z is measured parallel to the local direction of gravity.

not a simple matter; and the reader is referred to the discussion in Landau and Lifschitz (1987) and Yih (1965).

To the set (2.1.1)–(2.1.5), it is often useful to append a statement of conservation of a general tracer,

$$\frac{d(\rho C)}{dt} = q \tag{2.1.6}$$

where C represents any one of a variety of fields, such as oxygen, one of the nutrients, tritium, etc., and q represents any internal sources or sinks for the tracer, and so *conservation* is something of a misnomer but rarely causes confusion. Equation (2.1.5) is a special case with $C = 1,\ q = 0$.

For present purposes, an adequate thermodynamic statement is that, given the temperature, θ, and salt content per unit mass, S, of a parcel of fluid, its density can be computed as a function of the pressure from an empirically measured equation of state,

$$\rho(\rho, \phi, \lambda, t) = \rho(\theta(\phi, \lambda, t),\ S(\phi, \lambda, t)\,, p)\,. \tag{2.1.7}$$

Concentration evolution equations like (2.1.6) may be written separately for θ, S. The issues discussed by Landau and Lifschitz (1987) and by Yih (1965) concern the form of the right-hand side of (2.1.6) if temperature and salinity are permitted to behave diffusively.

A technical complication of ocean dynamics arises from the strong dependence in (2.1.7) of the density field on the pressure. Thus a parcel of fluid near-surface has a significantly different mass per unit volume, at the same temperature and salinity, than one at a depth of several thousand meters. Oceanographers therefore distinguish the density in situ from that evaluated at some fixed pressure, often $p = 0$, and called the *potential density*. For the momentum equations (2.1.1)–(2.1.3) expressing Newton's first law, the inertial mass is implied and the in situ density is to be used. But because mass is neither created nor destroyed by changing the pressure, a potential density is involved in equations such as (2.1.5). Detailed discussion of the effects of the equation of state of seawater on oceanic motions can become extremely complex (e.g., McDougall, 1987a,b), but qualitatively useful results can be obtained by ignoring the distinction. One must remain alert to quantitative difficulties, however.

2.1.2 Geostrophy

Observation suggests that for fluid flows which evolve on time scales longer than about one day, over spatial scales exceeding about 10 km, and away from the immediate vicinity of the seasurface and the equator, the dominant terms in the horizontal momentum equations (2.1.1) and (2.1.2) are the fourth term on the left and the first term on the right, that is, geostrophic balance,

$$-2\Omega \sin\phi\rho v \approx -\frac{\partial p}{a\cos\phi\partial\lambda} \tag{2.1.8}$$

$$2\Omega \sin\phi\rho u \approx -\frac{\partial p}{a\partial\phi}\,. \tag{2.1.9}$$

The vertical momentum equation (2.1.3) is also accurately given by hydrostatic balance,

$$0 = -g\rho - \frac{\partial p}{\partial z}\,. \tag{2.1.10}$$

Scaling arguments applied to the full Equations (2.1.1)–(2.1.3) based upon what is observed are completely consistent with the result (2.1.8)–(2.1.10). But one could imagine an ocean in a different dynamical regime, for which geostrophic, hydrostatic balance did not apply to lowest order.[2]

Local Cartesian approximations to (2.1.8)–(2.1.10) are adequate for many purposes. The simplest approximation can be derived systematically from spherical coordinates only if the origin of the y-coordinate is taken as the equator. At midlatitudes, there is is some geometric distortion, but it was a great contribution of Rossby and collaborators (1939) that the midlatitude "β-plane" nonetheless captures the essential physics of many oceanographic

[2] In fact, there are exceptions to our sweeping assertion of the validity of this balance, but they apply in special regions of small area–for example, in the motions corresponding to active wintertime convection.

(and atmospheric) motions while escaping from the inessential intricacies of the spherical coordinate system. Local geostrophic, hydrostatic balance for steady flow in Cartesian coordinates is written

$$-\rho f v = -\frac{\partial p}{\partial x} \tag{2.1.11}$$

$$\rho f u = -\frac{\partial p}{\partial y} \tag{2.1.12}$$

$$0 = -\frac{\partial p}{\partial z} - g\rho \tag{2.1.13}$$

where $f = 2\Omega \sin\phi$ is the *Coriolis parameter* and is usually written $f = f_0 + \beta y$ to account for its dependence on latitude. Mass conservation is

$$\frac{\partial u}{\partial x} + \frac{\partial v}{\partial y} + \frac{\partial w}{\partial z} = 0\,. \tag{2.1.14}$$

Conservation of density is

$$u\frac{\partial \rho}{\partial x} + v\frac{\partial \rho}{\partial y} + w\frac{\partial \rho}{\partial z} = 0\,, \tag{2.1.15}$$

or if combined with (2.1.14), is

$$\frac{\partial(\rho u)}{\partial x} + \frac{\partial(\rho v)}{\partial y} + \frac{\partial(\rho w)}{\partial z} = 0\,. \tag{2.1.16}$$

The equation of state is simply expressed in terms of the local coordinate system. The tracer conservation equation becomes

$$u\frac{\partial(\rho C)}{\partial x} + v\frac{\partial(\rho C)}{\partial y} + w\frac{\partial(\rho C)}{\partial z} = q(x, y, z, t)\,, \tag{2.1.17}$$

or if combined with (2.1.14), is

$$\frac{\partial(\rho C u)}{\partial x} + \frac{\partial(\rho C v)}{\partial y} + \frac{\partial(\rho C w)}{\partial z} = q(x, y, z, t)\,. \tag{2.1.18}$$

If f is treated as constant, cross-differentiation of (2.1.11) and (2.1.12) and use of (2.1.14) produces the well-known result

$$\frac{\partial w}{\partial z} = 0\,, \tag{2.1.19}$$

that is, *purely geostrophic motion is horizontally nondivergent* (this result is sometimes known as the Taylor-Proudman theorem). If the variation of f is accounted for, we obtain instead

$$\beta v = f\frac{\partial w}{\partial z}\,, \tag{2.1.20}$$

the *linear vorticity equation.*

The *thermal wind equations* are obtained from (2.1.11)–(2.1.13) by eliminating the pressure in terms of the density field, to give

$$-f\frac{\partial(\rho v)}{\partial z} = g\frac{\partial \rho}{\partial x} \tag{2.1.21}$$

$$f\frac{\partial(\rho u)}{\partial z} = g\frac{\partial \rho}{\partial y}, \tag{2.1.22}$$

which can be integrated in the vertical to produce

$$\rho u(x,y,z,t) = \frac{g}{f}\int_{z_0}^{z}\frac{\partial \rho}{\partial y}dz + \rho c(x,y,t,z_0) \equiv \rho(u_R + c) \tag{2.1.23}$$

$$\rho v(x,y,z,t) = -\frac{g}{f}\int_{z_0}^{z}\frac{\partial \rho}{\partial x}dz + \rho b(x,y,t,z_0) \equiv \rho(v_R + b)\,. \tag{2.1.24}$$

(The time, t, has been placed into the argument as a reminder that this balance can be a very good one even in a time-evolving field. Normally we will omit it.) The depth, z_0, is usually called the *reference depth* and is arbitrary: The two constants of integration, $b\,(x, y, t, z_0)$, $c\,(x, y, t, z_0)$ will change accordingly if z_0 is shifted, but usually we will suppress the explicit dependence upon it; u_R, v_R are called the *relative velocities*, and the integration constants b, c are the *reference-level velocities.* The density is the in situ value.

The two thermal wind equations have been central to physical oceanography for 100 years because of the realities of observations at sea. Historically, the only significant measurements that could be obtained at sea were made using ships. In the pre–solid-state electronics era,[3] the only routine measurements possible were the temperature, θ, and salinity, S, of the water at depth. The equation of state for seawater is determined by empirical laboratory measurements, that is, (2.1.7) permits estimates of the density at a given pressure from measurements of temperature and salinity. Thus, reasonable approximations to the density field can be obtained from observations made from ships and Equations (2.1.23)–(2.1.24) (using finite differences and approximating sums rather than analytical derivatives and integrals). Apart from the integration constants, the geostrophic flow is then calculable from shipboard measurements alone. The integration constants–that is, the reference level velocities–are mathematically trivial, but the inability to determine them plagued oceanography for 100 years and

[3] Vacuum tube electronics worked very badly at sea, and it was not until the advent of the transistor and its solid-state descendants that oceanographers were able to abandon the clever mechanical measurement systems employed until then. Baker (1981b) gives a good summary account of the evolution of oceanographic observation systems.

has been one of the major obstacles to understanding the ocean circulation. Much of this book is about their determination and use.

The transport of properties by a geostrophic flow is readily computed. Let the concentration of a property be given by $C(x, y, z)$ per unit mass. Then if the geostrophic flow is known, the flux of C between $x = x_1$ and $x = x_2$ lying between two depths $z_1(x)$ and $z_2(x)$ in the meridional direction is

$$V_c(y) \equiv \int_{x_1}^{x_2} \int_{z_1(x)}^{z_2(x)} \rho(v_R(x, y, z) + b(x, y))\, C(x, y, z) dz dx\,, \qquad (2.1.25)$$

which is made up of a part owing to the thermal wind and a part owing to the reference-level velocity. There is a corresponding expression for the flux in the zonal direction. When the property is mass itself, $C = 1$.

If one seeks the transport of mass between two fixed depths, z_1, z_2 not dependent upon x, then because the thermal wind component is computed from the density field, Equation (2.1.25) can be rewritten as

$$\begin{aligned} V_c(y) &= \int_{x_1}^{x_2} \int_{z_1}^{z_2} \left(\int_{z_0}^{z'} \frac{-g}{f} \frac{\partial \rho}{\partial x} dz + \rho b(x, y) \right) dz' dx \\ &= \frac{-g}{f} \int_{z_1}^{z_2} \int_{z_0}^{z'} (\rho(x_2, y, z) - \rho(x_1, y, z)) dz dz' \\ &\quad + \int_{x_1}^{x_2} \int_{z_1}^{z_2} \rho(x, y, z)\, b(x, y) dz dx\,, \end{aligned} \qquad (2.1.26)$$

which shows that the thermal wind contribution to the mass flux depends only upon the density of the end points x_1, x_2 and not upon the intermediate values (but only if z_1, z_2 are rigorously independent of x, precluding, for example, the situation in which the bottom might be encountered in the course of integration). The value of ρ differs little from a constant, $\bar{\rho} \approx 1.035\ \mathrm{gm/cm^3} = 1.035 \times 10^3\ \mathrm{kg/m^3}$, and if we treat it as a constant in the second term of (2.1.26), the mass transport is very nearly

$$\begin{aligned} V_c &= \frac{-g}{f} \int_{z1}^{z_2} \int_{z_0}^{z'} (\rho(x_2, y, z) - \rho(x_1, y, z)) dz dz' \\ &\quad + (z_2 - z_1)\bar{\rho} \int_{x_1}^{x_2} b(x, y) dx \end{aligned} \qquad (2.1.27)$$

in which the first term depends only upon the end-point values. Such simplifications of the transport are not possible in general with an arbitrary tracer $C \neq 1$.

Figure 2–2 displays for our future use a number of measurements of temperature, salinity, and oxygen as well as other properties, as obtained from ships, across some representative parts of the ocean.

A SMALL PRACTICAL ASIDE. It is often convenient to measure oceanic depths in pressure units. The original motivation arose because mechanical systems were devised to measure the pressures at which data were being collected by instruments on cables hung from ships. The hydrostatic relationship (2.1.13) provides a linear connection between pressure, p, and z. Because ρ differs little from 1.035 gm/cm^3 (although the small variations are *dynamically* extremely important), one can obviously interchange z and p up to a nearly constant factor. Thus, 10 meters of water exerts a pressure very nearly equal to that of 1 atmosphere (1 bar $= 10^5$ Pascals). By measuring pressure in the (non-SI) units of *decibars*, one has a unit that is numerically close to the pressure exerted by one meter of water. Saunders (1981) provides a more accurate conversion formula, but for many practical purposes, the interchange of meter and pressure depth in decibars is sufficient.

Water Masses. Oceanic water is labeled in a complex way by a number of measurable scalar properties: its temperature, salinity, oxygen, and silicate concentration, its depth, latitude, longitude, volume, etc. If there are n such properties, the water at any place in the ocean can be thought of as defining a point in an n-dimensional cartesian vector space. Because paper limits one to the representations of two-dimensions, much use has been made of depictions of oceanic water along selected two-dimensional subspaces or two-dimensional projections of three-dimensional subspaces of the n-space. Temperature and salinity are the water properties most easily and commonly measured, and so-called temperature-salinity diagrams have been much exploited. Figure 2–3 is Worthington's (1981) estimate, drawn as a histogram, for the temperature and salinity properties of the world ocean. Figures 2–2a-j are of course, latitude-longitude, or longitude-depth property diagrams, and one often encounters other choices of property; for example, Figure 2–4a is an oxygen-salinity diagram.

It was noticed long ago that there is a tendency for water to lie in restricted regions of particular property diagrams. In Figure 2–3 the single large peak, centered near 1.15°C and 34.685, represents a large fraction of the total water in the entire ocean.[4] Much of the remaining water lies on or near restricted curves in the temperature-salinity domain. Similar behavior can be seen on a regional basis in Figures 2–4. This behavior of oceanic water led early in the history of the subject to the idea of water masses. The concept has become somewhat blurred through varying usage. For our

[4] Salt content is measured on the practical salinity scale and is dimensionless; see Lewis and Perkin (1978) or Pickard and Emery (1982).

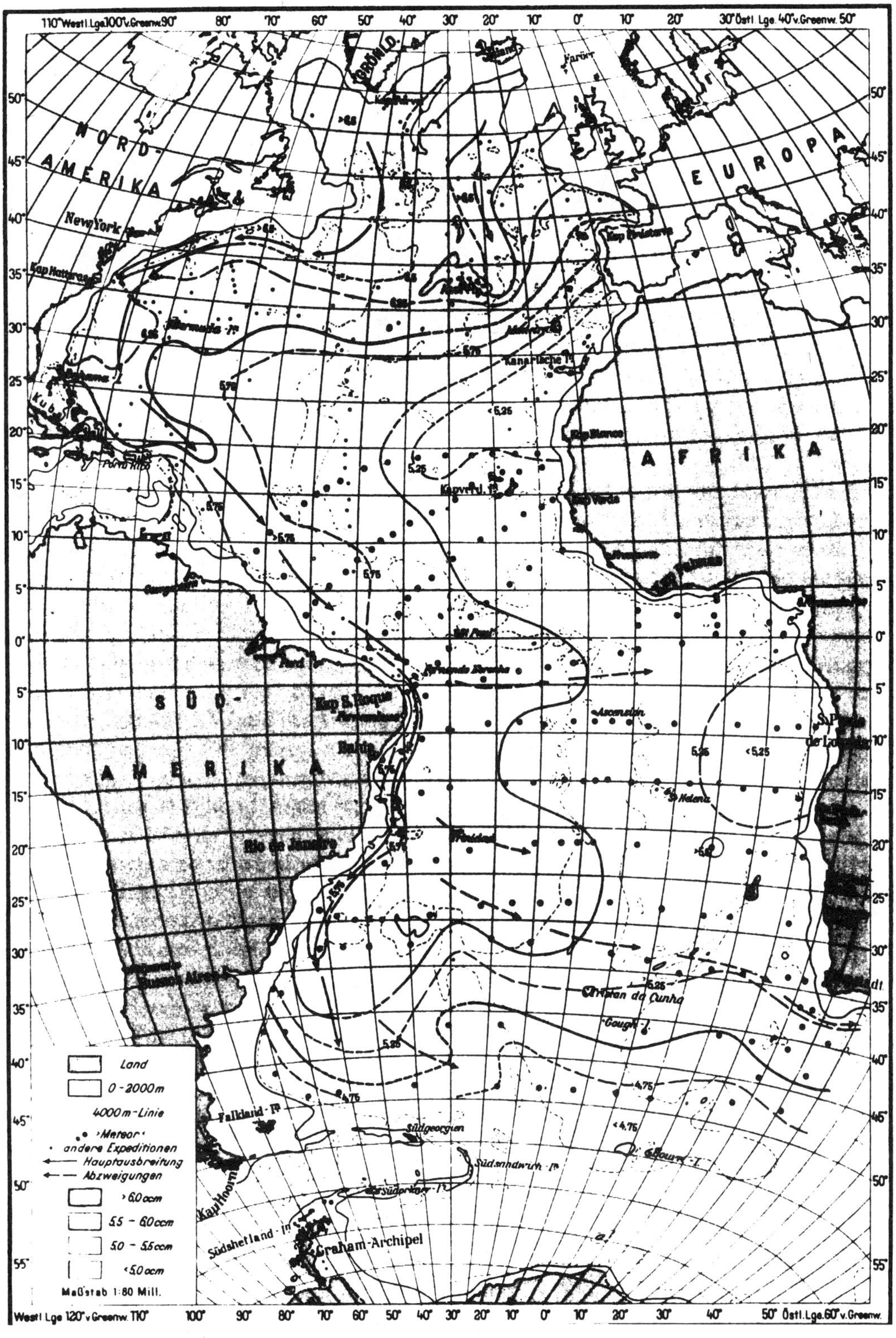

Figure 2–2a. The Atlantic oxygen concentration at the so-called intermediate oxygen maximum (in the vicinity of 2500–3000 m over much of the basin). Note the obvious tongue at the western boundary of the ocean. From Wüst (1935). Units are ml/l. (Figure 2–2a is also shown in the color insert.)

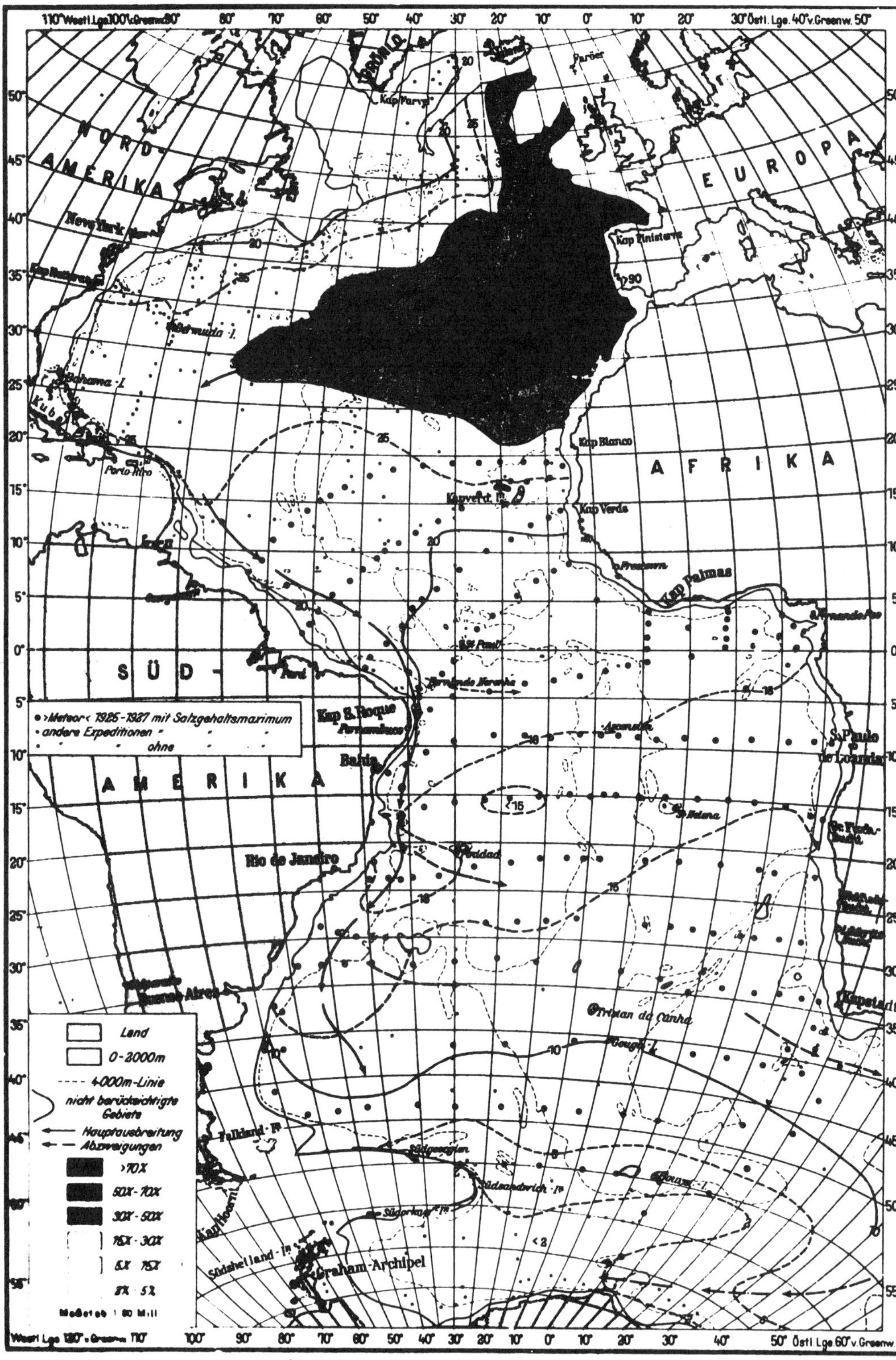

Figure 2–2b. Salinity in parts per thousand minus 35 at the "intermediate salinity maximum" in the Atlantic (the depth ranges from about 1000–2000 m). The most conspicuous feature is the Mediterranean salt tongue (Wüst, 1935). (Figure 2–2b is also shown in the color insert.)

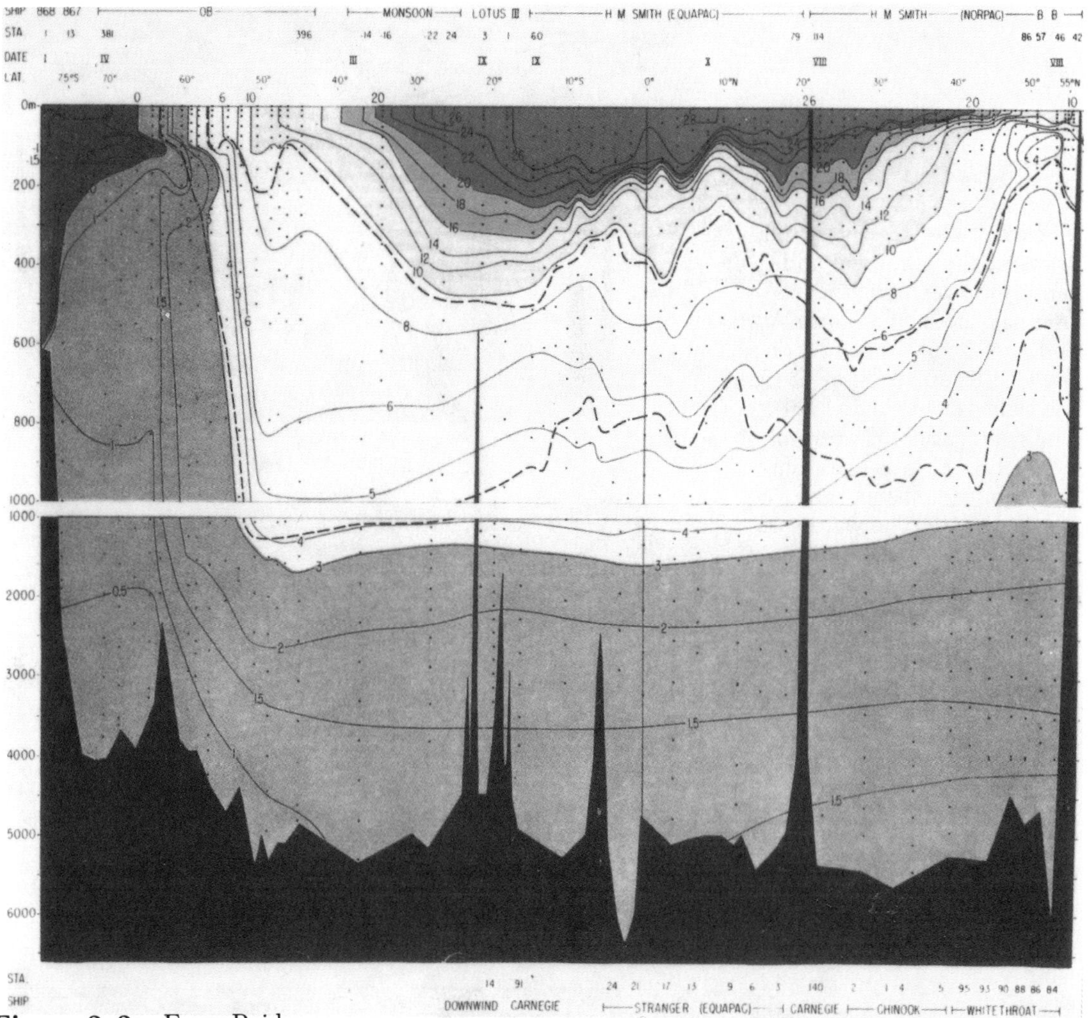

Figure 2–2c. From Reid (1965). Temperature section down the central Pacific Ocean. Dashed lines denote approximate location of two surfaces of constant density. Expedition names, station numbers, dates, and latitudes are indicated along the top. Outcrops of the isotherms at high latitudes are regions where dense water is formed, which sinks to mid-depths. (Figure 2–2c is also shown in the color insert.)

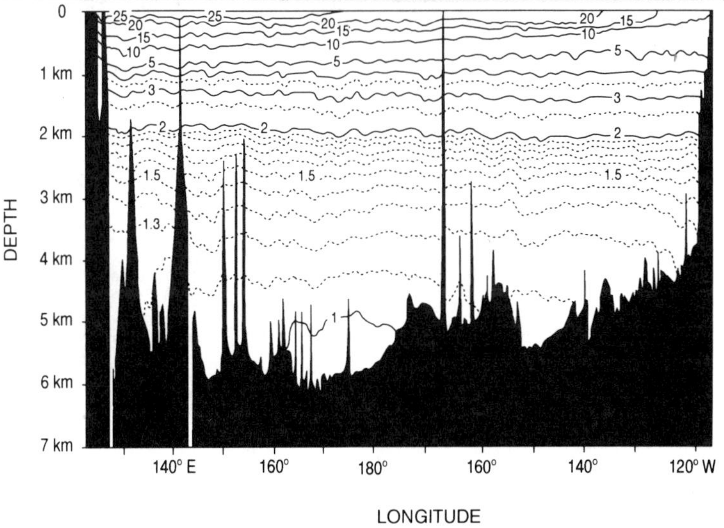

Figure 2–2d. Temperature section (°C), from Roemmich and McCallister (1989), along 24°N in the Pacific Ocean. Note the upward slope of the thermocline from west to east.

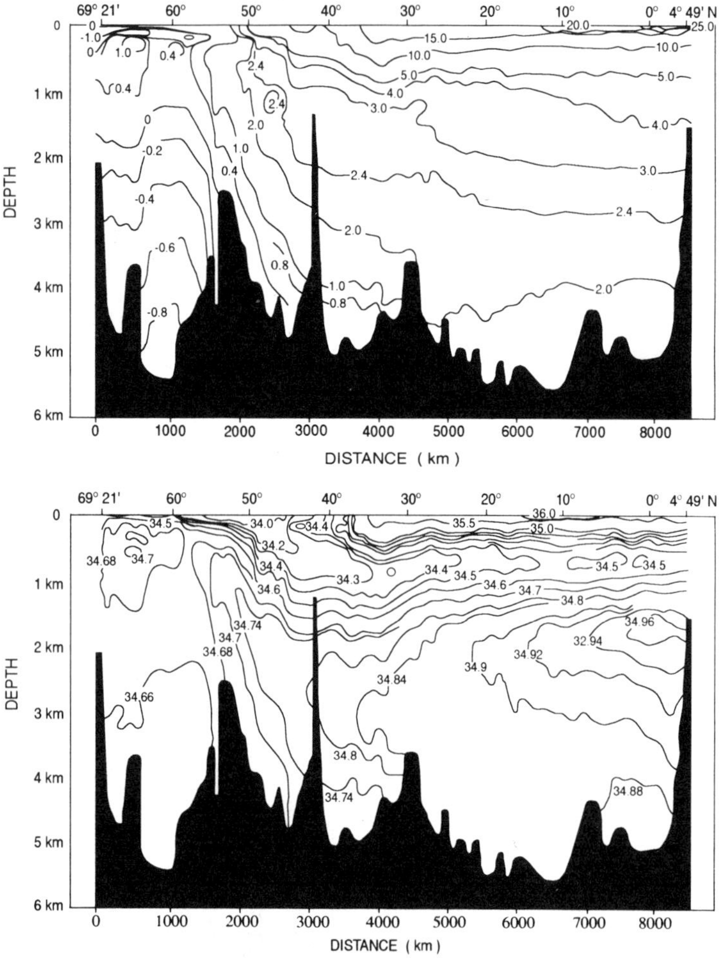

Figure 2–2e. Potential temperature (°C) (upper) and salinity (lower) along a section near 0° in the South Atlantic with the Antarctic on the left, terminating at 5° on the right (from Reid, 1989). The penetration of water of Antarctic origin is conspicuous in both fields at the southern end. The waters of North Atlantic origin are visible most plainly in the relative salinity maximum near 2000 m. But the inference of water movement is indirect; it is only a computation of the pressure field, which makes one conclude that these pictures do not represent a purely static (that is, resting) ocean in which the property tongues were formed long ago with no movement since.

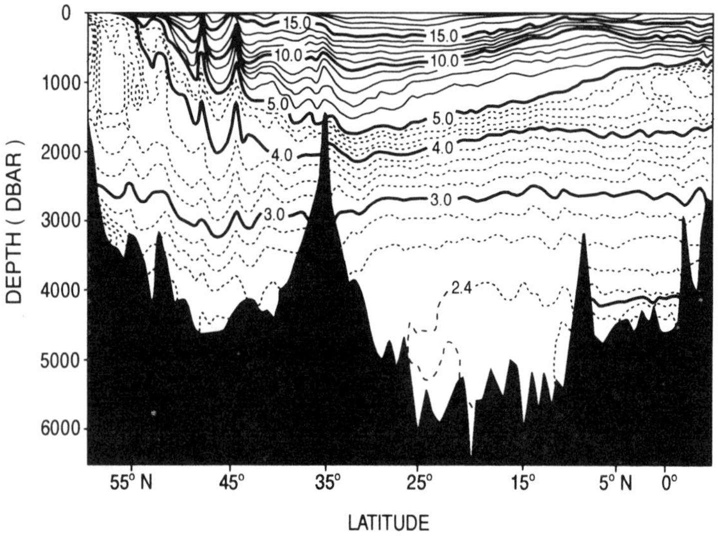

Figure 2–2f. In situ temperature (°C) along 35°W in the North Atlantic (Fukumori et al., 1991), running from Greenland to South America. The conspicuous bowl shape of the thermocline, with a long-slow shoaling toward the equator, is generally characteristic of the midlatitude interior ocean and is one of the features that the thermocline theories endeavor to explain.

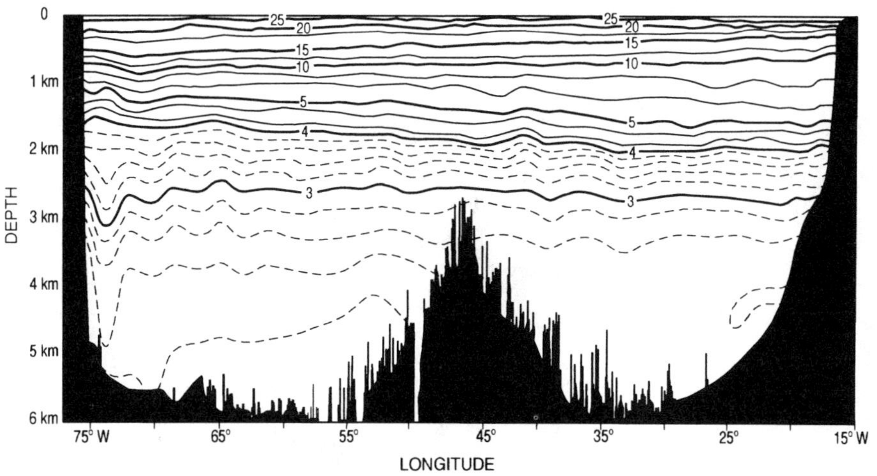

Figure 2–2g. In situ temperature (°C) along 24°N in the North Atlantic as measured in 1957 (Fuglister, 1960). Although there is a general shoaling of the thermocline toward the east–another feature in need of explanation–this particular section is remarkable for its nearly "flat" isotherms–a feature that is important in the discussion of the temperature flux across it taken up in Chapter 4. The spatial sampling employed here aliases some of the eddy field (unknown in 1957) so that the section appears to have more low-wavenumber energy than is correct (aliasing is discussed in Chapter 3).

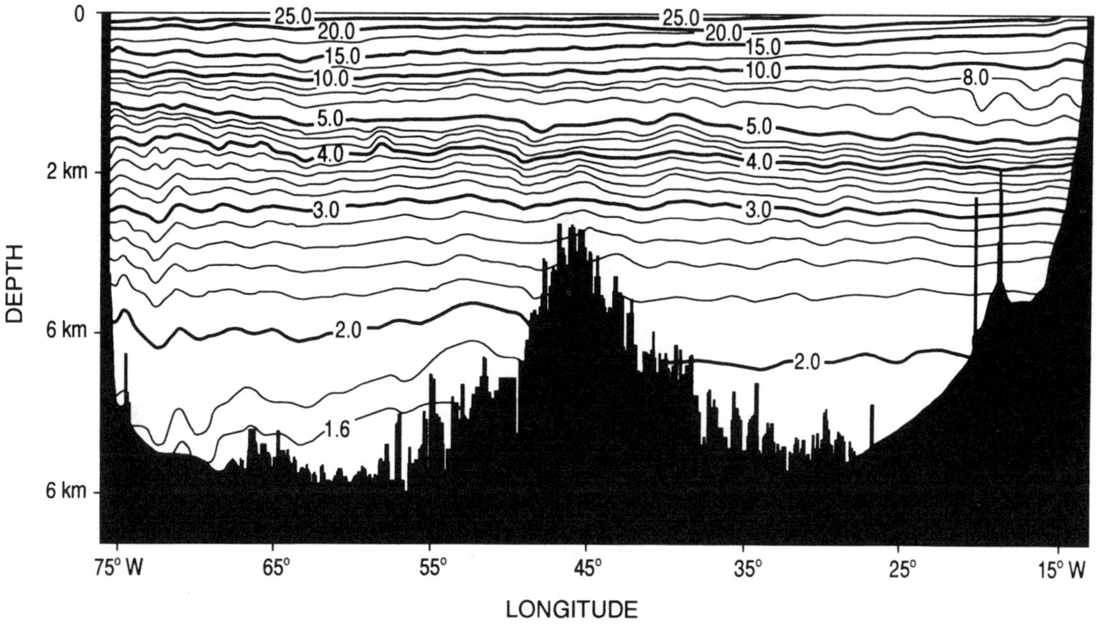

Figure 2–2h. Potential temperature along nearly the same section as seen in Figure 2–2g, but measured in 1981 (Roemmich & Wunsch, 1985). The comparatively slight differences between the sections (apart from the higher sampling rate in 1981 and the use of potential temperatures, which produce lower values in the deepest water) is strongly suggestive of the oceanic stability on the large scale over decades and longer.

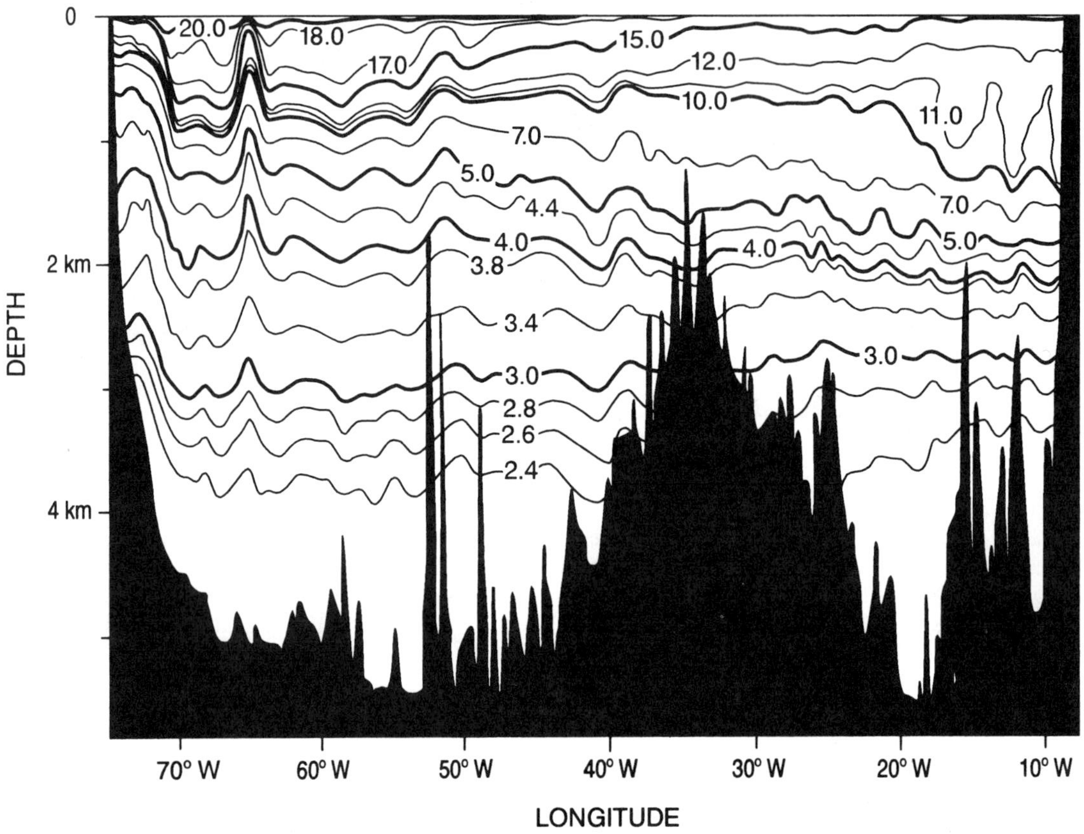

Figure 2–2i. Same as Figure 2–2h, except along 36°N, showing the interior Gulf Stream and associated flows. Data shown here and in Figure 2–2h are used as an example in Chapter 4.

purposes, a water mass can be defined as any n-dimensional vector of measurable scalar property, for example (θ, S, Si, $O, \ldots$), where Si, O, etc., are the silicate, oxygen, etc., concentrations; by convention, the geographical properties of depth, latitude, etc., are omitted. *Important* water masses are typically those for which a large quantity of water exists within a small distance of some central value θ_0, $S_0, \ldots$ in the n-space, or for which the n-tuple lies at the extreme end point of the existing curves–for example, at the tips of the "y-shape" seen in Figure 2–3. The reasons for the particular water masses and abundances actually seen in the ocean remain a comparatively neglected part of theoretical understanding (e.g., why is there such a large peak at the particular value observed in Figure 2–3?). Speer and Tziperman (1992) make a start on the problem.

In recent years, the awkwardness of representing n-dimensional spaces

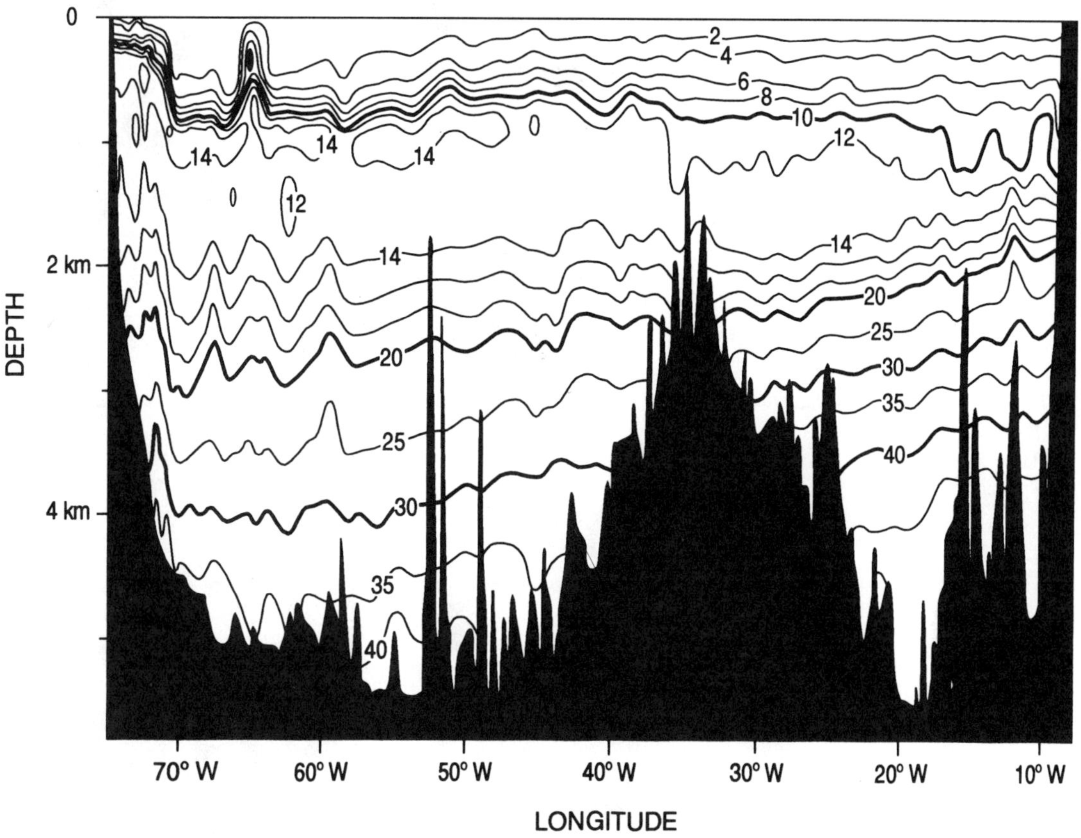

Figure 2–2j. Silicate (μ-gm at/l) along the same section as shown in Figure 2–2h. Like other nutrients, midlatitude silicate tends to be depleted near the surface and enhanced near the bottom–as a consequence of the biogeochemical cycling, which makes it a potentially valuable complement to temperature and salinity in conservation constraints, but greatly complicated by uncertainty over the interior source/sink terms.

in two dimensions has led to attempts to deal directly with the n-tuple representation in a more algebraic manner (e.g., Mackas, Denman, & Bennett, 1987; Tomczak & Large, 1989; Fukumori & Wunsch, 1991; Hamann & Swift, 1991). These efforts are briefly described in Chapter 5.

2.1.3 Ekman Theory

After a contentious nineteenth-century debate, a consensus emerged that the ocean circulation was driven primarily by the wind field and only secondarily by thermodynamic (buoyancy) forces. With the benefit of hindsight, and

Figure 2–2k. From Kawase and Sarmiento (1985). Nitrate distribution on potential density surface $\sigma_\theta = 36.85$ (reference pressure = 0). They interpret this figure as showing an equatorial tongue of nitrate, but the upwelling off Africa makes the interpretation ambiguous.

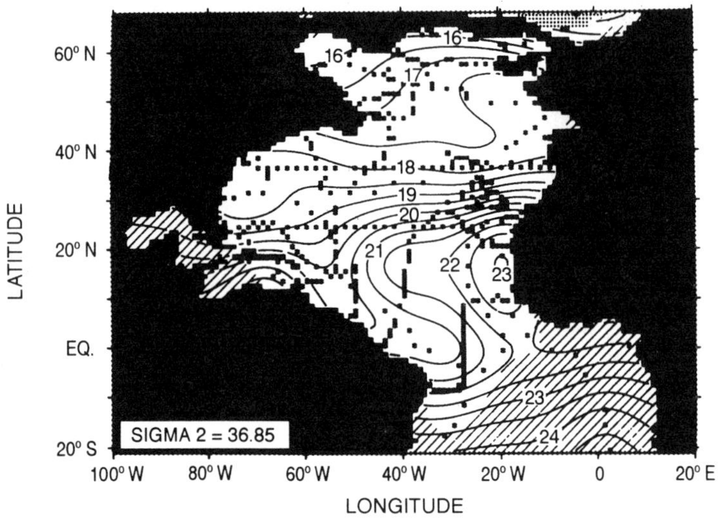

a

b

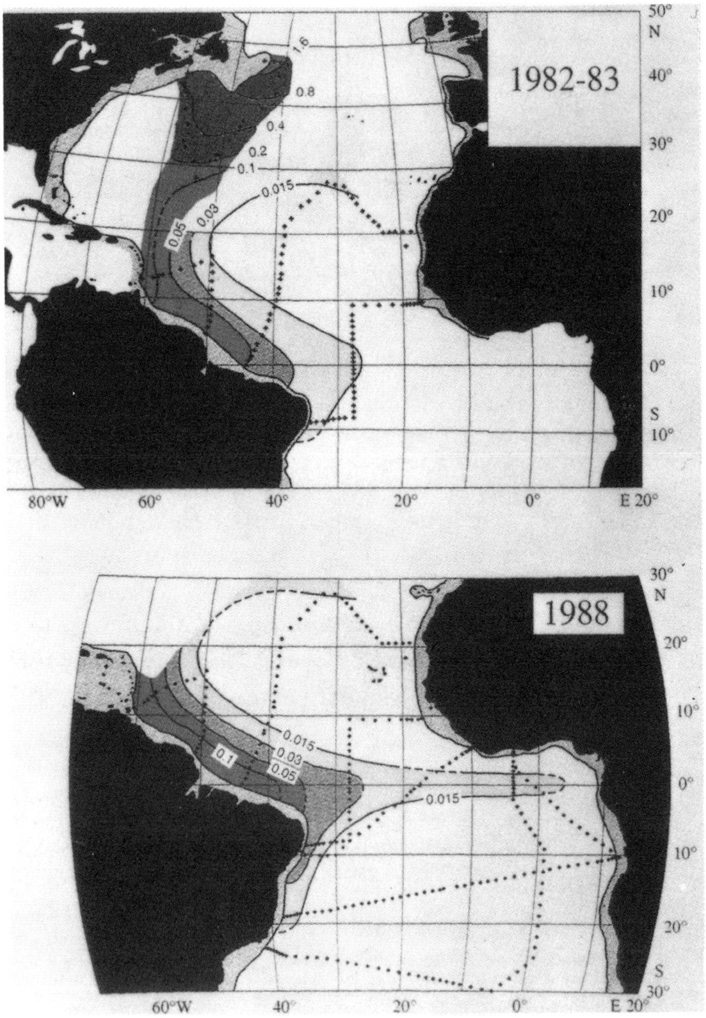

Figure 2–2l. Concentration (in picomoles/kg) of a fluorocarbon (CFC–11) on the density surface $\sigma_{1.5} = 34.63$ (i.e., reference pressure of 1500 decibars) in 1982–83 (a) and in 1988 (b) (Weiss et al., 1985, 1993, and personal communication, 1994). This surface lies near 1500–2000 meters over much of the subtropical Atlantic. CFCs are artificial compounds which, having been injected into the atmosphere, dissolve at the seasurface. They are then carried into the deep ocean through high-latitude convective processes. (Parts a and b are also shown in the color insert.)

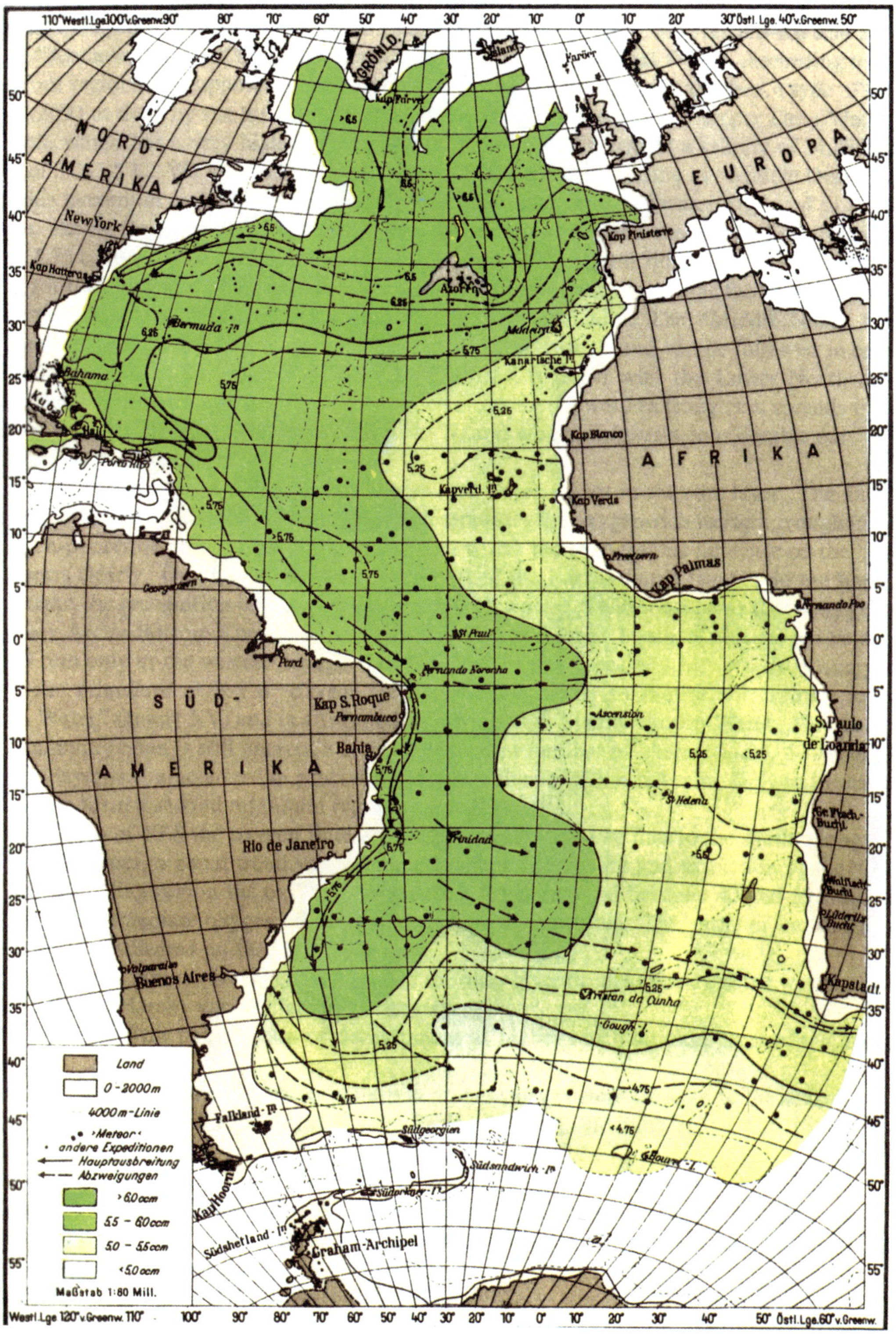

Figure 2–2a. The Atlantic oxygen concentration at the so-called intermediate oxygen maximum (in the vicinity of 2500–3000 m over much of the basin). Note the obvious tongue at the western boundary of the ocean. From Wüst (1935). Units are ml/l.

Figure 1–1b. Surface elevation of the ocean in cms during 10–20 March 1993 as seen by the TOPEX/POSEIDON altimeter satellite (after Stammer & Wunsch, 1994). The elevation is relative to a 2–year mean, has been averaged over 2° squares, and the 10–day "window" blurs the most rapidly changing features, therefore rendering the result somewhat simpler than a true instantaneous picture. Nonetheless, both data and model confirm the essentially turbulent nature of the circulation.

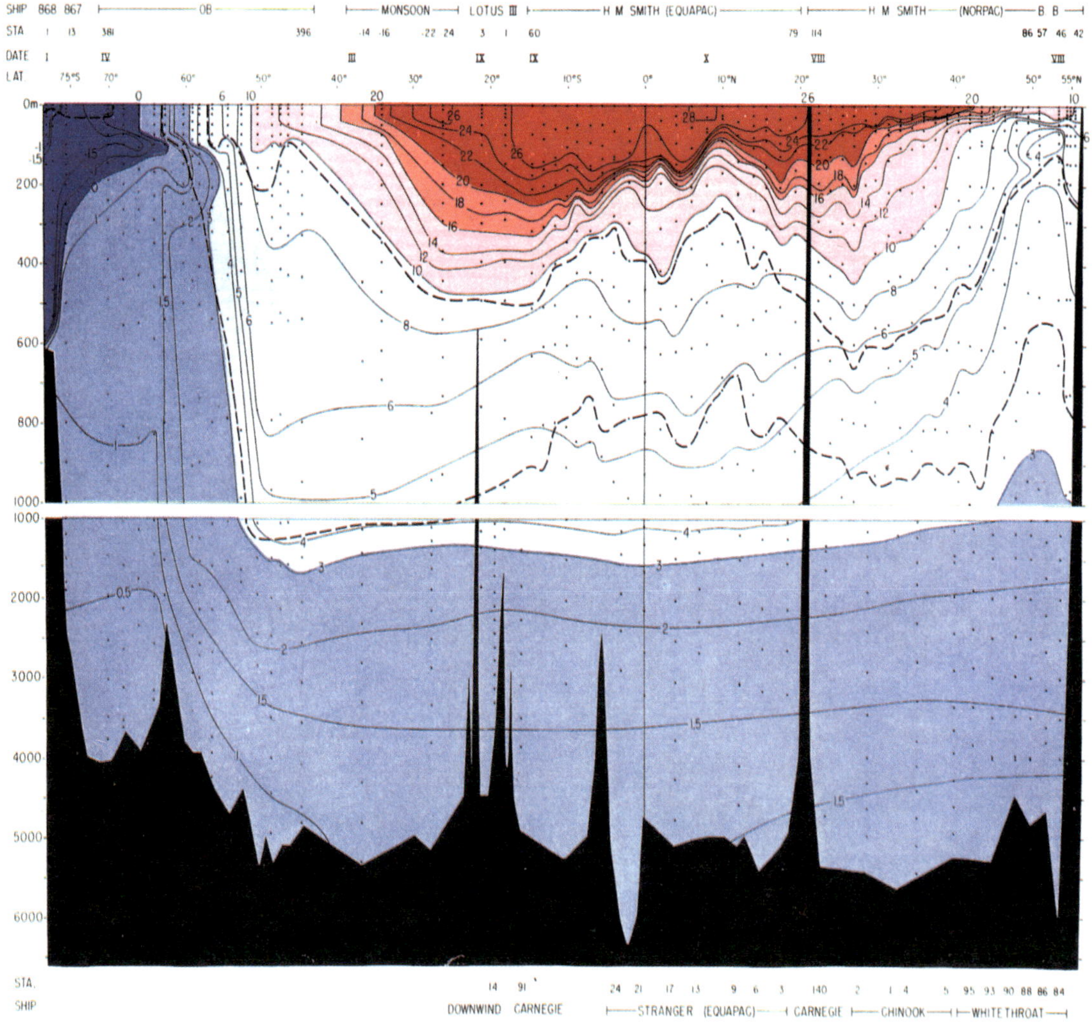

Figure 2–2c. From Reid (1965). Temperature section down the central Pacific Ocean. Dashed lines denote approximate location of two surfaces of constant density. Expedition names, station numbers, dates, and latitudes are indicated along the top. Outcrops of the isotherms at high latitudes are regions where dense water is formed, which sinks to mid-depths.

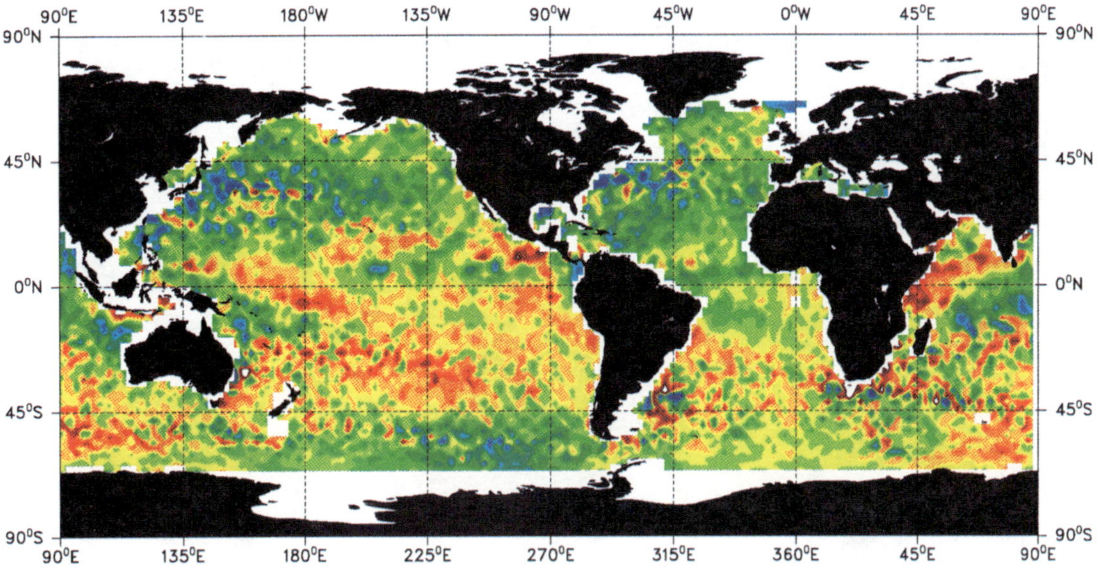

Figure 1–1b.

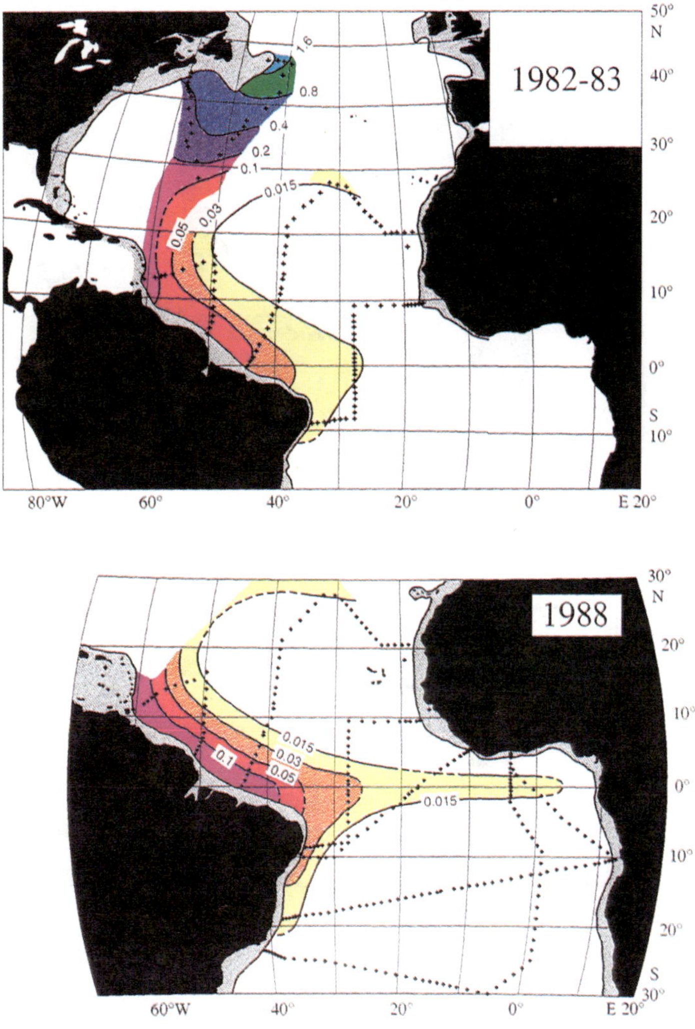

Figure 2–2l. Concentration (in picomoles/kg) of a fluorocarbon (CFC–11) on the density surface $\sigma_{1.5} = 34.63$ (i.e., reference pressure of 1500 decibars) in 1982–83 (a) and in 1988 (b) (Weiss et al., 1985, 1993, and personal communication, 1994). This surface lies near 1500–2000 meters over much of the subtropical Atlantic. CFCs are artificial compounds which, having been injected into the atmosphere, dissolve at the seasurface. They are then carried into the deep ocean through high-latitude convective processes.

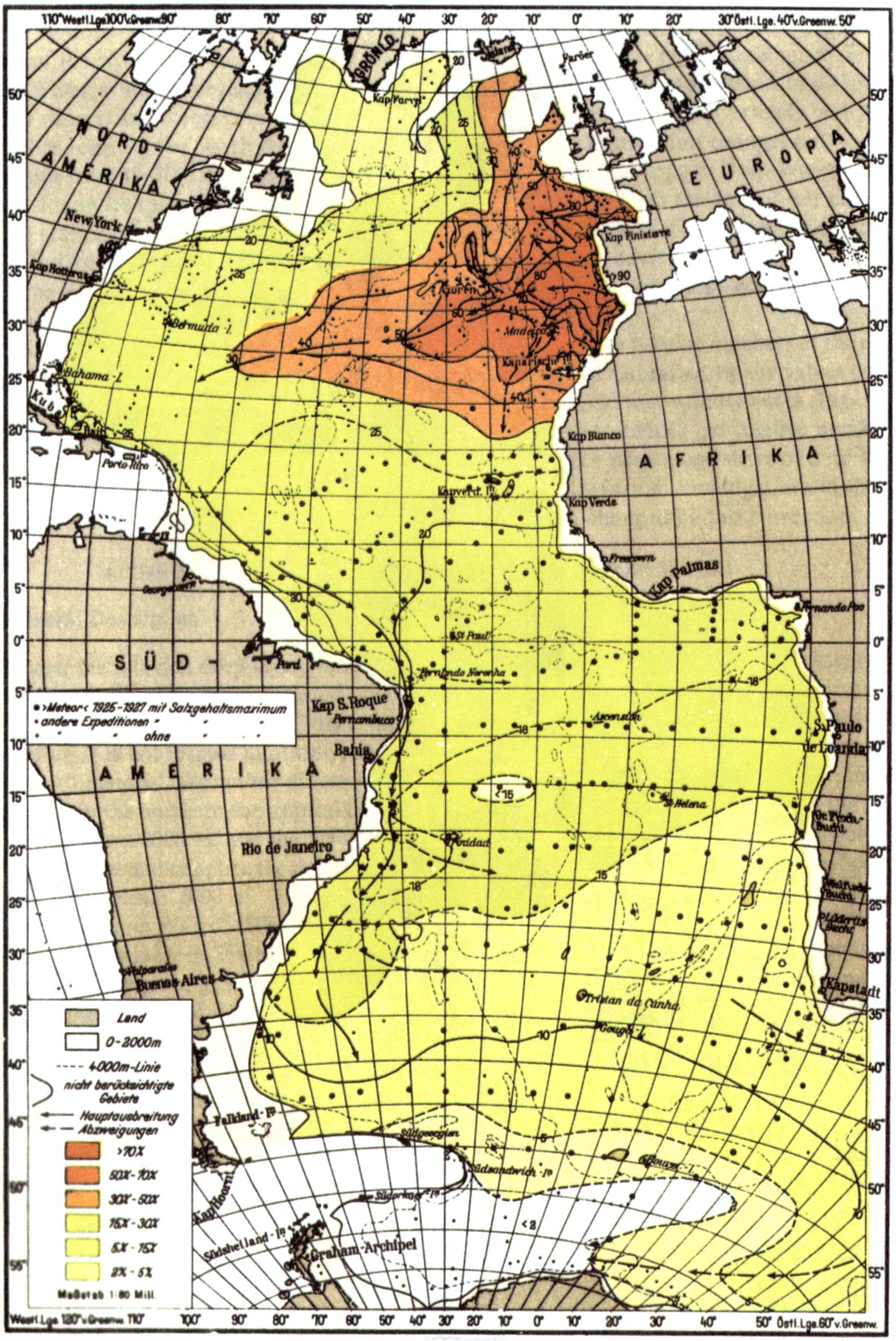

Figure 2–2b. Salinity in parts per thousand minus 35 at the "intermediate salinity maximum" in the Atlantic (the depth ranges from about 1000–2000 m). The most conspicuous feature is the Mediterranean salt tongue (Wüst, 1935).

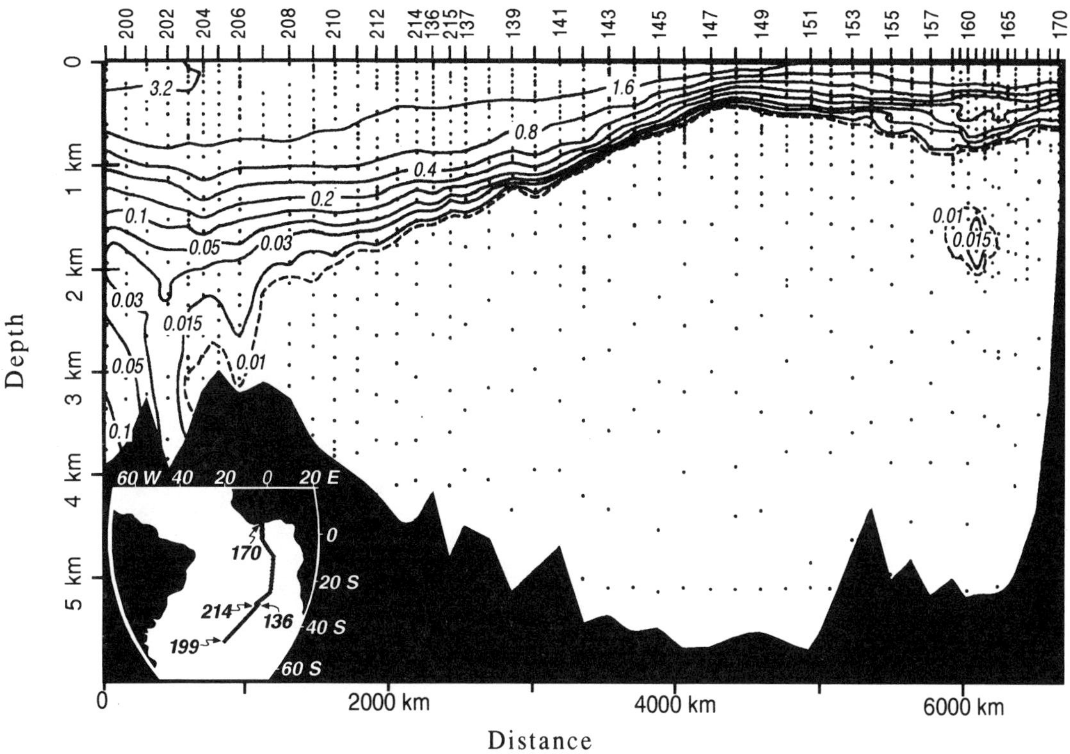

Figure 2–21. Caption on preceding page. Apart from demonstrating again the existence of a deep-western boundary current, the most remarkable feature is the long-equatorial tongue (c)–shown in cross-section near 5°W during 1988. The tongue is astonishingly thin and narrow, suggesting qualitatively how little mixing apparently takes place in the oceanic interior. (The inset shows the station positions of the cross-section.)

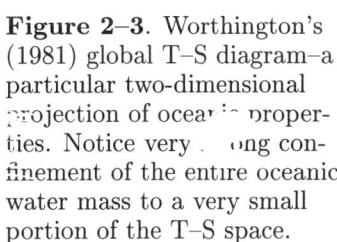

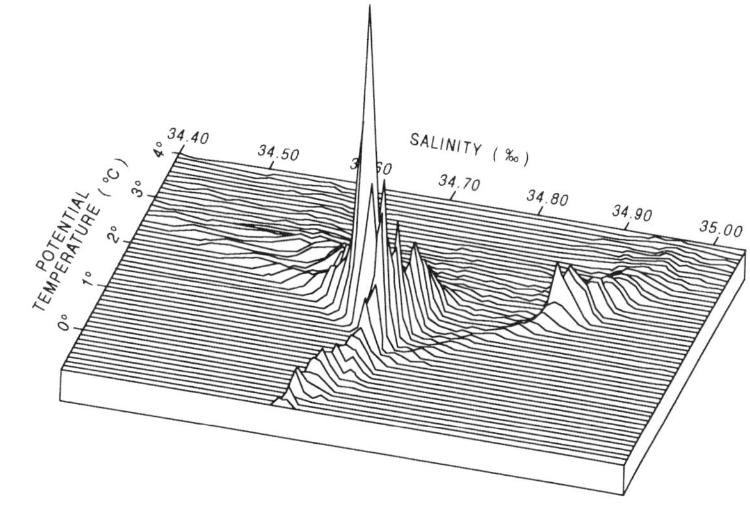

Figure 2–3. Worthington's (1981) global T–S diagram–a particular two-dimensional projection of ocea[illegible] proper-ties. Notice very [illegible]ong confinement of the entire oceanic water mass to a very small portion of the T–S space.

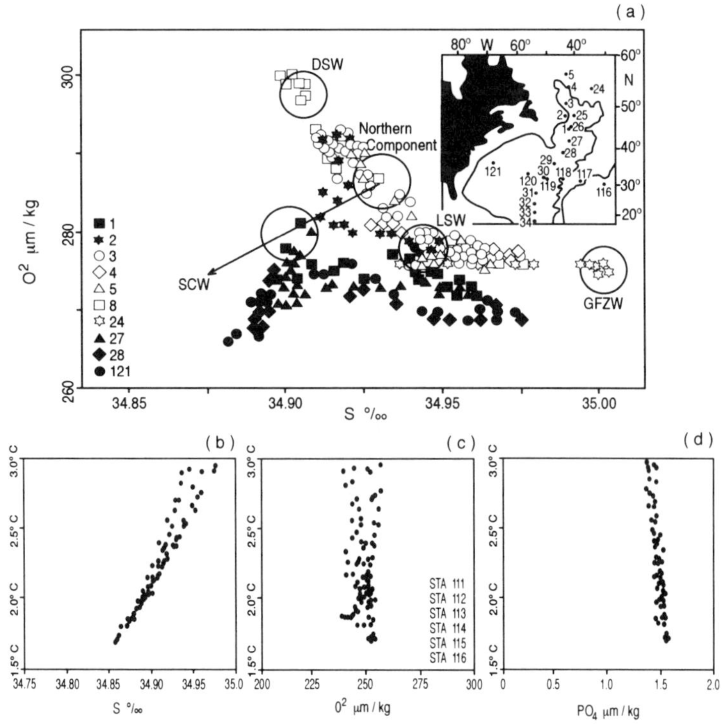

Figure 2–4. Collection of property-property diagrams, after Broecker, Takahashi, and Li (1976), showing various two-dimensional property projections: (a) oxygen-salinity, (b) (potential) temperature-salinity, (c) temperature-oxygen, (d) temperature-phosphate. The number of possible such pairs is very large. Various names are attached to water occurring in different parts of the property spaces. Here, there are Gibbs Fracture Zone Water (GFZW), Denmark Straits Water (DSW), Labrador Sea Water (LSW), and Southern Component Water (SCW). The different symbols denote varying locations (seen in the inset chart), which is an attempt to introduce an equivalent of a geographical dimension into the two-dimensional plane.

a clearer understanding of the behavior of nonlinear systems, we know that such a separation is not possible. But it is sensible nonetheless to try and first understand the motion of an ocean driven purely by the wind.

To represent the stress terms F_u, F_v of (2.1.1)–(2.1.2), Ekman (1905) appended eddy coefficient terms to Equations (2.1.11)–(2.1.12), producing a new system

$$-\rho f v = -\frac{\partial p}{\partial x} + A_v \frac{\partial^2 u}{\partial z^2} \tag{2.1.28}$$

$$\rho f u = -\frac{\partial p}{\partial y} + A_v \frac{\partial^2 v}{\partial z^2} \tag{2.1.29}$$

$$0 = -\frac{\partial p}{\partial z} - g\rho \,. \tag{2.1.30}$$

The eddy coefficients, A_v, are taken to be constants, with the dissipational terms written to mimic a molecular diffusion process. That is, molecular diffusion in an ordinary fluid produces stress terms of the form $\nu_m \nabla^2(u, v, w)$ in the x, y, z directions, respectively, where ν_m is the molecular diffusion coefficient. In what is an argument by analogy, the much more intense turbulent eddies are supposed to act on the larger scale flow as though they were molecular processes with a coefficient A_v that is many orders of magnitude larger than ν_m. Provision is made for different turbulent intensities in the vertical and horizontal directions and for a much more rapid variation of the larger-scale motions in the vertical dimension than in the horizontal dimension. In (2.1.28)–(2.1.29), it is therefore supposed that for the large-scale motions being described, the lateral turbulent motions are negligible compared to the vertical ones. The use of an eddy coefficient to represent the effects of scales smaller than the ones in which the investigator is interested, is a matter of mathematical convenience and tradition and is perhaps "a triumph of hope over experience." That it can fail badly in general was documented by Starr (1968). The wind at the surface being supposed to exert a vector stress $\boldsymbol{\tau} = (\tau_x(x, y),\ \tau_y(x, y))$, in analogy to the stress boundary conditions for molecular viscosity, the boundary conditions on the solutions to Equations (2.1.28)–(2.1.29) are

$$A_v \left(\frac{\partial u}{\partial z}, \frac{\partial v}{\partial z} \right)_{z=0} = (\tau_x, \tau_y), \quad w(0) = 0\,. \tag{2.1.31}$$

The boundary layer solution is Ekman's famous spiral of form

$$u_{BL} = \frac{\exp^{\sqrt{f/2A_v}z}}{\sqrt{2A_v f}} \left\{ (\tau_x + \tau_y) \cos\sqrt{f/2A_v}z + (\tau_x - \tau_y) \sin\sqrt{f/2A_v}z \right\} \tag{2.1.32}$$

$$v_{BL} = \frac{\exp^{\sqrt{f/2A_v}z}}{\sqrt{2A_v f}} \left\{ (\tau_y - \tau_x) \cos\sqrt{f/2A_v}z + (\tau_x + \tau_y) \sin\sqrt{f/2A_v}z \right\} \tag{2.1.33}$$

so that $(u_{BL}, v_{BL} \to 0$ as $z \to -\infty$, where the velocities have been labeled with the subscript BL as a reminder that they apply only in the surface boundary layer. Note a few salient features of this solution:

- The direct effects of the wind are confined to the Ekman depth $d_E = \sqrt{2A_v/f}$. Since the use of eddy coefficients is dubious, and the numerical values one is advised to use for A_v vary widely, it is fortunate that the square root reduces the sensitivity of the Ekman depth to the numerical choice. Extreme values of A_v lead to estimates of d_E ranging over about 25–200 m.
- The flow is *not* geostrophic, representing a balance primarily between the Coriolis forces and the stress terms.

- Consequently, to the order of the approximations leading to the balance (2.1.32)–(2.1.33), there is no pressure signature of the Ekman layer (see Pedlosky, 1987a). Pressure fields do not ordinarily form boundary layers, and we do not expect to see any pressure field associated with the Ekman layer itself. (Formally, the pressure terms in Equations (2.1.28)–(2.1.29) are set to zero in finding the solution (2.1.32)–(2.1.33). One confirms a posteriori that the pressure field owing to the Ekman spiral is very weak.)
- The transport of water in this boundary layer is

$$U = \int_{-\infty}^{0} \rho u dz = \tau_y / f \tag{2.1.34}$$

$$V = \int_{-\infty}^{0} \rho v dz = -\tau_x / f \tag{2.1.35}$$

directed–in the northern hemisphere–to the right of the wind and is independent of the value of A_v–a result of greatest importance, again because of the artificial nature of an eddy coefficient. If we take plausible numerical estimates for the wind stress as 0.1 Pascals (about 1 dyne/cm^2) and a latitude of 30°, the amount of fluid predicted to move within the Ekman layer is approximately 1.4×10^3 kg/m/s. Integrating over 1000 km, the mass flux is about 10^9 kg/s (volume flux of about 10^6 m^3/s, one "Sverdrup").[5]
- Away from the surface boundary layer, the stress terms become negligible, and the flow field reduces to a geostrophic "interior," satisfying (2.1.11)–(2.1.13), (2.1.20) whose flows we will label u_I, v_I, p_I.
- Apart from the slight variation in ρ, Equations (2.1.11)–(2.1.13) are linear, and we can write the total solution, Ekman plus interior, as

$$u = u_{BL} + u_I, \quad v = v_{BL} + v_I, \quad p = 0 + p_I\,. \tag{2.1.36}$$

That is, the flow within the boundary layer can usefully be viewed as being the *sum of the Ekman flow plus the geostrophic interior.* The pressure field seen near the seasurface is that owing to the interior flow alone, a result of great importance in the use of satellite altimetry.
- Because the lateral transport of fluid within the Ekman layer is a function of position, owing to variations in the value of f, and the strength of the wind, there must be an exchange of fluid with the region below if the flow

5 Because the density of seawater differs hardly at all from 1.03×10^3 kg/m^3, there is a nearly 3% error in converting 1 Sverdrup to 10^9 kg/s. But there are few flux properties in the ocean known to 3% accuracy, and the slightly sloppy conversion is often perfectly acceptable.

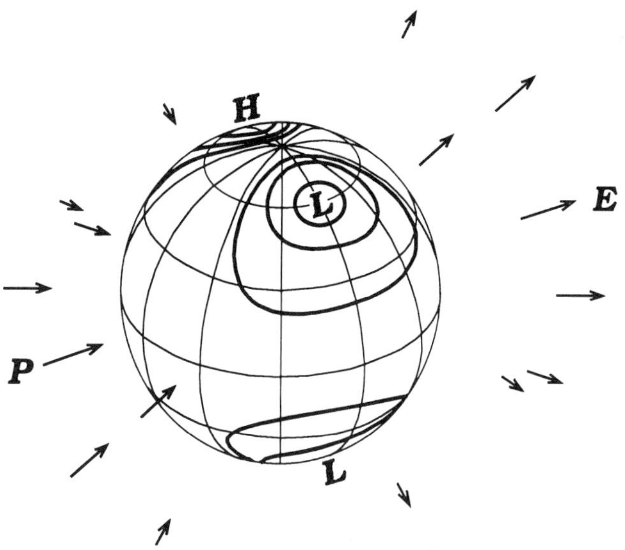

Figure 2–5. Stommel's (1957) Goldsbrough-type circulation in which Ekman pumping is replaced by evaporation (E) and precipitation (P) patterns to emphasize the flux divergence property of the Ekman layer. A geostrophic circulation around the highs and lows is established.

(and the seasurface) are to remain steady in time. The vertical velocity associated with this exchange is easily found to be

$$w_E = \frac{\partial}{\partial x}\left(\frac{\tau_y}{f}\right) - \frac{\partial}{\partial y}\left(\frac{\tau_x}{f}\right) \tag{2.1.37}$$

and this flux must be absorbed or provided by the underlying fluid, which is presumed to be in geostrophic balance, and which sets this deeper fluid into motion.

The wind-driven general circulation of the ocean, away from the immediate vicinity of the equator, is driven by this mechanism–that is, not by the wind or the Ekman layer but by the regional variations of the amount of fluid moving horizontally within the Ekman layer, and whose pumping sets the interior geostrophic fluid into motion. The underlying physics are probably best appreciated from the point of view taken by Stommel (1957, 1984), who avoided the Ekman layer entirely and discussed how geostrophic flows could be driven by patterns of pseudo-rainfall and evaporation (Figure 2–5). Rain acts analogously to the way Ekman pumping does, with evaporation equivalent to Ekman suction. The Ekman layer itself is of secondary interest in the general circulation, although its horizontal transport of mass and other properties are important at low latitudes. But its properties as a pump are the crucial element in understanding the general circulation.

Other boundary layers exist in the theory of the ocean circulation, including a weaker Ekman layer at the seafloor, plus complicated balances at sidewalls including the major western boundary currents. In one point of view, the region immediately encompassing the equator can be regarded as another boundary layer region. Nonetheless, the picture we have painted of an Ekman layer overlying and pumping an otherwise geostrophic interior remains helpful as a framework for discussion of the vast oceanic interior, subject to the warning that it is not by any means a complete depiction.

2.1.4 Sverdrup Balance

The dynamical connection between the wind field and the geostrophic movement of the interior fluid below the Ekman boundary layer was studied by Sverdrup (1947), although he did not consider the Ekman layer explicitly. He started with the Equations (2.1.8)–(2.1.30) and integrated them from a depth $z_S < 0$ to the seasurface, applying the boundary condition (2.1.31). The result is

$$-f\int_{z_s}^{\zeta} \rho v dz = -\int_{z_s}^{\zeta} \frac{\partial p}{\partial x} dz + \tau_x(\zeta) - \tau_x(z_s) \tag{2.1.38}$$

$$f\int_{z_s}^{\zeta} \rho u dz = -\int_{z_s}^{\zeta} \frac{\partial p}{\partial y} dz + \tau_y(\zeta) - \tau_y(z_s) \tag{2.1.39}$$

where $\zeta(x, y)$ is the seasurface.

The general manipulation of these equations is surprisingly complex; Fofonoff (1962) gives a complete account, of which a simplified form may be found in Pond and Pickard (1983, pp. 127–33). Here, we sketch a version of Sverdrup's argument so that we can discuss its validity in the next section. He supposed that if one went deep enough, a value z_S could be found such that $u, v = 0$, $z < z_S$, implying that both the horizontal pressure gradient and the stress vanish there as well. Define

$$(U, V) \equiv \int_{z_S}^{0} \rho(u, v) dz \tag{2.1.40}$$

$$P = \int_{z_S}^{0} p dz \tag{2.1.41}$$

and the equations become

$$-fV = -\frac{\partial P}{\partial x} + p(\zeta)\frac{\partial \zeta}{\partial x} - p(z_S)\frac{\partial z_S}{\partial x} + \tau_x(0) \tag{2.1.42}$$

$$fU = -\frac{\partial P}{\partial y} + p(\zeta)\frac{\partial \zeta}{\partial y} - p(z_S)\frac{\partial z_S}{\partial y} + \tau_y(0)\,, \tag{2.1.43}$$

where $p(\zeta)$ is atmospheric pressure and can be set safely to zero. Assuming *in addition* that the slopes of z_S are negligible (Sverdrup used a constant value for z_S), cross-differentiating these last two equations produces

$$\beta V + \left(\frac{\partial U}{\partial x} + \frac{\partial V}{\partial y}\right) = \frac{\partial \tau_y}{\partial x} - \frac{\partial \tau_x}{\partial y}\,. \tag{2.1.44}$$

Now integrate the continuity equation vertically,

$$\frac{1}{\rho}\left(\frac{\partial U}{\partial x} + \frac{\partial V}{\partial y}\right) + w(\zeta) - u(\zeta)\frac{\partial \zeta}{\partial x} - v(\zeta)\frac{\partial \zeta}{\partial y} - w(z_S) + u(z_S)\frac{\partial z_S}{\partial x} + v(z_S)\frac{\partial z_S}{\partial y} = 0\,. \tag{2.1.45}$$

But

$$w(\zeta) = u(\zeta)\frac{\partial \zeta}{\partial x} + v(\zeta)\frac{\partial \zeta}{\partial y} \tag{2.1.46}$$

is the condition that the seasurface not be moving in time. The sum of the last three terms of (2.1.45),

$$-w(z_S) + u(z_S)\frac{\partial z_S}{\partial x} + v(z_S)\frac{\partial z_S}{\partial y}\,, \tag{2.1.47}$$

vanish if (a) the individual terms vanish at z_S, or (b) z_S is the seafloor, or (c) z_S is a material surface (such as an isopycnal, i.e., a surface of constant density) across which there can be no fluid flow, in which case the sum of the three terms is zero. Assuming that one of (a), (b), or (c) obtains, we have

$$\frac{\partial U}{\partial x} + \frac{\partial V}{\partial y} = 0\,, \tag{2.1.48}$$

which when combined with (2.1.44) gives the famous *Sverdrup relation*

$$\beta V = \hat{\mathbf{k}} \cdot \boldsymbol{\nabla} \times \boldsymbol{\tau}\,, \tag{2.1.49}$$

whose importance in the history of the understanding of the oceanic general circulation is difficult to overestimate.

Equation (2.1.48) permits the introduction of a transport stream function, for example

$$U = -\frac{\partial \Psi}{\partial y}, \quad V = \frac{\partial \Psi}{\partial x}\,. \tag{2.1.50}$$

Substituting into (2.1.49), we obtain

$$\beta \Psi(x,y) = \int_{x_0}^{x} \hat{\mathbf{k}} \cdot \boldsymbol{\nabla} \times \boldsymbol{\tau} dx + G(y)\,. \tag{2.1.51}$$

The integration constant $G(y)$ can be used to set Ψ to zero on any boundary $x = L$, so as to assure no flow into the wall by putting

$$G(y) = -\int_{x_0}^{L} \hat{\mathbf{k}} \cdot \boldsymbol{\nabla} \times \boldsymbol{\tau} dx\,. \tag{2.1.52}$$

Simple mathematical considerations (e.g., Stommel, 1965; Pedlosky, 1987a) dictate that the boundary so chosen can only be an eastern boundary (assuming the physical boundary really is meridional). With this choice U, V are known, producing an estimate of the complete transport field in the ocean from the windfield alone. [Because there are no undetermined components of this solution, no additional boundary conditions can be satisfied, and it can be anticipated that the theory must break down in the vicinity of northern, southern, and western boundaries. This inability to satisfy all the boundary conditions is an immediate consequence of the reduction of the original Equations (2.1.1)–(2.1.5) to geostrophic balance. Whether this breakdown destroys the validity of the Sverdrup balance is a partially open issue, with the answer dependent upon where one is in which ocean basin.]

If Sverdrup's assumptions are valid, they produce a result of great value, for (2.1.49) or (2.1.51) permit one to connect the windfield at a single point (strictly the curl of the windstress) to the meridional transport of fluid over the whole water column. The power and simplicity of Sverdrup's result has led to its nearly universal employment in all theories of the wind-driven ocean circulation. In the next section we will examine the observational evidence for its validity.

The Sverdrup relationship is a transport theory, representing the sum of the horizontal Ekman layer transport plus that of the geostrophic interior, with the Ekman layer divergence owing to the β-effect exactly canceled by that of the interior, leaving the net flux of (2.1.51). Under the assumption of a geostrophic interior, it is easy to pull the vertically integrated horizontal transport apart, into an Ekman flow and an interior flow (e.g., Stommel, 1965). The interior meridional geostrophic flow is associated with a vertical velocity given by (2.1.20) that must accept the Ekman-pumped fluid; this Ekman pumping drives the meridional geostrophic interior flow v. [There is nothing to prevent one from turning this argument around and asserting that if the interior is set into motion with a meridional velocity component, then the vertical velocity forced through (2.1.20) must be absorbed by a near-surface Ekman layer, and satisfying a no-stress surface boundary condition if there is no wind; but this boundary layer would be a weak one.]

2.1.5 Stommel-Arons Theory

The question of how the ocean responds to buoyancy forcing through its interactions with the atmosphere has a long history, although it is thin by comparison to studies of the wind-driven circulation. Defant (1961, p. 489) outlines some of the early work, with particular emphasis upon a dispute between Sandström (1908) and Jeffreys (1925), although contrary to Defant, Jeffreys' arguments are the more correct. Schematically, one knows that at low latitudes, the ocean is warmed by the atmosphere, and at high latitudes, the ocean returns the heat. Getting beyond this simple statement is not so easy. The classical problem of the response of a fluid to heating goes under the name of the *Rayleigh-Benard problem*; Chandrasekhar (1961) and the collection by Saltzman (1962) are useful summaries of the older literature.

The oceanographic problem differs from the classical one in a number of important ways: (1) The system is both heated and cooled at the same upper boundary and is not heated from below and cooled from above as in the classical problem. The implications of this geometry were the center of the discussion by Sandström and Jeffreys. (2) The system is rotating and the apparent resulting circulation takes place over such a large meridional extent that the nonuniformity of the rotation is a central dynamical feature. (3) The ocean is stratified in both heat and salt; heat can be transferred to and from the atmosphere, but salt cannot (although freshwater can). This two-component convective system is potentially much more complex than a one-component system. (4) The boundary condition at the seasurface is not externally controlled; rather, the transfer of heat and moisture is a result of the *interaction* between the atmosphere and ocean, and the rates and signs can shift depending upon exactly what the oceanic motions are. (5) The convectively driven motion occurs in the presence of a wind-driven motion. The Equations of motion (2.1.1)–(2.1.5) are nonlinear, and it is not obvious that the convective motion can be understood without simultaneous understanding of the wind-driven elements.

Warren (1981) provides an interesting and helpful historical discussion of the convectively driven flow. Stommel (1962) pointed out what is perhaps the most important qualitative feature of large-scale oceanic convective motions–that the regions of sinking to great depth, owing to high latitude cooling and evaporation, are extremely small compared to those regions where the return upwelling is inferred to occur. (The issue of whether one really knows where the return flow occurs is taken up presently.) This apparent asymmetry (see Figure 2–6) is in striking contrast to the situation in Rayleigh-Benard convection, where there is at least a rough equality of the

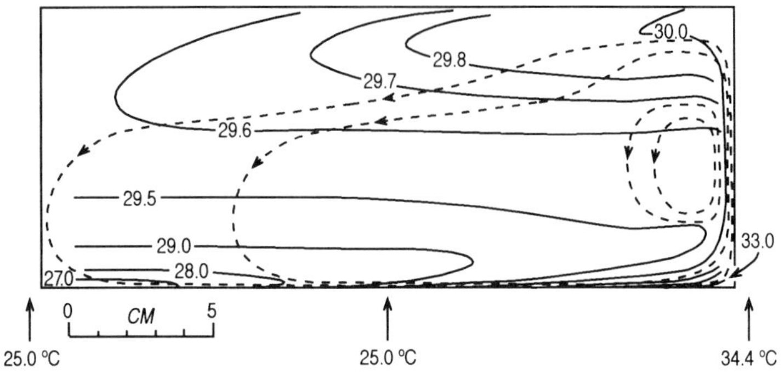

Figure 2–6a. Asymmetry of a circulation driven by heating and cooling at a fixed horizontal surface, here shown in Rossby's (1965) laboratory experiment. For technical reasons, the heating and cooling were applied from below and the roles of heating and cooling at the surface are interchanged. The warm, rising, branch against the right wall is much more confined than the broad cool one. Solid curves are the observed temperature field (°C), and the dashed lines show an estimate of the flow field inferred from Rossby's streak photographs. **b,c** (figures on facing page). Tritium injected into the ocean through nuclear weapons tests in the atmosphere, mostly in the early 1960s, is confined mainly to the oceanic surface, except where it reaches the seafloor at high northern latitudes in an analogue of the behavior in Figure 2–6a. (b) The condition in the northern part of the North Atlantic about 1971, and (c) conditions about 1981 (after Östlund & Rooth, 1990). Ticks along the bottom axis denote observation locations.

areas occupied by the ascending and descending elements. Stommel (1962) gave a qualitative explanation of the oceanic asymmetry in terms of the relative efficiency by which thermal convection is able to carry cold fluid to the floor of the ocean, and of the inefficiency by which the returning fluid could only be warmed by mixing processes. The latter phenomenon would thus require a far larger area to compensate the downward motions. Rossby (1965) produced a nonrotating, single component, laboratory analogue, and Beardsley and Festa (1972) extended it numerically. Nelkin (1987) summarizes the sparse more recent work.

Without producing a quantitative theory of the convection itself, Stommel and Arons (1960a,b) and Arons and Stommel (1967) constructed a theory of the deep circulation, resulting from the existence of the small convective regions, which stands today as one of the great monuments of oceanographic insight and imagination. The theory is attractive in its combination of elegant simplicity and results counter to intuition. Its elements can be reduced to a few simple assumptions and conclusions:

1. Away from the boundaries, the motion is geostrophic, with the divergence given by the linear vorticity balance (2.1.20).
2. Downwelling occurs in a few isolated regions, which can be considered point sources of given mass flux.

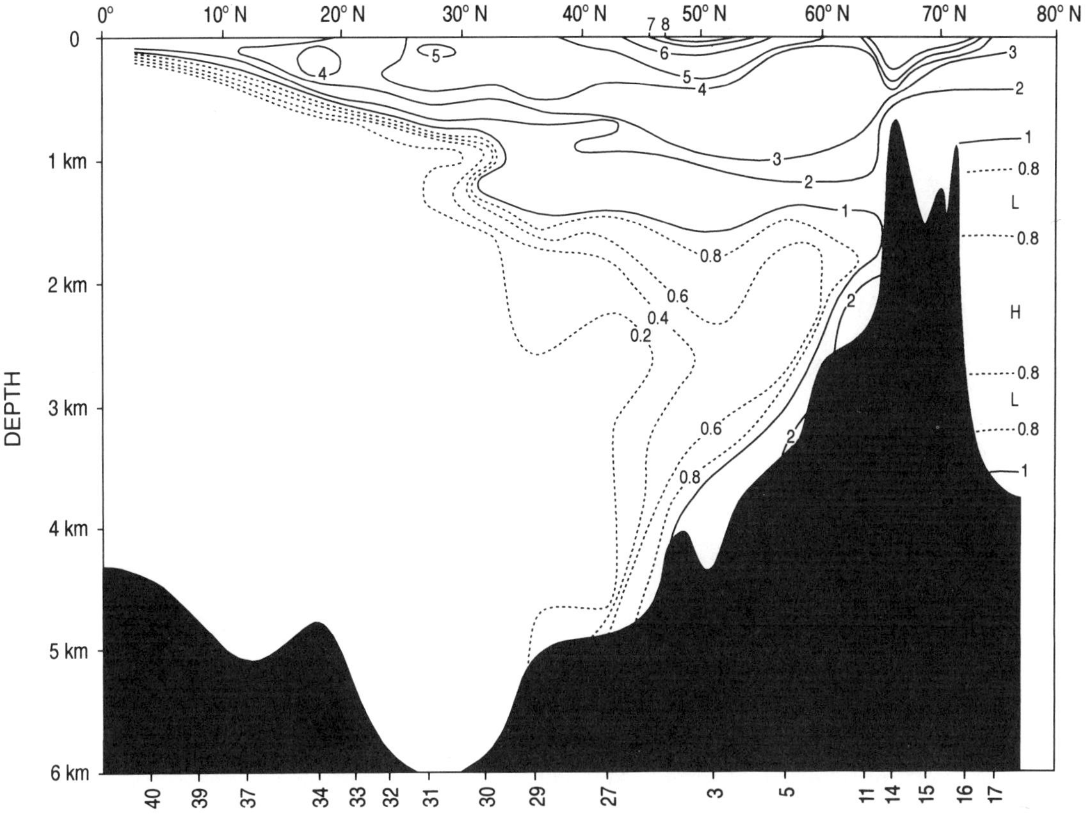
0°
10° N
20° N
30° N
40° N
50° N
60° N
70° N
80° N
0
1 km
2 km
3 km
4 km
5 km
6 km
DEPTH
L
H
L
40
39
37
34
33
32
31
30
29
27
3
5
11
14
15
16
17

b

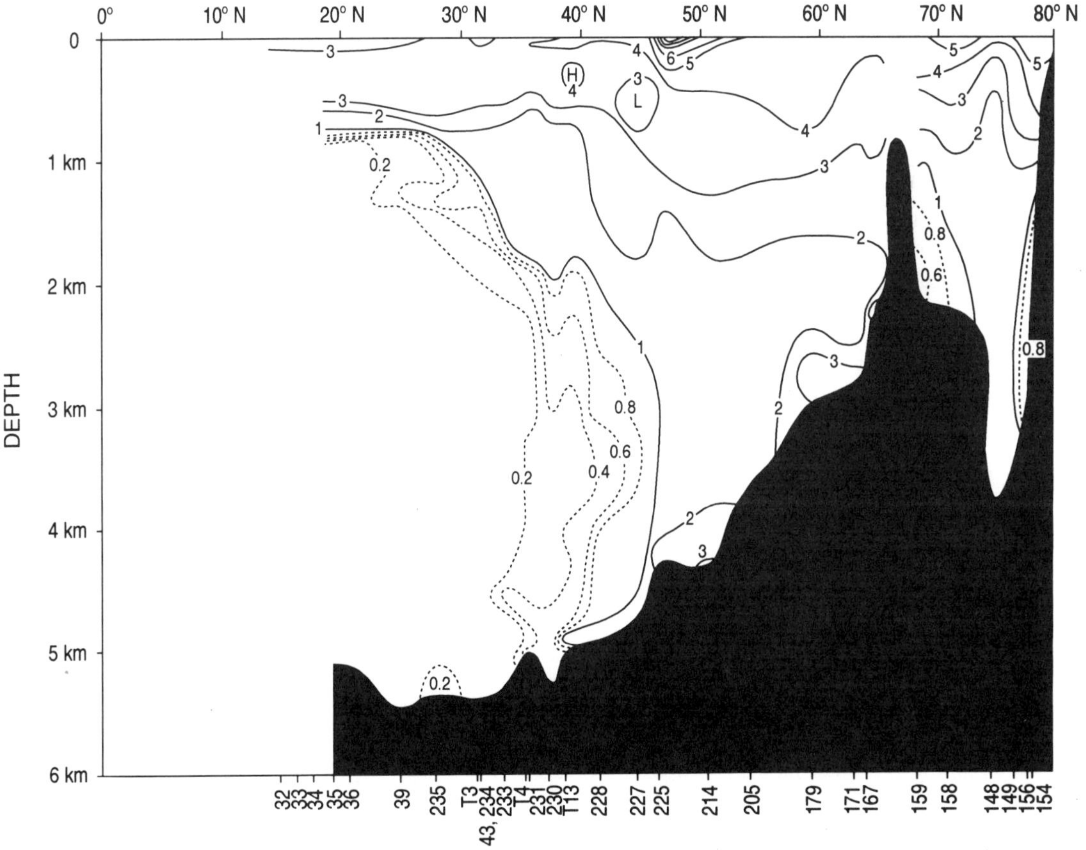
0°
10° N
20° N
30° N
40° N
50° N
60° N
70° N
80° N
0
1 km
2 km
3 km
4 km
5 km
6 km
DEPTH
H
L
32
33
34
35
36
39
235
T3
43, 234
233
T4
231
230
T13
228
227
225
214
205
179
171
167
159
158
148
149
156
154

c

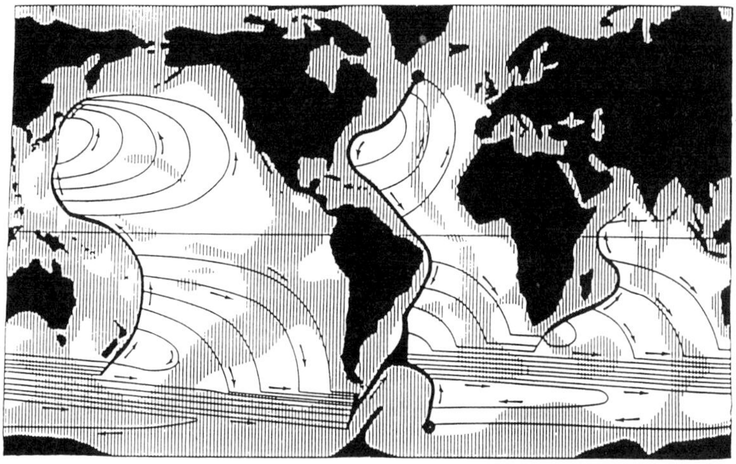

Figure 2–7. Abyssal circulation scheme constructed by Stommel (1958) under the assumptions of: mass sources in the northern Atlantic and in the Weddell Sea; uniform interior upwelling; and dynamically passive western boundary layers, which balance the local and global mass budgets.

3. The vertical return flow occurs as upwelling over the interior of the ocean. In the absence of additional information, the upwelling is assumed to be uniformly distributed. This uniformity assumption is not an essential ingredient but is arguably the simplest assumption that can be made.
4. The ocean is in steady state so that there must be mass balance over any oceanic volume. Any net meridional motion in the interior can be compensated in a western boundary current, whose dynamics are purely local. By "local" is meant that the boundary current structure and physics do not modify the interior physics given by (2.1.11)–(2.1.13), (2.1.20).

The consequence of these assumptions leads to some quite striking predictions, which were tested by laboratory analogue and reported by Stommel, Arons and Faller (1958); see also Faller (1981). The schematic circulation that results for the global ocean is depicted in Figure 2–7. We will take up later the observational evidence for the theory; here we remark only that it predicts the existence of a deep western boundary current (DWBC) in the North Atlantic (and analogous flows in all other oceans), a geostrophic interior motion that is eastward, out of the detraining DWBC, turning northward *toward* the mass source. The interior is everywhere upwelling, $w > 0$. Coupled with the picture described above, that the Ekman pumping in the subtropical gyre drives a downward flux of warm water, the Stommel-Arons upwelling of cold water suggests a picture of the thermocline as a layer sandwiched between these two competing flows. Indeed, the Stommel-Arons model led fairly quickly to the "old" thermocline theories (see Robinson & Stommel, 1959; Veronis, 1969; Welander, 1971; Needler, 1972). It is only recently that any significant extension of the Stommel-Arons picture has occurred, in the work of Kawase (1987; also see Cane, 1989), discussed in Section 2.2.

The simple physics laid out thus far–hydrostatic, geostrophic balance with an elementary Ekman layer superposed–does not do justice to the large and interesting subject of the physics of the ocean circulation. But textbooks (e.g., Pedlosky, 1987a) exist in which the theory is treated with great care. With some exceptions, taken up where needed later, the physics we have made explicit is adequate for discussion of most of the problems treated.

2.2 Observations

In this section we review some of the observations available to test the simple sketch of the physics of the ocean circulation and to understand, quantitatively, what we do and do not know for certain.

2.2.1 The Ekman Layer

It is possible to observe the Ekman spiral under laboratory conditions where the stress terms are, rigorously, the molecular ones (e.g., Tatro & Mollo-Christensen, 1967). But in the ocean, the Ekman spiral may be the most discussed and least observed of all theoretical constructs.

Over the years, many attempts have been made to verify the existence of Ekman layers at sea. The ensuing difficulties stand in stark contrast to the ease with which the layer is derived mathematically. Convincing observations of anything resembling a spiral are few. Ekman himself used his theory to explain why icebergs apparently moved at an angle to the right of the windfield. Hunkins (1966) showed a spiral under moving ice; Weller (1981) and Price, Weller, & Schudlich (1987), by averaging current meter observations over many days in a frame rotating with the wind, produced a spiral-like structure (Figure 2–8). These and other results confirm that spiral-like structures are indeed possible–welcome support for the general theoretical structure. What remains unclear is precisely how to connect the measured wind to the stress within the ocean–a stress meant to generate the mean transport to the right of the wind.[6] That is, even if an instantaneous estimate of the wind *stress* is available, what fraction of that stress is communicated to waves and Langmuir cells (Craik, 1985), and what fraction of that wave motion ultimately appears as the transport at the appropriate

[6] To the left of the wind in the southern hemisphere.

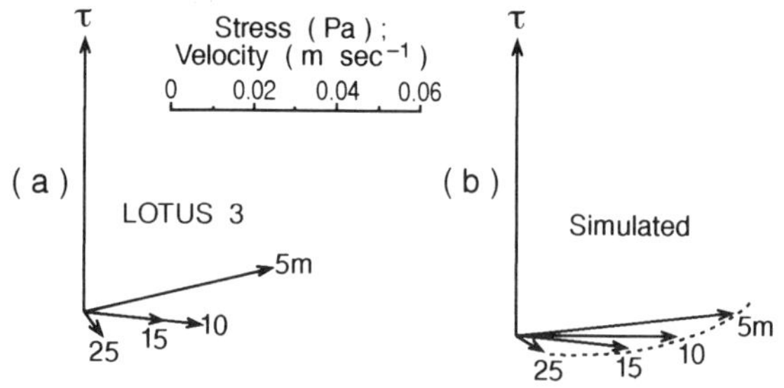

Figure 2–8. (a) Spiral-like structure observed in the surface layer (adapted from Price et al., 1987), as compared to (b), a theoretical Ekman spiral. This example is one of very few available in the ocean.

angle to the wind implied by (2.1.34)–(2.1.35), and how is it distributed with depth? A quantitative answer to this question remains beyond us.

Despite the difficulty in observing the actual Ekman structure at sea, and the absence of a quantitative connection with the observed wind, the transport result (2.1.34)–(2.1.35) and its independence of the detailed form of A_v leads most investigators to conclude that the fundamental result–Ekman divergence as the driver of the circulation–is valid. Its use in oceanography is thus mainly a consequence of the underlying simplicity of the physics, rather than any but the most indirect observational confirmation. Nonetheless, it must be admitted that the observational evidence for Ekman layers is uncomfortably thin. The main motivation for retaining it is that without the hypothesis (2.1.34)–(2.1.35), little or nothing could be done about the theory of the wind-driven ocean circulation.

But for a given windfield, quantitative understanding of just how much water will move on average to the right of the wind, as opposed to going into waves, or turbulence, remains inadequate. Very recently, Chereskin and Roemmich (1991), for a single trans-Atlantic section, managed to demonstrate some quantitative relationship between the basin-wide Ekman transport and direct observations from a ship–that is, they produced an *integral* demonstration of consistency.

What might the Ekman pumping look like if we assume the validity of the transport relation (2.1.34)–(2.1.35)? Leetmaa and Bunker (1978) took estimates of the windstress over the ocean and computed the divergence of the Ekman layer in the North Atlantic depicted in Figure 2–9a. Such a calculation involves differentiating (to obtain the curl) the estimated windstress field; no error estimates are given. Taking the result at face value for the moment, one sees a region of downwelling (pumping) south of the zero line that runs roughly from Cape Hatteras to Ireland. This region, of downward Ekman forcing, is usually referred to as the *subtropical gyre*, and the region to the north, of Ekman suction, is the *subpolar gyre*. Numerical values are order $10^{-5} - 10^{-4}$ cm/s, or 3–30 m/yr. The time variability of the windfields can be seen in Figure 2–9b,c.

Examination of Figures 2–2c,d,f,g lends a certain rough verisimilitude to this picture of Ekman pumping in the following sense. The subtropical gyre hydrography shows a bowl-shaped thermocline–that is, the thermocline is depressed relative to its values in the surrounding region, leading to a relative excess of warm water. This feature is often ascribed to the downward pumping by the Ekman layer of warm surface water, and indeed it is readily shown from thermocline theories (e.g., Needler, 1967; Welander, 1971; Luyten, Pedlosky, & Stommel, 1983) that such a feature emerges from a comparatively simple set of theoretical ideas.

Other indirect evidence also exists. For example, it is quite easy to see that if one blows a steady windfield parallel to the coastline (see Figure 2–10) in the configuration shown, the supply of water moving to the right of the wind can only be produced from below. The conclusion is that such "upwelling" regimes should occur when the wind drives the Ekman flux offshore (see, for example, Gill, 1982, for the theory and some of the many observations of upwelling regimes). Although again, there is little doubt that this physics actually does produce some elements of what is observed as upwelling, the situation in practice is far more complex than the simple static balance implied by Figure 2–10. Most observed upwelling contains strong time dependence (i.e., baroclinic wave-dominated motions, for example, see Gill & Clarke, 1974; Gill, 1982), and even the definition of upwelling has become unclear, some regarding it as the time-dependent vertical moving isotherms, others as the quasi-static situation depicted in Figure 2–10.

There is no reason to doubt that the physics contained in the Ekman layer, and that its zero-order consequence, a vertical pumping of the fluid interior, actually occurs. What is lacking generally is a quantitative connection between the windfield and the actual movement of water. Such a quantitative estimate would necessarily imply that one understood the space and time scales over which one had to average to produce the appropriate balance.

2.2.2 Geostrophic Balance

Wüst (1924) produced the classic test of geostrophy in the ocean. He took a set of remarkable current meter measurements made from an anchored ship in the Florida Current, reported by Pillsbury (1891), which when

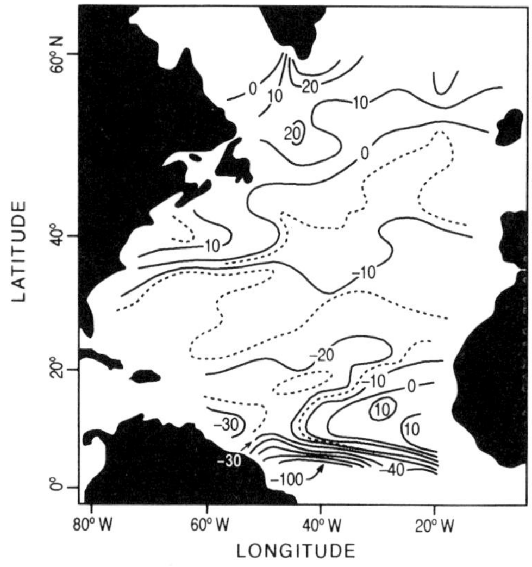

(a)

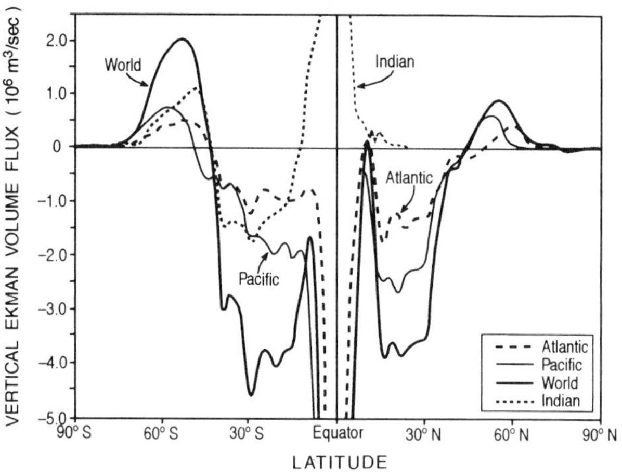

(b)

Figure 2–9. (a) Leetmaa and Bunker (1978) estimate of the Ekman pumping velocity in units of 10^{-5} cm/s. Contouring is incomplete near South America. The accuracy of such pictures is not known. (b) Zonally integrated Ekman fluxes as a function of latitude (from Levitus, 1988b) in different oceans and globally. The accuracy of these estimates is unclear, both because of sampling issues and suspected systematic errors in the underlying numbers, here derived from the Hellerman and Rosenstein (1983) climatology.

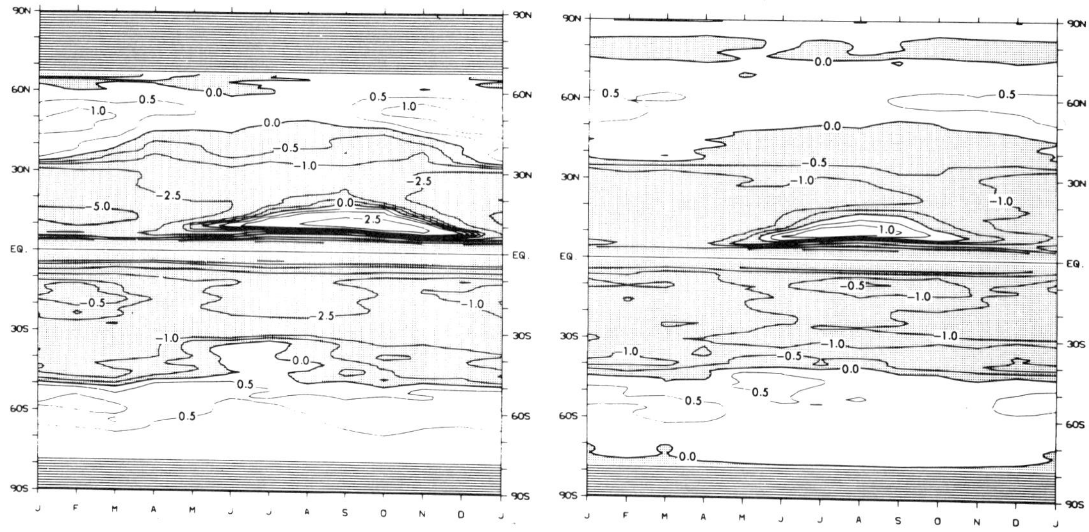

Figure 2–9c. Latitude-time diagrams of the zonally integrated Ekman flux throughout the year in the Pacific (left) and Atlantic (right), based upon climatological data (Levitus, 1988b). Notice the reversals in sign appearing at low latitudes and the fluctuations of the zero lines. Again, the reliability of these results and their interannual variability is unknown.

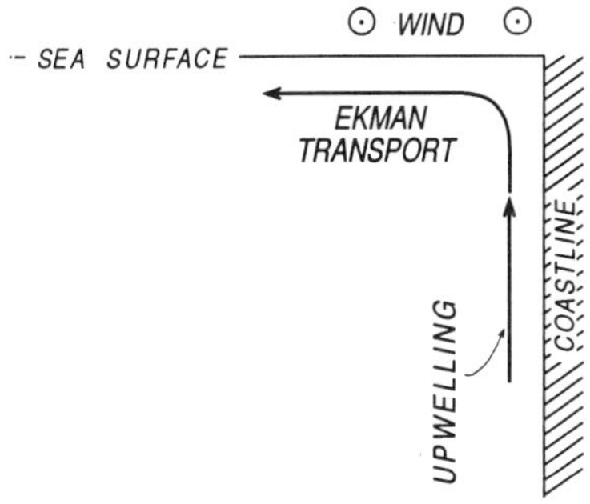

Figure 2–10 (above). Schematic of a simple, steady, upwelling situation with a wind blowing parallel to the coast, out of the page, in the northern hemisphere and the Ekman mass flux directed offshore. In practice, calculating the structure of the boundary layer next to the wall is a challenging mathematical problem. In observation, the situation is usually highly unsteady.

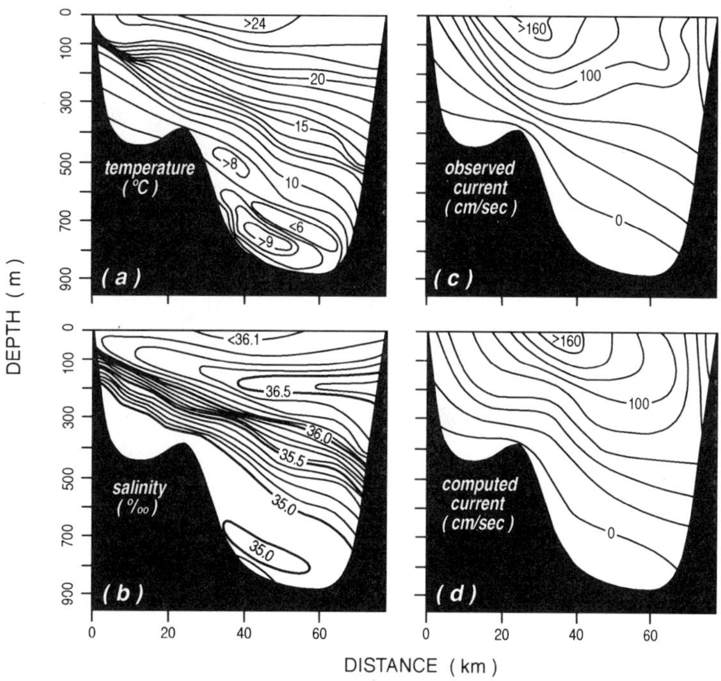

Figure 2–11. Florida Straits results as deduced by Wüst (1924) and reproduced in Sverdrup et al. (1942). Temperature (a) and salinity (b) sections were used by Wüst to compute the thermal wind (d) and compared to Pillsbury's measurements from an anchored ship (c), setting the level-of-no-motion to that calculated by Pillsbury. The result was taken to be a demonstration of the basic utility of the thermal wind relationships.

averaged produced a velocity profile as displayed in Figure 2–11b. He then took hydrographic measurements from 1878 and 1914 (Figures 2–11a,c) and computed the thermal wind (Figure 2–11d), setting the level-of-no-motion to coincide with that in Pillsbury's results. The agreement between the measured shear and that computed from the thermal wind was widely accepted as a proof of the utility of the dynamic method. Notice that the data were far from synoptic and that the test is made visually from comparison of Figures 2–11b,d.

Years later, Bryden (1977), Horton and Sturges (1979) using moored current meters, Swallow (1977) using neutrally buoyant floats, and others failed to detect any statistically significant deviation between the thermal wind shear computed from hydrography at point locations and corresponding time-average direct measurements. Theory suggests that detection of any such discrepancy by direct measurement of the difference between the thermal wind and directly measured shears should be very difficult; had any investigator found a significant deviation, it would have been extremely troublesome to our theoretical understanding of the ocean. Bryden and Fofonoff (1977) similarly failed to find any detectable deviation of the flow field from the *local* nondivergence demanded by Equation (2.1.19).

All of the above tests are based upon comparisons of the vertical shear of the geostrophic fields, not of the velocity field itself, owing to the missing integration constants. As already stated, the ocean cannot be strictly in geostrophic balance, but the required deviations are very small and evidently unmeasurable locally in space and time. We will see presently some nonlocal evidence. There is one clear example of small deviations from geostrophy: To the extent that the flow field and the mass field are observed to evolve in time, one is necessarily observing a nongeostrophic effect. Even so, at any given instant, the deviation from the balance (2.1.11)–(2.1.12) is typically unmeasurably slight.

2.2.3 Sverdrup Balance

Turning now to another pillar of oceanographic theory, Sverdrup balance, let us examine what is known of its validity. It provides an interesting way to discuss the difficulty of turning oceanographic observations and theory into the quantitative subject that is the overall goal of physical oceanography.

In his original derivation (Sverdrup, 1947), the circulation in the tropical Pacific computed from (2.1.51)–(2.1.52) was compared to that obtained from the dynamic method with a level-of-no-motion. He took the hydrography available to him, a level-of-no-motion at 500 decibars, and assumed the resulting flow was the correct one. With the best windfield estimate he could find, he drew the picture displayed in Figure 2–12 using (2.1.51) to compare the resulting wind-driven flow with that estimated from the hydrography and the level-of-no-motion. The appearance of the countercurrent, flowing directly into the prevailing winds, was an exciting result and a very convincing element of this theory.[7] On the other hand, the predicted values do not everywhere lie on top of the estimated values, no error estimates were provided, and one is still faced with the usual quantitative questions of whether the flow in the tropical Pacific is estimated to, say, 50% or to 95% accuracy by Sverdrup's theory, as measured by some criterion, for example, the mean kinetic energy; the time and space scales over which one

[7] See the description by Munk (1984) of Sverdrup's approach to this problem.

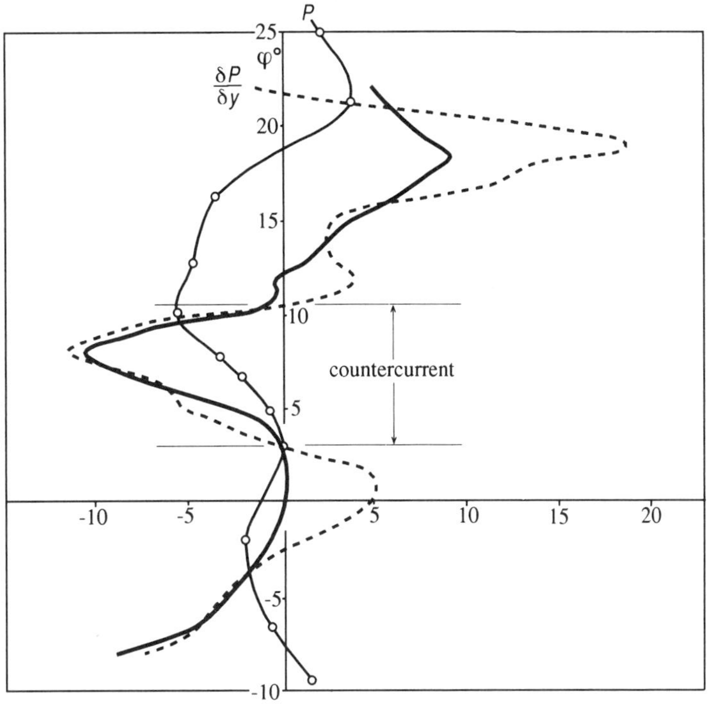

Figure 2–12. Picture, adapted from Sverdrup (1947), of the zonal transport computed from Equation (2.1.51) (solid line), determined directly from the thermal wind and a level-of-no-motion at 500 decibars (dashed line). Light solid line is the pressure field relative to the reference level, and the dashed line is its meridional derivative. The prediction of the countercurrent ($u \propto -\partial p/\partial y$) is quite convincing, but its sensitivity to the arbitrary assumption of a level-of-no-motion at 500 decibars is unclear. Also unknown is whether the quantitative discrepancies at latitudes away from the countercurrent are an indication of a physical problem or simply the differences expected given the data noise and sampling problems.

must average to achieve this accuracy; and whether the model has similar or different accuracies over any other part of the world ocean.

Leetmaa, McCreary, & Moore (1981), in reexamining this result, concluded that it was a lucky accident–depending upon Sverdrup's choice of 500 decibars as a level-of-no-motion and the particular wind estimates employed (October–November). Had any other choices been made, the result would have looked much less convincing, with who knows what consequences for the subsequent development of ocean dynamics theory.

Sverdrup's relationship is so central to theories of the ocean circulation that almost all discussions assume it to be valid without any comment at all and proceed to calculate its consequences for higher-order dynamics. It is not difficult to see the attraction–the Sverdrup relationship in the form (2.1.51) produces a simple connection between one of the forcing fields, the windstress, and the interior oceanic flow, a relationship that is purely local. Its simplicity has permitted theory to focus on the dynamically more interesting parts of the ocean such as the boundary currents. The absence of such a simple relationship between the interior flow and the other major forcing function, the buoyancy flux to and from the atmosphere, led to a neglect for decades of the theory of the thermodynamic response of the ocean.

So it is difficult to overestimate the importance of Sverdrup balance. But because it is so important to all putative understanding of the circulation, one should carefully examine the evidence for its actual applicability. In deriving it, several suppositions were made: (1) that the interior is geostrophic, and by inference, that the linear vorticity balance (2.1.20) is valid; (2) that there is a uniform depth where all three components of velocity vanish; and (3) that the Ekman transport is given by (2.1.34)–(2.1.35). Thus, a quantitative test of Sverdrup balance becomes a test of the simultaneous validity of several assumptions and conclusions.

In an interesting and influential paper, Leetmaa, Niiler, & Stommel (1977; I will refer to this paper as LNS) examined the Sverdrup relationship in the North Atlantic and inferred its validity as an integral relationship over the oceanic width. But Wunsch and Roemmich (1985) suggested that in fact much of the theory was being assumed rather than tested, and that to the contrary, the major oceanographic evidence was against its validity in the North Atlantic, at least on a large scale.

Let us examine the problem beginning with the issue of the windstress. Figure 2–13 shows two estimates of the climatological windstress over the North Atlantic, one from Hellerman and Rosenstein (1983), the other from Isemer and Hasse (1985), and the difference between them. Although the database employed in both estimates is essentially the same, the difference field is a considerable fraction of the variance of either of them. Böning, Döscher, & Isemer (1991) ascribe the difference between the results to a difference in the drag coefficient used to convert low windspeeds to stress estimates. In particular, the conversion between windspeed nominally measured at 10 m on the decks of ships and stress at the seasurface used in the climatologies is of the form

$$\boldsymbol{\tau} = C_D \rho |\mathbf{u}| \mathbf{u}$$
$$C_D = C_D(|\mathbf{u}|,\ \theta_a - \theta_w) \qquad (2.2.1)$$

where $\theta_a - \theta_w$ is a measure of the air-sea temperature difference. Because all ships' anemometers are not mounted at the same height, there is a correction made for the actual height. A further correction is necessary for the ship's motion and, in principle, for the actual superstructure disturbance

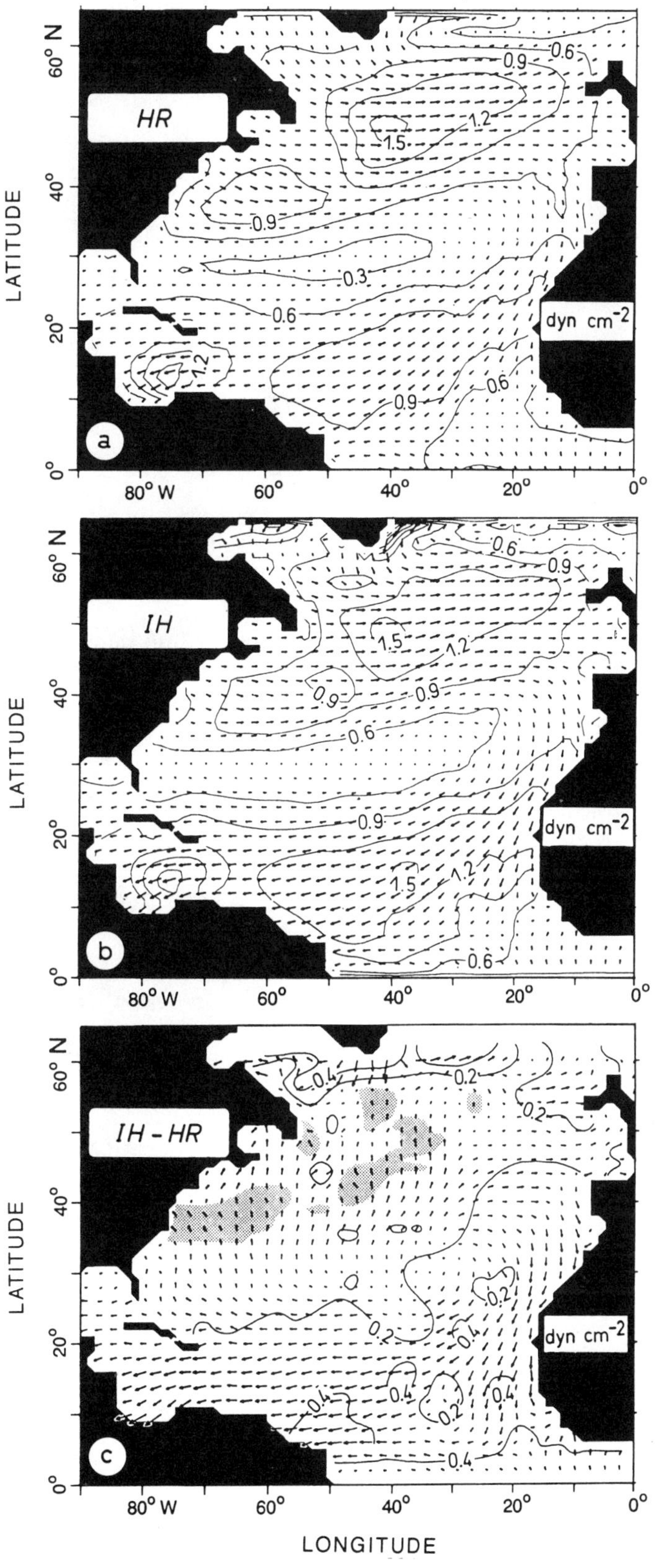

Fig. 2–13. Comparison by Böning et al. (1991) of two wind climatologies in the North Atlantic, that of Hellerman and Rosenstein (1983) and of Isemer and Hasse (1985). Difference is shown in the lowest panel; note that the contour interval is 0.2 in the lowest panel, but 0.3 in the upper. The data used were basically the same in both, so the differences are largely attributable to model errors.

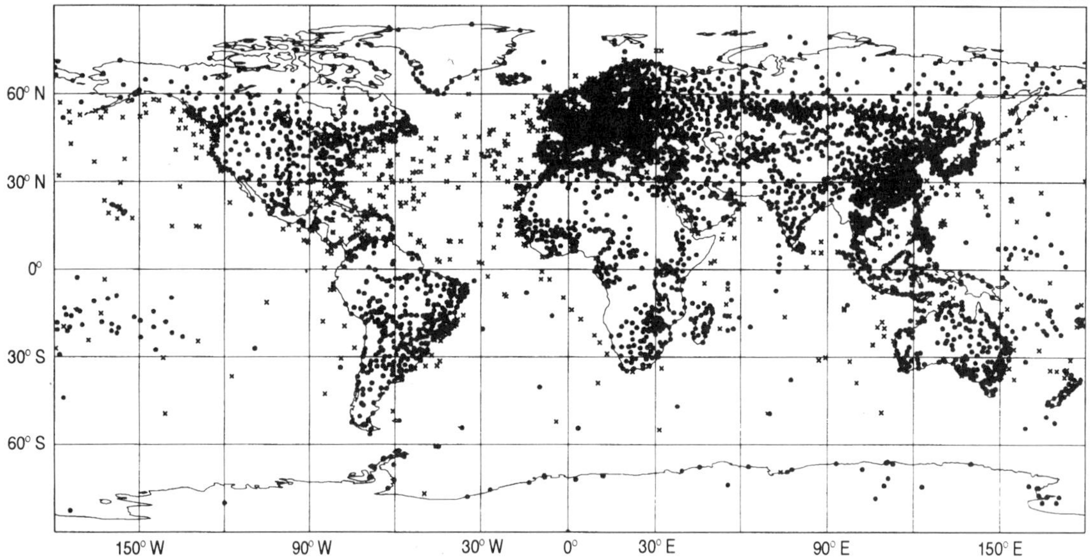

Figure 2–14. Distribution of ship and land observations available for meteorological analysis and forecasting on one particular, but typical, day. The majority of the oceanic observations are from ship reports (x) and represent much of the information available for estimating the oceanic windfield and oceanic seasurface temperatures. The great inhomogeneity and sparsity of the oceanic measurements means that the overocean fields are fundamentally undersampled, rendering gridding almost impossible. (From the European Centre for Medium Range Weather Forecasting, courtesy P. Viterbo and L. Illari.)

on the ship. Different investigators make substantially different choices of the function C_D and its dependence upon the windspeed. There is also an issue of the spatial distribution of ships. Figure 2–14 shows all the ship reports of wind during one day. Large areas of the world ocean are unmeasured even once, much less adequately to produce any reasonable average. Fundamentally, we do not know how good the climatologies in Figure 2–13 are or which, if either, is more accurate than the other. The subject is difficult and in an unsatisfactory state: Trenberth, Large, & Olson (1989a), Trenberth, Olson, & Large (1989b), Harrison (1989), Böning et al., (1991), and Halpern, Hollingsworth, & Wentz (1994), among others, have suggested that serious systematic errors exist in the various climatologies. Even if the Sverdrup balance were completely accurate, how well would the circulation be computable with the existing wind climatologies?

As with many elements of oceanographic physics, one can perform direct tests of a particular balance pointwise. We examined geostrophy and concluded that pointwise balance obtained to such a high accuracy that we were

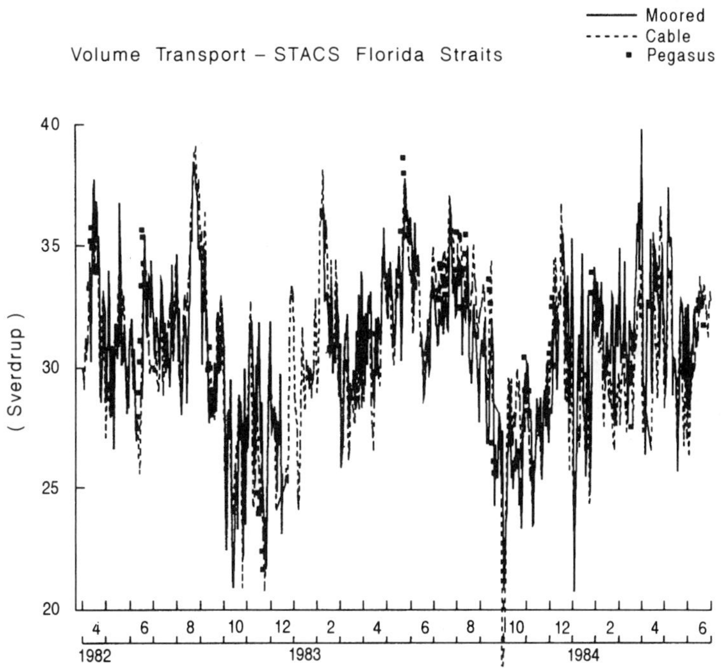

Figure 2–15. Time series of measurements of transport in Florida Current (Schott et al., 1988, and R. Zantopp, private communication, 1995). Measurements by three different techniques are shown: from moorings, electromagnetic cable, and a transport float (labeled *Pegasus*). The day-to-day variability is from less than 20 Sv to more than 40, including a superposed seasonal cycle. The problem of sampling such a flow in order to produce an adequate mean will be apparent, but this particular region is one of the few lending itself to long-term intensive measurements. Other currents that have not been adequately measured are often simply assumed to be stable in time.

frustrated in determining any deviations from it. The Sverdrup relation in the form (2.1.49) is a pointwise balance. No one has seriously attempted a test that extends beyond the sort that Sverdrup himself conducted. (There have been some claimed tests, but they were directed instead at the *time-varying* circulation, which is a very different problem.)

A weaker test is to show that the large-area-average of the physics is quantitatively correct, and this approach is the one taken by LNS, which we will now describe. First, there have been a long series of measurements of the transport of the Florida Current (the Gulf Stream) as it flows bounded at 25°N between Florida and the Bahamas. These measurements (for example, Schmitz & Richardson, 1968; Molinari, Wilson, & Leaman, 1985; Schott, Lee, & Zantopp, 1988) produce a stable average transport that is roughly 32 ± 3 Sv. (Figure 2–15). The wind climatologies, such as either one in Figure 2–13, can be used to compute $\hat{\mathbf{k}} \cdot \boldsymbol{\nabla} \times \boldsymbol{\tau}$. LNS show that the Leetmaa and Bunker (1978) climatology, when integrated westward across the Atlantic at 25°N, produces a value of nearly 30 Sv, with an estimated error of about 30%. Thus, one necessary integral consistency check is passed–that if (2.1.49) is valid everywhere across the Atlantic at 25°N latitude, the numerical value produced by the integrated wind-curl is roughly consistent with the northward transport by the Gulf Stream, a necessary agreement if mass is to be conserved. (Baker, 1981a, produced a similar consistency result in the Southern Ocean.)

It remains to demonstrate the validity of (2.1.51). As already described, there are two elements to the meridional integrated transport–the part owing to the Ekman flux and the part owing to the geostrophic interior. The wind climatology produces an Ekman transport of about 3 Sv (although again, an error estimate would be desirable).

An estimate of the geostrophic mass transport is required. LNS take the hydrographic section at 24.5°N displayed in Figure 2–2g and use the dynamic method. Here, one sees the beginning of some potential trouble: A time-averaged windfield has been used to estimate the right-hand side of (2.1.51); but a nearly synoptic (i.e., nonaveraged) hydrographic section is used to calculate the geostrophic velocity on the left-hand side. Does this make quantitative sense? We do not know enough about oceanic variability to be sure. An argument in its favor is that since we are going to integrate zonally across the Atlantic, the spatial averaging that results may be, through an ergodic assumption, equivalent to a time average along the same section. We can assume for the moment that this is true and see where it leads us. Therefore, the meridional velocity is computed from (2.1.24) in the conventional way. But there remains the problem of z_S (actually the problem of determining the integration constant in the thermal wind balance). LNS show that they can find a fixed value z_S taken as a level-of-no-motion, $z_S = z_0$, near 1500 decibars, for which the net mass transport, integrated westward from the eastern boundary produces the values shown in Figure 2–16.

One sees large oscillations near the western boundary. If the final integrated value is taken seriously, the net meridional transport is not the 32–33 Sv we require (ignoring the Ekman contribution), but closer to 16 Sv. LNS argue that the structure seen is ephemeral, being time-dependent eddies, and therefore extrapolate across this region as shown, to produce a final value of about 25 Sv, which within error bars, agrees with the required value. The need for the extrapolation is, of course, a statement that the ergodic hypothesis has failed. Whether these structures are indeed time-dependent, with zero time average, is a question to which we will return.

There are other difficulties. We have seen that in the Sverdrup model z_S must be a depth where the velocity vanishes. It would suffice to show either that $u, v = 0$, or since we are dealing only with a zonal section, that $\partial w/\partial z = 0$ there. Neither of these tests is possible with the data being

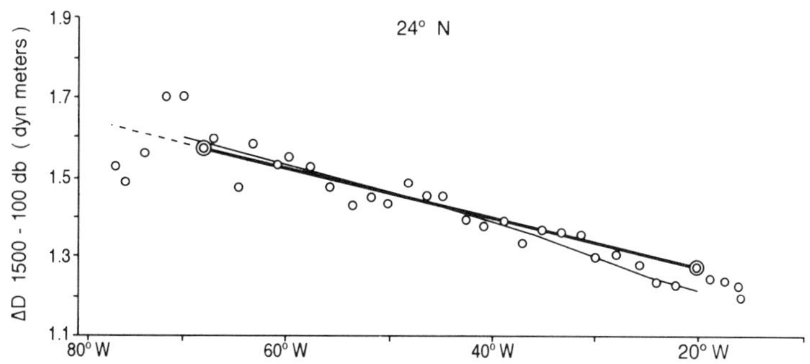

Figure 2–16. Total transport computed (Leetmaa et al., 1977) by integrating the dynamic topography from the eastern Atlantic along the 24°N section (Figure 2–2g) using a level-of-no-motion of 1500 decibars. The extrapolation on the western end (dashed) was justified by arguing that it represented transient eddies; but it probably actually represents the large-scale recirculation regime of the western boundary current system. The light line represents the dynamic topography estimated from the windstress curl.

used. We must *assume* the validity of this crucial assumption and can only examine its plausibility, to which we will return.

Thus, the test of Sverdrup balance as a basin-wide average can be described as dependent upon several critical *assumptions* that: a synoptic section can be combined with a climatological windfield; the climatology is an accurate reflection of the true wind; the extrapolation across the western portion of the hydrographic section is sensible, despite the failure of the ergodic hypothesis; and the chosen z_S is a level-of-no-horizontal motion.

When one draws conclusions about the behavior of some aspect of the ocean circulation, one should try to understand what the implications are for other aspects of the circulation. In that spirit, what are the implications if one assumed that the LNS conclusion was correct? Among other reasons for understanding the ocean circulation, the fluxes of properties that have climatological consequences by the circulation loom large. Consider the meridional transport of heat across this same section. In Chapter 4, we will examine a number of estimates of the heat flux across this section. There is a consensus that about $1.1 \pm 0.2 \times 10^{15}$W (1.1 PW) of heat is carried poleward across the section. Suppose, however, that the circulation were consistent with Sverdrup balance as just suggested–that is, that the 30 Sv carried northward by the Gulf Stream was returned above $z_S \approx 1500$ decibars. The mean temperature flux of the Florida Current has been estimated several times; Bryden and Hall (1980) suggest a value of about $18.8°\text{C} \times 29.5$ Sv $\approx$ 555°C Sv. The temperature flux of the 24°N section, with a level-of-no motion at about 1500 decibars, is about 540°C Sv. Because mass is assumed conserved, we can turn the temperature flux into a net northward heat flux across 24°N of about 15°C Sv $\approx 6.3 \times 10^{13}$W, a small fraction of what appears to be required in reality.

We thus have a conundrum: If we believe the estimate that the meridional heat flux is about 1.1 PW, as suggested by many authors, then there is a contradiction with zonally integrated Sverdrup balance. Which is to be taken more seriously? We must really leave this question to Chapter 4, but some comment can be made here. The conclusion that the geostrophic flow returned the requisite 30 Sv southward above z_S depended specifically upon the validity of the extrapolation across the last 10° of longitude on the basis that the large oscillations were time-dependent with zero true-time average. Figures 2–17c-e are various estimates that have been made of the horizontal transports of mass or volume in the North Atlantic. These estimates differ widely; what they tend to have in common is the suggestion that there are large-scale, very strong recirculations in the vicinity of the Gulf Stream. These recirculations are not the mesoscale eddy fields described in the last section. They may not be permanent in detail, but they appear to be always present in some form. The argument that they have a time-vanishing meridional circulation is far from proven.

We can examine yet another bit of evidence concerning this question. Figure 2–18 (Böning et al., 1991) shows an estimate of the Sverdrup transport stream function computed from (2.1.51)–(2.1.52)–that is, what the transport would be should we be able to find a z_S satisfying Sverdrup's assumption. Both wind climatologies shown agree that the zonal flow is into the western boundary current south of about 29°N latitude, and *out* of the western boundary current north of there. Note that this latitude is well south of Cape Hatteras.

Theories of the Gulf Stream (e.g., Stommel, 1965; Pedlosky, 1987a) and in particular Charney's (1955) inertial theory show that in order for the current to remain against the coast, it requires an inflow from the interior ocean. Figure 2–18 suggests that were the interior east of the Stream actually consistent with Sverdrup balance, the Gulf Stream would have to be losing water. The theories show that in this circumstance the Stream would leave the coast well before Cape Hatteras. Because one of the few things we are really sure of is that the Gulf Stream follows the coast to Cape Hatteras, either the theory is wrong or the interior flow adjacent to the Stream cannot be in Sverdrup balance. Given the structures always observed in this region, the latter appears to be the more compelling conclusion, and it seems unlikely that an extrapolation can be defended. But without the ex-

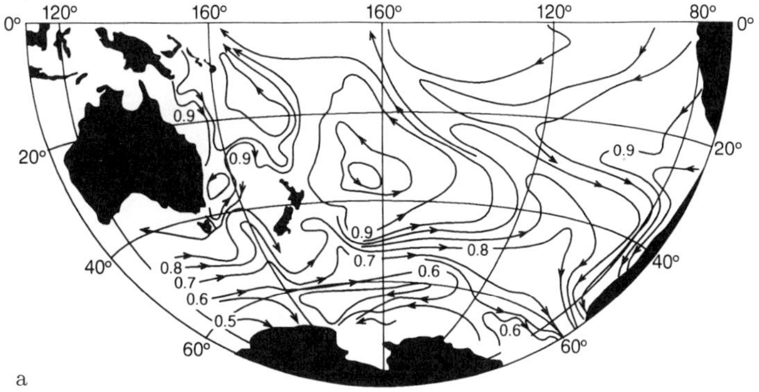

a

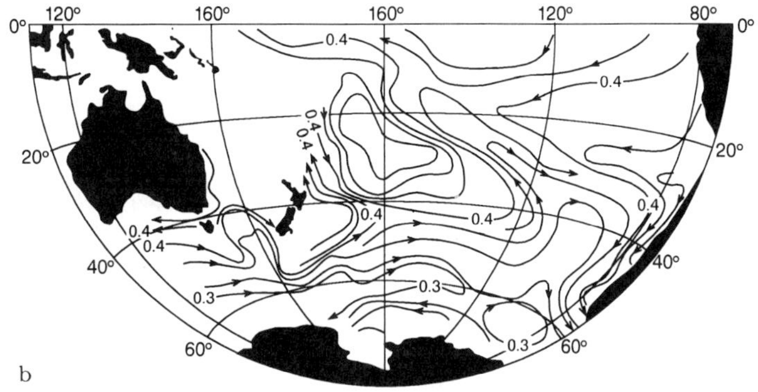

b

Figure 2–17. Representative circulation diagrams, at different levels and depths in different oceans. (a) and (b) are from Reid (1986) in the South Pacific Ocean and show his inferred absolute flows at 2000 and 3000 decibars, respectively. No simple reference level is used. (c) and (d) are from Stommel, Niiler, and Anati (1978) and depict the flows at 100 decibars, relative to a level-of-no-motion at 1500 decibars, and at 1500 decibar relative to 3000 decibars. Elevation units are dynamic millimeters (10^{-3} dynamic meters). (e) North Atlantic transport streamfunction between 250 and 3500 decibars from Fukumori et al. (1991), based upon a 3000-decibar reference level and a different data set from that used in (c,d). These diagrams are representative of attempts to infer the circulation by qualitative methods. Different data sets and different assumptions, such as levels-of-no-motion, produce varying results for which determining whether any is closer to the real ocean in some aspects than others has proved nearly impossible. Such pictures agree qualitatively but disagree quantitatively. Buried in them is the serious question of the degree to which it makes physical sense to combine nonsimultaneous data from a turbulent circulation. Over the years, differences between successive versions of such pictures have hardly diminished–suggesting that estimates of the ocean circulation are not converging.

trapolation, there is no longer agreement between Equation (2.1.51) and the Gulf Stream transport, and the last vestige of proof of integrated Sverdrup balance has disappeared.

Having lost the integrated balance possibility, we can still ask whether the pointwise balance might be valid, or perhaps would be so if integrated over some region in the North Atlantic. One is led to the eastern basin of the North Atlantic, as far as possible from the Gulf Stream recirculations. Wunsch and Roemmich (1985) examined this possibility by asking whether

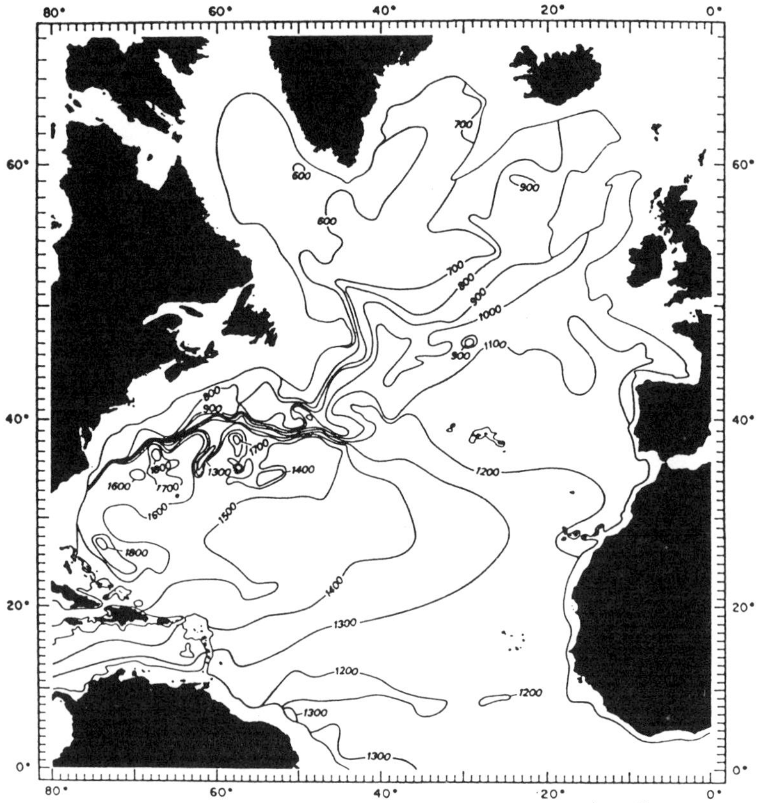

Figure 2–17c. Caption on preceding page.

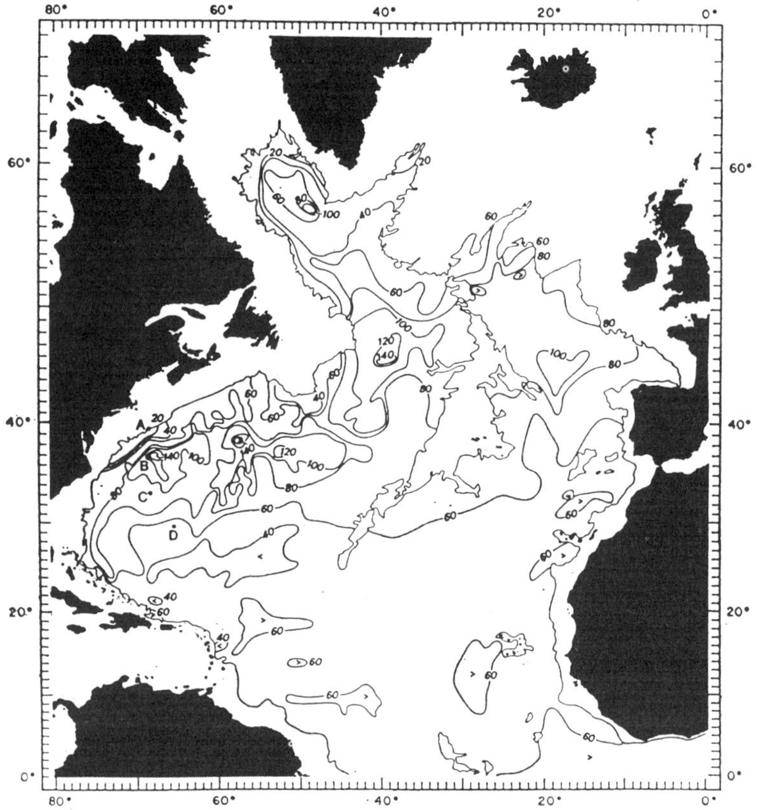

Figure 2–17d. Caption on preceding page.

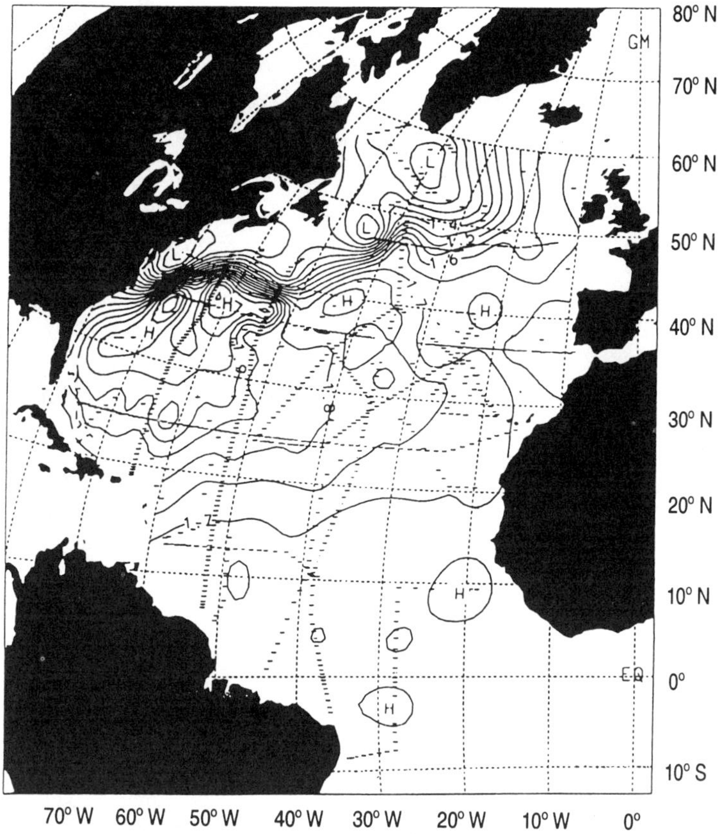

Figure 2–17e. Caption on p. 59.

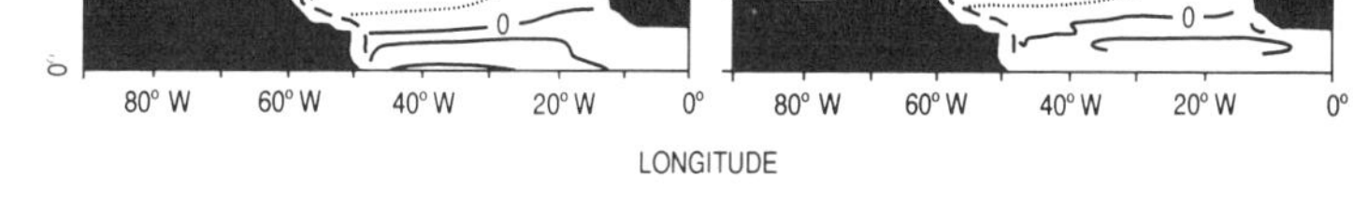

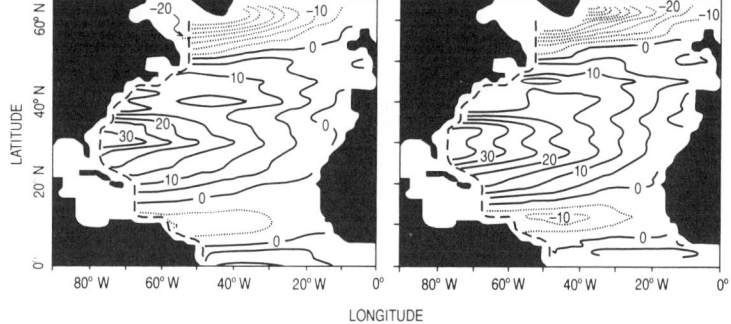

Figure 2–18. Sverdrup transports from Isemer and Hasse (1985) (left) and Hellerman and Rosenstein (1983) (right) wind climatologies (from Böning et al., 1991). In both cases, mass flux is predicted to be outward from the Gulf Stream well south of Cape Hatteras, in contradiction to most theories of that current. The suggestion is that the flow to the east of the Stream in this area cannot be produced by the Sverdrup relationship. There may well also be times when the windfield differs considerably from its climatology, and one has the same question as to whether the response of a turbulent fluid to a climatological windfield actually resembles the time-average response.

it was reasonable to suppose that one could find a depth z_S corresponding to Sverdrup's hypotheses. In particular, they noted two observational points. First, there is a considerable zonal structure in the density field in the deep eastern basin (see Figures 2–2h,i). If some z_S could be found somewhere in the middle of the water column satisfying Sverdrup's hypotheses, the thermal wind relationship would then imply a very substantial north-south velocity near the bottom. Second, there is a considerable north-south slope to the bottom here; the bottom boundary condition [(2.1.47) must vanish] then requires a large w because the flow must remain parallel to the bottom. Wunsch and Roemmich estimated that this bottom-induced w would dominate the w from the Ekman pumping (Bryden, 1980, produced a specific example). One then must try to understand a wind-driven ocean circulation in the presence of stratification and topography. It is possible, of course, that the observed near-bottom shear is time-dependent with zero mean and that a fluctuating flow generates no time-average contribution to w; we are not sure of either of these things. If there is a time-average w at the bottom, then the presence of a mid-depth location where $\partial w/\partial z = 0$ would have to be shown as consistent with the value of w at the base of the Ekman layer and at the bottom. No such theory exists in any completeness; it is not easy to say whether there should be such a mid-depth z_S. In a diagnostic calculation, Luyten, Stommel, & Wunsch (1985) found a flow field that differed radically from ordinary Sverdrup balance, producing transports far larger than one would obtain otherwise. See also the work of deSzoeke (1985) and Cessi and Pedlosky (1986).

The importance of the bottom-induced vertical velocities is consistent with the general circulation model results of Bryan, Böning, & Holland (1995), who concluded that over the North Atlantic, bottom torques were comparable to Ekman pumping over most of the basin, rendering inadvisable the use of Sverdrup balance to describe the flow field. Their model study was made complicated by the uncertainties in the windstress climatologies already alluded to and by considerable large-scale time dependencies in the model, rendering the instantaneous flows often far from any time-average one.

The purpose of this extended discussion has not been to disprove the validity of Sverdrup balance; rather, it was to emphasize the gap commonly existing in oceanography between a plausible and attractive theoretical idea and the ability to demonstrate its quantitative applicability to actual oceanic flow fields. Given, too, the very great regional variations in oceanic physics, one cannot arbitrarily assume that validity of a theoretical framework in one region necessarily implies its validity somewhere else. As we have seen, the evidence for Sverdrup balance in the North Atlantic is at best meager (which is not the same as a disproof), and there is ample reason to doubt its applicability over more than a fraction of the North Atlantic Ocean. In contrast, Hautala, Roemmich, & Schmitz (1994) have produced evidence from recent data suggesting that Sverdrup balance may indeed be the observed zero-order physical balance in much of the subtropical North Pacific Ocean. This result is plausible given the great physical differences between the Atlantic and Pacific (e.g., the absence of a strong, deep-water buoyancy-driven flow in the latter). On the other hand, they also demonstrated significant discrepancies from Sverdrup balance over much of the area.

2.2.4 Measurements: The Eddy Field

The discussion so far has mainly addressed the ocean circulation as though there were a simple large-scale flow, perhaps slowly changing through time. The situation is not so simple. A hydrographic section such as that in Figure 2–2f shows to the eye a large-scale structure that from nearly the beginning of oceanography was taken to represent a large-scale general circulation, and it is indeed true that this type of structure is the focus of the present book. Applying the thermal wind to structures such as those seen in Figure 2–2f over spatial scales of hundreds to thousands of kilometers leads to a picture of a large-scale flow, with magnitudes of a few cm/s, and varying hardly at all with time. It was thus greatly disconcerting, when beginning about 1960, neutrally buoyant floats and moored current meters became available and it was discovered that the flow was far from slowly varying in either time or space and the magnitudes were often 10–50 cm/s, not 1–5 cm/s; see Robinson (1983). This period marked the beginning of the study of the *mesoscale* (to preserve consistency with meteorological practice, the motions should have been called the *synoptic scale*, as the Russian literature does).

The mesoscale is visible as the small wiggles in the isotherms in Figure 2–2f; Equations (2.1.21)–(2.1.22) show that the thermal wind is proportional to the horizontal derivatives of the density field. When one takes the derivatives of a field such as those in Figure 2–2f or 2–2h, we remove (filter out) the large spatial scales and amplify the smaller scales (discussed further in Chapter 3). The result is a thermal windfield as displayed in Figure 2–19a, dominated by rapidly (in space) reversing eddies. Direct-velocity measuring devices such as current meters produce records dominated by these energetic small-scale motions. Figure 2–20 is a remarkable nine-year

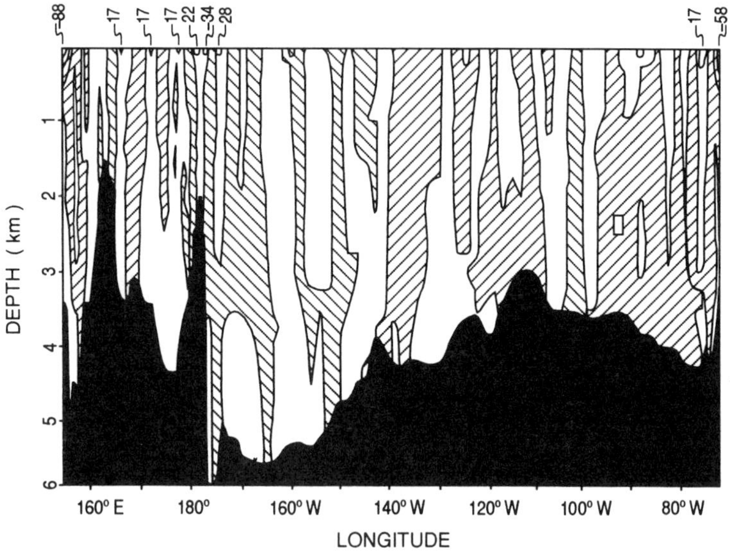

Figure 2–19a. Estimate (Wunsch, Hu, & Grant, 1983) of geostrophic flow in cm/s across 28°S in the South Pacific. Strong cellular flow is some combination of permanent and transient eddy features. In any case, it is consistent with what one anticipates occurring in any quasi-synoptic section–a layered ("Wüstian") climatological flow water mass movement is not expected. (This section is undersampled in longitude.)

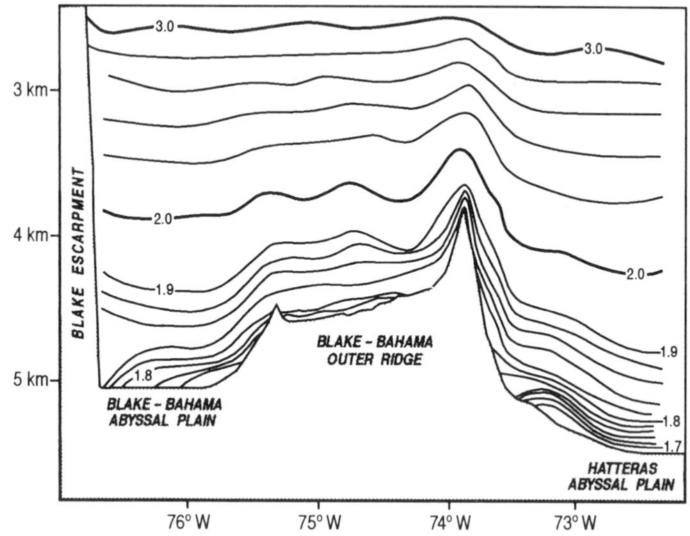

Figure 2–19b. Potential temperature isotherms overlying the Blake-Bahama Outer Ridge (from Amos, Gordon, & Schneider, 1971). Note extremely steep gradient to the east of the Ridge, carrying the DWBC, which would be missed completely if the region were not well resolved.

long record of Müller and Siedler (1992) from which no significant mean velocity emerges at any depth. Figure 1–1 shows the strong eddy features of a numerical model and of the ocean. We now understand that the kinetic energy of the ocean is dominated by the eddy velocities, while the potential energy is dominated by the much larger-scale general circulation. What this means is that velocity measuring devices are comparatively insensitive to the large spatial scales of interest.

The presence of an energetic eddy field has at least two distinct consequences. The first is purely kinematic. It is a sampling issue: One must

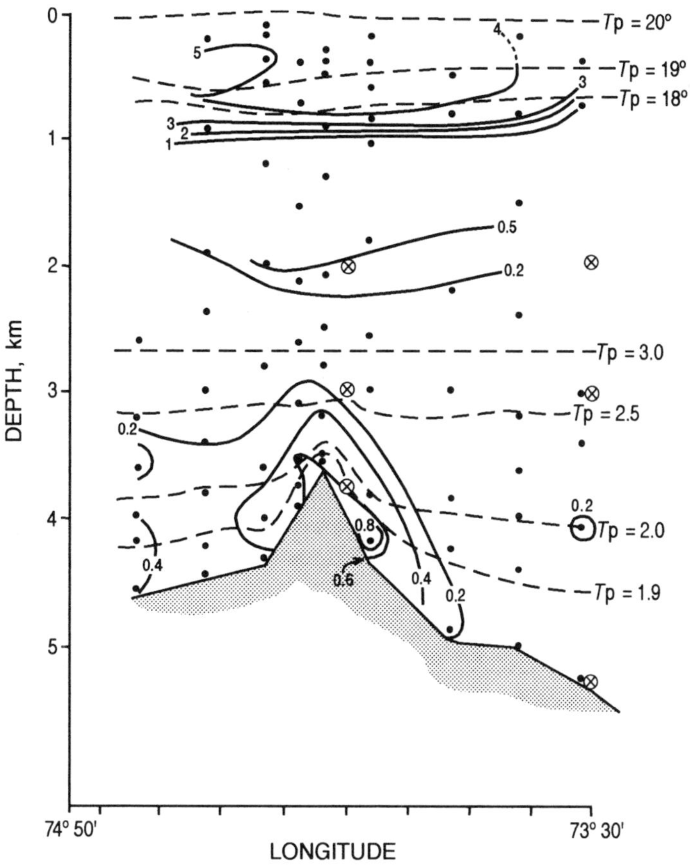

Figure 2–19c. Tritium concentration (solid line) and potential temperature isotherms (dashed line), from Jenkins and Rhines (1980), in the DWBC at the Blake-Bahama Outer Ridge. Notice that tritium maximum is near the core of the Current. Tritium and other property transports cannot be computed accurately without resolving both the velocity field and the tracer concentration on this extremely small scale. The coldest temperatures, as well as the tritium, tend to nearly coincide with the highest apparent velocities, which implies that useful transport computations of the temperature of tritium will be dependent upon having extremely dense spatial sampling of the fields (x denotes the position of current meter measurements described in the original paper).

measure the ocean often enough in space and time so that the eddies do not "alias" into the larger-scale motions (this subject is also taken up in Chapter 3). The resulting sampling requirements are onerous, but an undersampled ocean may look radically different from the real one. Second, because the ocean is a fluid, there can be dynamical consequences for the larger-scale motions of the presence of the smaller-scale eddies. These consequences have a rich variety of possibilities and are part of the large subject of understanding the small deviations from geostrophy that govern the circulation.

2.2.5 Stommel-Arons: The Abyssal Flows

The most notable prediction of the Stommel-Arons model described in Section 2.1.5 was that there should be deep western boundary currents

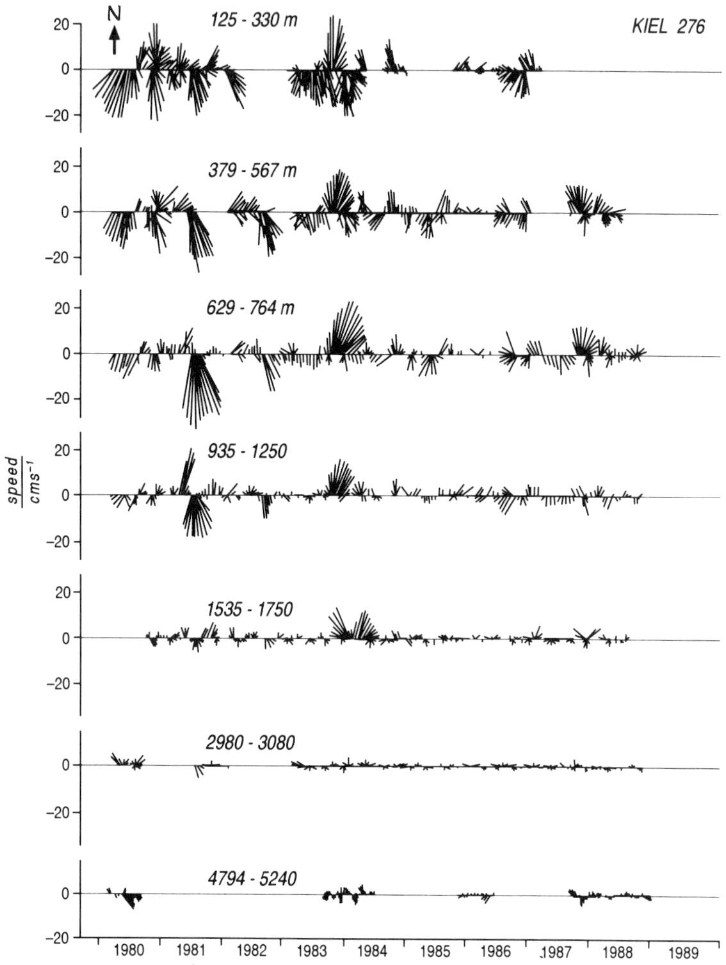

Figure 2–20. Subsampled daily mean velocity vectors from nearly nine years of records in the Canary Basin (Müller & Siedler, 1992). Even with such a long record, there is no statistically significant mean velocity. Because of mooring replacements, records were obtained over the depth ranges indicated.

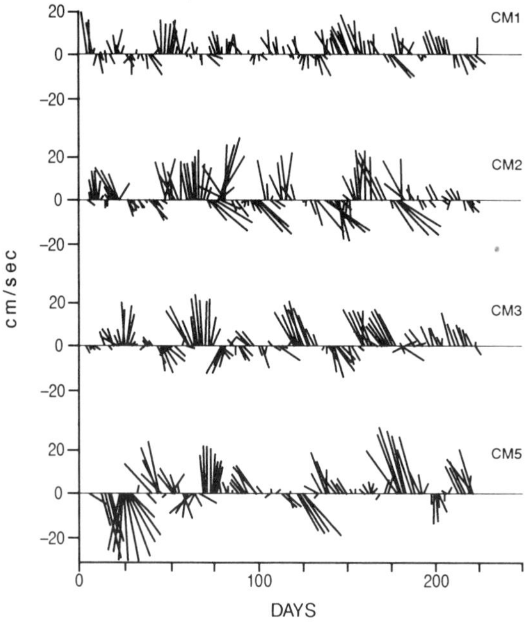

Figure 2–21. Forty-hour means of current meter records (Pickart & Watts, 1990), from within the North Atlantic Deep Western Boundary Current. Their somewhat unfortunate convention is that positive (upward) values are directed southwest and would correspond to the conventional flow direction of the DWBC. There are, however, extended periods when the flow is clearly to the north, and a short record is often unrepresentative of the mean conditions.

(DWBC) carrying fluid in specific directions in all the oceans. In the North Atlantic (then as now the most scientifically accessible of all oceans), it was predicted that the flow should be southward, opposite to the Gulf Stream. This prediction was soon tested by Swallow and Worthington (1961), who managed to insert some of the then-new neutrally buoyant floats (Swallow floats) into the predicted flow. The floats were seen to move rapidly southward, and the theory was triumphantly vindicated.

With hindsight, there is a problem: The longest float measurement lasted only four days. With the discovery of the eddy field, we recognize that there was a significant probability that the floats would have actually moved northward over four days (see, for example, Figure 2–21 from Pickart & Watts, 1990). Today, knowing what we do about variability, it would be regarded as ludicrous to claim that a four-day record proved the existence of a mean-flow. As with the good agreement that Sverdrup found with his theory, one wonders what the course of oceanographic research would have been had the Swallow floats moved northward?

Nonetheless, the existence of a DWBC in the North Atlantic can hardly be doubted. Indeed, Wüst's (1935) diagrams (Figure 2–2a) plainly show a property tongue down the western side of the ocean that is suggestive of a flow–but no proof in a time-dependent system. Warren (1981) discusses the general evidence for predicted DWBCs in the world's oceans. Although there are some specific difficulties, for example, in the North Pacific (see Warren & Owens, 1988), and the Indian Ocean ridge system presents some interpretive problems, the existence of the DWBC system is firmly established.

Tying down the quantitative aspects of the Stommel and Arons scheme has proved more elusive, however. The volume flux of the convective elements dense enough to be regarded as Deep Water is very poorly known, in part because the rate of formation is very uncertain, and because the convecting water entrains surrounding fluid both in the process of sinking (see Killworth, 1983b, for a discussion of the convective process) and as it flows in the DWBC. That is, the DWBCs evidently exchange fluid in a complex and poorly understood way with the interior fluid (e.g., Pickart, 1992). Furthermore, the convective process is believed to occur mainly in wintertime. Yet, there is no evidence (Dickson, Gmitrowicz, & Watson, 1990) for any seasonal oscillation in the DWBC mass flux, even quite far upstream toward the convective region.

Consider Figure 2–22a, which is an estimate of the annual average heat exchange between ocean and atmosphere (Isemer, Willebrand, & Hasse, 1989).

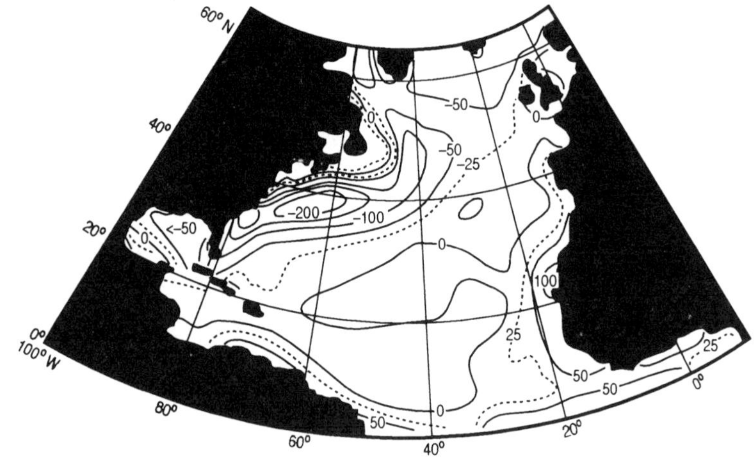

Figure 2–22a. Estimate by Isemer, Willebrand, and Hasse (1989) of annual average heat transfer (W/m^2) between the North Atlantic ocean and the atmosphere. The estimate is a combination of bulk formula methods with a constraint requiring 1 PW moving poleward across 25°N. Note that most of the North Atlantic is estimated to be losing heat on average to the atmosphere, even in the tropics. Errors here may be 50% or higher with large regional correlations.

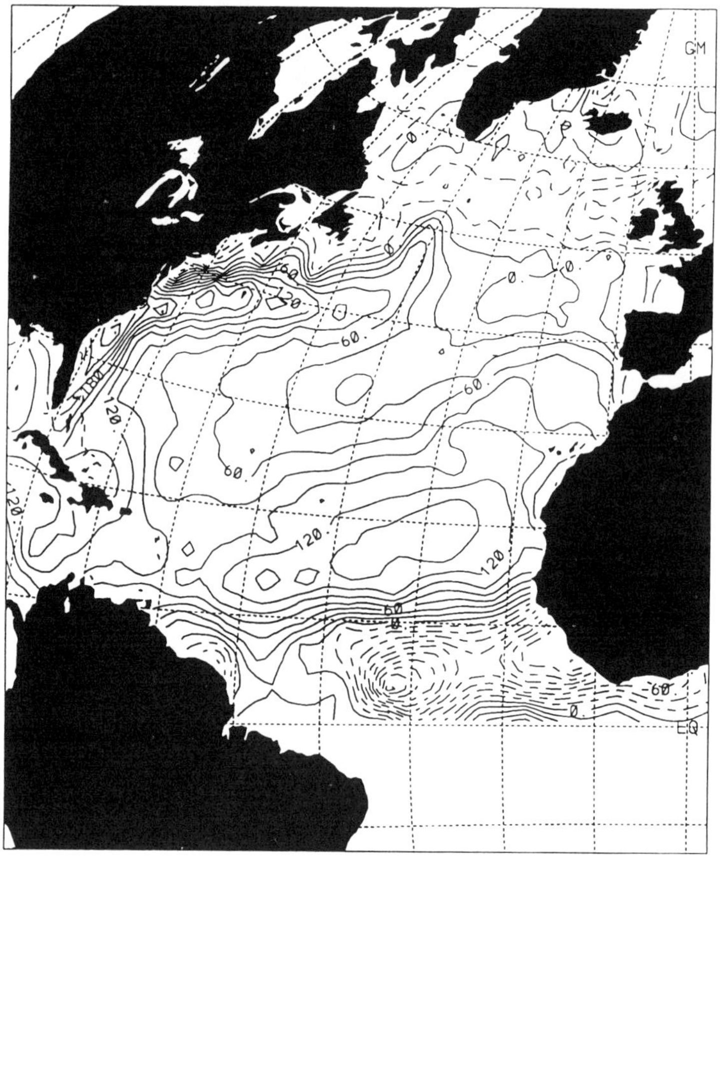

Figure 2–22b. Estimate (in cm/year) from Schmitt, Bogden, and Dorman (1989) of the climatological mean evaporation minus precipitation in the North Atlantic. Error estimate is not available but is almost certainly extremely large, perhaps approaching 100%. Compare to Figure 2–5.

Somewhat contrary to what one might have anticipated from the discussion thus far, the chief transfer of heat from ocean to atmosphere appears to occur not in the high-latitude regions of the Norwegian-Greenland Sea area but over the Gulf Stream system (thought to be the result of atmospheric cyclogenesis as cold continental air masses encounter the warm Stream waters). Indeed, convection *is* known to occur in this area; it is the region of the formation of so-called 18°C water; see Warren (1972). There is another region of known important North Atlantic convection, the so-called Labrador Sea Water being the result (Lazier, 1973). These latter water masses are not dense enough to reach the bottom of the ocean and hence have received less attention than the very special deep water masses. Indeed, only the high-latitude North Atlantic and the Weddell Sea area are believed to generate water masses heavy enough to reach the seafloor. The northern Indian Ocean does not convect except in the Arabian and Red Seas, in an evaporation-dominated system, and the North Pacific convection does not make water heavy enough to get below about 800 m (Pacific Intermediate Water, see Figures 2–2c). The problem is that we cannot produce a priori estimates leading to quantitative values for the rates and properties, including the density, of the convection elements. Until we can do so, there is no possibility of producing numerical values for the properties, including the mass fluxes, of the currents predicted by the Stommel-Arons theory.

What has been even more difficult has been the demonstration that the interior flows are consistent with the model, particularly in the South Pacific. There, the predicted DWBC had been clearly found (Stommel, Stroup, Reid, & Warren, 1973; Warren, 1981). The predictions for the South Pacific interior are shown as part of the global map in Figure 2–7. Figures 2–17a,b show Reid's estimates of the absolute flow at depth in the South Pacific. One need not accept the details of Reid's estimate of the deep flow field in the Pacific (there are objections to it), but visually there is little resemblance between the Stommel and Arons prediction and one of the very few existing estimates of the actual flow. Figure 2–19a, taken from Wunsch, Hu, and Grant (1983), shows another element of the problem. The meridional flow as computable using the dynamic method and hydrographic data is dominated by the eddy field. The Stommel and Arons picture implies a weak net southward meridional motion that is obviously difficult to find in the presence of such a noisy data set.

It is important to recall that the precise form of figures like 2–7 is dependent upon the Stommel and Arons assumption that the vertical upwelling was uniform over the oceanic interior. Stommel and Arons make it clear that they had no illusions about the truth of this assumption but that it was worth exploring the consequences of assuming its validity until a more complex structure could be justified. So one interpretation of the discrepancy between Figures 2–7 and 2–17a,b is that the actual distribution of vertical velocity in the South Pacific is not uniform. Using methods discussed in Chapter 4, Olbers, Wenzel, & Willebrand (1985) produced the estimates of w for the North Atlantic depicted in Figure 2–23a and Olbers and Wenzel (1989) the Southern Ocean values in Figure 2–23b. Again, one sees a complex system of upwelling and downwelling regions. Such a situation does not contradict the physics employed by Stommel and Arons; it merely suggests that the simple constant interior w assumption is not adequate. But without detailed knowledge of the actual w, one cannot predict the geostrophic interior flow and use the predictions to test the theory.

The first real step beyond the basic form of the Stommel and Arons theory was taken by Kawase (1987). He noticed (see Figure 2–2e) that in the Atlantic, the tracer properties seemingly associated with the DWBC extended, in part, zonally along the equator. These zonal extensions are reminiscent of the zonal structures that Reid shows in the South Pacific. The Stommel and Arons map, to the contrary, shows no particular equatorial structure. Kawase (1987) attempted to determine what would happen if the DWBC were time-dependent, permitting it to turn on or oscillate slowly. To understand the result, one needs a few elements of the theory of the time-dependent ocean, for which the reader is referred to Gill (1982), Pedlosky (1987a), and Moore and Philander (1977), with the latter being focused on the equatorial ocean. A nearly purely verbal discussion will suffice here.

The equations of motion (2.1.1)–(2.1.5) permit the existence of a wave trapped against the walls of the ocean, traveling (in the northern hemisphere) so that the wall is on the right-side of the wave and with the pressure gradient normal to the wall in geostrophic balance. This Kelvin wave travels southward toward the equator on the western side of the ocean and northward away from it on the eastern side (the wall is to the left in the southern hemisphere, and so the wave also travels toward the equator on the western side there, etc.). One can think of this Kelvin wave as the time-dependent counterpart to the steady DWBC, which is trapped against the western wall. Now it is one of the peculiarities of the change in sign of the Coriolis parameter at the equator that the equator can support a Kelvin wave, moving eastward (see Figure 2–24). [In an unpublished thesis, Moore (1968; see also Moore & Philander, 1977) showed that indeed one could make up a complete mode of oscillation consisting, in part, of a Kelvin

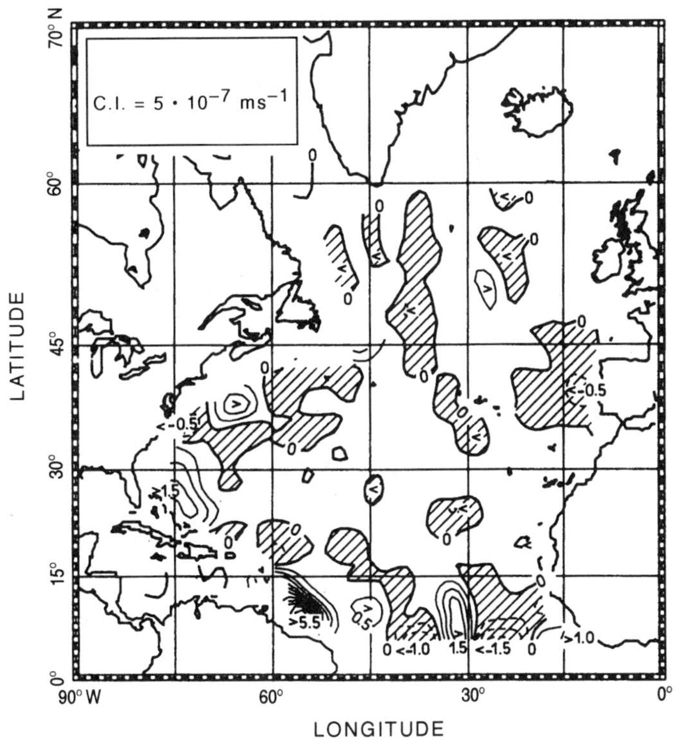

Figure 2–23xa. Estimate, from Olbers et al. (1985), of the vertical velocity at 2000 m in the North Atlantic. Regions of downwelling are shaded. Such figures suggest that the Stommel and Arons assumption of a uniform interior oceanic upwelling is too simple to produce the observed circulation, but the estimates suffer from the need to work with climatologies.

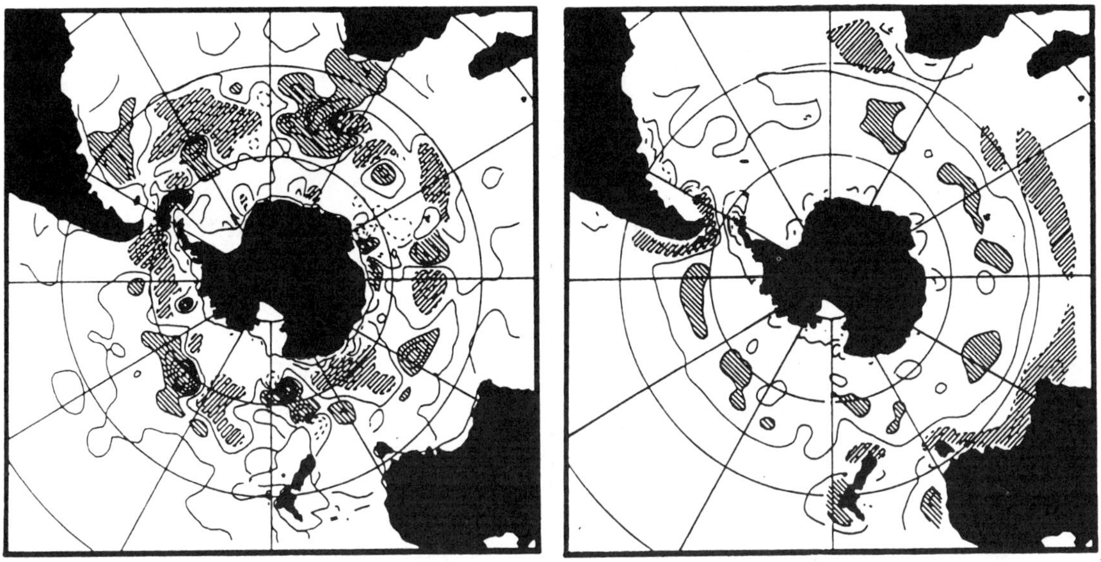

Figure 2–23b. Estimate, from Olbers and Wenzel (1989), of upwelling at 50 (left) and 2000 m (right) in the Southern Ocean. Contour interval is 10^{-6} cm/s, with negative contours dashed. Shaded areas show extreme values (above $|10^{-6}|$ cm/s on the left and above $|2 \times 10^{-6}|$ on the right).

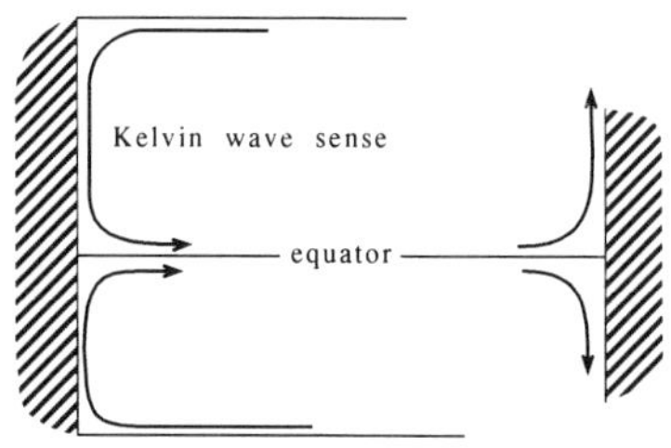

Figure 2–24. Schematic of movement of Kelvin waves around a closed basin: counterclockwise in the Northern hemisphere, clockwise in the Southern. The Kelvin wave, although not appearing in the so-called quasi-geostrophic approximation, appears to be essential to understanding of the time-dependent adjustment of the ocean.

wave running clear around a closed ocean.] Kawase (1987) made a model ocean in which the mass source was confined to an abyssal layer underlying another much thicker layer. The major step was to formulate the exchange between layers (i.e., the vertical velocity) as a simple rule dependent upon the time-dependent thickness of the lower layer. Thus, the vertical velocity was both time-dependent and, in its asymptotic steady limit, was not uniform but would be a function of the actual flow. The mass source effectively excites Kelvin waves, which carry information about the time-dependence around the basin, but because the waves are baroclinic,[8] the signal velocity is small. If the mass source were held constant after an initial transient, the asymptotic steady state would not be reached for a long time.

The reader is referred to Kawase (1987) for details; the calculation suggests that the Stommel and Arons theory can be regarded as the steady-state limit of an ocean that has had a very long time to adjust to the strength of its deep mass sources. But any dissipation, or slow time dependence, would prevent this particular steady state from being reached, and one could generate complex along-equatorial flows and zonal property jets. This theory is quite incomplete, the artificiality of the vertical velocity computation being only one problem (for example, entrainment and detrainment from the western wall flow, see Figures 2–18, needs to be accounted for). One might wonder if the Dickson et al. (1990) suggestion that the overflows are remarkably steady is in contradiction to the importance of time dependence in the Kawase model; but the measurements span only three years, may be untypical of the longer-term behavior, the interpretation is based upon some simplifying assumptions, and they may not preclude oscillations further downstream. (The crucial role of the Kelvin wave in the oceanic response suggests that approximations to the equations of motion which filter out this mode–for example, so-called quasi-geostrophic approximations–do not correctly model the time-dependent response. Milliff and McWilliams, 1994, grapple with this issue.)

[8] That is, dependent for the restoring force on the small interior density gradients.

2.3 The Classical Problem

From the preceding discussion, it is apparent that much of the uncertainty surrounding the physical basis of the observed circulation could be removed if the nature of the geostrophic interior circulation alone were known accurately. For example, if the meridional velocity, v, were known, then the linear vorticity Equation (2.1.20) could be integrated to produce the vertical profile of w (up to a function of x, y) and a direct test made of the relative contributions of Ekman pumping and bottom-velocity-driven flows. Or one could directly compare the result with the predictions of Equations (2.1.51)–(2.1.52) or those of the Stommel-Arons flows.

This message is critical: The ocean circulation is observed to be–in almost all places, at all times–in near-geostrophic equilibrium. One recognizes that the structure and intensity of the geostrophic flows are determined by the ageostrophic components. But a determination of the geostrophically balanced flows permits very strong inferences about what those ageostrophic flows and corresponding forces must be or have been in the past.

Consider that the nongeostrophic oceanic boundary layers may well control the interior flow field, defining its structures and intensities. If we can determine what that interior flow field is, we can infer many of the properties of the boundary layers themselves–for example, how much and where they are exchanging fluid with the oceanic interior.

One should not confuse this statement with an assertion that the boundary layers are therefore passive, as sometimes assumed in theory, meaning that the interior geostrophic solution can be made to satisfy whatever boundary conditions are required, through the presence of boundary layers which simply absorb or supply the flows required by the interior. [This assumption underlies the Stommel (1948), Munk (1950), and Stommel and Arons (1960a,b) models, for example.] Even if the interior flow is *determined* by active boundary layer behavior, once the interior flow has been established, its strength and shape having been set, it remains nearly geostrophic, can be determined by our methods, and permits inferences about the boundary layers. Consider the flow fields in Figure 2–18; if the apparent outflow south of Cape Hatteras is real, then determination of the interior flow permits a determination of the rate with which the Gulf Stream is losing fluid through a nongeostrophic process.

Many oceanographers would describe as the "classical problem of physical oceanography" the question of how to compute the geostrophic flow of water in the deep sea from observations easily made from ships. In practice, one uses the thermal wind relations, and the major issue is reduced to the missing integration constants. (This statement of the problem is historically correct; today, however, determination of the absolute flow alone would not be regarded as the end of the matter. Given knowledge of the presence of mesoscale and other variability, the problem of defining a useful time average of the absolute flow is equally essential.)

2.3.1 Dynamic Calculations

If f is locally constant, there is nothing in Equations (2.1.23)–(2.1.24) to distinguish x from y, and we will arbitrarily (for the moment) assign x as the along-ship-track dimension and work solely with (2.1.24). The entirety of the classical oceanographic problem alluded to above can be reduced to the question of "what is the integration constant b"?–a seemingly trivial issue but one that perplexed generations of oceanographers and greatly hindered the understanding of the ocean circulation. Such integration constants are often called *reference-level velocities* because they represent the absolute velocity at the *reference-level depth*, z_0. It is important *not* to call this, as some authors do, the *barotropic velocity*.[9] If the integration depth z_0 is chosen to be $z = 0$, then from (2.1.11) we have immediately

$$\rho_s v(x, y, z = 0) = \frac{1}{f}\frac{\partial p_0}{\partial x} \,. \tag{2.3.1}$$

The hydrostatic Equation (2.1.13) can be integrated in the vertical from the seasurface, $z = \zeta(x, y, t)$, to produce

$$p(z, x, y, t) = g \int_z^{z=\zeta} \rho(x, y, z', t) dz' \,. \tag{2.3.2}$$

If ζ is small, then near $z \approx 0$,

$$p(x, y, z = 0, t) \approx g\rho(x, y, z = 0)\zeta(x, y, z, t) \,. \tag{2.3.3}$$

Thus, $p\,(x, y, 0, t)$ is the elevation of the seasurface relative to the earth's equipotentials (see Figure 2–25), and the total pressure at depth is

$$p(x, y, z, t) = g\rho(x, y, 0, t)\zeta(x, y, t) + \int_z^{o} g\rho(x, y, z', t) dz' \tag{2.3.4}$$

[9] Some authors use it to mean the bottom velocity; some, the vertically averaged velocity; and some, the reference-level velocity. None of these definitions is unique, no one knows what the word means anymore, and it should be banished from the oceanographic vocabulary except in the special context of the theoretician's flat-bottomed, dynamically linear ocean.

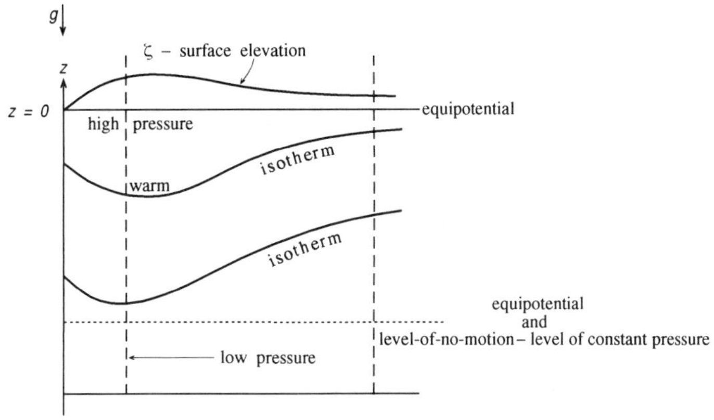

Figure 2–25. Schematic of the relationship between isopycnals (assumed identical to the isotherms) and a true level-of-no-motion (zero pressure gradient at a fixed gravitational equipotential). As depicted, the excess of fluid leading to the surface elevation is just compensated deep in the water column because the entire fluid column is assumed to be warmer than the neighboring fluid. If the warmth persists below the resulting level of zero pressure gradient, the sign of the pressure gradient will be reversed there.

where the second term is readily measured by a ship and is the pressure as a function of depth *relative to the seasurface.* If the shape of the seasurface were known, relative to a gravitational equipotential, as depicted in Figure 2–25, the flow field could be computed at all depths from the geostrophic relations. Knowledge of the shape of the seasurface relative to the gravitational equipotentials in effect fixes the absolute pressure at a reference level of $z = 0$.

Suppose shipborne measurements are made at two different hydrographic stations, as depicted in Figure 2–25, so that the second term in (2.3.4) is known at two locations and the necessary horizontal derivative can be estimated. The problem is that the seasurface shape is not known, and comparatively small elevations or valleys in the seasurface relative to a near-surface gravitational equipotential can compensate observed horizontal variations in density. As depicted in the figure, the common pressure value, p', might occur at the same deep gravitational equipotential (isostatic compensation) where the locally horizontal pressure gradient vanishes, and then so does the flow field. Fixing the absolute pressure or pressure gradient on any gravitational equipotential would clearly be equivalent to determining the shape of the seasurface.

What about measuring the slope of the seasurface? A rough numerical estimate is easily made. Taking a typical seasurface geostrophic velocity v_s

to be 1 cm/sec, and a latitude of 30°,

$$g\frac{\partial \zeta}{\partial x} \approx f v_s \,, \tag{2.3.5}$$

$$\frac{\partial \zeta}{\partial x} \approx \frac{7.3 \times 10^{-5}\ \mathrm{sec}^{-1} \times 1\ \mathrm{cm/s}}{980\ \mathrm{cm/sec}^2} \approx 7 \times 10^{-8} \,,$$

which is about 1 meter change in 10,000 km. Until very recently, the idea of measuring such slopes at sea was hopeless; technology (satellite altimetry) has, however, evolved to the point where it has become practical; see Fu, Christensen, Yamarone, Lefebvre, Menard, Dorrer, & Escudier (1994).

The same problem does *not* arise in meteorology: One can determine the elevation of an observer and the equipotentials relative to the center of the earth with accuracies of centimeters. Ordinary barometers then produce absolute pressures at known elevations and the absolute pressure gradients can be computed quite accurately at a reference surface. This difference between the two subjects is just one of many divergent observational problems.

What about measuring v at some depth, thus fixing b? Here, the central difficulty lies in the presence of the background variability of internal waves and the mesoscale. Months and years are required to produce stable mean velocities in the presence of this energetic noise. Even if stable means were apparent after a year or two, such measurements are impractical for ships; the internal wave band all by itself would require the ship to stay in one place for several days to average out these fluctuations. One could never afford to tie up ships in such a manner on a routine basis. (Pillsbury's 1893 measurements in the Florida Current remain a remarkable tour de force.)

The difficulty of obtaining stable mean flows by direct methods might make one wonder very sensibly about the entire notion that the ocean has a large-scale, time-average flow field. For the moment, note only that in some places there are obvious large-scale nearly unidirectional stable flows (e.g., the Gulf Stream).

2.3.2 The Reference-Level Problem: Early Attempts

The inability to determine the reference-level velocity distorted and stymied the study of oceanography for many decades. So as to avoid complete paralysis, oceanographers made the assumption that if one went deeply enough into the sea, the fluid movement would become so weak as to be negligible. This not-unreasonable assumption of a "level-of-motion," as in Figure

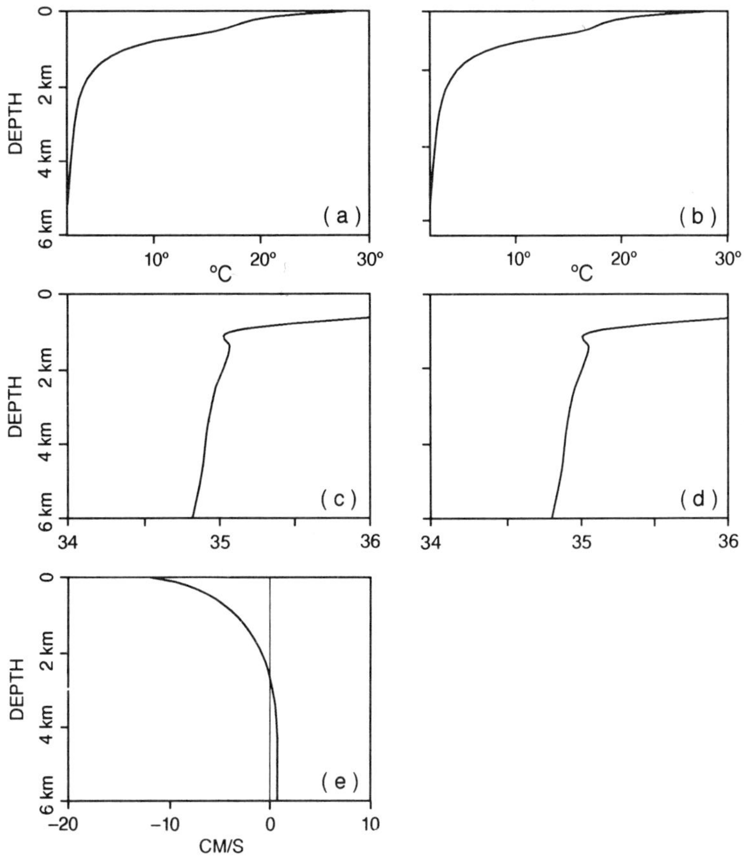

Figure 2–26. Pair of nearby hydrographic stations taken from the 24°N section (Figure 2–2h) (61.18°W, 24.50°N and 60.49°W, 24.52°N). The temperature profiles are shown in (a), (b), and the corresponding salinity profiles are in (c), (d). These result in a thermal wind, which is shown relative to a 3000-decibar level in (e).

2–25, permitted one to discuss the surface circulation in quantitative terms. It had the unfortunate practical consequence of removing any incentive to make observations below the putative level-of-no-motion, observations which are very difficult, not to speak of time-consuming and expensive. Figure 2–26e displays the vertical profile of the horizontal velocity as computed from the data displayed in Figure 2–26a-d, with the reference level placed at $z_0 = 3000$ decibars. One observes that vertical displacement of the level of zero motion over several thousand meters in the vertical makes little change in the near-surface velocity estimates, simply because the vertical derivative of v, the shear, is very weak in the abyss. To the extent that one is concerned only with the upper ocean velocities, the assumption of levels-of-no-motion appears reasonable, with the results insensitive to the particular choice. The *transport* properties of the ocean velocity, representing integrals of the velocities, are, of course, very sensitive to the level-of-no-motion chosen.

Reid (1981) has reviewed many circulation schemes based on levels-of-no-motion. The idea of such a level became so ingrained in oceanographic thinking that it tended to become routinely accepted procedure. But many were still bothered by its necessity. Two of the important German oceanographers who flourished between the two World Wars, Wüst and Defant, tackled the problem in different ways. Wüst's notions have been more enduring, and we examine them first.

Property sections through the ocean (see Figure 2–2) strongly suggest to the eye that these "dyes," as one can think of them, delineate movements of water from a region where the properties are most intense into the interior ocean. An appealing picture is that the property tongues indicate water mass origins–for example, in the Circumpolar region, or the high northern Atlantic, or the mouth of the Mediterranean. One then argues that the flow must be from the source regions along the axis of the tongue into the oceanic interior. With some oversimplification, this is Wüst's *core-layer* idea, the tongue axes denoting the core-layers. Wüst then drew arrows along these tongues (see Figure 2–2a,b) and discussed the "spreading" ("*Ausbreitung*" in German) of the water. His arrows are all of the same length, and he recognized that such ideas did not permit computation of flow magnitudes or of the distinction between flow and "mixing" (we return to mixing later).

But Wüst's ideas had a powerful impact on those who sought to employ geostrophic balance and levels-of-no-motion. His picture supported the idea that the oceanic circulation was essentially layerlike; if in figures like 2–2e one identifies one of the water masses (e.g., the North Atlantic Deep Water) as moving from the Arctic regions, and one just overlying it, the Circumpolar Intermediate Water, as coming from the Antarctic, then it is eminently reasonable to think that there must be a line of zero velocity between them. Such arguments have been widely employed but finally broke down in two separate ways: in Worthington's (1976) attempted synthesis of the circulation of the North Atlantic Ocean, which will be described later, and in the clearer recognition that even if the core-layers were appropriate in a climatological sense, they might be irrelevant to the use of synoptic or even modern data. Thus, for example, figures like 2–2b suggest that the massive salinity tongue emanating from the Mediterranean is the result of a fluid flow from the Mediterranean, as depicted by Wüst's arrows. But the picture by itself provides no information as to whether the required flow did not actually take place long ago (millennia?), with either no flow having taken place since, or with the flow continuing to this day. (Indeed, present-day flows appear to be mainly *around* the tongue rather than along its axis.) The problem is to distinguish between the physics that produced the "standing crop"–that is,

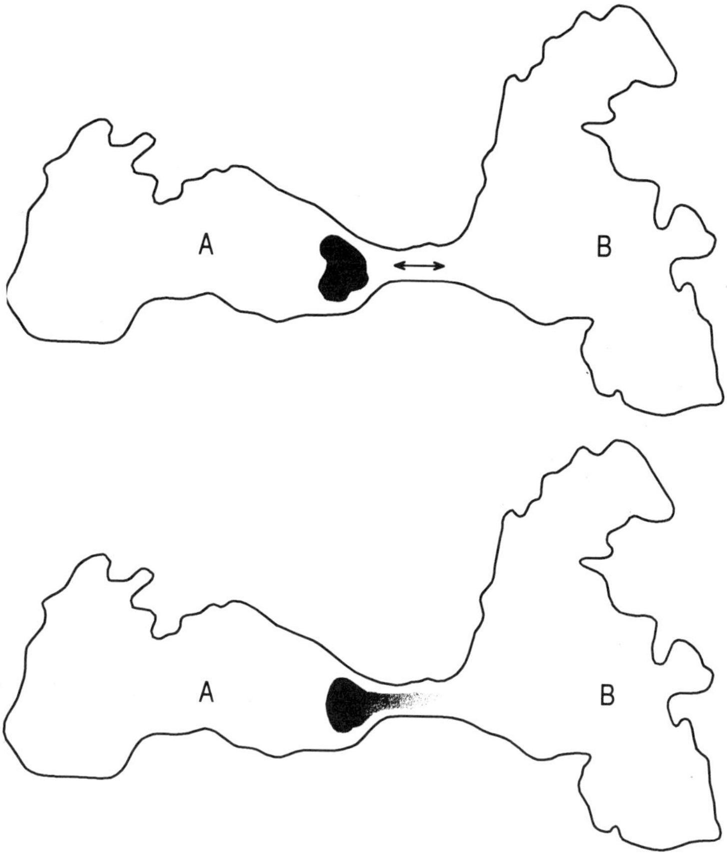

Figure 2–27. Schematic of the movement of a tracer in the ocean. Initially (top), basin A contains a dye, and basin B is devoid of the tracer. An oscillatory flow occurs from A to B such that the time-average mass flux between the basins is zero. But because the particle trajectories from A to B are not identical to those from B to A, dye is effectively mixed with the fluid in basin B. Following a number of oscillations of the flow, a tracer tongue is found in B (bottom), which could be interpreted, erroneously, as implying a net mass flux from A to B.

the concentrations observed in the property charts–and the property fluxes required to maintain what is already there. The property tongues by themselves require no present-day flow at all! Dynamically passive tracers, with no significant interior sources or sinks, such as oxygen below the zone of active photosynthesis and biological processes, are incapable by themselves of saying anything about absolute rates of flow in the ocean; they lack the built-in "clock" necessary to determine the circulation.

The introduction of time variability into the discussion of the general circulation greatly complicates the entire Wüstian conceptual framework. If the flow fluctuates, there is no longer a simple well-defined general circulation that describes the movement of mass and associated properties; instead, one must carefully define the property whose circulation is of interest and specify in detail the time span of interest. A simple example may be useful. Consider the situation in Figure 2–27. Two ocean basins are depicted, connected by a narrow strait (perhaps the Drake Passage or the passages of the Indonesian archipelago). The mass flow connecting the two basins is taken to be purely periodic so that the long-time average, $\overline{\rho u} = 0$.

It is supposed, however, that the fluid in basin A is dyed with some chemical (e.g., tritium), C, and that initially the concentration of C vanishes in basin B. When the flow is from A to B, dye is carried into B; but unless the flow field is completely reversible–no mixing at all with the fluid in B–some dye is left behind when the flow reverses. Over many cycles of oscillation, a large inventory (a plume) of C will build up in basin B so that $\overline{\rho u C} \neq 0$. But inference from the plume of the existence of a net mass flux from A to B is clearly incorrect. Different tracers, distributed somewhat differently in the fluid exchanged between A and B, can lead to complex, varying distributions, and with transports from A to B that are quite different. Thus, one must carefully specify not only the property whose circulation is under discussion but also the time interval over which the transport is to be determined (averages over fractions of the oscillation period, which could be annual or interannual, can be very different).

At the present time, there have been inadequate data to clearly separate the transport properties of different chemical tracers and of mass in the ocean circulation. But there are enough data to preclude the blind conclusion that property tongues can be simply converted to inferences about mass fluxes. Compare Figure 2–2a to Figure 2–27. Only measurements directly connected to the dynamical flows (e.g., the pressure gradients derived from the density field) can be relied upon for deductions of actual water movement, and ultimately, the introduction of time variations into the picture becomes essential.

Defant (1941) discussed a rule for determining the level-of-no-motion, which was permitted to slope across the ocean (in contrast to most practical circulation attempts in precomputer days of assigning one deep pressure or depth level as that of no motion). He chose to use the level of minimum vertical geostrophic shear. The arguments he provided to justify this choice are difficult to understand. I suggested (Wunsch, 1978) that the choice is a rational one in the absence of any other criterion–the depth of minimum geostrophic shear is the depth for which the resulting velocities and transports are least sensitive to perturbation in the level-of-no-motion. Although Defant's arguments do not seem rigorous, the pictures he drew (see Figure 2–28) are qualitatively reminiscent of much more recent estimates of the North Atlantic circulation.

Hidaka (1940a,b), a Japanese contemporary of Wüst and Defant, tackled the problem in a very different way. Consider the hydrographic stations

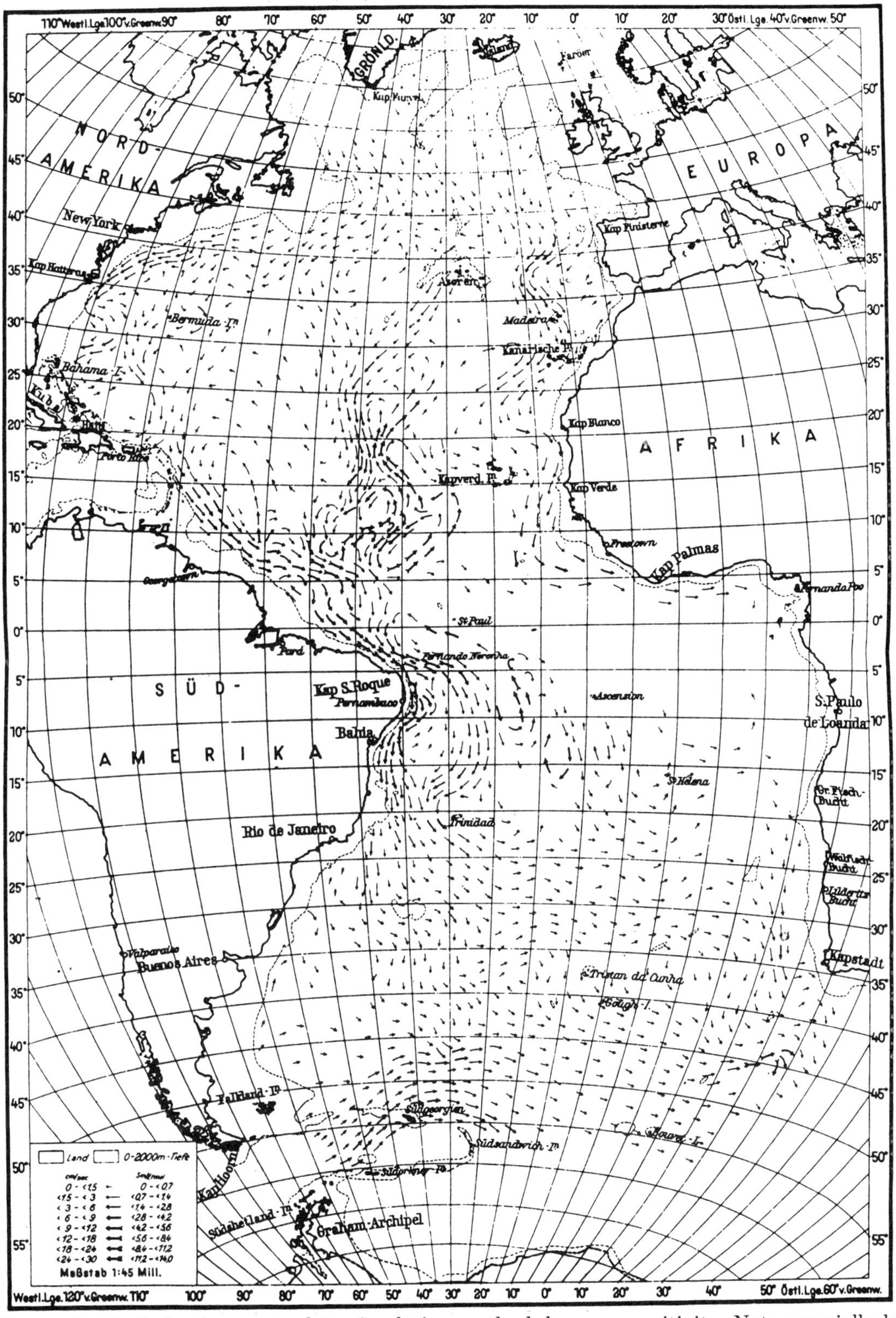

Figure 2–28. Defant's (1941) circulation scheme at 2000 m based upon his absolute circulation method described in the text–essentially using a level of minimum sensitivity. Note especially his inference of a strong deep western boundary current.

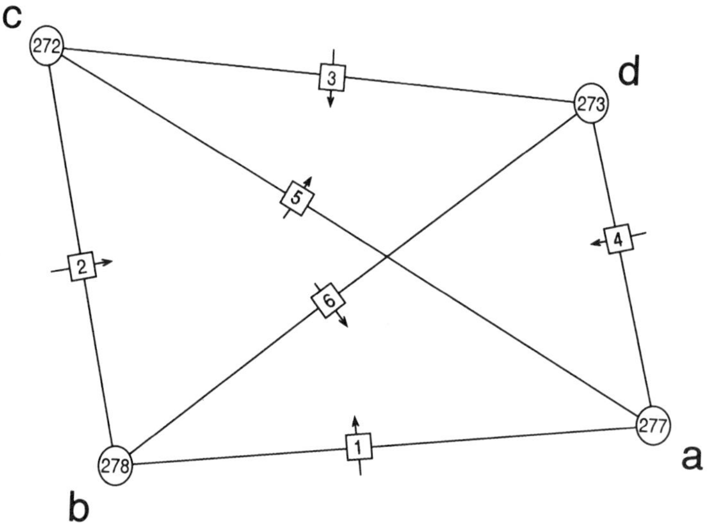

Figure 2–29. Geometry used by Hidaka (1940b) to estimate reference-level velocities by mass and salt conservation in defining triangular volumes. Circled numbers are station identifiers, and boxed integers are interface labels used to identify the flows between volumes.

depicted in Figure 2–29. He argued that with each pair of stations, one could compute the mass and salt flux into the volumes depicted by the straight lines. Let the reference-level velocity in each station pair along each line be denoted $c_i, i = 1$ to 6, all of which are unknowns. We can require sensibly that total mass and salt must be conserved–what enters the volumes must leave again. One recognizes that Hidaka attempted to make use of the mass and salt conservation Equations (2.1.15)–(2.1.18), which are not otherwise used in the dynamic method. Writing conservation equations for each of the boxes depicted in Figure 2–29, a simple count produces six equations in six unknowns, which Hidaka solved. Unfortunately, Defant (1961) demolished this idea of Hidaka's for computing reference-level velocities by demonstrating unequivocally that the system of equations was ill-conditioned and that the numerical values produced by Hidaka (1940b) were meaningless. The method then sank from view (apart from a mention in passing on p. 457 of Sverdrup, Johnson, & Fleming, 1942, and its discussion as a failed idea in Defant's textbook). There is an echo of it in Riley's notable (1951) paper, but that, too, produced equivocal results.

2.4 Hidaka's Problem and the Algebraic Formulation

Despite the best efforts of a large number of people over many decades, no generally adequate method was found for the determination of the reference-level velocity on the space and time scales of a section obtainable from a ship. A number of ad hoc assumptions were made (see, for example, Sverdrup et al., 1942; Defant, 1961; Reid, 1981). No procedure that was proposed could be derived from first principles, and none became widely accepted as generally applicable. The consequence was that most discussions of the large-scale ocean circulation in fact, if not in name, became based upon the assumption that somewhere at depth in the ocean the horizontal velocity became so small as to be negligible. Much was learned this way, and there was little reason to be troubled by the arbitrariness of the level-of-no-motion assumption as long as most hydrographic data did not extend below one or two thousand meters, and as long as the chief interest lay in qualitative descriptions of the upper-ocean velocity field. Many published circulation schemes today are still based upon this idea.

But eventually it became possible to nearly routinely make hydrographic observations extending to thousands of meters, right to the seafloor. At this point, attempts to use the deep level-of-motion assumption led to great difficulties. Though moderate vertical shifts of the assumed level made little difference to the near-surface velocities, they made very great changes in the estimated deep velocities. Consider, for example, that a deep western boundary current (Figure 2–19c) would have the sense of its flow reversed by shifting the level-of-no-motion from where it is shown in the figure to the seafloor. Furthermore, as the hydrographic database expanded, attempts were being made to compute not just the water velocities but the net *flux* of properties: mass, salt, heat, etc. Vertical shifts, when integrated over large areas of ocean, make very large differences in the property fluxes. It was Worthington (1976) who perhaps most forcefully called the attention of the oceanographic community to these problems and who proposed a very drastic remedy.

Worthington had struggled for many years to calculate the transports of water masses in the North Atlantic using the thermal wind equations and the assumption of simply definable levels-of-no-motion (he did permit them to deviate from simple horizontal depth/pressure surfaces). Unlike many earlier investigators, Worthington took seriously the abyssal flows in the ocean, for example, trying to calculate what happens below the level-of-no-motion under the assumption that these water masses should be largely conserved. After years of grappling with the problem, he concluded that there was something fundamentally wrong with the assumptions and published a complete North Atlantic circulation scheme (Worthington, 1976). This scheme has been widely reproduced and has been very influential. Unfortunately, one must read the book very carefully (see the footnote on his p. 36) to understand that the assumption that Worthington chose to remove was geostrophic balance. In particular, the Gulf Stream recirculation flows

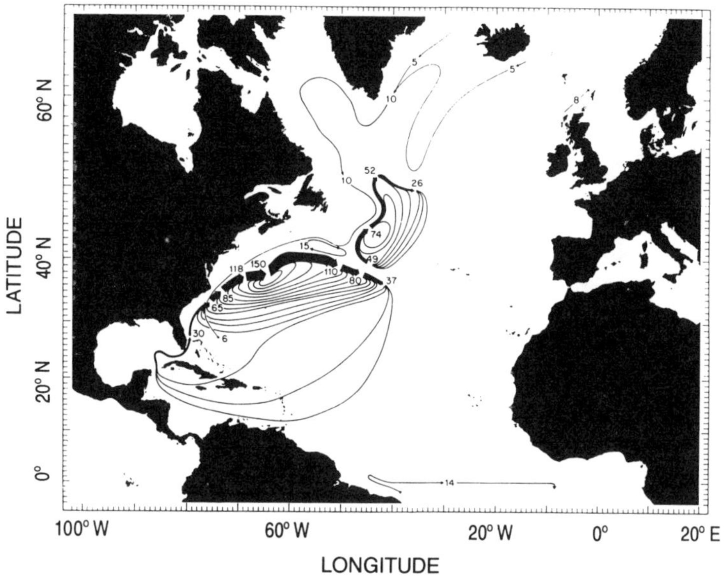

Figure 2–30. Worthington's (1976) steady-circulation scheme showing putative top-to-bottom flows. Transport values are in Sverdrups. Geostrophy is violated both in the Gulf Stream return flow and in the division between the two gyres. The widespread reproduction of this scheme, despite its physical impossibility, is a tribute to its striking visual power and elements of scientific discourse best left to sociologists. But note that Defant in 1941 (see Figure 271 in Defant, 1961) had suggested a very tight Gulf Stream recirculation and a separate gyre in the Labrador Sea. (See Krauss, 1986, for comments by an experienced, but clearly exasperated, oceanographer on the relationship between Worthington's picture and previous knowledge of the ocean circulation.)

and the region of splitting of the North Atlantic circulation into two distinct gyres, which attracted so much attention to the proposed circulation, are not consistent with geostrophy (Figure 2–30).

When one obtains a contradiction between a model (and Worthington had a very specific model) and data, one is led to modify the model. But the art of it comes into the question of which assumptions are the likely candidates for change. Among the assumptions of Worthington's model (he did not lay them out in such formal fashion) we can list:

1. Geostrophic, hydrostatic balance;
2. Conservation of water mass volumes when not in contact with the atmosphere;
3. The circulation is kinematically steady–that is, a section across the Gulf Stream in, say, 1958 represents the same ocean as a section in 1970; and
4. The circulation has simply prescribed deep levels-of-no-motion.

The problem with taking geostrophic balance in (1) as the assumption to be modified first is that it is an extremely drastic step. Geostrophy is derived from the Navier-Stokes equations, which are the fluid form of

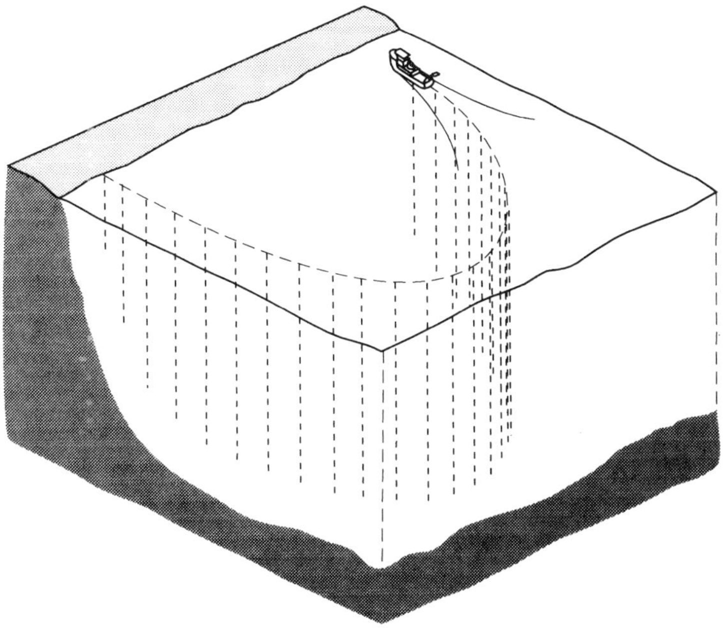

Figure 2–31. Schematic of a closed volume of fluid defined by a land boundary and a single hydrographic survey. Inferences by inverse methods are made from constraints expressing conservation laws for fluid properties within the volume.

Newton's laws of motion; as already discussed, the approximation can be systematically and rigorously justified as the major fluid dynamical balance for the time/space scales under consideration. The flow field in Worthington's transport picture changed with depth to conserve various fluid properties; but the relationship between the vertical structure in the flow field and the structure of the density field in (2.1.21)–(2.1.22) is a consequence of geostrophy. If these relations are not adhered to, the pressure field in (2.1.11)–(2.1.12) will not balance the Coriolis forces; the inferred flows are unsupported by the pressure gradients or anything else and are inconsistent with Newton's Laws.

To analyze the issues further, let us write an algebraic/numerical statement of Worthington's model. Consider that the fixing of a level-of-no-motion, z_0, is a local decision. Because we know so little, we are free to let it vary arbitrarily from one pair of hydrographic stations to another, removing assumption (4). We now follow the lead provided by Hidaka (1940a,b) and write down the requirements that the flow into and out of a volume of ocean, perhaps that shown in Figure 2–31, should conserve mass and salt.

Make the convention that velocities and transports are positive to the north and/or east and that the sign of the unit normal for a closed volume is positive *inward.* Choose a reference depth $z_0(s_j)$ where s is an arc length along the volume periphery and j denotes the station pair number, and compute the thermal wind relative to this reference level for each station pair. Consider first the total amount of fluid moving geostrophically into and out of the closed volume shown in the figure:

$$\sum_{j}^{N}\sum_{q}^{Q}\rho_j(q)(v_{Rj}(q)+b_j)\delta_j\Delta a_j(q)\approx 0 \tag{2.4.1}$$

where we have denoted by $v_{Rj}(q)$ the thermal wind (relative velocity) in station pair j at depth interval q, b_j is the reference-level velocity, $\Delta a_j(q)$ is the differential area in depth interval q in station pair j, and δ_j is the unit normal (± 1) for pair j for the volume under consideration (it is supposed, for the moment, that the Ekman flow is negligible). Equation (2.4.1) is a discrete approximation to the areal integrals over the boundary section and can be carried out in a variety of different approximations. Everything is known in (2.4.1) except for the reference-level velocities. Equation (2.4.1) has been written as approximately equal to zero, rather than precisely so, in anticipation of the need to grapple with errors in the various terms of the sum. [The presence of ρ in (2.4.1) is mainly cosmetic–the variation of density is so slight that numerically one cannot distinguish mass conservation from volume conservation–consistent with Equation (2.1.4) being the appropriate mass conservation equation.]

Equation (2.4.1) is one equation in N unknowns, b_j, and the addition of some further constraints–that is, equations–would be helpful. One possibility is to require that the salt content of the volume should also be conserved. Let S be the salinity (salt/unit mass). Then we might require

$$\sum_{j=1}^{N}\sum_{q=1}^{Q}\rho_j(q)S_j(q)(v_{Rj}(q)+b_j)\delta_j\Delta a_j(q)\approx 0\,. \tag{2.4.2}$$

The historical hydrography, as in Figure 2–2g,h, provides some reason to believe that the volume of water in the ocean lying in fixed density intervals does not change significantly over a few years. Define the depth of fluid of density ρ_i as $z(\rho_i,x,y)\equiv q_i(x,y)$. Then conservation of mass between density intervals $\rho_i\le\rho\le\rho_{i+1}$ is

$$\sum_{j=1}^{N}\sum_{q_i(j)}^{q_{i+1}(j)}\rho_j(q)(v_{Rj}(q)+b_j)\delta_j\Delta a_j(q)\approx 0\,. \tag{2.4.3}$$

Since $v_{Rj}(q)$ are assumed known, (2.4.3) can be rewritten as

$$\sum_{j=1}^{N}\sum_{q_i(j)}^{q_{i+1}(j)}\rho_j(q)b_j\delta_j\Delta a_j(q)\approx-\sum_{j=1}^{N}\sum_{q_i(j)}^{q_{i+1}(j)}\rho_j(q)\delta_j v_{Rj}(q)\Delta a_j(q) \tag{2.4.4}$$

and similarly for (2.4.1) and (2.4.2).

One can go on like this (and Chapter 4 is devoted in part to a discussion of what reasonable requirements on the flow are). The collection of all such relations can be written as

$$\mathbf{Ex} + \mathbf{n} = \mathbf{y} \tag{2.4.5}$$

where $\mathbf{x}$ is the vector of unknown reference-level velocities b_j. The noise vector, $\mathbf{n}$, has been introduced to permit the writing of these equations as equalities, in presence of errors in $\mathbf{y}$ and in the elements of the coefficient matrix $\mathbf{E}$.

If we could write down enough constraints, and if they were accurate enough, one might try to calculate the values of $\mathbf{x}$ from (2.4.5). As a practical matter we are faced with a number of troubling issues in doing so. First, we are usually unable to write down as many or more equations as there are x_i. Second, the values of $\mathbf{y}$, here the fluxes computed from the relative-velocity contributions, are in practice very noisy. Third, there are errors in the coefficient matrix $\mathbf{E}$. In general terms, the ill-conditioning problems that Defant noticed in Hidaka's calculation will be present.

Despite the anticipation of trouble, the heart of the geostrophic inverse problem is the extraction of useful information from equation sets like (2.4.5). Chapter 3 is devoted to providing some basic background to the mathematical problem, and Chapter 4 discusses the practical application of the mathematical machinery. We will see that discussion and solution of this problem has very general application to a wide variety of estimation problems. Along the way, we will discuss variants of the box model just formulated, including the so-called β-spiral, and its relatives. As a byproduct, it will be seen that the problem Worthington posed, far from having no solution, formally has an infinite number of them–turning his difficulty on its head.

Before moving on, some perspective is useful. The conventional dynamic method does not fully use the Equation set (2.1.11)–(2.1.18), in particular making no use of Equations (2.1.14)–(2.1.18). Hidaka (1940a,b) added the conservation of mass and salt equations to those normally used by the dynamic method. Riley (1951) added conservation of oxygen and other traditional oceanographic tracers. With hindsight, we know that their approach was an effective one, but they lacked knowledge of the mathematics required to work with noisy ill-conditioned sets of equations (although that mathematics already existed in the published literature). It was not until the mid–1970s that it began to be fully recognized that the complete set of equations does permit estimation of the absolute velocity field–the reference-level velocity plus the thermal wind. That this is so is the basis of the β-spiral (Stommel & Schott, 1977; Schott & Stommel, 1978), the geostrophic inverse method (Wunsch, 1977, 1978), and various combinations and variations of these ideas, for example, Killworth (1983a) and Olbers and Willebrand (1984). Development of computers was an obvious and essential ingredient in making the methodologies practical. The next chapter is devoted to developing the machinery required to proceed with these methods. In Chapter 5, however, we will show that the underlying problem is really nonlinear and the linear methods of the β-spiral and geostrophic box balances are not the full story.

2.5 The Absolute Velocity Problem in Retrospect

The problem of determining the geostrophic interior flow of the ocean circulation is one of *estimation*, because it involves the combination of noisy observations with a dynamical and kinematical model (geostrophic, hydrostatic balance, property conservation). It is vital to distinguish this problem from the mathematical one: *Apart from observational noise and sampling issues, the so-called level-of-no-motion problem is completely solved.*

The reference-level velocity problem arises only if one restricts the use of dynamics and kinematics to the hydrostatic, geostrophic Equations (2.1.11)–(2.1.13), which lead to the thermal wind relations (2.1.23)–(2.1.24). Surprisingly, it was only after the development of the geostrophic inverse and β-spiral methods that it was recognized that inclusion of the continuity and mass conservation Equations [(2.1.14) plus any version of (2.1.15)–(2.1.18)] produced a unique determination of the full, three-dimensional, absolute velocity.

The most succinct display of the solution is due to Needler (1985), who manipulated Equations (2.1.11)–(2.1.18) into an expression for the three-dimensional velocity:

$$\rho \mathbf{v}(x, y, z) = g \frac{\hat{\mathbf{k}} \cdot (\nabla \rho \times \nabla q)}{\nabla (f q_z) \cdot (\nabla \rho \times \nabla q)} \nabla \rho \times \nabla q \tag{2.5.1}$$

where $q = f \partial \rho / \partial z$, $q_z = \partial q / \partial z$, and assuming $\nabla \rho \times \nabla q \neq 0$. (Obtaining this expression is not so easy; a brief derivation appears in the appendix to this chapter.)

Equation (2.5.1) is exact. What it shows is that if one knew the actual density field, the three-dimensional, absolute velocity could be computed and there is no reference-level velocity problem. But since ρ is never known, but only some noisy, undersampled approximation to it, we have a serious reference-level velocity estimation problem.

2A. Appendix. Conservation Laws

The equations of motion can be manipulated into various useful forms, including statements of property conservation. The finding of conserved dynamical quantities, and of Needler's (1985) expression, Equation (2.5.1) from (2.1.1)–(2.1.18), is not obvious, and a derivation is offered here. For more detail and discussion, the reader is referred to Welander (1971), Killworth (1979), Needler (1985), Pedlosky (1987a,b) and Wunsch (1994). For some purposes, spherical coordinates should be used and not the Cartesian version we employ.

Equation (2.1.15) is an assertion that to the order of our approximations, density is conserved, $d\rho/dt = 0$ (the system is steady and $d/dt \equiv u\partial/\partial x + v\partial/\partial y + w\partial/\partial z$). Differentiate this equation with respect to z,

$$\frac{\partial u}{\partial z}\frac{\partial \rho}{\partial x} + u\frac{\partial^2 \rho}{\partial x \partial z} + \frac{\partial v}{\partial z}\frac{\partial \rho}{\partial y} + v\frac{\partial^2 \rho}{\partial y \partial z} + \frac{\partial w}{\partial z}\frac{\partial \rho}{\partial z} + w\frac{\partial^2 \rho}{\partial z^2} = 0\,, \tag{2.5.2}$$

which is

$$\mathbf{v}\cdot\nabla\left(\frac{\partial \rho}{\partial z}\right) + \frac{\partial u}{\partial z}\frac{\partial \rho}{\partial x} + \frac{\partial v}{\partial z}\frac{\partial \rho}{\partial y} + \frac{\partial w}{\partial z}\frac{\partial \rho}{\partial z} = 0\,. \tag{2.5.3}$$

Substituting for the thermal wind [Equations (2.1.21)–(2.1.22)], but treating ρ as constant except where multiplied by g–the Boussinesq approximation–produces cancellation of the second and third terms, and

$$\mathbf{v}\cdot\nabla\left(\frac{\partial \rho}{\partial z}\right) + \frac{\partial w}{\partial z}\frac{\partial \rho}{\partial z} = 0\,. \tag{2.5.4}$$

But the linear vorticity Equation (2.1.20) can be used to eliminate w from (2.5.4), producing

$$\mathbf{v}\cdot\nabla\frac{\partial \rho}{\partial z} + \frac{\beta v}{f}\frac{\partial \rho}{\partial z} = 0$$

or

$$\mathbf{v}\cdot\nabla\left(f\frac{\partial \rho}{\partial z}\right) = 0\,, \tag{2.5.5}$$

that is, the quantity

$$q = f\frac{\partial \rho}{\partial z}\,, \tag{2.5.6}$$

the planetary potential vorticity, is conserved.

Another quantity, the Bernoulli function, $B = p + g\rho z$, is also conserved (e.g., Welander, 1971), as can be seen by writing

$$\frac{dp}{dt} = u\frac{\partial p}{\partial x} + v\frac{\partial p}{\partial y} + w\frac{\partial p}{\partial z}$$

$$= -wg\rho\,,$$

using the geostrophic and hydrostatic balance equations. Thus, since

$$\frac{d(g\rho z)}{dt} = gz\frac{d\rho}{dt} + g\rho w = g\rho w\,,$$

we have

$$\frac{dB}{dt} \equiv \frac{d(p + gz\rho)}{dt} = 0\,. \tag{2.5.7}$$

Needler's (1985) expression (2.5.1) follows from conservation of density and potential vorticity. He notes that the absolute velocity must be normal to both $\nabla\rho$, ∇q and hence

$$\rho\mathbf{v} = \alpha(x, y, z)\nabla\rho \times \nabla q \tag{2.5.8}$$

where α is a scalar to be determined. The velocity field is made up of its horizontal and vertical vector components,

$$\mathbf{v} = \mathbf{v}_H + \hat{\mathbf{k}}w \tag{2.5.9}$$

where $\hat{\mathbf{k}}$ is the vertical unit vector. The vertical shear is

$$\frac{\partial\mathbf{v}}{\partial z} = \frac{\partial\mathbf{v}_H}{\partial z} + \hat{\mathbf{k}}\frac{\partial w}{\partial z} = \frac{\partial\mathbf{v}_H}{\partial z} + \hat{\mathbf{k}}\frac{\beta}{f}v \tag{2.5.10}$$

where the last equality follows from the linear vorticity equation. The thermal wind Equations (2.1.21)–(2.1.22) are

$$\rho\frac{\partial\mathbf{v}_H}{\partial z} = -\frac{g}{f}\hat{\mathbf{k}} \times \nabla\rho\,,$$

which when substituted into (2.5.10) produces

$$\rho\frac{\partial\mathbf{v}}{\partial z} = -\frac{g}{f}\hat{\mathbf{k}} \times \nabla\rho + \hat{\mathbf{k}}\frac{\beta}{f}\alpha\hat{\boldsymbol{\jmath}}\cdot(\nabla\rho \times \nabla q) \tag{2.5.11}$$

where $\hat{\boldsymbol{\jmath}}$ is the unit vector in the y-direction and (2.5.8) was used for $\mathbf{v}$. Now the z-derivative of (2.5.8) must be the same as (2.5.11) (within the Boussinesq approximation), and setting the two equal results in

$$\begin{aligned}\alpha\nabla\rho_z \times \nabla q + \alpha\nabla\rho \times \nabla q_z + \alpha_z\nabla\rho \times \nabla q\\ = -\frac{g}{f}\hat{\mathbf{k}} \times \nabla\rho + \hat{\mathbf{k}}\frac{\beta}{f}\alpha\hat{\boldsymbol{\jmath}}\cdot(\nabla\rho \times \nabla q)\,,\end{aligned} \tag{2.5.12}$$

that is,

$$\begin{aligned}\alpha\left\{\nabla\rho_z \times \nabla q + \nabla\rho \times \nabla q_z - \hat{\mathbf{k}}\frac{\beta}{f}\hat{\boldsymbol{\jmath}}\cdot(\nabla\rho \times \nabla q)\right\}\\ + \alpha_z(\nabla\rho \times \nabla q) = -\frac{g}{f}\hat{\mathbf{k}} \times \nabla\rho\end{aligned} \tag{2.5.13}$$

(the subscript here denotes the partial derivative). Taking the dot product of this last expression with ∇q eliminates the term in α_z and

$$\alpha\left\{(\nabla\rho\times\nabla q_z\cdot\nabla q)-\hat{\mathbf{k}}\cdot\nabla q\frac{\beta}{f}\hat{\boldsymbol{\jmath}}\cdot(\nabla\rho\times\nabla q)\right\}=-\frac{g}{f}\hat{\mathbf{k}}\times\nabla\rho\cdot\nabla q\,. \quad (2.5.14)$$

From an elementary identity,[10]

$$(\nabla\rho\times\nabla q_z)\cdot\nabla q=-\nabla q_z\cdot(\nabla\rho\times\nabla q)\,,$$

and

$$\frac{g}{f}\hat{\mathbf{k}}\times\nabla\rho\cdot\nabla q=\frac{g}{f}\hat{\mathbf{k}}\cdot(\nabla\rho\times\nabla q)\,. \quad (2.5.15)$$

Furthermore,

$$\hat{\mathbf{k}}\cdot\nabla q\frac{\beta}{f}\hat{\boldsymbol{\jmath}}\cdot(\nabla\rho\times\nabla q)=q_z\frac{\beta}{f}\hat{\boldsymbol{\jmath}}\cdot(\nabla\rho\times\nabla q)\,,$$

so

$$\nabla(fq_z)\cdot(\nabla\rho\times\nabla q)=f\nabla q_z\cdot(\nabla\rho\times\nabla q)+q_z\beta\hat{\boldsymbol{\jmath}}\cdot(\nabla\rho\times\nabla q)\,,$$

and (2.5.14) finally results in

$$\alpha=g\frac{\hat{\mathbf{k}}\cdot(\nabla\rho\times\nabla q)}{\nabla(fq_z)\cdot(\nabla\rho\times\nabla q)}\,,\qquad\nabla\rho\times\nabla q\neq0\,, \quad (2.5.16)$$

which if substituted into (2.5.8) produces Needler's expression (2.5.1).

These conservation laws are extremely important in understanding the theory of the ocean circulation and are the basis for many estimation procedures. But one cannot expect that direct substitution of observed density fields–with all their distortions–will produce sensible results.

[10] $\mathbf{a}\cdot\mathbf{b}\times\mathbf{c}=\mathbf{b}\cdot\mathbf{c}\times\mathbf{a}=\mathbf{c}\cdot\mathbf{a}\times\mathbf{b}$ for arbitrary vectors $\mathbf{a},\mathbf{b},\mathbf{c}$.

3

Basic Machinery

The purpose of this chapter is to record a number of results that are essential tools for the discussion of the problems already described. Much of this material is elementary and is discussed here primarily to produce a consistent notation for later use. Reference will be made to some of the good available textbooks. But some of the material is given what may be an unfamiliar interpretation, and I urge everyone to at least skim the chapter.

Our basic tools are those of matrix and vector algebra as they relate to the solution of simultaneous equations, and some elementary statistical ideas mainly concerning covariance, correlation, and dispersion. Least squares is reviewed, with an emphasis placed upon the arbitrariness of the distinction between knowns, unknowns, and noise. The singular-value decomposition is a central building block, producing the clearest understanding of least squares and related formulations. I introduce the Gauss-Markov theorem and its use in making property maps, as an alternative method for obtaining solutions to simultaneous equations, and show its relation to and distinction from least squares. The chapter ends with a brief discussion of recursive least squares and estimation as essential background for the time-dependent methods of Chapter 6.

3.1 Matrix and Vector Algebra

This subject is very large and well developed, and it is not my intention to repeat material better found elsewhere (e.g., Noble & Daniel, 1977; Strang, 1988). Only a brief survey of central results is provided.

A matrix is an $M \times N$ array of elements of the form

$$\mathbf{A} = \{A_{ij}\},\ 1 \le i \le M,\ 1 \le j \le N\,.$$

Normally a matrix is denoted by a boldface capital letter. A vector is a special case of an $M \times 1$ matrix, written as a boldface lower-case letter, for example, $\mathbf{q}$. Corresponding capital or lower-case letters for Greek symbols are also indicated in boldface. Unless otherwise stipulated, vectors are understood to be column vectors. The transpose of a matrix interchanges its rows and columns. Transposition applied to vectors is sometimes used to save space in printing, for example, $\mathbf{q}^T = [q_1 \; q_2 \ldots q_N]^T$ is the same as

$$\mathbf{q} = \begin{bmatrix} q_1 \\ q_2 \\ \vdots \\ q_N \end{bmatrix} .$$

3.1.1 Matrices and Vectors

The inner, or dot, product between two $L \times 1$ vectors $\mathbf{a}$, $\mathbf{b}$ is written $\mathbf{a}^T\mathbf{b} \equiv \mathbf{a} \cdot \mathbf{b} = \sum_{i=1}^{L} a_i b_i$ and is a scalar. Such an inner product is the projection of $\mathbf{a}$ onto $\mathbf{b}$ (or vice versa). The magnitude of this projection can be measured as

$$\mathbf{a}^T\mathbf{b} = |\mathbf{a}||\mathbf{b}| \cos \phi$$

where $\cos \phi$ ranges between zero, when the vectors are orthogonal, and one, when they are parallel.

Suppose we have a collection of N vectors, $\mathbf{e}_i$, each of dimension N. If it is possible to represent perfectly an arbitrary N–dimensional vector $\mathbf{f}$ as the linear sum

$$\mathbf{f} = \sum_{i=1}^{N} \alpha_i \mathbf{e}_i \,, \tag{3.1.1}$$

then $\mathbf{e}_i$ are said to be a spanning set. A necessary and sufficient condition for them to be a spanning set is that they should be independent–that is, no one of them can be perfectly representable by the others:

$$\mathbf{e}_{j_0} - \sum_{i=1,\, i \neq j_0}^{N} \beta_i \mathbf{e}_i \neq 0, \quad 1 \leq j_0 \leq N \,, \tag{3.1.2}$$

for any choice of β_i.

The expansion coefficients α_i in (3.1.1) are obtained by taking the dot product of (3.1.1) with each of the vectors in turn:

$$\sum_{i=1}^{N} \alpha_i \mathbf{e}_k^T \mathbf{e}_i = \mathbf{e}_k^T \mathbf{f}, \quad 1 \leq k \leq N \,, \tag{3.1.3}$$

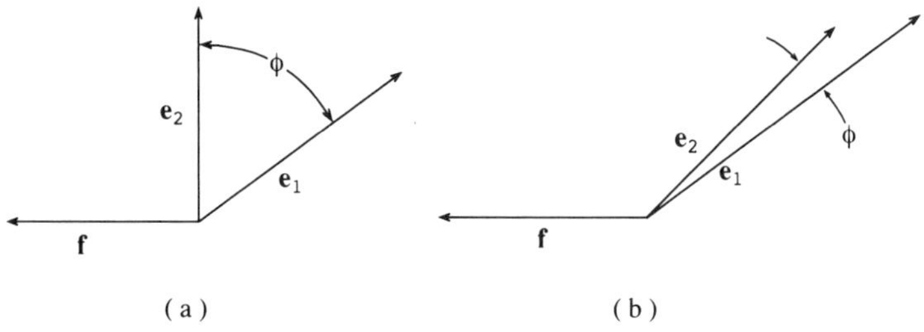

Figure 3–1. An arbitrary two-dimensional vector **f** can be expanded exactly in any two nonparallel vectors $\mathbf{e}_1$, $\mathbf{e}_2$, as in (a). In (b), the angle ϕ is not actually zero, but as it becomes arbitrarily small, it is readily confirmed that the slightest errors in knowledge of **f** render unstable calculation of the expansion coefficients.

a system of N equations in N unknowns. The α_i are most readily found if the $\mathbf{e}_i$ are a mutually orthonormal set–that is, if

$$\mathbf{e}_i^T \mathbf{e}_j = \delta_{ij} \,,$$

but this requirement is not necessary for a spanning set. With a spanning set, the information contained in the set of projections, $\mathbf{e}_i^T \mathbf{f} = \mathbf{f}^T \mathbf{e}_i$, is adequate then to determine the α_i and thus all the information required to reconstruct **f**.

The concept of nearly dependent vectors is helpful and can be understood heuristically. Consider Figure 3–1a, in which the space is two-dimensional. Then the two vectors $\mathbf{e}_1$, $\mathbf{e}_2$, as depicted there, are independent and can be used to expand an arbitrary two-dimensional vector **f** in the plane. But if the vectors become nearly parallel, as in Figure 3–1b, as long as they are not exactly parallel, they can still be used mathematically to represent **f** perfectly. However, one anticipates, and we find in practice, that as the angle ϕ between them becomes very small, they are *almost dependent*, and numerical problems arise in finding the expansion coefficients α_1, α_2. The generalization to higher dimensions is left to the reader's intuition.

It has been found convenient and fruitful to define multiplication of two matrices **A**, **B** by the operation $\mathbf{C} = \mathbf{AB}$, such that

$$C_{ij} = \sum_{p=1}^{P} A_{ip} B_{pj} \,. \tag{3.1.4}$$

For the definition (3.1.4) to make sense, **A** must be an $M \times P$ matrix and **B** must be $P \times N$ (including the special case of $P \times 1$, a column vector). That is, the two matrices must be *conformable*. If two matrices are multiplied, or a matrix and a vector are multiplied, conformability is implied; otherwise one can be assured that an error has been made. Note that $\mathbf{AB} \neq \mathbf{BA}$ even where both products exist, except under special circumstances. For both products to exist, $\mathbf{A}$ and $\mathbf{B}$ must be square and of the same dimension.

The mathematical operation in (3.1.4) may appear arbitrary, but a physical interpretation is available: Matrix multiplication is the dot product of all of the rows of $\mathbf{A}$ with all the columns of $\mathbf{B}$.

If we define a matrix, $\mathbf{E}$, each of whose columns is the corresponding vector $\mathbf{e}_i$, and a vector, $\boldsymbol{\alpha} = \{\alpha_i\}$, in the same order, the expansion (3.1.1) can be written in the compact form

$$\mathbf{f} = \mathbf{E}\boldsymbol{\alpha}\,. \tag{3.1.5}$$

The transpose of a matrix $\mathbf{A}$ is written $\mathbf{A}^T$ and is defined as $\{A_{ij}\}^T = A_{ji}$, an interchange of the rows and columns of $\mathbf{A}$. A *symmetric matrix* is one for which $\mathbf{A}^T = \mathbf{A}$. The product $\mathbf{A}^T\mathbf{A}$ represents the array of all the dot products of the columns of $\mathbf{A}$ with themselves, and similarly, $\mathbf{A}\mathbf{A}^T$ represents the set of all dot products of all the rows of $\mathbf{A}$ with themselves. It follows that $(\mathbf{AB})^T = \mathbf{B}^T\mathbf{A}^T$. Because we have $(\mathbf{A}\mathbf{A}^T)^T = \mathbf{A}\mathbf{A}^T$, $(\mathbf{A}^T\mathbf{A})^T = \mathbf{A}^T\mathbf{A}$, both these matrices are symmetric ones. [We used $(\mathbf{A}^T)^T = \mathbf{A}$.]

The *trace* of a square $M \times M$ matrix $\mathbf{A}$ is defined as $\mathrm{trace}(\mathbf{A}) = \sum_i^M A_{ii}$. A *diagonal matrix* is square and zero except for the terms along the main diagonal. The operator $\mathrm{diag}(\mathbf{q})$ makes a square diagonal matrix with $\mathbf{q}$ along the main diagonal.

The special $L \times L$ diagonal matrix $\mathbf{I}_L$, with $I_{ii} = 1$, is the *identity*. Usually, when the dimension of $\mathbf{I}_L$ is clear from the context, the subscript is omitted. If there is a matrix $\mathbf{B}$, such that $\mathbf{BE} = \mathbf{I}$, then $\mathbf{B}$ is the *left-inverse* of $\mathbf{E}$. If $\mathbf{B}$ is the left inverse of $\mathbf{E}$ and $\mathbf{E}$ is square, a standard result is that it must also be a right inverse: $\mathbf{EB} = \mathbf{I}$, $\mathbf{B}$ is then called *the inverse of* $\mathbf{E}$ and is usually written $\mathbf{E}^{-1}$. If $\mathbf{E}$ is not square, such an inverse cannot exist, and special inverses, like a left inverse, are sometimes written $\mathbf{E}^{+}$ and referred to as *generalized inverses*. Some of them will be encountered later. A useful result is that $(\mathbf{AB})^{-1} = \mathbf{B}^{-1}\mathbf{A}^{-1}$ if the inverses exist. Square matrices with inverses are *nonsingular*.

We need the idea of the *length*, or norm, of a vector. Several choices are possible; for present purposes, the conventional l_2 norm,

$$\|\mathbf{f}\|_2 \equiv (\mathbf{f}^T\mathbf{f})^{1/2} = \left(\sum_{i=1}^{N} f_i^2\right)^{1/2} \tag{3.1.6}$$

is most useful; often the subscript will be omitted. This definition leads in turn to the measure of distance between two vectors, $\mathbf{a}$, $\mathbf{b}$ as

$$\|\mathbf{a} - \mathbf{b}\|_2 = \sqrt{(\mathbf{a} - \mathbf{b})^T (\mathbf{a} - \mathbf{b})}\,, \tag{3.1.7}$$

the familiar Cartesian distance. Distances can also be measured in such a way that deviations of certain elements of $\mathbf{c} = \mathbf{a} - \mathbf{b}$ count for more than others–that is, a metric, or set of weights can be introduced with a definition

$$\|\mathbf{c}\| = \sqrt{\sum_i c_i W_{ii} c_i}\,, \tag{3.1.8}$$

depending upon the importance to be attached to magnitudes of different elements, stretching and shrinking various coordinates. Finally, in the most general form, distance can be measured in a coordinate system both stretched and rotated relative to the original one

$$\|\mathbf{c}\|_W = \sqrt{\mathbf{c}^T \mathbf{W} \mathbf{c}} \tag{3.1.9}$$

where $\mathbf{W}$ is an arbitrary matrix (but usually, for physical reasons, symmetric and positive definite[1]).

Consider a set of M linear equations in N unknowns,

$$\mathbf{E}\mathbf{x} = \mathbf{y}\,. \tag{3.1.10}$$

Because of the appearance of simultaneous equations in situations in which the y_i are observed, and where $\mathbf{x}$ are parameters that we wish to determine, it is often convenient to refer to (3.1.10) as a set of measurements of $\mathbf{x}$ which produced the observations or data, $\mathbf{y}$. If $M > N$, the system is said to be *overdetermined*, or *formally overdetermined*. If $M < N$, it is *underdetermined*, and if $M = N$, it is *just-determined* or *formally just-determined*. The use of the word *formally* has a purpose we will come to later. Knowledge of the matrix inverse to $\mathbf{E}$ would make it easy to solve a set of L equations in L unknowns by left-multiplying (3.1.10) by $\mathbf{E}^{-1}$. The reader is cautioned that although matrix inverses are a very powerful tool, one is usually ill advised to solve large sets of simultaneous equations by inverting the coefficient matrix (e.g., Golub & Van Loan, 1989).

There are several ways to view the meaning of any set of linear simultaneous equations. If the columns of $\mathbf{E}$ continue to be denoted $\mathbf{e}_i$, but without necessarily stipulating that they are either a spanning set or orthogonal, then (3.1.10) is of the form,

$$x_1\mathbf{e}_1 + x_2\mathbf{e}_2 + \cdots + x_n\mathbf{e}_N = \mathbf{y}\,. \tag{3.1.11}$$

The ability to so describe an arbitrary $\mathbf{y}$, or to solve the equations, would thus depend upon whether the $M \times 1$ vector $\mathbf{y}$ can be specified by a sum of N column vectors, $\mathbf{e}_i$–that is, it would depend upon their being a spanning set. In this view, the elements of $\mathbf{x}$ are simply the corresponding expansion

[1] *Positive definite* will be defined later.

coefficients. Depending upon the ratio of M to N–that is, the number of equations to unknown elements–one faces the possibility that there are fewer expansion vectors $\mathbf{e}_i$ than elements of $\mathbf{y}$ $(M > N)$, or that there are more expansion vectors available than elements of $\mathbf{y}$ $(M < N)$. Thus, the overdetermined case corresponds to having *fewer* expansion vectors, and the underdetermined case corresponds to having *more* expansion vectors, than the dimension of $\mathbf{y}$. It is possible that in the overdetermined case, the too-few expansion vectors are not actually independent, so that there are even fewer vectors available than is first apparent. Similarly, in the underdetermined case, there is the possibility that although it appears we have more expansion vectors than required, fewer may be independent than the number of elements of $\mathbf{y}$, and the consequences of that case need to be understood as well.

Alternatively, if the rows of $\mathbf{E}$ are denoted $\mathbf{r}_i^T$, $1 \le i \le M$, (3.1.10) is a set of M-inner products,

$$\mathbf{r}_i^T \mathbf{x} = y_i, \quad 1 \le i \le M\,. \tag{3.1.12}$$

That is, the set of simultaneous equations is equivalent to being provided with the value of M–dot products of the N–dimensional unknown vector, $\mathbf{x}$, with M known vectors, $\mathbf{r}_i$. Whether that is sufficient information to determine $\mathbf{x}$ depends upon whether the $\mathbf{r}_i$ are a spanning set. In this view, in the overdetermined case, one has *more* dot products available than unknown elements x_i, and in the underdetermined case, there are *fewer* such values than unknowns. (These statements are particularly transparent if the rows or columns happen to be orthonormal vectors, and the reader is urged to examine the relative determinancy in that special situation.)

3.1.2 Identities, Differentiation, and So Forth

Here are some identities and matrix/vector definitions that prove useful.

A square positive definite matrix $\mathbf{A}$ is one for which the scalar quadratic form,

$$J = \mathbf{x}^T \mathbf{A} \mathbf{x}\,, \tag{3.1.13}$$

is positive for all vectors $\mathbf{x}$. (It suffices to consider only symmetric $\mathbf{A}$ because for a general matrix, $\mathbf{x}^T\mathbf{A}\mathbf{x} = \mathbf{x}^T[(\mathbf{A} + \mathbf{A}^T)/2]\mathbf{x}$, which follows from the scalar property of the quadratic form.) If $J \ge 0$ for all $\mathbf{x}$, $\mathbf{A}$ is positive semidefinite, or nonnegative definite. Linear algebra books show that a necessary and sufficient requirement for positive definiteness is that $\mathbf{A}$ have all positive eigenvalues and a semidefinite one must have all nonnegative eigenvalues.

Nothing has been said about actually finding the numerical values of either the matrix inverse or the eigenvectors and eigenvalues. Computational algorithms for obtaining them have been developed by experts and are discussed in many good textbooks (Lawson & Hanson, 1974; Golub & van Loan, 1989; Press, Flannery, Teukolsky, & Vetterling, 1992; etc.), and software systems like MATLAB implement them in easy-to-use form. For purposes of this book, we assume the reader has at least a rudimentary knowledge of these techniques and access to a good software implementation.

We end up doing a certain amount of differentiation and other operations with respect to matrices and vectors. A number of formulas are very helpful and save a lot of writing. They are all demonstrated by doing the derivatives term-by-term. Let $\mathbf{q}$, $\mathbf{r}$ be $N \times 1$ column vectors, and $\mathbf{A}$, $\mathbf{B}$, $\mathbf{C}$ be matrices. Then if s is any scalar,

$$\frac{\partial s}{\partial \mathbf{q}} = \mathbf{b} = \left\{ \begin{matrix} \frac{\partial s}{\partial q_1} \\ \cdot \\ \cdot \\ \cdot \\ \frac{\partial s}{\partial q_N} \end{matrix} \right\} \tag{3.1.14}$$

is a vector (the gradient). The second derivative of a scalar,

$$\frac{\partial^2 s}{\partial \mathbf{q}^2} = \left\{ \frac{\partial}{\partial q_i} \frac{\partial s}{\partial q_j} \right\} = \left\{ \begin{matrix} \frac{\partial^2 s}{\partial q_1^2} & \frac{\partial^2 s}{\partial q_1 q_2} & \cdot & \cdot & \frac{\partial^2 s}{\partial q_1 q_N} \\ \cdot & \cdot & \cdot & \cdot & \cdot \\ \frac{\partial^2 s}{\partial q_N q_1} & \cdot & \cdot & \cdot & \frac{\partial^2 s}{\partial q_N^2} \end{matrix} \right\}, \tag{3.1.15}$$

is the *Hessian* of s.

The derivative of one vector by another is a matrix:

$$\frac{\partial \mathbf{r}}{\partial \mathbf{q}} = \left\{ \frac{\partial r_j}{\partial q_j} \right\} = \left\{ \begin{matrix} \frac{\partial r_1}{\partial q_1} & \frac{\partial r_2}{\partial q_1} & \cdot & \frac{\partial r_M}{\partial q_1} \\ \frac{\partial r_1}{\partial q_2} & \cdot & \cdot & \frac{\partial r_M}{\partial q_2} \\ \cdot & \cdot & \cdot & \cdot \\ \frac{\partial r_1}{\partial q_N} & \cdot & \cdot & \frac{\partial r_M}{\partial q_N} \end{matrix} \right\} \equiv \mathbf{P}\,. \tag{3.1.16}$$

If $\mathbf{r}$, $\mathbf{q}$ are of the same dimension, the determinant of $\mathbf{P}$ is the *Jacobian* of $\mathbf{r}$.

Assuming conformability, the inner product

$$\mathbf{r}^T \mathbf{q} = \mathbf{q}^T \mathbf{r}$$

is a scalar, and

$$\frac{\partial(\mathbf{q}^T\mathbf{r})}{\partial\mathbf{q}} = \frac{\partial(\mathbf{r}^T\mathbf{q})}{\partial\mathbf{q}} = \mathbf{r}\,, \tag{3.1.17}$$

$$\frac{\partial(\mathbf{q}^T\mathbf{q})}{\partial\mathbf{q}} = 2\mathbf{q}\,. \tag{3.1.18}$$

For a quadratic form,

$$J = \mathbf{q}^T\mathbf{A}\mathbf{q}$$

$$\frac{\partial J}{\partial\mathbf{q}} = (\mathbf{A}+\mathbf{A}^T)\mathbf{q}\,, \tag{3.1.19}$$

and its Hessian is $\mathbf{A}+\mathbf{A}^T$.

Let $\mathbf{A}$, $\mathbf{B}$, $\mathbf{C}$ be square. Then

$$\frac{\partial \mathrm{trace}\mathbf{A}}{\partial\mathbf{A}} = \mathbf{I}\,, \tag{3.1.20}$$

$$\frac{\partial \mathrm{trace}(\mathbf{BAC})}{\partial\mathbf{A}} = \mathbf{B}^T\mathbf{C}^T\,, \tag{3.1.21}$$

$$\frac{\partial \mathrm{trace}(\mathbf{ABA}^T)}{\partial\mathbf{A}} = \mathbf{A}(\mathbf{B}+\mathbf{B}^T)\,. \tag{3.1.22}$$

Rogers (1980) is an entire volume of matrix derivative identities, and many other useful properties are discussed by Magnus and Neudecker (1988).

There are a few, unfortunately unintuitive, matrix inversion identities that are essential to some of the later chapters. Liebelt (1967, Section 1–19) derives them by considering the square, partitioned matrix

$$\begin{Bmatrix} \mathbf{A} & \mathbf{B} \\ \mathbf{B}^T & \mathbf{C} \end{Bmatrix} \tag{3.1.23}$$

where $\mathbf{A}^T=\mathbf{A}$, $\mathbf{C}^T=\mathbf{C}$, but $\mathbf{B}$ can be rectangular of conformable dimensions in (3.1.23). The most important of the identities, sometimes called the *matrix inversion lemma* is, in one form,

$$\{\mathbf{C}-\mathbf{B}^T\mathbf{A}^{-1}\mathbf{B}\}^{-1} = \mathbf{C}^{-1} - \mathbf{C}^{-1}\mathbf{B}^T(\mathbf{B}\mathbf{C}^{-1}\mathbf{B}^T-\mathbf{A})^{-1}\mathbf{B}\mathbf{C}^{-1} \tag{3.1.24}$$

where it is assumed that the inverses exist. (The history of this not-very-obvious identity is discussed by Haykin, 1986, p. 385.) A variant (Liebelt's equation 1–51) is

$$\mathbf{A}\mathbf{B}^T(\mathbf{C}+\mathbf{B}\mathbf{A}\mathbf{B}^T)^{-1} = (\mathbf{A}^{-1}+\mathbf{B}^T\mathbf{C}^{-1}\mathbf{B})^{-1}\mathbf{B}^T\mathbf{C}^{-1}\,. \tag{3.1.25}$$

Both (3.1.24)–(3.1.25) are readily confirmed by direct multiplication, for example, by showing that $\mathbf{A}\mathbf{B}^T(\mathbf{C}+\mathbf{B}\mathbf{A}\mathbf{B}^T)^{-1}$ times the right-hand side of (3.1.25) is the identity.

Another identity, found by completing the square, is demonstrated by directly multiplying it out and requires $\mathbf{C} = \mathbf{C}^T$ ($\mathbf{A}$ is unrestricted, but the matrices must be conformable as shown):

$$\mathbf{ACA}^T - \mathbf{BA}^T - \mathbf{AB}^T = (\mathbf{A} - \mathbf{BC}^{-1})\mathbf{C}(\mathbf{A} - \mathbf{BC}^{-1})^T - \mathbf{BC}^{-1}\mathbf{B}^T . \quad (3.1.26)$$

A number of useful definitions of a matrix norm exist. For present purposes the so-called spectral norm or 2–norm defined as

$$\|\mathbf{A}\|_2 = \sqrt{\text{maximum eigenvalue of } (\mathbf{A}^T\mathbf{A})} \quad (3.1.27)$$

is usually adequate. Without difficulty (e.g., Haykin, 1986, p. 61), it may be seen that this definition is equivalent to

$$\|\mathbf{A}\|_2 = \max \frac{\mathbf{x}^T\mathbf{A}^T\mathbf{A}\mathbf{x}}{\mathbf{x}^T\mathbf{x}} = \max \frac{\|\mathbf{A}\mathbf{x}\|_2}{\|\mathbf{x}\|_2} \quad (3.1.28)$$

where the maximum is defined over all vectors $\mathbf{x}$. Another useful measure is the Frobenius norm,

$$\|\mathbf{A}\|_F = \sqrt{\sum_{i=1}^{M} \sum_{j=1}^{N} A_{ij}^2} = \sqrt{\text{trace}(\mathbf{A}^T\mathbf{A})} . \quad (3.1.29)$$

Neither definition requires $\mathbf{A}$ to be square.

These norms permit one to derive various useful results. Consider one illustration. $\mathbf{Q}$ is square, and $\|\mathbf{Q}\| < 1$, then

$$(\mathbf{I} + \mathbf{Q})^{-1} = \mathbf{I} - \mathbf{Q} + \mathbf{Q}^2 - \cdots , \quad (3.1.30)$$

which may be verified by multiplying both sides by $\mathbf{I} + \mathbf{Q}$, doing term-by-term multiplication and measuring the remainders with either norm.

3.1.3 Gram-Schmidt Process

One often has a set of p-independent but nonorthonormal vectors $\mathbf{h}_i$, and it is convenient to find a new set $\mathbf{g}_i$, which are orthonormal. The *Gram-Schmidt process* operates by induction. Suppose we have orthonormalized the first k of the $\mathbf{h}_i$ to a new set, $\mathbf{g}_i$, and wish to generate the $k+1$st. Let

$$\mathbf{g}_{k+1} = \mathbf{h}_{k+1} - \sum_{j}^{k} \gamma_j \mathbf{g}_j . \quad (3.1.31)$$

Because $\mathbf{g}_{k+1}$ must be orthogonal to the preceding $\mathbf{g}_i$, $i = 1, k$, we take the dot products of (3.1.31) with each of these vectors, producing a set of simultaneous equations for determining the unknown γ_j. The resulting $\mathbf{g}_{k+1}$ is easily given unit norm by division by its length.

If one has the first k of N necessary vectors, one needs an additional $N-k$ independent vectors h_i. There are several possibilities. One might simply generate the necessary vectors by filling their elements with random numbers. Or one might take a very simple trial set like $\mathbf{h}_{k+1} = [1 \ \ 0 \ \ 0 \ \ \ldots \ \ 0]^T$, $\mathbf{h}_{k+2} = [0 \ \ 1 \ \ . \ \ . \ \ 0], \ldots$. If one is unlucky, the set might prove not to be independent. But a simple numerical perturbation usually suffices to render them so. In practice, the algorithm is changed to what is usually called the *modified Gram-Schmidt process* (see Lawson & Hanson, 1974) for purposes of numerical stability.

3.2 Simple Statistics; Regression

Some statistical ideas are required, but the discussion is confined to stating some basic notions and to developing a notation. A statistics text such as Cramér (1946), or one on regression such as Seber (1977), should be consulted for real understanding.

We require the idea of a probability density for a random variable x. This subject is a very deep one–as described, for example, by Feller (1957) and Jeffreys (1961)–but our approach will be heuristic. Suppose that an arbitrarily large number of experiments can be conducted for the determination of the values of x, denoted X_i, $1 \le i$, and a histogram of the experimental values found. The frequency function, or probability density, will be defined as the limit, supposing it exists, of the histogram per bin width of an arbitrarily large number of experiments divided into bins of arbitrarily small value ranges. Let the corresponding limiting density or frequency function be denoted $p_x(X)$. (This notation distinguishes between a random variable, x, and the numerical values it assumes, X. The distinction is not always preserved where the context prevents confusion.) The *average*, or *mean*, or *expected value* is denoted $< x >$ and defined as

$$< x > \equiv \int_{\text{all } X} X p_x(X) dX \tag{3.2.1}$$

and is the center of mass of $p_x(X)$. Using the definition of a frequency function, it is easy to show heuristically that the *sample average* or *mean*,

$$< x >_N \equiv \frac{1}{N} \sum_{i=1}^{N} X_i , \tag{3.2.2}$$

when it exists, will usually asymptotically approach $< x >$ in the limit as N approaches infinity. Knowledge of the true mean value of a random variable is commonly all that we are willing to assume known. If forced to forecast the numerical value of x under such circumstances, often the best we can do is to employ $< x >$. If the deviation from the true mean is denoted x' so that $x = < x > + x'$, such a forecast has the virtue that we are assured the average forecast error, $< x' >$, would be zero if many such forecasts are made. The bracket operation is very important throughout this book; it has the property that if a is a nonrandom quantity, $< ax >= a < x >$.

The idea of a frequency function generalizes easily to two or more random variables, x, y. We can in concept do an arbitrarily large number of experiments in which we count the occurrences of differing pair values, (X_i, Y_i), of x, y and make a histogram, dividing by the bin area, and taking the limit to produce a joint probability density, $p_{xy}(X, Y)$. A simple example would be the simultaneous measurement by a current meter of the two components of horizontal velocity.

An important use of joint probability densities is in what is known as *conditional probability.* Suppose that the joint probability density for x, y is known and furthermore, $y = Y$–that is, information is available concerning the actual value of y. What then is the probability density for x given that a particular value for y is known to have occurred? This new frequency function is usually written as $p_{x|y}(X|Y)$ and is read as "the probability of x, given that y has occurred with value Y." It follows immediately from the definition of the probability density that

$$p_{x|y}(X|Y) = \frac{p_{xy}(X, Y)}{p_y(Y)}. \tag{3.2.3}$$

(This equation is readily interpreted by going back to the original experimental concept and understanding the restriction on x given that y is known to lie within one of the bins.)

If one finds that $p_{xy}(X, Y) = p_x(X)p_y(Y)$, then x, y are said to be *independent.* Using the joint frequency function, define the average product as

$$< xy >= \int\int_{\text{all } X,Y} XY p_{xy}(X, Y) dX dY. \tag{3.2.4}$$

Should $< (x- < x >)(y - < y >) > \neq 0$, x, y are said to *covary* or to be *correlated.* From the definition of frequency function and the bracket operation, if x, y are independent, then $< (x- < x >)(y - < y >) > = 0$. Under these circumstances x, y are *uncorrelated* or do not covary. Independence implies lack of correlation, but the reverse is not necessarily true. If the two variables are independent, then (3.2.3) is

$$p_{x|y}(X|Y) = p_x(X); \tag{3.2.5}$$

that is, knowledge of the value of y does not change the probability density for x–a sensible result–and there is then no predictive power for one variable given knowledge of the other.

We need the idea of dispersion–the expected or average squared value of some quantity about some interesting value, like its mean. The most familiar measure of dispersion is the variance, already used above, the expected fluctuation of a random variable about its mean:

$$\sigma_x^2 =< (x - < x >)^2 > .$$

More generally, define the dispersion of any random variable x as

$$D^2(x) = < x^2 > .$$

Thus, $\sigma_x^2 = D^2(x - < x >)$.

Sample estimates of quantities like the mean and other properties of random variables made from observations occur throughout science. In the case of the sample mean, it is possible to show without difficulty that the expected value of $< x >_N$ is the true average–that is, $<< x >_N > = < x >$. The interpretation is that for finite N, we do not expect that the sample mean will equal the true mean, but that if we could produce sample averages from distinct groups of observations, the sample averages would themselves have an average that would fluctuate about the true mean.

The variance of the sample mean (3.2.2) is easily shown to be

$$D^2(< x >_N - << x >_N>) = \frac{\sigma_x^2}{N} , \tag{3.2.6}$$

and thus the dispersion diminishes with N.

There are many sample estimates, however, some of which we encounter, where the expected value of the sample estimate is not equal to the true estimate. Such an estimator is said to be *biased*. Otherwise, it is *unbiased*. The simplest example of a biased estimator is the sample variance if defined as

$$s^2 \equiv \frac{1}{N} \sum_i^N (X_i - < x >_N)^2 . \tag{3.2.7}$$

For simplicity, but without loss of generality, assume the true mean of x is zero, $< x > = 0$ (a nonzero mean can be removed first if necessary). Equation (3.2.7) is

$$s^2 = \frac{1}{N} \left(\sum_i^N x_i^2 \right) - \frac{< x >_N^2}{N} ,$$

whose expected value is

$$< s^2 > = \sigma_x^2 - \frac{\sigma_x^2}{N} = \sigma_x^2 \frac{N-1}{N}\,, \tag{3.2.8}$$

using (3.2.6) and the zero true mean. Thus, in the definition (3.2.8) the expected value of the sample variance is not the correct value but is rather $(N-1)/N$ times it. To remove the bias, one often redefines the sample variance as

$$s^2 = \frac{1}{N-1} \sum_{i=1}^{N} (X_i - < x >_N)^2\,. \tag{3.2.9}$$

Suppose there are two random variables x, y between which there is anticipated to be some linear relationship

$$x = ay + n \tag{3.2.10}$$

where n represents any contributions to x that remain unknown despite knowledge of y. Then

$$< x > = a < y > + < n >\,, \tag{3.2.11}$$

and (3.2.10) shows

$$x - < x > = a(y - < y >) + (n - < n >)\,,$$

or

$$x' = ay' + n', \quad x' = x - < x >, \quad \text{etc.} \tag{3.2.12}$$

From this last equation,

$$a = \frac{< x'y' >}{< y'^2 >} = \frac{< x'y' >}{(< y'^2 >< x'^2 >)^{1/2}} \frac{< x'^2 >^{1/2}}{< y'^2 >^{1/2}} = \rho \frac{< x'^2 >^{1/2}}{< y'^2 >^{1/2}} \tag{3.2.13}$$

where it was supposed $< y'n' > = 0$, thus defining n. The quantity

$$\rho \equiv \frac{< x'y' >}{(< y'^2 >< x'^2 >)^{1/2}} \tag{3.2.14}$$

is the *correlation coefficient* and is easily shown to have the property $|\rho| \leq 1$. If ρ should vanish, then so does a. If a vanishes, then knowledge of y' carries no information about the value of x'. If $\rho = \pm 1$, then it follows from the definitions that $n = 0$, and knowledge of a permits perfect prediction of x' from knowledge of y' (because probabilities are being used, rigorous usage would state "perfect prediction almost always," but this distinction will be ignored).

A measure of how well the prediction of x from y will work can be obtained in terms of the variance of x'. We have

$$< x'^2 > = a^2 < y'^2 > + < n'^2 > = \rho^2 < x'^2 > + < n'^2 >$$

or

$$(1-\rho^2) < x'^2 > = < n'^2 > ; \tag{3.2.15}$$

that is, the fraction of the variance in x' that is unpredictable by y' is $(1-\rho^2) < x'^2 >$ and is the *unpredictable power.* Conversely, $\rho^2 < x'^2 >$ is the *predictable power.* The limits as $\rho \to 0, 1$ are readily apparent.

Thus, we interpret the statement that two variables x', y' are correlated or covary to mean that knowledge of one permits at least a partial prediction of the other, the expected success of the prediction depending upon the size of ρ. This result represents an implementation of the statement that if two variables are not independent, then knowledge of one permits some skill in the prediction of the other. If two variables do not covary but are also not independent, a linear model like (3.2.10) would not be useful and some nonlinear one would be required. Such nonlinear methods are possible and are touched on briefly later. The idea that correlation or covariance between various physical quantities carries useful predictive skill between them is an essential ingredient of many of the methods taken up in this book.

If a sequence of pairs of values x_i, y_i is measured so that we have a set of simultaneous equations

$$ay_i + n_i = x_i \,, \tag{3.2.16}$$

we might think to use these equations to determine $< x >$, $< y >$, a, $< n >$, n', etc. This leads into the huge subject of regression analysis (see, for example, Seber, 1977; or Draper & Smith, 1982), which is necessary to understand the connection between the theoretical values of quantities like $< x >$ and their sample values computed from objects like $< x >_N$. Some more machinery is required to do so, which we will eventually obtain in part.

In the absence of other information, the Gaussian, or normal, probability density is often invoked to describe observations. Apart from its comparatively simple mathematical properties, justification for the assumption of normality lies with the so-called Central Limit Theorem (Cramér, 1946). This theorem, which can be proven under hypotheses of varying strength, shows that under general circumstances, phenomena that are the result of summing many independent stochastic phenomena will tend toward a normal distribution. But not all physical phenomena conform to the assumptions of the Central Limit Theorem; in particular, the ocean is demonstrably nonnormal in many respects, making the Gaussian hypothesis a dangerous one. Nonetheless, it is worth recalling the fundamental properties of the normal probability density. For a single variable x, it is defined as

$$p_x(X) = \frac{1}{\sqrt{2\pi}\sigma_x} \exp\left(-\frac{(X-m_x)^2}{2\sigma_x^2}\right)$$

[sometimes abbreviated as $G(m_x, \sigma_x)$]. It is readily confirmed that $< x > = m_x$, $< (x - < x >)^2 > = \sigma_x^2$. Suppose that x, y are *independent* Gaussian variables $G(m_x, \sigma_x)$, $G(m_y, \sigma_y)$. Then their joint probability density is just the product of the two individual densities,

$$p_{xy}(X,Y) = \frac{1}{2\pi\sigma_x\sigma_y} \exp\left(-\frac{(X-m_x)^2}{2\sigma_x^2} - \frac{(Y-m_y)^2}{2\sigma_y^2}\right). \tag{3.2.17}$$

We need to consider the probability density for normal variables that are correlated. Let two new random variables, ξ_1, ξ_2, be defined as a linear combination of x, y,

$$\xi_1 = a_{11}(x-m_x) + a_{12}(y-m_y) + m_{\xi_1}$$
$$\xi_2 = a_{21}(x-m_x) + a_{22}(y-m_y) + m_{\xi_2}$$

or

$$\boldsymbol{\xi} = \mathbf{A}(\mathbf{x} - \mathbf{m}_x) + \mathbf{m}_\xi \tag{3.2.18}$$

where $\mathbf{x} = \{x, y\}^T$, $\mathbf{m}_x = [m_x, m_y]^T$, $\mathbf{m}_\xi = [m_{\xi_1}, m_{\xi_2}]^T$. What is the probability density for these new variables? The general rule for changes of variable in probability densities follows from area conservation in mapping from x, y space to ξ_1, ξ_2 space–that is,

$$p_{\xi_1\xi_2}(\Xi_1, \Xi_2) = p_{xy}(X(\Xi_1,\Xi_2),\, Y(\Xi_1,\Xi_2)) \frac{\partial(X,Y)}{\partial(\Xi_1,\Xi_2)} \tag{3.2.19}$$

where $\partial(X,Y)/\partial(\Xi_1,\Xi_2)$ is the Jacobian of the transformation between the two variable sets, and the numerical values satisfy the functional relations,

$$\Xi_1 = a_{11}(X-m_x) + a_{12}(Y-m_y) + m_{\xi_1}\,,$$

etc. Suppose that the relationship (3.2.18) is invertible–that is, we can solve for

$$x = b_{11}(\xi_1 - m_{\xi_1}) + b_{12}(\xi_2 - m_{\xi_2}) + m_x$$
$$y = b_{21}(\xi_1 - m_{\xi_1}) + b_{22}(\xi_2 - m_{\xi_2}) + m_y$$

or

$$\mathbf{x} = \mathbf{B}(\boldsymbol{\xi} - \mathbf{m}_\xi) + \mathbf{m}_x\,. \tag{3.2.20}$$

Then the Jacobian of the transformation is

$$\frac{\partial(X,Y)}{\partial(\Xi_1,\Xi_2)} = b_{11}b_{22} - b_{12}b_{21} = \det(\mathbf{B}) \tag{3.2.21}$$

[det(**B**) is the determinant of **B**]. Equation (3.2.18) produces

$$\begin{aligned}
< \xi_1 > &= m_{\xi_1} \\
< \xi_2 > &= m_{\xi_2} \\
< (\xi_1 - < \xi_1 >)^2 > &= a_{11}^2\sigma_x^2 + a_{12}\sigma_y^2 \\
< (\xi_1 - < \xi_1 >)(\xi_2 - < \xi_2 >) > &= a_{11}a_{21}\sigma_x^2 + a_{12}a_{22}\sigma_y^2 \neq 0\,.
\end{aligned} \tag{3.2.22}$$

In the special case,

$$\mathbf{A} = \begin{Bmatrix} \cos\phi & \sin\phi \\ -\sin\phi & \cos\phi \end{Bmatrix}, \quad \mathbf{B} = \begin{Bmatrix} \cos\phi & -\sin\phi \\ \sin\phi & \cos\phi \end{Bmatrix}, \tag{3.2.23}$$

the transformation (3.2.23) is a simple coordinate rotation through angle ϕ, and the Jacobian is 1. The second-order moments in (3.2.22) then become

$$< (\xi_1 - < \xi_1 >)^2 >= \sigma_{\xi_1}^2 = \cos^2\phi\sigma_x^2 + \sin^2\phi\sigma_y^2\,, \tag{3.2.24}$$

$$< (\xi_2 - < \xi_2 >)^2 >= \sigma_{\xi_2}^2 = \sin^2\phi\sigma_x^2 + \cos^2\phi\sigma_y^2\,, \tag{3.2.25}$$

$$< (\xi_1 - < \xi_1 >)(\xi_2 - < \xi_2 >) >\equiv \mu_{\xi_1\xi_2} = (\sigma_y^2 - \sigma_x^2)\cos\phi\sin\phi\,. \tag{3.2.26}$$

The new probability density is

$$p_{\xi_1\xi_2}(\Xi_1,\Xi_2) = \frac{1}{2\pi\sigma_{\xi_1}\sigma_{\xi_2}\sqrt{1-\rho_\xi^2}} \times \tag{3.2.27}$$

$$\exp\left\{-\frac{1}{2\sqrt{1-\rho_\xi^2}}\left[\frac{(\Xi_1 - m_{\xi_1})^2}{\sigma_{\xi_1}^2} - \frac{2\rho_\xi(\Xi_1 - m_{\xi_1})(\Xi_2 - m_{\xi_2})}{\sigma_{\xi_1}\sigma_{\xi_2}} + \frac{(\Xi_2 - m_{\xi_2})^2}{\sigma_{\xi_2}^2}\right]\right\}$$

where $\rho_\xi = (\sigma_y^2 - \sigma_x^2)\sin\phi\cos\phi/\left(\sigma_{\xi_1} + \sigma_{\xi_2}\right)^{1/2} = \mu_{\xi_1\xi_2}/\sigma_{\xi_1}\sigma_{\xi_2}$ is the correlation coefficient of the new variables. A probability density derived through a linear transformation from two independent variables that are Gaussian will be said to be jointly Gaussian, and (3.2.27) is a canonical form. Because a coordinate rotation is invertible, it is important to note that if we had two random variables ξ_1, ξ_2 that were jointly Gaussian with $\rho \neq 1$, then we could find a pure rotation (3.2.23), which produces two other variables x, y that are uncorrelated and therefore *independent*. Notice that (3.2.26) shows that two such uncorrelated variables x, y will necessarily have different variances; otherwise, ξ_1, ξ_2 would have zero correlation, too.

It follows that two uncorrelated jointly Gaussian random variables are also independent. This property is one of the reasons Gaussians are so nice to work with.

3.2.1 Vector Random Processes

Simultaneous discussion of two random processes, x, y can be regarded as discussion of a vector random process $[x, y]^T$ and suggests a generalization to N dimensions. Let us label N random processes as $x(i)$ and define them as the elements of a vector $\mathbf{x}^T = [x(1), x(2), \ldots, x(N)]^T$. Then the mean is a vector: $< \mathbf{x} > = \mathbf{m}_x$, and the covariance is a matrix:

$$\mathbf{C}_{xx} = D^2(\mathbf{x}- < \mathbf{x} >) = < (\mathbf{x}- < \mathbf{x} >)(\mathbf{x}- < \mathbf{x} >)^T >, \qquad (3.2.28)$$

which is necessarily symmetric and positive semidefinite. The cross-covariance of two processes $\mathbf{x}$, $\mathbf{y}$ is

$$\mathbf{C}_{xy} = < (\mathbf{x}- < \mathbf{x} >)(\mathbf{y}- < \mathbf{y} >)^T > \qquad (3.2.29)$$

and $\mathbf{C}_{xy} = \mathbf{C}_{yx}^T$.

It proves convenient to introduce two further moment matrices in addition to the covariance matrices for which the dispersion is measured about the mean. The *second-moment* matrices will be defined as

$$\mathbf{R}_{xx} \equiv D^2(\mathbf{x}) = < \mathbf{x}\mathbf{x}^T >, \quad \mathbf{R}_{xy} = < \mathbf{x}\mathbf{y}^T >$$

($\mathbf{R}_{xy} = \mathbf{R}_{yx}^T$, etc.). Let $\tilde{\mathbf{x}}$ be an estimate of the true value, $\mathbf{x}$. Then the dispersion of $\tilde{\mathbf{x}}$ about the true value will be called the *uncertainty* (sometimes it is called the *error covariance*) and is

$$\mathbf{P} \equiv D^2(\tilde{\mathbf{x}} - \mathbf{x}) = < (\tilde{\mathbf{x}} - \mathbf{x})(\tilde{\mathbf{x}} - \mathbf{x})^T > . \qquad (3.2.30)$$

An intuitively pleasing requirement for an estimator $\mathbf{x}$ is that it should minimize the variance about the true value–that is, minimize the diagonal elements of $\mathbf{P}$. This choice is an aesthetic one, but it is the one we will use extensively later.

If there are N variables, ξ_i, $1 \le i \le N$, they will be said to have an "N-dimensional jointly normal probability density" if it is of the form

$$p_{\xi_1,\ldots,\xi_N}(\Xi_1, \ldots, \Xi_N) = \frac{\exp -\frac{1}{2}(\boldsymbol{\Xi} - \mathbf{m})^T \mathbf{C}_{\xi\xi}^{-1}(\boldsymbol{\Xi} - \mathbf{m})}{(2\pi)^{N/2}\sqrt{\det(\mathbf{C}_{\xi\xi})}} . \qquad (3.2.31)$$

It is readily demonstrated that $<\boldsymbol{\xi}>=\mathbf{m}$, $<(\boldsymbol{\xi}-\mathbf{m})(\boldsymbol{\xi}-\mathbf{m})^T>=\mathbf{C}_{\xi\xi}$. Equation (3.2.27) is a special case of (3.2.31) for $N=2$.

Positive definite symmetric matrices can be factored as

$$\mathbf{C}_{\xi\xi}=\mathbf{C}_{\xi\xi}^{T/2}\mathbf{C}_{\xi\xi}^{1/2}, \tag{3.2.32}$$

called the *Cholesky decomposition*, where $\mathbf{C}_{\xi\xi}^{1/2}$ is upper triangular and nonsingular. Numerical schemes for finding $\mathbf{C}_{\xi\xi}^{1/2}$ are described by Lawson and Hanson (1974) and Golub and Van Loan (1989). It follows that the transformation (a rotation and stretching),

$$\mathbf{x}=\mathbf{C}_{\xi\xi}^{-T/2}(\boldsymbol{\xi}-\mathbf{m}), \tag{3.2.33}$$

produces new variables $\mathbf{x}$ of zero mean, and identity covariance–that is, a probability density

$$\begin{aligned} p_{x_1,\ldots,x_N}(X_1,\ldots,X_N) &= \frac{\exp-\frac{1}{2}(X_1^2+\cdots X_N^2)}{(2\pi)^{N/2}} \\ &= \frac{\exp\left(-\frac{1}{2}X_1^2\right)}{(2\pi)^{1/2}}\cdots\frac{\exp\left(-\frac{1}{2}X_N^2\right)}{(2\pi)^{1/2}}, \end{aligned} \tag{3.2.34}$$

which factors into N independent, normal variates of zero mean and unit variance ($\mathbf{C}_{xx}=\mathbf{R}_{xx}=\mathbf{I}$). Such a process is often denoted *white noise.* (Cramér, 1946, discusses what happens when the determinant of $\mathbf{C}_{\xi\xi}$ vanishes–that is, if $\mathbf{C}_{\xi\xi}$ is singular.)

3.2.2 Functions of Random Variables

If the probability density of x is $p_x(x)$, then the mean of a function of x, $g(x)$ is just

$$<g(x)>=\int_{-\infty}^{\infty} g(X)p_x(X)dX, \tag{3.2.35}$$

which follows from the definition of the probability density as the limit of the outcome of a number of trials. The probability density for g regarded as a new random variable is given by (3.2.19) as

$$p_g(G)=p_x(X(G))\frac{dx}{dg} \tag{3.2.36}$$

where the Jacobian is just dx/dg for a one-dimensional transformation.

An important special case is $g=x^2$ where x is Gaussian of zero mean and unit variance (any Gaussian variable z of mean m and variance σ^2 can be transformed to one of zero mean and unit variance by the transformation

$$x = \frac{z - m}{\sigma},$$

whose Jacobian is very simple). Then the probability density of g is

$$p_g(G) = \frac{1}{G^{1/2}\sqrt{2\pi}} \exp(-G/2), \qquad G \geq 0, \tag{3.2.37}$$

a probability density usually denoted as χ_1^2 (chi-squared), and $< g > = 1$, $D^2(g- < g >) = 2$.

3.2.3 Sums of Random Variables

It is often helpful to be able to compute the probability density of sums of independent random variables. The procedure for doing so is based upon (3.2.35). Let x be a random variable, and consider the expected value of the function e^{ixt}:

$$< e^{ixt} > = \int_{-\infty}^{\infty} e^{iXt} p_x(X) dX \equiv \phi_x(t), \tag{3.2.38}$$

which is also the Fourier transform of $p_x(X)$; $\phi_x(t)$ is usually termed the *characteristic function* of x. Now consider the sum of two independent random variables x, y with probability densities p_x, p_y, respectively, and define a new random variable $z = x + y$. What is the probability density of z? A method for finding it is based upon first determining the characteristic function, $\phi_z(t)$ for z and then using the Fourier inversion theorem to obtain $p_x(Z)$. To obtain ϕ_z,

$$\phi_z(t) = < e^{izt} > = < e^{i(x+y)t} > = < e^{ixt} >< e^{iyt} >$$

where the last step depends upon the independence assumption. This last equation shows

$$\phi_z(t) = \phi_x(t)\phi_y(t). \tag{3.2.39}$$

That is, the characteristic function for a sum of two independent variables is the product of the characteristic functions. The *convolution theorem* (see, for example, Bracewell, 1978) asserts that the Fourier transform (forward or inverse) of a product of two functions is the convolution of the Fourier transforms of the two functions. We will not explore this relation in any detail, leaving the reader to pursue the subject in the references (e.g., Cramér, 1946). But it follows immediately that the multiplication of the characteristic functions of a sum of independent Gaussian variables produces a new variable, which is also Gaussian, with a mean equal to the sum of the means and a variance that is the sum of the variances ("sums of Gaussians are Gaussian"). It also follows immediately from the convolution theorem that if a variable ξ is defined as

$$\xi = x_1^2 + x_2^2 + \cdots + x_v^2 \tag{3.2.40}$$

where each x_i is Gaussian of zero mean and unit variance, the probability density for ξ is

$$p_\xi(\Xi) = \frac{\Xi^{\nu/2-1}}{2^{\nu/2}\Gamma\left(\frac{\nu}{2}\right)} \exp(-\Xi/2)\,, \tag{3.2.41}$$

known as χ_v^2, "chi-square with ν degrees of freedom." The chi-square probability density is central to the discussion of the sizes of vectors, such as $\tilde{\mathbf{n}}$, measured as $\tilde{\mathbf{n}}^T\tilde{\mathbf{n}} = \sum_i \tilde{\mathbf{n}}_i^2$ if the elements of $\tilde{\mathbf{n}}$ can be assumed to be independent and Gaussian. Equation (3.2.37) is the special case $\nu = 1$. One has,

$$<\xi> = \nu\,, \qquad D^2(\xi - <\xi>) = 2\nu\,. \tag{3.2.42}$$

3.2.4 Degrees of Freedom

The number of independent variables described by a probability density is usually called the *number of degrees of freedom.* Thus, the densities in (3.2.31) and (3.2.34) have N degrees of freedom, and (3.2.41) has ν of them. If a sample average (3.2.2) is formed, it is said to have N degrees of freedom if each of the x_j is independent. But what if the x_j have a covariance $\mathbf{C}_{xx}$ that is nondiagonal? This question of how to interpret averages of correlated variables will be explicitly discussed in Section 3.5.

Consider for the moment only the special case of the sample variance (3.2.9), with divisor $N-1$ rather than N as might be expected. The reason is that even if the sample values $x(i)(\equiv x_i)$ are independent [we are not distinguishing here between $x(i)$ and $X(i)$], the presence of the sample average in the sample variance means that there are only $N-1$ independent terms in the sum. That this is so is most readily seen by examining the two-term case. Two samples produce a sample mean, $< x >_2 = (x_1 + x_2)/2$. A two-term sample variance is

$$s^2 = \tfrac{1}{2}((x_1 - < x >_2)^2 + (x_2 - < x >_2)^2)\,,$$

but knowledge of x_1 and the sample average permits perfect prediction of x_2 and thus of the second term in the sample variance, and there is just one independent piece of information in the two-term sample variance.

3.2.5 Stationarity

Consider a vector random variable with elements $x_i = x(i)$ where the argument i denotes a position in time or space. Then $x(i)$, $x(j)$ denote two different random variables–for example, the temperature at two different positions in the ocean, or the temperature at two different times at the same position. If the physics governing these two different random variables are independent of the parameter i (i.e., independent of time or space), then $x(i)$ is said to be *stationary*, meaning that the underlying statistics are independent of i. Specifically, $< x(i) > = < x(j) > \equiv < x >$, $D^2(x(i)) = D^2(x(j))$, etc. Furthermore, $x(i)$, $x(j)$ have a covariance

$$C_{xx}(i,j) = < (x(i) - < x(i) >)(x(j) - < x(j) >) > = C_{xx}(|i-j|)\,, \tag{3.2.43}$$

that is, independent of i, j, and depending only upon the difference $|i-j|$; $|i-j|$ is often called the *lag*. Then

$$< (x(i) - < x >)(x(j) - < x >)^T > = \{C_{xx}(i,j)\} = \{C_{xx}(|i-j|)\}$$

is called the *autocovariance* of $\mathbf{x}$ or just the covariance, because we now regard $x(i)$, $x(j)$ as intrinsically the same process.[2] If C_{xx} does not vanish, then by the discussion above, knowledge of the numerical value of $x(i)$ implies some predictive skill for $x(j)$ and vice versa–a result of great importance when we examine map making and objective analysis. A jointly normal stationary time series would have probability density (3.2.31) in which all the elements of $\mathbf{m}$ are identical, and the ij-th elements of $\mathbf{C}_{\xi\xi}$ depend only upon $i-j$.

3.2.6 Maximum Likelihood

Given a set of observations with known joint probability density, one can base a method for estimating various sample parameters upon a principle of maximum likelihood, which finds those parameters that render the actual observations to be the most probable ones. Consider one simple example for an uncorrelated jointly normal stationary time series,

$$< x(i) > = m, \; < (x(i) - m)(x(j) - m) > = \sigma_x^2 \delta_{ij}\,,$$

[2] If the means and variances are independent of i, j and the first cross-moment is dependent only upon $|i-j|$, the process x is said to be stationary in the *wide sense*. If all higher moments also depend only on $|i-j|$, the process is said to be stationary in the *strict sense*, or more simply, just *stationary*. A Gaussian process has the unusual property that wide-sense stationarity implies strict-sense stationarity.

with corresponding joint probability density

$$p_{\mathbf{x}}(x(1), x(2), x(3), \ldots, x(N)) = \frac{1}{(2\pi)^{N/2}\sigma_x^N} \times$$
$$\exp\left(-\frac{1}{2\sigma_x^2}\left[(x(1)-m)^2 + (x(2)-m)^2 + \cdots + (x(N)-m)^2\right]\right). \quad (3.2.44)$$

Substitution of the observed values into (3.2.44)[3] permits evaluation of the probability that these particular values occurred. Denote this probability as L. One can demand those values of m, σ_x, rendering the probability a maximum of L for all possible series mean and standard deviations. The probability can be maximized by minimizing the exponent in (3.2.44)–that is, minimizing

$$\log L = \frac{1}{2\sigma_x^2}\sum (x(i)-m)^2 - \frac{1}{2}N\log(2\sigma_x) - \frac{1}{2}N\log(2\pi), \quad (3.2.45)$$

the log-likelihood function. Maximizing log L with respect to m, σ_x produces

$$\tilde{m} = \frac{1}{N}\sum_1^N x(i), \ \tilde{\sigma}_x^2 \equiv s^2 = \frac{1}{N}\sum_1^N (x(i)-\tilde{m})^2, \quad (3.2.46)$$

and the result is the usual sample average and the biased estimate of the sample variance (3.2.7). A likelihood function derived from (3.2.31) provides a straightforward generalization to covarying variables.

A complete methodology for most of what follows in this book can be built upon the general ideas of maximum likelihood estimation, but it is not the course I choose to follow. Extended discussions can be found in numerous places, including Van Trees (1968), who carries the idea all the way through the material found here in Chapter 6.

3.3 Least Squares

Much of what follows in this book can be described using very elegant and powerful mathematical tools. On the other hand, by restricting ourselves to discrete models and finite numbers of measurements, almost everything can also be viewed as a form of ordinary least squares. It is thus useful to go back and review what "everyone knows" about this most-familiar of all approximation methods.

[3] Strictly speaking, we should work with the conditional probability (3.2.3), $-p_{x|m,\sigma}\big(x(1), x(2) \ldots x(N)|m, \sigma\big)$.

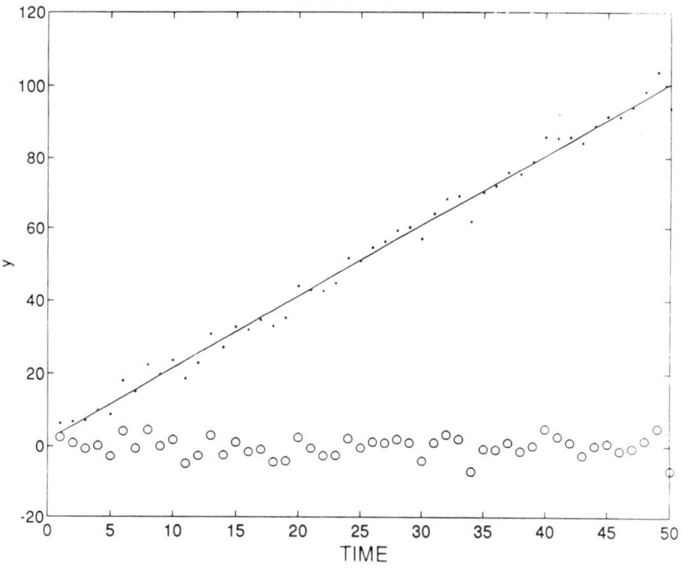

Figure 3–2. "Data" generated through the rule $y = 1 + 2t + n$, where $< n > = 0$, $< n_i n_j > = 9\delta_{ij}$, denoted by small dots. Solid line shows a least-squares fit to the data, which is $\tilde{y} = 1.9 \pm 0.8 + (1.96 \pm 0.03)t$ with $< \tilde{a}\tilde{b} > = -0.02$; open circles denote the residuals of the fit, which appear to be qualitatively white noise in character.

3.3.1 Basic Formulation

Consider the elementary problem motivated by the "data" shown in Figure 3–2; t is supposed to be an independent variable, which could be time, or a spatial coordinate or just an index. Some physical variable, call it $\theta(t)$, perhaps temperature at a point in the ocean, has been measured at times $t = t_i$, $1 \leq i \leq M$, as depicted in the figure.

We have reason to believe that there is a linear relationship between $\theta(t)$ and t in the form $\theta(t) = at + b$ so that the measurements are

$$y(t) = \theta(t) + n(t) = a + bt + n(t) \tag{3.3.1}$$

where $n(t)$ is the inevitable measurement noise. We want to determine a, b.

The set of observations can be written in the general standard form,

$$\mathbf{Ex} + \mathbf{n} = \mathbf{y} \tag{3.3.2}$$

where in the present special case,

$$\mathbf{E} = \begin{Bmatrix} 1 & t_1 \\ 1 & t_2 \\ \cdot & \cdot \\ \cdot & \cdot \\ 1 & t_M \end{Bmatrix}, \quad \mathbf{x} = \begin{bmatrix} a \\ b \end{bmatrix}, \quad \mathbf{y} = \begin{bmatrix} y(t_1) \\ y(t_2) \\ \cdot \\ \cdot \\ y(t_M) \end{bmatrix}, \tag{3.3.3}$$

and $\mathbf{n}(t)$ is the noise vector. Equation sets like (3.3.2) appear in many practical situations, including the ones described in Chapter 2.

One sometimes sees (3.3.2) written as

$$\mathbf{Ex} \sim \mathbf{y}$$

or even

$$\mathbf{Ex} = \mathbf{y}\,.$$

But Equation (3.3.2) is preferable, because it explicitly recognizes that $\mathbf{n} = \mathbf{0}$ is exceptional. Sometimes, by happenstance or arrangement, one finds that $M = N$ and that $\mathbf{E}$ has an inverse. But the obvious solution, $\mathbf{x} = \mathbf{E}^{-1}\mathbf{y}$, leads to the conclusion, $\mathbf{n} = \mathbf{0}$, probably regarded as unacceptable if the $\mathbf{y}$ are the result of measurements. We will need to return to this case, but for now, let us consider the problem where $M > N$.

Commonly, then, one sees a *best possible* solution–defined as producing the smallest possible value of $\mathbf{n}^T\mathbf{n}$–that is, the one producing the minimum of

$$J = \sum_{i=1}^{M} (a + bt_i - y(t_i))^2 \equiv \sum_{i=1}^{M} n_i^2 = \mathbf{n}^T\mathbf{n} = (\mathbf{Ex} - \mathbf{y})^T(\mathbf{Ex} - \mathbf{y})\,. \quad (3.3.4)$$

Differentiating (3.3.4) with respect to a, b or $\mathbf{x}$ [using (3.1.17) and (3.1.19)] and by setting $dJ = \sum (\partial J/\partial x_i)\, dx_i = 0$, term-by-term (anticipating a minimum rather than a maximum), leads to the system called the *normal equations*,

$$\mathbf{E}^T\mathbf{Ex} = \mathbf{E}^T\mathbf{y}\,. \quad (3.3.5)$$

Making the sometimes valid assumption that $(\mathbf{E}^T\mathbf{E})^{-1}$ exists,

$$\tilde{\mathbf{x}} = (\mathbf{E}^T\mathbf{E})^{-1}\mathbf{E}^T\mathbf{y}\,. \quad (3.3.6)$$

The solution is written as $\tilde{\mathbf{x}}$ rather than as $\mathbf{x}$ because the relationship between (3.3.6) and the correct value is not clear. The fit is displayed in Figure 3–2, as are the residuals,

$$\tilde{\mathbf{n}} = \mathbf{y} - \mathbf{E}\tilde{\mathbf{x}} = \mathbf{y} - \mathbf{E}(\mathbf{E}^T\mathbf{E})^{-1}\mathbf{E}^T\mathbf{y} = (\mathbf{I} - \mathbf{E}(\mathbf{E}^T\mathbf{E})^{-1}\mathbf{E}^T)\mathbf{y}\,. \quad (3.3.7)$$

That is, the M equations have been used to estimate N values, $\tilde{x}_i$, and M values $\tilde{n}_i$, or $M + N$ altogether.

All this is easy and familiar and applies to any set of simultaneous equations, not just the straight-line example. Before proceeding, let us apply some of the statistical machinery to understanding (3.3.6). Notice that no statistics were used in obtaining (3.3.6), but we can nonetheless ask the extent to which this value for $\tilde{\mathbf{x}}$ is affected by the random elements, the noise in $\mathbf{y}$. Let $\mathbf{y}_0$ be the value of $\mathbf{y}$ that would be obtained in the hypothetical situation for which $\mathbf{n} = 0$. Assume further that $<\mathbf{n}> = 0$ and that $\mathbf{R}_{nn} = \mathbf{C}_{nn} = <\mathbf{n}\mathbf{n}^T>$ is known. Then the expected value of $\tilde{\mathbf{x}}$ is

$$<\tilde{\mathbf{x}}> = (\mathbf{E}^T\mathbf{E})^{-1}\mathbf{E}^T\mathbf{y}_0 . \tag{3.3.8}$$

If the matrix inverse exists, then in many situations, including the problem of fitting a straight line to data, perfect observations would produce the correct answer, and (3.3.6) is an unbiased estimate of the true solution, $<\tilde{\mathbf{x}}> = \mathbf{x}$. On the other hand, if the data were actually produced from physics governed, for example, by a quadratic rule, $\theta(t) = a + ct^2$, then fitting the linear rule to such observations, even if they are perfect, could never produce the right answer, and the solution would be biased. An example of such a fit is shown in Figure 3–4. Such errors are distinguishable from the noise of observation and are properly labeled *model errors*. Assume, however, that the correct model is being used and therefore that $<\tilde{\mathbf{x}}> = \mathbf{x}$. Then the uncertainty of the solution is the same as the variance about the mean and is

$$\begin{aligned}\mathbf{P} = \mathbf{C}_{\tilde{\mathbf{x}}\tilde{\mathbf{x}}} &=< (\tilde{\mathbf{x}} - \mathbf{x})(\tilde{\mathbf{x}} - \mathbf{x})^T > \\ &= (\mathbf{E}^T\mathbf{E})^{-1}\mathbf{E}^T < \mathbf{n}\mathbf{n}^T > \mathbf{E}(\mathbf{E}^T\mathbf{E})^{-1} \\ &= (\mathbf{E}^T\mathbf{E})^{-1}\mathbf{E}^T\mathbf{R}_{nn}\mathbf{E}(\mathbf{E}^T\mathbf{E})^{-1} .\end{aligned} \tag{3.3.9}$$

In the special case, $\mathbf{R}_{nn} = \sigma_n^2\mathbf{I}$–that is, there is no correlation between the noise in different equations (often called *white noise*)–then (3.3.9) simplifies to

$$\mathbf{P} = \sigma_n^2(\mathbf{E}^T\mathbf{E})^{-1} . \tag{3.3.10}$$

If we are not confident that $<\tilde{\mathbf{x}}> = \mathbf{x}$, (3.3.9)–(3.3.10) are still interpretable but as $C_{\tilde{\mathbf{x}}\tilde{\mathbf{x}}} = D^2(\tilde{\mathbf{x}} - <\tilde{\mathbf{x}}>)$ – the covariance of $\tilde{\mathbf{x}}$. The *standard error* of $\tilde{x}_i$ is usually defined to be $\pm\sqrt{C_{\tilde{\mathbf{x}}\tilde{\mathbf{x}}}}$ and is used to understand the adequacy of data for distinguishing different possible estimates of $\tilde{\mathbf{x}}$. If applied to the straight line fit of Figure 3–2, we obtain an estimate as $\tilde{\mathbf{x}}^T = [a \quad b]^T = [1.9 \pm 0.8,\ 1.96 \pm 0.03]^T$. If the noise in $\mathbf{y}$ is Gaussian, it follows that the probability density of $\tilde{\mathbf{x}}$ is also Gaussian, with mean $<\tilde{\mathbf{x}}>$ and covariance $\mathbf{C}_{\tilde{\mathbf{x}}\tilde{\mathbf{x}}}$. Of course, if $\mathbf{n}$ is not Gaussian, then neither will be the estimate, and one must be wary of the accuracy of standard error estimates.

The uncertainty of the residual estimates is

$$\begin{aligned}\mathbf{P}_{\tilde{n}\tilde{n}} &\equiv< (\tilde{\mathbf{n}} - \mathbf{n})(\tilde{\mathbf{n}} - \mathbf{n})^T > = (\mathbf{I} - \mathbf{E}(\mathbf{E}^T\mathbf{E})^{-1}\mathbf{E}^T)\mathbf{R}_{nn}(\mathbf{I} - \mathbf{E}(\mathbf{E}^T\mathbf{E})^{-1}\mathbf{E}^T)^T \\ &= \sigma_n^2(\mathbf{I} - \mathbf{E}(\mathbf{E}^T\mathbf{E})^{-1}\mathbf{E}^T)^2 = \sigma_n^2(\mathbf{I} - \mathbf{E}(\mathbf{E}^T\mathbf{E})^{-1}\mathbf{E}^T)\end{aligned} \tag{3.3.11}$$

where the second line is valid for white-noise residuals, and where $< \mathbf{n} > = \mathbf{0}$ is assumed to be correct.

The fit of a straight line to observations demonstrates many of the issues involved in making inferences from real, noisy data that appear in more complex situations. In Figure 3–3, the correct model used to generate the data was the same as in Figure 3–2, but the noise level is very high. The parameters $[\tilde{a}, \tilde{b}] = [1.5 \pm 2.8,\ 2.0 \pm 0.1]$–that is, $\tilde{a}$ is numerically incorrect, and formally indistinguishable from zero (but consistent within one standard error with the correct value). One sometimes reads, in such situations, that "least squares failed," but such a statement represents a fundamental confusion of the methodology with the lack of data adequate to demonstrate a hypothesis. Least squares functions exactly as intended, and one could conclude legitimately either that (1) there is no evidence that a straight-line rule explains the data, or (2) the data are consistent with the hypothesis $a = 1$, $b = 2$, and there is no reason to change such a prior estimate. In Figure 3–5, the quadratic model of Figure 3–4 was used to generate the numbers, but with enough additional data supplied that the residuals now clearly fail to satisfy a hypothesis of being white noise. Modeling a quadratic field with a linear model produces a systematic or model error. In contrast, the fit of a quadratic rule $y = a + bt + ct^2$, shown in Figure 3–6, does leave small, random appearing residuals. But if the true noise were not random, one might well erroneously deduce the presence of a quadratic model; such possibilities strongly suggest that the residuals had better be examined at least as closely as the model parameter estimates–and the need to do so is a constant theme in this book.

Visual tests for randomness of residuals have obvious limitations, and elaborate statistical tests help to determine objectively whether one should accept or reject the hypothesis that no significant structure remains in a sequence of numbers. Books on regression analysis (e.g., Seber, 1977, or Box & Jenkins, 1978) should be consulted for general methodologies. As an indication of what can be done, Figures 3–7a and b show the sample autocovariance,

$$\tilde{R}_{nn}(\tau) = \frac{1}{M} \sum_{i=1}^{M-|\tau|} \tilde{n}_i \tilde{n}_{i+\tau} \,, \tag{3.3.12}$$

for the residuals of the fits shown in Figures 3–5 and 3–6. [$\tilde{R}_{nn}(\tau)$ is an estimate of $< n_i n_{i+\tau} >$.] If the residuals were truly uncorrelated, $< \tilde{n}_i \, \tilde{n}_{i+\tau} > = 0$, $\tau \neq 0$, and one expects $\tilde{R}_{nn}(\tau)$ to approach a delta function at $\tau = 0$. Tests are available to determine if the nonzero values

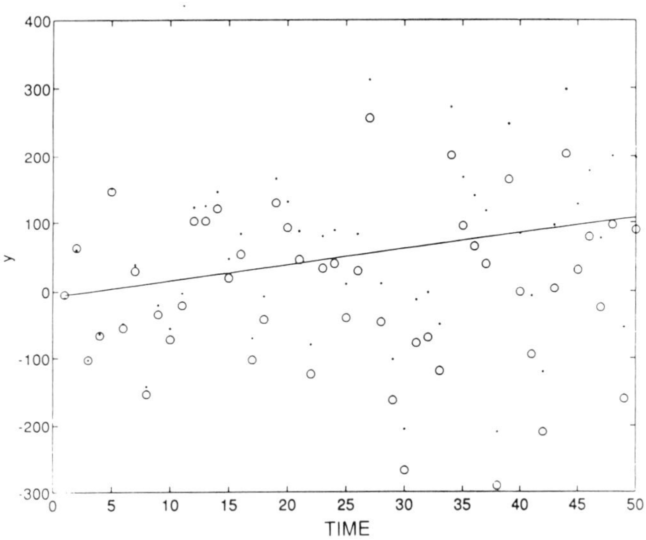

Figure 3–3. The same situation as described in Figure 3–2 except $< n > = 0$, $< n_i n_j > = 100^2 \delta_{ij}$, meaning the "data" were very noisy. The fit is now $\tilde{y} = 1.5 \pm 2.75 + (2.02 \pm 0.09)t$.

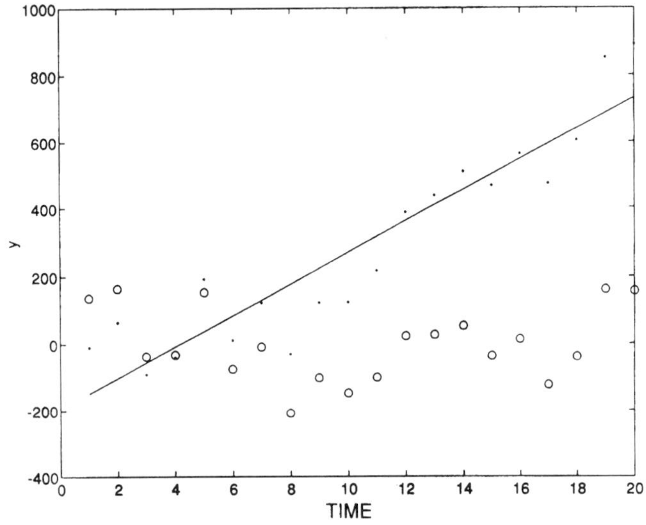

Figure 3–4. Here, the "data" were generated through a quadratic rule, $y = 1 + t^2 + n$, $< n > = 0$, $< n_i n_j > = 100^2 \delta_{ij}$. But a linear fit was nonetheless made that produces $y = -194 \pm 1.6 + (46.4 \pm 0.1)t$. To the eye, at least, it is a reasonably good fit, and one might have great difficulty in rejecting the hypothesis that a straight-line rule is valid.

for $\tau \neq 0$ are significantly nonzero (Box & Jenkins, 1978, Ch. 2). Here, we merely note that the sample autocovariance behavior again confirms visually what we already know–that the fit in Figure 3–6 is adequate, and in Figure 3–5 it is not.

3.3.2 Weighted and Tapered Least Squares

The least-squares solution (3.3.6)–(3.3.7) was derived by minimizing the objective function (3.3.4), in which each residual element is given equal

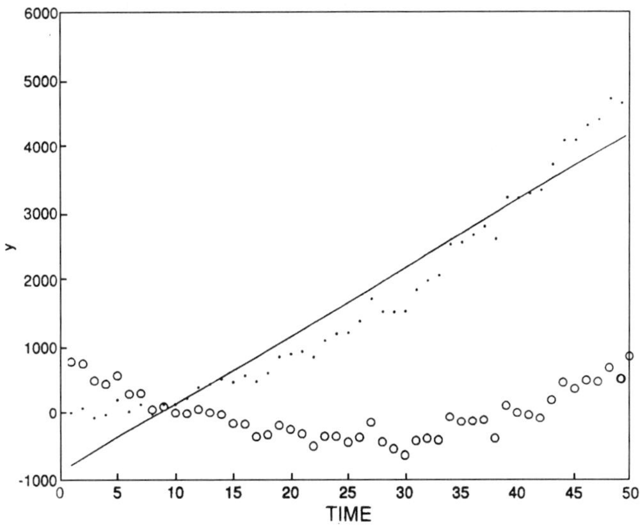

Figure 3–5. The same situation as in Figure 3–4 except that the duration was extended. Now the linear fit leaves obvious residuals that are nonrandom, producing a strong indication that a linear model is inadequate, or that the noise is not white, or both.

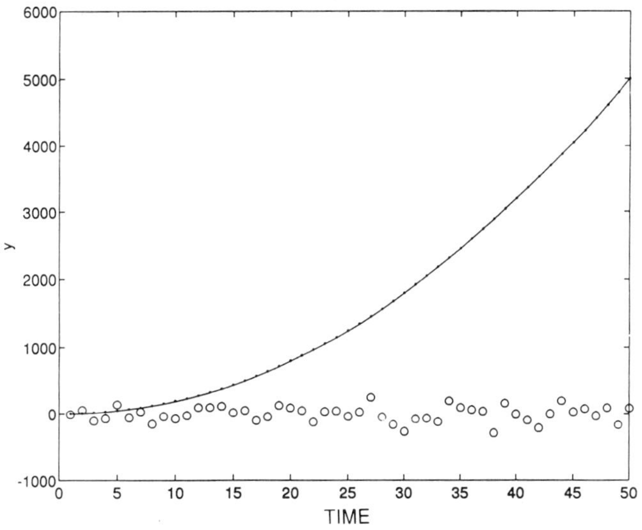

Figure 3–6. The same situation as in Figure 3–5 except that now a quadratic model, $y = a + bt + ct^2$, was fit, resulting in a solution $y = -2 \pm 5.2 + (-0.26 \pm 4.7)t + (2.0 \pm 3.9)t^2$ and leaving small, apparently random, residuals.

weight. An important feature of least squares is that we can give whatever emphasis we please to minimizing individual equation residuals, for example, by introducing an objective function

$$J = \sum_i W_{ii} n_i^2 \tag{3.3.13}$$

where W_{ii} are any numbers desired. The choice $W_{ii} = 1$ might be reasonable, but it is clearly an arbitrary one that without further justification does not produce a solution with any special claim to significance. In the least

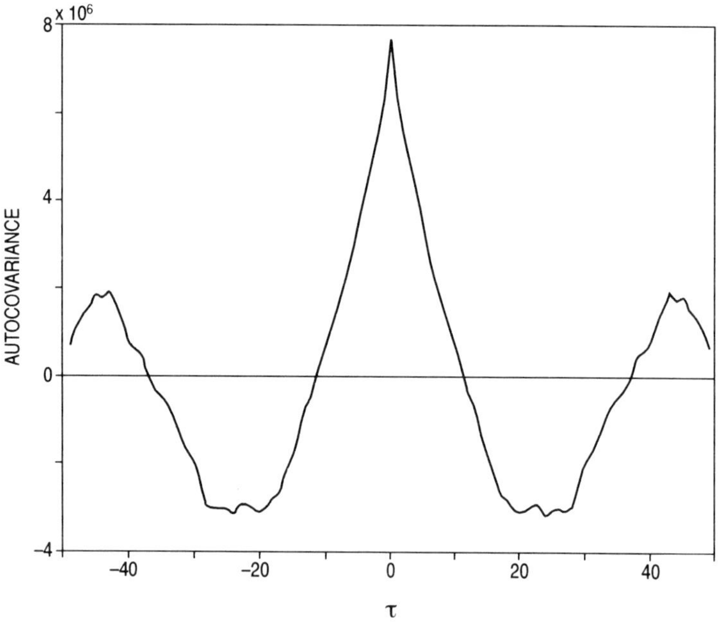

Figure 3–7a. Autocovariance of the residuals from Figure 3–5, the autocovariance of a non-white process. This is an example of a test for adequacy of a model. We would probably reject the model.

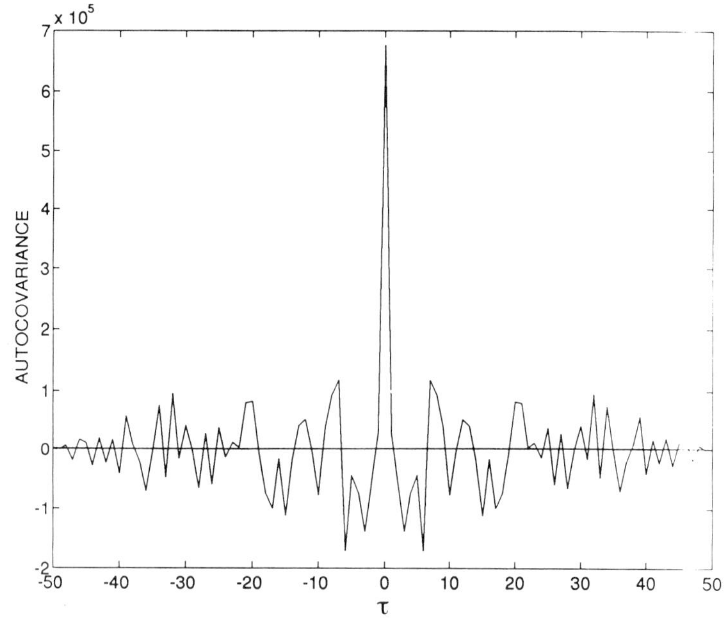

Figure 3–7b. Autocovariance of fit in Figure 3–6, when a quadratic was fit. Here, the result is indistinguishable from white noise, and the model would be acceptable.

squares context, we are free to make any other reasonable choice, including demanding that some residuals should be much larger than others, perhaps just to determine if it is possible.

A general formalism is obtained by defining a diagonal weight matrix,

$\mathbf{W} = \mathrm{diag}([W_{ii}])$.[4] Divide each equation in (3.3.2) by $\sqrt{W_{ii}}$,

$$W_{ii}^{-1/2} \sum_j E_{ij} x_j + W_{ii}^{-1/2} n_i = W_{ii}^{-1/2} y_i \tag{3.3.14}$$

or

$$\begin{gathered} \mathbf{E}'\mathbf{x} + \mathbf{n}' = \mathbf{y}' \\ \mathbf{E}' = \mathbf{W}^{-T/2}\mathbf{E},\ \mathbf{n}' = \mathbf{W}^{-T/2}\mathbf{n},\ \mathbf{y}' = \mathbf{W}^{-T/2}\mathbf{y} \end{gathered} \tag{3.3.15}$$

where we used the fact that the square root of a diagonal matrix is its element-by-element square roots. Such a matrix is its own transpose, and the purpose of writing $\mathbf{W}^{-T/2}$ will become clear below. The operation in (3.3.14) or (3.3.15) is usually called *row scaling* because it operates on the rows of $\mathbf{E}$ (as well as on $\mathbf{n}$, $\mathbf{y}$).

For the new equations (3.3.15), the objective function

$$J = \mathbf{n}'^T\mathbf{n}' = (\mathbf{y}' - \mathbf{E}'\mathbf{x})^T(\mathbf{y}' - \mathbf{E}'\mathbf{x}) = \mathbf{n}^T\mathbf{W}^{-1}\mathbf{n} = (\mathbf{y} - \mathbf{E}\mathbf{x})^T\mathbf{W}^{-1}(\mathbf{y} - \mathbf{E}\mathbf{x}) \tag{3.3.16}$$

weights the residuals as desired. If for some reason, $\mathbf{W}$ is nondiagonal but symmetric and positive-definite, then it has a Cholesky decomposition,

$$\mathbf{W} = \mathbf{W}^{T/2}\mathbf{W}^{1/2},$$

and (3.3.15) remains useful more generally.

The values $\tilde{\mathbf{x}}$, $\tilde{\mathbf{n}}$, minimizing (3.3.16), are

$$\begin{aligned} \tilde{\mathbf{x}} &= (\mathbf{E}'^T\mathbf{E}')^{-1}\mathbf{E}'^T\mathbf{y}' = (\mathbf{E}^T\mathbf{W}^{-1}\mathbf{E})^{-1}\mathbf{E}^T\mathbf{W}^{-1}\mathbf{y} \\ \tilde{\mathbf{n}} &= \mathbf{W}^{T/2}\mathbf{n}' = \left\{\mathbf{I} - \mathbf{E}(\mathbf{E}^T\mathbf{W}^{-1}\mathbf{E})^{-1}\mathbf{E}^T\mathbf{W}^{-1}\right\}\mathbf{y} \end{aligned} \tag{3.3.17}$$

and

$$\mathbf{P} = \mathbf{C}_{\tilde{x}\tilde{x}} = (\mathbf{E}^T\mathbf{W}^{-1}\mathbf{E})^{-1}\mathbf{E}^T\mathbf{W}^{-1}\mathbf{R}_{nn}\mathbf{W}^{-1}\mathbf{E}(\mathbf{E}^T\mathbf{W}^{-1}\mathbf{E})^{-1}. \tag{3.3.18}$$

Uniform diagonal weights are a special case. The rationale for choosing differing diagonal weights or a nondiagonal $\mathbf{W}$ is probably not very obvious to the reader. There is one common situation in which $\mathbf{W} = \mathbf{R}_{nn} = \{ < n_i n_j > \}$, that is, the weight matrix is chosen to be the expected second-moment matrix of the residuals. Then (3.3.18) simplifies to

$$\mathbf{P} = \mathbf{C}_{\tilde{x}\tilde{x}} = (\mathbf{E}^T\mathbf{R}_{nn}^{-1}\mathbf{E})^{-1}. \tag{3.3.19}$$

Here, the weighting (3.3.15) has a ready interpretation: The equations (and hence the residuals) are rotated and stretched so that in the new coordinate

[4] If $\mathbf{q}$ is a vector, the operator $\mathrm{diag}(\mathbf{q})$ forms a square diagonal matrix, whose elements are q_i.

system, the covariance of n_i is diagonal and the variances $< n_i^2 >$ are all unity. In this space, the simple, original objective functions (3.3.4) make physical sense. But we emphasize that this choice of $\mathbf{W}$ is a very special one and has confused many users of inverse methods. To emphasize again: Least squares is a deterministic process in which $\mathbf{W}$ is a set of weights wholly at the disposal of the investigator; setting $\mathbf{W} = \mathbf{R}_{nn}$ is a special case.[5]

If it is intended that $\mathbf{W}$ should reflect the actual expected noise variance, one should confirm after the fact that substitution of $\tilde{\mathbf{x}}$ into (3.3.16) produces a value of J consistent with the hypothesis. That is, because

$$< J > = < \mathbf{n}^T \mathbf{R}_{nn}^{-1} \mathbf{n} > = \sum_1^M < n_i^2 > = M - K\,, \tag{3.3.20}$$

one should obtain (3.2.42),

$$\tilde{J} = \tilde{\mathbf{n}}^T \mathbf{R}_{nn}^{-1} \tilde{\mathbf{n}} \simeq M - K\,. \tag{3.3.21}$$

$M - K$ degrees of freedom (here, $K = N$) are anticipated because the residuals are not independent but are related by (3.3.7). [That there are N degrees of freedom removed by (3.3.7) becomes obvious later on.] The degree of approximation required to $\tilde{J} = M - K$ is then readily determined from χ^2_{M-K}, assuming the $\tilde{n}_i$ are approximately Gaussian. As an illustration, 30 equations in 15 unknowns were constructed for which $(\mathbf{E}^T\mathbf{E})^{-1}$ existed. Then with $\mathbf{x}$ known, an ensemble of 50 values of $\mathbf{y}$ was generated by forming

$$\mathbf{y} = \mathbf{E}\mathbf{x} + \mathbf{n}$$

and by generating $\mathbf{n}$ with a pseudorandom number generator. The system of equations was solved for $\tilde{\mathbf{x}}$ by Equation (3.3.6), producing 50 different estimates of both $\tilde{\mathbf{x}}$, $\tilde{\mathbf{n}}$ and the resulting value of $\tilde{J}$ formed for each and plotted in Figure 3–8a. By expression (3.3.20) we would expect the mean value of $\tilde{J}$ to be near 15, as the figure suggests is approximately correct. Figure 3–8b shows the empirical frequency function of $\tilde{J}$ as compared to χ^2_{15}. The study of the deviations between the expected and computed distributions is the basis of hypothesis testing for the validity of the results, but we leave this discussion to the large literature on the subject. Similarly, the individual elements whose sum is $\tilde{J}$ should have a probability density consistent with χ^2_1.

Whether the equations are scaled or not, the previous limitations of the

[5] In maximum likelihood estimation, least-squares is used to find the likelihood function extreme, and $\mathbf{W} = \mathbf{R}_{nn}$ emerges as a natural choice (e.g., Cramér, 1946; Van Trees, 1968). But the logic of this process is distinct from least-squares as we are employing it here.

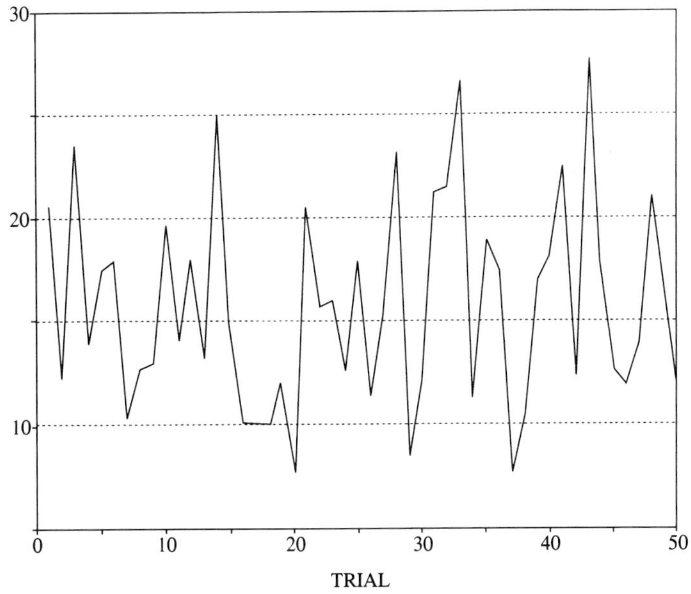

Figure 3–8a. Example in which a system of 30 equations in 15 unknowns was solved for 50 different noise realizations in **y**, showing the different values of $\tilde{J}$ [equation (3.3.20)] for each resulting solution. The mean-square residuals had a value of 15.9.

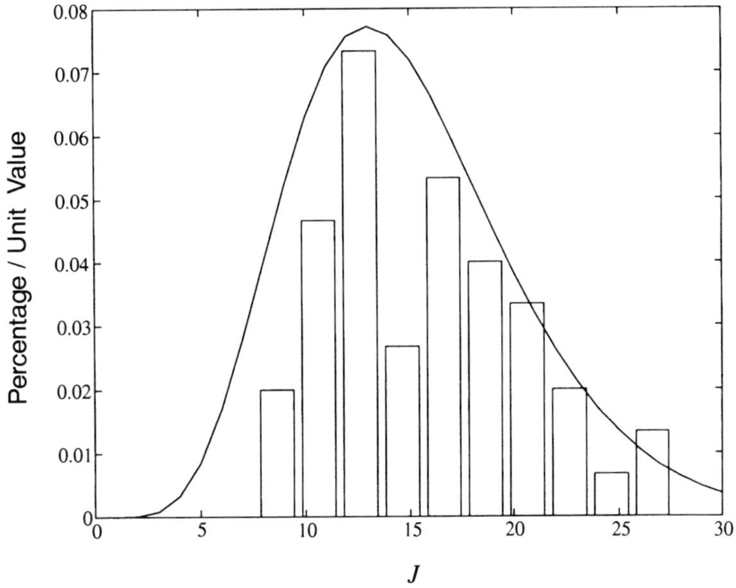

Figure 3–8b. Histogram of values of $\tilde{J}$ shown in Figure 3–8a, compared to the probability density χ^2_{15}.

simple least-squares solutions remain. In particular, we still have the problem that the solution may produce solution elements $\tilde{\mathbf{x}}$, $\tilde{\mathbf{n}}$, whose relative values are not in accord with expected or reasonable behavior, and the solution uncertainty or variances could be unusably large. These quantities are all determined, mechanically and automatically, from combinations such as $(\mathbf{E}^T\mathbf{W}^{-1}\mathbf{E})^{-1}$, an operator that is neither controllable nor very easy to understand and that may not even exist if the matrix is singular.

Suppose, without loss of generality, that any necessary weight matrix $\mathbf{W}$ has already been applied to the equations so that (3.3.4) is a reasonable objective function. It was long ago recognized that some control over the magnitudes of $\tilde{\mathbf{x}}$, $\tilde{\mathbf{n}}$, $\mathbf{C}_{\tilde{x}\tilde{x}}$ could be obtained in the simple least-squares context by modifying the objective function (3.3.4) to have an additional term:

$$J' = \mathbf{n}^T\mathbf{n} + \alpha^2\mathbf{x}^T\mathbf{x} = (\mathbf{y} - \mathbf{E}\mathbf{x})^T(\mathbf{y} - \mathbf{E}\mathbf{x}) + \alpha^2\mathbf{x}^T\mathbf{x} \tag{3.3.22}$$

in which α^2 is a given positive constant.

If the minimum of (3.3.22) is sought by setting to zero the derivatives with respect to $\mathbf{x}$, the resulting normal equations produce

$$\tilde{\mathbf{x}} = (\mathbf{E}^T\mathbf{E} + \alpha^2\mathbf{I})^{-1}\mathbf{E}^T\mathbf{y} \tag{3.3.23}$$

$$\tilde{\mathbf{n}} = \left\{\mathbf{I} - \mathbf{E}(\mathbf{E}^T\mathbf{E} + \alpha^2\mathbf{I})^{-1}\mathbf{E}^T\right\}\mathbf{y} \tag{3.3.24}$$

$$\mathbf{C}_{\tilde{x}\tilde{x}} = (\mathbf{E}^T\mathbf{E} + \alpha^2\mathbf{I})^{-1}\mathbf{E}^T\mathbf{R}_{nn}\mathbf{E}(\mathbf{E}^T\mathbf{E} + \alpha^2\mathbf{I})^{-1} \tag{3.3.25}$$

$$\mathbf{P}_{\tilde{n}\tilde{n}} = \left\{\mathbf{I} - \mathbf{E}(\mathbf{E}^T\mathbf{E} + \alpha^2\mathbf{I})^{-1}\mathbf{E}^T\right\}\mathbf{R}_{nn} \times \left\{\mathbf{I} - \mathbf{E}(\mathbf{E}^T\mathbf{E} + \alpha^2\mathbf{I})^{-1}\mathbf{E}^T\right\}^{-1}. \tag{3.3.26}$$

By letting $\alpha^2 \to 0$, the solution (3.3.6)–(3.3.7), (3.3.9) is recovered, and if $\alpha^2 \to \infty$, $\|\tilde{\mathbf{x}}\|_2 \to 0$, $\tilde{\mathbf{n}} \to \mathbf{y}$, α^2 is called a *trade-off parameter*, because it trades the magnitude of $\tilde{\mathbf{x}}$ against that of $\tilde{\mathbf{n}}$. By varying the size of α^2, we gain some influence over the norm of the residuals relative to that of $\tilde{\mathbf{x}}$. The expected value of $\tilde{\mathbf{x}}$ is now

$$< \tilde{\mathbf{x}} > = (\mathbf{E}^T\mathbf{E} + \alpha^2\mathbf{I})^{-1}\mathbf{E}^T\mathbf{y}_0\,. \tag{3.3.27}$$

If the true solution is believed to be (3.3.8), then this new solution is biased. But the variance of $\tilde{\mathbf{x}}$ (3.3.25) has been reduced by introduction of $\alpha^2 > 0$–that is, the acceptance of a bias reduces the variance. Equations (3.3.23)–(3.3.26) are sometimes known as the *tapered least-squares* solution, a label whose implication becomes clear later.

A physical motivation for the modified objective function (3.3.22) is obtained by noticing that it would be the simplest one to use if the equations being solved consisted of (3.3.2), augmented with a second set asserting $\mathbf{x} \approx \mathbf{0}$–that is, a combined set

$$\mathbf{E}\mathbf{x} + \mathbf{n} = \mathbf{y}$$
$$\alpha^2(\mathbf{x} + \mathbf{n}_1) = \mathbf{0}$$

or

$$\mathbf{E}_1\mathbf{x} + \mathbf{n}_2 = \mathbf{y}_2$$

$$\mathbf{E}_1 = \begin{Bmatrix} \mathbf{E} \\ \alpha^2\mathbf{I} \end{Bmatrix}, \ \mathbf{n}_2^T = [\mathbf{n}^T \ \alpha^2\mathbf{n}_1^T], \ \mathbf{y}_2^T = [\mathbf{y}^T \quad \mathbf{0}^T], \tag{3.3.28}$$

and in which α^2 expresses a preference for fitting the first or second sets more accurately. It then comes as no surprise that the solution covariance depends upon the relative weight given to the second set of equations. A preference that $\mathbf{x} \approx \mathbf{x}_0$ is readily imposed instead, with an obvious change in (3.3.22).

Note the important points, to be shown later, that the matrix inverses in (3.3.23)–(3.3.26) will always exist as long as $\alpha^2 > 0$ and, furthermore, that the expressions remain valid even if $M < N$.

Tapered least squares produce some control over the sum of squares of the relative norms of $\tilde{\mathbf{x}}$, $\tilde{\mathbf{n}}$ but still does not produce control over the individual elements $\tilde{x}_i$. In analogy to the control of the elements of $\tilde{n}_i$ obtained by using a weight matrix $\mathbf{W}$, we can further generalize the objective function by introducing another $N \times N$ weight matrix, $\mathbf{S}$, and using

$$J = \mathbf{n}^T\mathbf{n} + \mathbf{x}^T\mathbf{S}^{-1}\mathbf{x} = (\mathbf{y} - \mathbf{E}\mathbf{x})^T(\mathbf{y} - \mathbf{E}\mathbf{x}) + \mathbf{x}^T\mathbf{S}^{-1}\mathbf{x}\,. \tag{3.3.29}$$

Setting the derivatives with respect to $\mathbf{x}$ to zero results in

$$\tilde{\mathbf{x}} = (\mathbf{E}^T\mathbf{E} + \mathbf{S}^{-1})^{-1}\mathbf{E}^T\mathbf{y} \tag{3.3.30}$$

$$\tilde{\mathbf{n}} = \left\{\mathbf{I} - \mathbf{E}(\mathbf{E}^T\mathbf{E} + \mathbf{S}^{-1})^{-1}\mathbf{E}^T\right\}\mathbf{y} \tag{3.3.31}$$

$$\mathbf{C}_{\tilde{x}\tilde{x}} = (\mathbf{E}^T\mathbf{E} + \mathbf{S}^{-1})^{-1}\mathbf{E}^T\mathbf{R}_{nn}\mathbf{E}(\mathbf{E}^T\mathbf{E} + \mathbf{S}^{-1})^{-1}\,. \tag{3.3.32}$$

The only restriction is that the matrix inverses must exist. Tapered least squares is a special case in which $\mathbf{S}^{-1} = \alpha^2\mathbf{I}_N$, and plain least squares further sets $\alpha^2 = 0$. Like $\mathbf{W}$, $\mathbf{S}$ is often diagonal, in which the numerical values simply assert a preference for making individual terms of $\tilde{x}_i$ large or small.

Suppose that $\mathbf{S}$ is positive definite and symmetric and thus has a Cholesky decomposition. If the equations are scaled as (sometimes called *column scaling* because it weights the columns of $\mathbf{E}$),

$$\mathbf{E}\mathbf{S}^{T/2}\mathbf{S}^{-T/2}\mathbf{x}+\mathbf{n}=\mathbf{y}$$
$$\mathbf{E}'\mathbf{x}'+\mathbf{n}=\mathbf{y}$$
$$\mathbf{E}'=\mathbf{E}\mathbf{S}^{T/2}\,,\qquad \mathbf{x}'=\mathbf{S}^{-T/2}\mathbf{x}\,, \tag{3.3.33}$$

then the objective function (3.3.22), with $\alpha^2 = 1$, is a plausible one in the new coordinate system of $\mathbf{x}'$. Like $\mathbf{W}$, one is completely free to choose $\mathbf{S}$ as one pleases. A common example is to write $\mathbf{S} = \mathbf{F}^T\mathbf{F}$, where $\mathbf{F}$ is $N \times N$,

$$\mathbf{F}=\alpha\begin{Bmatrix} 1 & -1 & 0 & \cdots & 0\\ 0 & 1 & -1 & \cdots & 0\\ \vdots & \vdots & \vdots & \vdots & \vdots\\ 0 & \cdots & \cdots & 0 & 1\end{Bmatrix}, \tag{3.3.34}$$

whose effect is to minimize a term $\alpha^2\sum_i(x_i - x_{i+1})^2$, which can be regarded as a smoothest solution, and using α^2 to trade smoothness against the size of $\|\tilde{\mathbf{n}}\|_2$. $\alpha\mathbf{F}$ is the Cholesky decomposition of $\mathbf{S}$. Another common choice is $\mathbf{S} = \mathbf{R}_{xx}$–that is, the second moments of the solution where known. In this special situation, the x'_i would be uncorrelated with unit variance. Usually it is assumed that both row and column scaling have been done:

$$\mathbf{W}^{-T/2}\mathbf{E}\mathbf{S}^{T/2}\mathbf{S}^{-T/2}\mathbf{x}+\mathbf{W}^{-T/2}\mathbf{n}=\mathbf{W}^{-T/2}\mathbf{y}$$

$$\mathbf{E}'\mathbf{x}'+\mathbf{n}'=\mathbf{y}'$$

$$\mathbf{E}'=\mathbf{W}^{-T/2}\mathbf{E}\mathbf{S}^{T/2},\ \mathbf{x}'=\mathbf{S}^{-T/2}\mathbf{x},\ \mathbf{n}'=\mathbf{W}^{-T/2}\mathbf{n},\ \mathbf{y}'=\mathbf{W}^{-T/2}\mathbf{y}\,, \tag{3.3.35}$$

at which point the plain objective function

$$J=\mathbf{n}'^T\mathbf{n}'+\mathbf{x}'^T\mathbf{x}' \tag{3.3.36}$$

is used (α^2 has been absorbed into $\mathbf{S}$, but it is often convenient to carry it as a separate parameter). If the primes are dropped, Equations (3.3.23)–(3.3.26) result, with $\alpha^2 = 1$. If the original variables $\mathbf{E}$, $\mathbf{x}$, $\mathbf{n}$, $\mathbf{y}$ are restored, we obtain the most general row- and column-scaled form, which is, for future reference,

$$J = \mathbf{n}\mathbf{W}^{-1}\mathbf{n} + \mathbf{x}^T\mathbf{S}^{-1}\mathbf{x} \tag{3.3.37}$$

$$\tilde{\mathbf{x}} = (\mathbf{E}^T\mathbf{W}^{-1}\mathbf{E} + \mathbf{S}^{-1})^{-1}\mathbf{E}^T\mathbf{W}^{-1}\mathbf{y} \tag{3.3.38}$$

$$\tilde{\mathbf{n}} = \left\{\mathbf{I} - \mathbf{E}(\mathbf{E}^T\mathbf{W}^{-1}\mathbf{E} + \mathbf{S}^{-1})^{-1}\mathbf{E}^T\mathbf{W}^{-1}\right\}\mathbf{y} \tag{3.3.39}$$

$$\mathbf{C}_{\tilde{x}\tilde{x}} = (\mathbf{E}^T\mathbf{W}^{-1}\mathbf{E} + \mathbf{S}^{-1})^{-1}\mathbf{E}^T\mathbf{W}^{-1}\mathbf{R}_{nn} \times \mathbf{W}^{-1}\mathbf{E}(\mathbf{E}^T\mathbf{W}^{-1}\mathbf{E} + \mathbf{S}^{-1})^{-1} \tag{3.3.40}$$

$$\mathbf{P}_{\tilde{n}\tilde{n}} = \left\{\mathbf{I} - \mathbf{E}(\mathbf{E}^T\mathbf{W}^{-1}\mathbf{E} + \mathbf{S}^{-1})^{-1}\mathbf{E}^T\mathbf{W}^{-1}\right\}^{-1}\mathbf{R}_{nn} \times \left\{\mathbf{I} - \mathbf{E}(\mathbf{E}^T\mathbf{W}^{-1}\mathbf{E} + \mathbf{S}^{-1})\mathbf{E}^T\mathbf{W}^{-1}\right\}^T . \tag{3.3.41}$$

So far, all of this is conventional. But we have made a special point of displaying explicitly not only the elements $\tilde{\mathbf{x}}$ but those of the residuals $\tilde{\mathbf{n}}$. Notice that although we have considered only the formally overdetermined system, $M > N$, we *always* determine not only the N elements of $\tilde{\mathbf{x}}$ but also the M elements of $\tilde{\mathbf{n}}$, for a total of $M + N$ values, extracted from the M equations. It is apparent that any change in any element $\tilde{n}_i$ forces changes in $\tilde{\mathbf{x}}$. In this view, to which we adhere, systems of equations involving observations *always* contain more unknowns than knowns. There is compelling reason, therefore, to rewrite (3.3.2) as

$$\mathbf{E}_1\boldsymbol{\xi} = \mathbf{y}$$

$$\mathbf{E}_1 = \{\mathbf{E} \quad \mathbf{I}_M\}, \quad \boldsymbol{\xi}^T = [\mathbf{x} \quad \mathbf{n}]^T , \tag{3.3.42}$$

which is to be solved exactly, as an underdetermined system. That even overdetermined observed systems are of necessity actually underdetermined, leads to taking a first look at formal underdetermination.

3.3.3 Undetermined Systems–A First Discussion

What does one do when the number of equations is less than the number of unknowns, and no more observations are possible? One often attempts in such a situation to reduce the number of unknowns so that the formal overdeterminism is restored. Such a parameter reduction procedure may be sensible, but there are pitfalls. Consider data produced from a law

$$y = 1 + a_M t^M + n(t) , \tag{3.3.43}$$

which might be deduced by fitting a parameter set $[a_0, \ldots, a_M]$. If there are fewer than M observations, an attempt to fit with fewer parameters,

$$y = \sum_{i=0}^{Q} a_i t^i , \quad Q < M , \tag{3.3.44}$$

may give a good, even perfect fit, but it would be wrong. The reduction in model parameters in such a case biases the result. One is better off retaining the underdetermined system and making inferences concerning the possible values of a_i rather than using the form (3.3.44), in which any possibility of learning something about a_M has been eliminated.

In more general terms (discussed by Wunsch & Minster, 1982), parameter reduction can lead to model errors or biases that can produce wholly illusory results. A specific example was provided by Wunsch (1988a); a two-dimensional ocean circulation model was used to calculate values for the apparent oxygen utilization rate (AOUR). But when the parameterization was made more realistic (a three-dimensional model), it was found that AOUR was indeterminate to within any useful range. The conclusions from the underparameterized model are erroneous; the second model produces the useful information that the database was inadequate to estimate AOUR, and one avoids drawing incorrect conclusions.

Another example is Munk's (1966) well-known discussion of the property fields of the abyssal Pacific Ocean (see Figure 4–29). He fit the observations to the solutions of one-dimensional vertical balance equations

$$w \frac{\partial C}{\partial z} - \kappa \frac{\partial^2 C}{\partial z^2} = \text{sinks} \tag{3.3.45}$$

where C is temperature, salinity, or radiocarbon for the vertical velocity w and vertical mixing coefficient κ. The fit was quite good and has been cherished by a generation of chemical oceanographers as showing that the ocean is one-dimensional and steady. But such an ocean circulation is impossible, and one is misled by the good fit of an underparameterized model.

A general approach to solving underdetermined problems is to render them unique by minimizing an objective function, subject to satisfaction of the linear constraints. To see how this can work, suppose that (3.3.2) are indeed formally underdetermined–that is, $M < N$–and seek the solution that exactly satisfies the equations and simultaneously renders the objective function, $J = \mathbf{x}^T \mathbf{x}$, as small as possible. Direct minimization of J leads to

$$dJ = \left(\frac{\partial J}{\partial \mathbf{x}} \right)^T d\mathbf{x} = 2\mathbf{x}^T d\mathbf{x} = 0 , \tag{3.3.46}$$

but the coefficients of the individual dx_i can no longer be separately set to zero (i.e., $\mathbf{x} = 0$ is incorrect) because the dx_i no longer vary independently but are restricted to values satisfying $\mathbf{Ex} = \mathbf{y}$. One approach is to use the known dependencies to reduce the problem to a new one in which the differentials are independent. For example, suppose that there are general functional relationships

$$\begin{bmatrix} x_1 \\ \vdots \\ x_L \end{bmatrix} = \begin{bmatrix} \xi_1(x_{L+1}, \ldots, x_N) \\ \vdots \\ \xi_L(x_{L+1}, \ldots, x_N) \end{bmatrix}.$$

Then the first L elements of x_i may be eliminated, and the cost function becomes

$$J = \xi_1^2 + \cdots + \xi_L^2 + x_{L+1}^2 + \cdots + x_N^2$$

in which the remaining x_i, $L+1 \le i \le N$ are independently varying. But an explicit solution for L elements of $\mathbf{x}$ in terms of the remaining ones may be difficult to find.

When it is inconvenient to find such an explicit representation eliminating some variables in favor of others, a standard procedure for finding the constrained minimum is to introduce a new vector *Lagrange multiplier*, $\boldsymbol{\mu}$, of M unknown elements, to make a new objective function

$$J' = J - 2\boldsymbol{\mu}^T(\mathbf{Ex} - \mathbf{y}) = \mathbf{x}^T\mathbf{x} - 2\boldsymbol{\mu}^T(\mathbf{Ex} - \mathbf{y}) \tag{3.3.47}$$

and ask for its stationary point, treating both $\boldsymbol{\mu}$ and $\mathbf{x}$ as independently varying unknowns. The numerical 2 is introduced solely for notational tidiness. The rationale for this procedure is straightforward (e.g., Morse & Feshbach, 1953, p. 238; Strang, 1986): $\mathbf{Ex} = \mathbf{y}$ requires that

$$\mathbf{E}d\mathbf{x} = \mathbf{e}_1 dx_1 + \mathbf{e}_2 dx_2 + \cdots + \mathbf{e}_N dx_N = 0$$

where the $\mathbf{e}_i$ are the column vectors of $\mathbf{E}$. A constant, $-2\boldsymbol{\mu}^T$, times this last expression can be added to $dJ = 0$ so that

$$\begin{aligned} dJ - 2\boldsymbol{\mu}^T\mathbf{E}d\mathbf{x} = \left(\frac{\partial J}{\partial x_1} - 2\boldsymbol{\mu}^T\mathbf{e}_1\right) dx_1 + \left(\frac{\partial J}{\partial x_2} - 2\boldsymbol{\mu}^T\mathbf{e}_2\right) dx_2 + \cdots \\ + \left(\frac{\partial J}{\partial x_N} - 2\boldsymbol{\mu}^T\mathbf{e}_N\right) dx_N = 0\,. \end{aligned} \tag{3.3.48}$$

In this form, there are M elements of $\boldsymbol{\mu}$ that can be used to set any M of the coefficients of the dx_i to zero, leaving coefficients of $N - M$ of the remaining dx_i, which can be treated as independent variables. If the objective function J' is differentiated with respect to $\boldsymbol{\mu}$, $\mathbf{x}$ and is set to zero, it is readily found that the result is a set of simultaneous equations equivalent to the vanishing of each coefficient in (3.3.48), plus the model. With (3.3.47), this recipe produces

$$\frac{1}{2}\frac{\partial J'}{\partial \boldsymbol{\mu}} = \mathbf{Ex} - \mathbf{y} = \mathbf{0} \tag{3.3.49}$$

$$\frac{1}{2}\frac{\partial J'}{\partial \mathbf{x}} = \mathbf{x} - \mathbf{E}^T\boldsymbol{\mu} = \mathbf{0} \tag{3.3.50}$$

where the first of these are just the original equations and the second are the coefficients in (3.3.48) whose solution must therefore be equivalent to setting the individual terms of (3.3.48) to zero as required, subject to the model. Because the original equations emerge, the second term of J' will vanish at the stationary point. The convenience of being able to treat all the x_i as independently varying is offset by the increase in problem dimensions by the introduction of the unknown μ_i.

Equation (3.3.50) gives

$$\mathbf{E}^T\boldsymbol{\mu} = \mathbf{x}\,, \tag{3.3.51}$$

and substituting for $\mathbf{x}$ into (3.3.49),

$$\mathbf{E}\mathbf{E}^T\boldsymbol{\mu} = \mathbf{y}\,,$$

$$\tilde{\boldsymbol{\mu}} = (\mathbf{E}\mathbf{E}^T)^{-1}\mathbf{y}\,, \tag{3.3.52}$$

assuming the inverse exists, and

$$\tilde{\mathbf{x}} = \mathbf{E}^T(\mathbf{E}\mathbf{E}^T)^{-1}\mathbf{y} \tag{3.3.53}$$

$$\tilde{\mathbf{n}} = \mathbf{0} \tag{3.3.54}$$

$$\mathbf{C}_{\tilde{x}\tilde{x}} = \mathbf{0} \tag{3.3.55}$$

($\mathbf{C}_{\tilde{x}\tilde{x}} = 0$ because formally we estimate $\tilde{\mathbf{n}} = \mathbf{0}$).

Equations (3.3.51) for $\boldsymbol{\mu}$ in terms of $\mathbf{x}$ involves the coefficient matrix $\mathbf{E}^T$. An intimate connection exists between matrix transposes and adjoints of differential equations (see especially, Lanczos, 1961; or Morse & Feshbach, 1953), and thus $\boldsymbol{\mu}$ is sometimes called the *adjoint solution*, with $\mathbf{E}^T$ being the adjoint model.[6] The original Equations (3.3.2) were assumed formally underdetermined, and thus the adjoint model equations in (3.3.51) are necessarily formally overdetermined. The physical interpretation of $\boldsymbol{\mu}$ comes from the result

$$\frac{\partial J'}{\partial \mathbf{y}} = 2\boldsymbol{\mu}\,; \tag{3.3.56}$$

[6] But the matrix transpose should not be confused with the adjoint matrix, which is quite different.

the Lagrange multipliers represent the sensitivity of the minimum of J' to the perturbations in the data $\mathbf{y}$.

Equation (3.3.53) is the classical solution of minimum norm of $\mathbf{x}$, satisfying the constraints exactly while minimizing the solution length. That a minimum is achieved can be verified by evaluating the second derivatives of J' at the solution point. The minimum occurs at a saddle point in $\mathbf{x}$, $\boldsymbol{\mu}$ space (see Sewell, 1987, for an interesting discussion) and where the term proportional to $\boldsymbol{\mu}$ necessarily vanishes. The operator $\mathbf{E}^T(\mathbf{E}\mathbf{E}^T)^{-1}$ is sometimes called a *Moore-Penrose inverse.*

If the Equations (3.3.2) are first column-scaled using $\mathbf{S}^{-T/2}$, Equations (3.3.53)–(3.3.55) are in the primed variables, the solution in the original variables is

$$\tilde{\mathbf{x}} = \mathbf{S}\mathbf{E}^T(\mathbf{E}\mathbf{S}\mathbf{E}^T)^{-1}\mathbf{y} \tag{3.3.57}$$

$$\tilde{\mathbf{n}} = \mathbf{0} \tag{3.3.58}$$

$$\mathbf{C}_{\tilde{x}\tilde{x}} = \mathbf{0}, \tag{3.3.59}$$

and the result depends directly upon $\mathbf{S}$. If a row-scaling with $\mathbf{W}^{-T/2}$ is used, it is readily shown that $\mathbf{W}$ disappears from the solution and has no effect on it. Equations (3.3.57)–(3.3.59) are a valid solution, but there is a potentially fatal defect–$\tilde{\mathbf{n}} = \mathbf{0}$ is rarely acceptable when $\mathbf{y}$ are observations. Furthermore, $\|\tilde{\mathbf{x}}\|$ is again uncontrolled, and $\mathbf{E}^T\mathbf{E}$ may not have an inverse.

We have been emphasizing that $\mathbf{n}$ must be regarded as fully an element of the solution, as much as $\mathbf{x}$, that equations representing observations can always be written as (3.3.42) and can be solved exactly. Therefore, we use a modified objective function

$$J = \alpha^2\mathbf{x}^T\mathbf{x} + \mathbf{n}^T\mathbf{n} - 2\boldsymbol{\mu}^T(\mathbf{E}\mathbf{x} + \mathbf{n} - \mathbf{y}), \tag{3.3.60}$$

with both $\mathbf{x}$, $\mathbf{n}$ appearing in the objective function. Setting the derivatives of (3.3.60) with respect to $\mathbf{x}$, $\mathbf{n}$, $\boldsymbol{\mu}$ to zero, and solving the resulting normal equations produces

$$\tilde{\mathbf{x}} = \mathbf{E}^T(\mathbf{E}\mathbf{E}^T + \alpha^2\mathbf{I})^{-1}\mathbf{y} \tag{3.3.61}$$
$$\tilde{\mathbf{n}} = \left\{\mathbf{I} - \mathbf{E}\mathbf{E}^T(\mathbf{E}\mathbf{E}^T + \alpha^2\mathbf{I})^{-1}\right\}\mathbf{y} \tag{3.3.62}$$
$$\mathbf{C}_{\tilde{x}\tilde{x}} = \mathbf{E}^T(\mathbf{E}\mathbf{E}^T + \alpha^2\mathbf{I})^{-1}\mathbf{R}_{nn}(\mathbf{E}\mathbf{E}^T + \alpha^2\mathbf{I})^{-1}\mathbf{E} \tag{3.3.63}$$
$$\tilde{\boldsymbol{\mu}} = \tilde{\mathbf{n}} \tag{3.3.64}$$
$$\mathbf{P}_{\tilde{n}\tilde{n}} = \left\{\mathbf{I} - \mathbf{E}\mathbf{E}^T(\mathbf{E}\mathbf{E}^T + \alpha^2\mathbf{I})^{-1}\right\}\mathbf{R}_{nn} \times \left\{\mathbf{I} - \mathbf{E}\mathbf{E}^T(\mathbf{E}\mathbf{E}^T + \alpha^2\mathbf{I})^{-1}\right\}, \tag{3.3.65}$$

and as before, we could employ α^2 as a means to control the relative norms of $\tilde{\mathbf{x}}$, $\tilde{\mathbf{n}}$ and the elements of $\mathbf{C}_{\tilde{x}\tilde{x}}$. If we suppose that weights $\mathbf{W}^{T/2}$, $\mathbf{S}^{T/2}$ were applied to the equations prior to forming J, then the solution (3.3.61)–(3.3.65) is in the primed variables, and in terms of the original variables is

$$\tilde{\mathbf{x}} = \mathbf{S}\mathbf{E}^T(\mathbf{E}\mathbf{S}\mathbf{E}^T + \mathbf{W})^{-1}\mathbf{y} \tag{3.3.66}$$
$$\tilde{\mathbf{n}} = \left\{\mathbf{I} - \mathbf{E}\mathbf{S}\mathbf{E}^T(\mathbf{E}\mathbf{S}\mathbf{E}^T + \mathbf{W})^{-1}\right\}\mathbf{y} \tag{3.3.67}$$
$$\mathbf{C}_{\tilde{x}\tilde{x}} = \mathbf{S}\mathbf{E}^T(\mathbf{E}\mathbf{S}\mathbf{E}^T + \mathbf{W})^{-1}\mathbf{R}_{nn}(\mathbf{E}\mathbf{S}\mathbf{E}^T + \mathbf{W})^{-1}\mathbf{E}\mathbf{S} \tag{3.3.68}$$
$$\tilde{\boldsymbol{\mu}} = \mathbf{W}^{-T/2}\tilde{\mathbf{n}} \tag{3.3.69}$$
$$\mathbf{P}_{\tilde{n}\tilde{n}} = \left\{\mathbf{I} - \mathbf{E}\mathbf{S}\mathbf{E}^T(\mathbf{E}\mathbf{S}\mathbf{E}^T + \mathbf{W})^{-1}\right\}\mathbf{R}_{nn} \times \left\{\mathbf{I} - \mathbf{E}\mathbf{S}\mathbf{E}^T(\mathbf{E}\mathbf{S}\mathbf{E}^T + \mathbf{W})^{-1}\right\}, \tag{3.3.70}$$

with α^2 absorbed into $\mathbf{S}$. Despite the different form, we claim that (3.3.66)–(3.3.68) are identical to (3.3.38)–(3.3.41)–their identity is readily shown by using the matrix inversion lemma in the form (3.1.25). A choice between the two forms is often made on the basis of the dimensionality of the matrices being inverted: $\mathbf{E}^T\mathbf{W}^{-1}\mathbf{E}$ is $N \times N$ and $\mathbf{E}\mathbf{S}\mathbf{E}^T$ is $M \times M$. But even this criterion is ambiguous, for example, because $\mathbf{W}$ is $M \times M$, and if it is not actually diagonal, or its inverse otherwise known, one would have to invert it.

Equations (3.3.38)–(3.3.40) and (3.3.66)–(3.3.70) result from two very different appearing objective functions–one in which the equations are imposed in the mean square (3.3.38)–(3.3.40), and one in which they are imposed exactly (3.3.66)–(3.3.70), using Lagrange multipliers. In the terminology of Sasaki (1970) and others, exact relationships are called *strong* constraints, and those imposed in the mean-square are *weak* ones. A preferable terminology, which we will sometimes use, is *hard* and *soft* constraints. But in the present situation in particular, the distinction is illusory: Although (3.3.2) are being imposed exactly, it is only the presence of the error term, $\mathbf{n}$, that permits the equations to be written as equalities and thus as hard constraints. The hard and soft constraints here produce an identical solution. In some circumstances, which we will discuss briefly below, one may wish to impose exact constraints upon the elements of $\tilde{x}_i$; these are often model constraints, for example, that the flow should be exactly geostrophic. But it is actually rare that one's models are exactly correct, and even geostrophy is always violated slightly. The solution (3.3.53)–(3.3.55) was derived from a true hard constraint, $\mathbf{Ex} = \mathbf{y}$, but we ended by rejecting it as generally inapplicable.

It should be ever more clear that $\mathbf{n}$ is only by convention discussed separately from $\mathbf{x}$ and is fully a part of the solution. The combined form (3.3.42), which literally treats $\mathbf{x}$, $\mathbf{n}$ as the solution, is imposed through a hard constraint on the objective function,

$$J = \boldsymbol{\xi}^T\boldsymbol{\xi} - 2\boldsymbol{\mu}^T(\mathbf{E}_1\boldsymbol{\xi} - \mathbf{y}_1), \qquad (3.3.71)$$

which is (3.3.60) with $\alpha^2 = 1$. (There are numerical advantages, however, in working with objects in two spaces of dimensions M and N rather than a single space of dimension $M + N$.)

3.4 The Singular Vector Expansion

Least squares is a very powerful, very useful method for finding solutions of linear simultaneous equations of any dimensionality, and one might wonder why it is necessary to discuss any other form of solution. But in the simplest form of least squares, the solution is dependent upon the existence of inverses of $\mathbf{E}^T\mathbf{E}$, or $\mathbf{E}\mathbf{E}^T$. In practice, their existence cannot be guaranteed, and we need to understand first what that means, the extent to which solutions can be found when the inverses do not exist, and the effect of introducing weight matrices $\mathbf{W}$, $\mathbf{S}$. This problem is intimately related to the issue of controlling solution and residual norms. Second, the relationship between the equations and the solutions is somewhat impenetrable, in the sense that structures in the solutions are not easily related to particular elements of the data y_i. For many purposes, particularly physical insight, understanding the structure of the solution is essential.

3.4.1 Simple Vector Expansions

Consider again the elementary problem (3.1.1) of representing an L-dimensional vector $\mathbf{f}$ as a sum of a complete set of L-orthonormal vectors $\mathbf{g}_i$, $1 \le i \le L$, $\mathbf{g}_i^T \mathbf{g}_j = \delta_{ij}$. Without error,

$$\mathbf{f} = \sum_{j=1}^{L} a_j \mathbf{g}_j,\ a_j = \mathbf{g}_j^T \mathbf{f}\,. \tag{3.4.1}$$

But if for some reason only the first K coefficients a_j are known, we can only approximate $\mathbf{f}$ by its first K terms:

$$\begin{aligned}\tilde{\mathbf{f}} &\approx \sum_{j=1}^{K} a_j \mathbf{g}_j \\ &= \mathbf{f} + \delta\mathbf{f}_1\,,\end{aligned} \tag{3.4.2}$$

and there is an error, $\delta\mathbf{f}_1$. From the orthogonality of the $\mathbf{g}_i$, it follows that $\delta\mathbf{f}_1$ will have minimum l_2 norm if and only if it is orthogonal to the K vectors retained in the approximation, and if and only if a_j are given by (3.4.1). The only way the error could be reduced is by increasing K.

Define an $L \times K$ matrix $\mathbf{G}_K$ whose columns are the first K of the $\mathbf{g}_j$. Then $\mathbf{a} = \mathbf{G}_K^T \mathbf{f}$ is the vector of coefficients $a_j = \mathbf{g}_j^T \mathbf{f}$, $1 \le j \le K$, and the finite representation (3.4.2) is (one should write it out)

$$\tilde{\mathbf{f}} = \mathbf{G}_K \mathbf{a} = \mathbf{G}_K (\mathbf{G}_K^T \mathbf{f}) = (\mathbf{G}_K \mathbf{G}_K^T)\mathbf{f}\,, \quad \mathbf{a} = \{a_i\} \tag{3.4.3}$$

where the third equality follows from the associative properties of matrix multiplication. This expression shows that a *representation of a vector in an incomplete orthonormal set produces a resulting approximation that is a simple linear combination of the elements of the correct values* (i.e., a weighted average, or *filtered* version of them).

Because the columns of $\mathbf{G}_K$ are orthonormal, $\mathbf{G}_K^T \mathbf{G}_K = \mathbf{I}_K$–that is, the $K \times K$ identity matrix; but $\mathbf{G}_K \mathbf{G}_K^T \ne \mathbf{I}_L$ unless $K = L$ (that $\mathbf{G}_L \mathbf{G}_L^T = \mathbf{I}_L$ for $K = L$ follows from the theorem for square matrices, which show a left inverse is also a right inverse; see any book on linear algebra). If $K < L$, $\mathbf{G}_K$ is semi-orthogonal. If $K = L$, it is orthogonal; in this case, $\mathbf{G}_L^{-1} = \mathbf{G}_L^T$. If it is only semi-orthogonal, $\mathbf{G}_K^T$ is a left inverse but not a right inverse.

$\mathbf{G}_K \mathbf{G}_K^T$ is known as a *resolution matrix*, with a simple interpretation. Suppose the true value of $\mathbf{f}$ were

$$\mathbf{f}_{j_0} = [0 \;\; 0 \;\; 0 \;\; . \;\; . \;\; . \;\; 0 \;\; 1 \;\; 0 \;\; . \;\; 0 \;\; .. \;\; 0]^T\,,$$

that is, a Kronecker delta with unity in element j_0. Then the incomplete

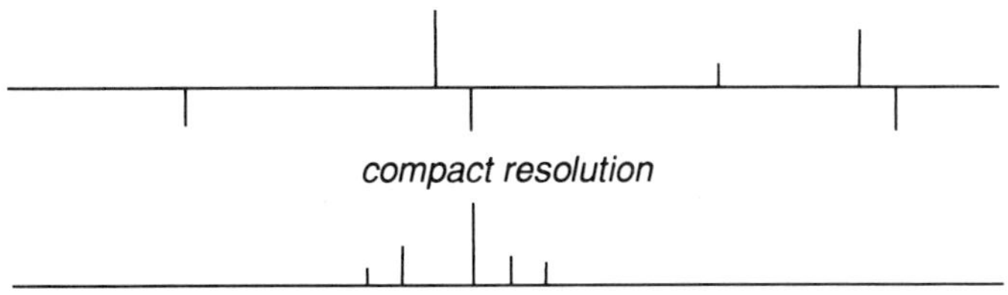

Figure 3–9. Incomplete vector expansions produce solutions that are linear combinations of the elements of correct ones. One can distinguish *compact* resolution where the linear combinations are a simple neighborhood weighted average (where *neighborhood* has a physical interpretation in time or space) from *non-compact* resolution where the averaging involves physically distant elements (although they are typically close in some other space, e.g., as measured in water mass properties). The figure shows, schematically, how the weights differ in the two cases.

expansion (3.4.2) or (3.4.3) would not reproduce the delta function but rather

$$\tilde{\mathbf{f}}_{j_0} = \mathbf{G}_K \mathbf{G}_K^T \mathbf{f}_{j_0} \,, \qquad (3.4.4)$$

which is row (or column, because it is symmetric) j_0 of $\mathbf{G}_K \mathbf{G}_K^T$. The j_0-th row of the resolution matrix tells one what the corresponding form of the vector would be if its true form were a delta function at position j_0.

To form a Kronecker delta function requires a complete set of vectors. An analogous elementary result of Fourier analysis shows that a Dirac delta function demands contributions from all frequencies to arrange for a narrow, very high pulse. Removal of some of the requisite vectors (sinusoids) produces broadening and sidelobes. Here, depending upon the precise structure of the $\mathbf{g}_i$, the broadening and sidelobes can be complicated. If one is lucky, the effect could be a simple broadening (schematically shown in Figure 3–9) without distant sidelobes (Wiggins, 1972, who has a good discussion, calls this *compact resolution*), leading to the tidy interpretation of the result as a local average of the true values.

A resolution matrix has the property

$$\text{trace}(\mathbf{G}_K \mathbf{G}_K^T) = K \,, \qquad (3.4.5)$$

which follows from noting that

$$\text{trace}(\mathbf{G}_K^T \mathbf{G}_K) = \text{trace}(\mathbf{I}_K) = K \,,$$

and by direct evaluation,

$$\text{trace}(\mathbf{G}_K \mathbf{G}_K^T) = \text{trace}(\mathbf{G}_K^T \mathbf{G}_K) \,.$$

Orthogonal vector expansions are particularly simple to use and interpret, but their relevance to solving a set of simultaneous equations may be obscure. What we will show, however, is that we can always find sets of orthonormal vectors to simplify greatly the job of solving simultaneous equations. To do so, we digress to recall the basic elements of the eigenvector/eigenvalue problem.

Consider a *square*, $M \times M$ matrix $\mathbf{E}$ and the simultaneous equations

$$\mathbf{E}\mathbf{g}_i = \lambda_i \mathbf{g}_i, \quad 1 \leq i \leq M, \tag{3.4.6}$$

that is, the problem of finding a set of vectors $\mathbf{g}_i$ whose dot products with the rows of $\mathbf{E}$ are proportional to themselves. Such vectors are *eigenvectors*, and the constants of proportionality are the *eigenvalues*. Under special circumstances, the eigenvectors form an orthonormal spanning set. Textbooks show that if $\mathbf{E}$ is square and symmetric, such a result is guaranteed. Suppose for the moment that we have such a special case, and recall how eigenvectors can be used to solve (3.1.10). With an orthonormal, spanning set, both the known $\mathbf{y}$ and the unknown $\mathbf{x}$ can be written as

$$\mathbf{x} = \sum_{i=1}^{M} \alpha_i \mathbf{g}_i, \; \alpha_i = \mathbf{g}_i^T \mathbf{x}, \tag{3.4.7}$$

$$\mathbf{y} = \sum_{i=1}^{M} \beta_i \mathbf{g}_i, \; \beta_i = \mathbf{g}_i^T \mathbf{y}. \tag{3.4.8}$$

By convention, $\mathbf{y}$ is known, and therefore the β_i can be regarded as given. If we could find the α_i, $\mathbf{x}$ would be known.

Substitute (3.4.7) into (3.1.10), and using the eigenvector property,

$$\mathbf{E}\sum_{i=1}^{M} \alpha_i \mathbf{g}_i = \sum_{i=1}^{M} \left(\mathbf{g}_i^T \mathbf{y}\right) \mathbf{g}_i$$

or

$$\sum_{i=1}^{M} \alpha_i \lambda_i \mathbf{g}_i = \sum_{i}^{M} \left(\mathbf{g}_i^T \mathbf{y}\right) \mathbf{g}_i. \tag{3.4.9}$$

But the expansion vectors are orthonormal, and so

$$\lambda_i \alpha_i = \mathbf{g}_i^T \mathbf{y} \tag{3.4.10}$$

$$\alpha_i = \frac{\mathbf{g}_i^T \mathbf{y}}{\lambda_i} \tag{3.4.11}$$

$$\mathbf{x} = \sum_{i=1}^{N} \frac{\mathbf{g}_i^T \mathbf{y}}{\lambda_i} \mathbf{g}_i. \tag{3.4.12}$$

Apart from an evident difficulty if any eigenvalue vanishes, the problem is now completely solved. If we define a diagonal matrix, $\boldsymbol{\Lambda}$, with elements,

λ_i, ordered by convention in descending numerical value, and the matrix $\mathbf{G}$, whose columns are the corresponding $\mathbf{g}_i$ in the same order, the solution to (3.1.10) can be written [from (3.4.7), (3.4.10)–(3.4.12)] as

$$\boldsymbol{\alpha} = \boldsymbol{\Lambda}^{-1}\mathbf{G}^T\mathbf{y} \tag{3.4.13}$$

$$\mathbf{x} = \mathbf{G}\boldsymbol{\Lambda}^{-1}\mathbf{G}^T\mathbf{y} \tag{3.4.14}$$

where $\boldsymbol{\Lambda}^{-1} = \mathrm{diag}(1/\lambda_i)$.

Vanishing eigenvalues, $i = i_0$, cause trouble, and we must consider them. Let the corresponding eigenvectors be $\mathbf{g}_{i_0}$. Then any part of the solution that is proportional to such an eigenvector is annihilated by $\mathbf{E}$–that is, $\mathbf{g}_{i_0}$ is orthogonal to all the rows of $\mathbf{E}$. Such a result means that there is no possibility that anything in $\mathbf{y}$ could provide any information about the coefficient α_{i_0}. If $\mathbf{y}$ corresponds to a set of observations (data), then $\mathbf{E}$ represents the connection (mapping) between system unknowns and observations. The existence of zero eigenvalues shows that the act of observation of $\mathbf{x}$ removes certain structures in the solution that are then indeterminate. Vectors $\mathbf{g}_{i_0}$ (and there may be many of them) are said to lie in the nullspace of $\mathbf{E}$. Eigenvectors corresponding to nonzero eigenvalues lie in its range. The simplest example is given by the observations

$$x_1 + x_2 = 3\,,$$

$$x_1 + x_2 = 3\,.$$

Any structure in $\mathbf{x}$ such that $x_1 = -x_2$ is destroyed by this observation, and by inspection, the nullspace vector must be $\mathbf{g}_2 = [1 \quad -1]^T/\sqrt{2}$ (the purpose of showing the observation twice is to produce an $\mathbf{E}$ that is square).

Suppose there are $K < M$ nonzero λ_i. Then for $i > K$, Equation (3.4.10) is

$$0\alpha_i = \mathbf{g}_i^T\mathbf{y}, \quad K+1 \le i \le M\,, \tag{3.4.15}$$

and two cases must be distinguished.

Case (1):

$$\mathbf{g}_i^T\mathbf{y} = 0\,, \qquad K+1 \le i \le M\,. \tag{3.4.16}$$

We could then put $\alpha_i = 0$, $K+1 \le i \le M$, and the solution can be written

$$\tilde{\mathbf{x}} = \sum_{i=1}^{K} \frac{\mathbf{g}_i^T\mathbf{y}}{\lambda_i}\mathbf{g}_i \tag{3.4.17}$$

and $\mathbf{E}\tilde{\mathbf{x}} = \mathbf{y}$, *exactly*. We have put a tilde over $\mathbf{x}$ because a solution of the form

$$\tilde{\mathbf{x}} = \sum_{i=1}^{K} \frac{\mathbf{g}_i^T \mathbf{y}}{\lambda_i} \mathbf{g}_i + \sum_{i=K+1}^{M} \alpha_i \mathbf{g}_i \,, \tag{3.4.18}$$

with the remaining α_i taking on arbitrary values, also satisfies the equations exactly. That is to say, the true value of $\mathbf{x}$ *could* contain structures proportional to the nullspace vectors of $\mathbf{E}$, but the equations (3.1.10) neither require their presence nor provide the information necessary to determine their amplitudes. We thus have a situation with a *solution nullspace.* If the matrix $\mathbf{G}_K$ is $M \times K$, carrying only the first K of the $\mathbf{g}_i$–that is, the range vectors–$\boldsymbol{\Lambda}_K$ is $K \times K$ with only the first K, nonzero eigenvalues, and the columns of $\mathbf{Q}_G$ are the M-K nullspace vectors [it is $M \times (M - K)$], then the solutions (3.4.17) and (3.4.18) are

$$\tilde{\mathbf{x}} = \mathbf{G}_K \boldsymbol{\Lambda}_K^{-1} \mathbf{G}_K^T \mathbf{y} \,, \tag{3.4.19}$$

$$\tilde{\mathbf{x}} = \mathbf{G}_K \boldsymbol{\Lambda}_K^{-1} \mathbf{G}_K^T \mathbf{y} + \mathbf{Q}_G \boldsymbol{\alpha}_G \tag{3.4.20}$$

where $\boldsymbol{\alpha}_G$ is the vector of unknown nullspace coefficients, respectively. Equation (3.4.16) is often known as a *solvability condition.* The solution (3.4.19) with no nullspace contribution will be called the *particular* solution.

If $\mathbf{G}$ is written as a partitioned matrix,

$$\mathbf{G} = \{\mathbf{G}_K \quad \mathbf{Q}_G\} \,,$$

it follows from the column orthonormality that

$$\mathbf{G}\mathbf{G}^T = \mathbf{I}_L = \mathbf{G}_K \mathbf{G}_K^T + \mathbf{Q}_G \mathbf{Q}_G^T \tag{3.4.21}$$

or

$$\mathbf{Q}_G \mathbf{Q}_G^T = \mathbf{I}_L - \mathbf{G}_K \mathbf{G}_K^T \,. \tag{3.4.22}$$

Case (2):

$$\mathbf{g}_i^T \mathbf{y} \neq 0 \,, \qquad i > K \,, \tag{3.4.23}$$

for one or more of the nullspace vectors. In this case, Equation (3.4.10) is the contradiction

$$0\alpha_i \neq 0 \,,$$

and Equation (3.4.9) is actually

$$\sum_{i=1}^{K} \lambda_i \alpha_i \mathbf{g}_i = \sum_{i=1}^{M} (\mathbf{g}_i^T \mathbf{y}) \mathbf{g}_i, \; K < M \,, \tag{3.4.24}$$

that is, with differing upper limits on the sums. Owing to the orthonormality of the $\mathbf{g}_i$, there is no choice of α_i, $1 \leq i \leq K$ on the left that can match the last M-K terms on the right. Evidently there is no solution in the conventional sense unless (3.4.16) is satisfied, hence the name *solvability condition.* What is the best we might do? Define *best* to mean that the solution $\tilde{\mathbf{x}}$ should be chosen such that

$$\mathbf{E}\tilde{\mathbf{x}} = \tilde{\mathbf{y}}$$

where the difference, $\tilde{\mathbf{n}} = \mathbf{y} - \tilde{\mathbf{y}}$, which we call the *residual*, should be as small as possible (in the l_2 norm). If this choice is made, then the orthogonality of the $\mathbf{g}_i$ shows immediately that the best choice is still (3.4.11), $1 \leq i \leq K$. No choice of nullspace vector coefficients, nor any other value of the coefficients of the range vectors, can reduce the norm of $\mathbf{n}$. The best solution is then also (3.4.17) or (3.4.19).

In this situation, we are no longer solving the equations (3.1.10) but rather are dealing with a set that could be written

$$\mathbf{E}\mathbf{x} \sim \mathbf{y} \tag{3.4.25}$$

where the demand is for a solution that is the best possible, in the sense just defined. Such statements of approximation are awkward, and it is more useful to always rewrite (3.4.25) as

$$\mathbf{E}\mathbf{x} + \mathbf{n} = \mathbf{y} \tag{3.4.26}$$

where $\mathbf{n}$ is the residual. If $\tilde{\mathbf{x}}$ is given by (3.4.18), then

$$\tilde{\mathbf{n}} = \sum_{i=K+1}^{M} (\mathbf{g}_i^T \mathbf{y}) \mathbf{g}_i \tag{3.4.27}$$

by (3.4.24). Notice that $\tilde{\mathbf{n}}^T \tilde{\mathbf{y}} = 0$.

This situation, where we started with M equations in M unknowns, but found in practice that some structures of the solution could not actually be determined, is labeled *formally just-determined*, where the word *formally* alludes to the fact that the mere appearance of a just-determined system did not mean that the characterization was true in practice. One or more vanishing eigenvalues means that the rows and columns $\mathbf{E}$ are not spanning sets.

Some decision has to be made about the coefficients of the nullspace vectors in (3.4.18) or (3.4.20). We could use the form as it stands, regarding it as the *general solution.* The analogy with the solution of differential equations should be apparent–typically, such equations have particular and homogeneous solutions. In the present case, the homogeneous solution corresponds to the nullspace vectors. When solving a differential equation, determination of the magnitude of the homogeneous solution requires additional information, often provided by boundary or initial conditions; here, additional information is also necessary but is missing. Despite the presence of indeterminate elements in the solution, we know exactly what they are: proportional to the nullspace vectors. Depending upon the specific situation, we might conceivably be in a position to obtain more observations and would seriously consider observational strategies directed at detecting these missing structures. The reader is also reminded of the discussion of the Neumann problem in Section 1.3.

Another approach is to define a simplest solution, appealing to what is usually known as *Occam's Razor*, or the *principal of parsimony*, that in choosing between multiple explanations of a given phenomenon, the simplest one is usually the best. What is simplest can be debated, but here there is a compelling choice: The solution (3.4.17) or (3.4.19)–that is, without any nullspace contributions, is less structured than any other solution. [It is often but not always (again, recall the Neumann problem) true that the nullspace vectors are more "wiggily" than those in the range. In any case, including any vector not required by the data is arguably producing more structure than is required.] Setting all the unknown α_i to zero is thus one choice. It follows from the orthogonality of the $\mathbf{g}_i$ that this particular solution is also the one of minimum solution norm. Later, we will see some other choices for the nullspace vectors.

If the nullspace vector contributions are set to zero, the true solution has been expanded in an incomplete set of orthonormal vectors. Thus, $\mathbf{G}_K\mathbf{G}_K^T$ is the resolution matrix, and the relationship between the true solution and the particular one is just

$$\tilde{\mathbf{x}} = \mathbf{G}_K\mathbf{G}_K^T\mathbf{x} = \mathbf{x} - \mathbf{Q}_G\boldsymbol{\alpha}_G, \quad \tilde{\mathbf{y}} = \mathbf{G}_K\mathbf{G}_K^T\mathbf{y}, \quad \tilde{\mathbf{n}} = \mathbf{Q}_G\mathbf{Q}_G^T\mathbf{y}\,. \tag{3.4.28}$$

These results are so important, we recapitulate them: (3.4.18) or (3.4.20) is the general solution. There are three vectors involved–one of them, $\mathbf{y}$, is known, and two of them, $\mathbf{x}$, $\mathbf{n}$, are unknown. Because of the assumption that $\mathbf{E}$ has a complete orthonormal set of eigenvectors, all three of these vectors can be expanded, exactly, as

$$\mathbf{x} = \sum_{i=1}^{M}\alpha_i\mathbf{g}_i, \quad \mathbf{n} = \sum_{i=1}^{M}\gamma_i\mathbf{g}_i, \quad \mathbf{y} = \sum_{i=1}^{M}(\mathbf{y}^T\mathbf{g}_i)\mathbf{g}_i\,. \tag{3.4.29}$$

Substituting into (3.4.26), and using the eigenvector property produces

$$\sum_{i=1}^{M} \alpha_i \mathbf{E}\mathbf{g}_i + \sum_{i=1}^{M} \gamma_i \mathbf{g}_i = \sum_{i=1}^{M} (\mathbf{y}^T \mathbf{g}_i)\mathbf{g}_i$$
$$= \sum_{i=1}^{K} \lambda_i \alpha_i \mathbf{g}_i + \sum_{i=1}^{M} \gamma_i \mathbf{g}_i = \sum_{i=1}^{M} (\mathbf{y}^T \mathbf{g}_i)\mathbf{g}_i \,. \tag{3.4.30}$$

From the orthogonality property, we must have

$$\lambda_i \alpha_i + \gamma_i = \mathbf{y}^T \mathbf{g}_i, \quad 1 \le i \le K\,, \tag{3.4.31}$$
$$\gamma_i = \mathbf{y}^T \mathbf{g}_i, \quad K+1 \le i \le M\,. \tag{3.4.32}$$

In dealing with the first relationship, we must make a choice. If we set

$$\gamma_i = \mathbf{g}_i^T \mathbf{n} = 0\,, \quad 1 \le i \le K\,, \tag{3.4.33}$$

the residual norm is made as small as possible by completely eliminating the range vectors from the residual. This choice is motivated by the attempt to satisfy the equations as well as possible but is seen to have elements of arbitrariness. A decision about other possibilities depends upon knowing more about the system and will be the focus of considerable later attention.

It may be objected that this entire development is of little use, because the problems discussed in Chapter 2 produced $\mathbf{E}$ matrices that could not be guaranteed to have complete orthonormal sets of eigenvectors. Indeed, the problems considered produce matrices that are usually nonsquare and for which the eigenvector problem is not even defined.

For arbitrary *square* matrices, the question of when a complete orthonormal set of eigenvectors exists is not difficult to answer but becomes somewhat elaborate; it is treated in all texts on linear algebra. Brogan (1985) has a succinct discussion.

In the general situation, where an $N \times N - \mathbf{E}$ is not symmetric, one must consider cases in which there are N distinct eigenvalues and where some are repeated, and the general approach requires the so-called Jordan form. But we will find a way to avoid these intricacies and yet deal with sets of simultaneous equations of arbitrary dimensions, not just square ones. In the next several sections, a machinery is developed for doing exactly that. Although the mathematics are necessarily somewhat more complicated than is employed in solving the just-determined simultaneous linear equations using a complete orthonormal eigenvector set, this simplest problem provides full analogues to all of the issues in the more general case, and the reader will probably find it helpful to refer back to this situation for insight.

Before leaving this special case, note one more useful property of the eigenvectors/eigenvalues. For the moment, let $\mathbf{G}$ have all its columns, containing both the range and nullspace vectors, with the nullspace vectors being last. It is thus an $M \times M$ matrix. Correspondingly, let $\boldsymbol{\Lambda}$ contain all the eigenvalues on its diagonal, including the zero ones; it, too, is $M \times M$. Then the eigenvector definition (3.4.6) produces

$$\mathbf{EG} = \mathbf{G}\boldsymbol{\Lambda}\,. \tag{3.4.34}$$

Multiply both sides of (3.4.34) by $\mathbf{G}^T$:

$$\mathbf{G}^T\mathbf{EG} = \mathbf{G}^T\mathbf{G}\boldsymbol{\Lambda} = \boldsymbol{\Lambda} \tag{3.4.35}$$

using the orthogonality of $\mathbf{G}$; $\mathbf{G}$ is said to *diagonalize* $\mathbf{E}$. Now multiply both sides of (3.4.35) on the left by $\mathbf{G}$ and on the right by $\mathbf{G}^T$:

$$\mathbf{GG}^T\mathbf{EGG}^T = \mathbf{G}\boldsymbol{\Lambda}\mathbf{G}^T \tag{3.4.36}$$

or, using the orthogonality of $\mathbf{G}$ when it has all its columns,

$$\mathbf{E} = \mathbf{G}\boldsymbol{\Lambda}\mathbf{G}^T\,, \tag{3.4.37}$$

a useful decomposition of $\mathbf{G}$, consistent with the symmetry of $\mathbf{E}$. Recall that $\boldsymbol{\Lambda}$ has zeros on the diagonal corresponding to the zero eigenvalues, and the corresponding rows and columns are entirely zero. Writing out (3.4.37), these zero rows and columns multiply all the nullspace vector columns of $\mathbf{G}$ by zero, and it is found that the nullspace columns of $\mathbf{G}$ can be eliminated, $\boldsymbol{\Lambda}$ reduced to its $K \times K$ form, and the decomposition (3.4.37) is still exact in the form

$$\mathbf{E} = \mathbf{G}_K\boldsymbol{\Lambda}_K\mathbf{G}_K^T\,. \tag{3.4.38}$$

Then the simultaneous equations (3.4.26) are

$$\mathbf{G}_K\boldsymbol{\Lambda}_K\mathbf{G}_K^T\mathbf{x} + \mathbf{n} = \mathbf{y}\,. \tag{3.4.39}$$

Left multiply both sides by $\boldsymbol{\Lambda}_K^{-1}\mathbf{G}_K^T$ (existence of the inverse is guaranteed by the removal of zero eigenvalues), and

$$\mathbf{G}_K^T\mathbf{x} + \boldsymbol{\Lambda}_K^{-1}\mathbf{G}_K^T\mathbf{n} = \boldsymbol{\Lambda}_K^{-1}\mathbf{G}_K^T\mathbf{y}\,. \tag{3.4.40}$$

But $\mathbf{G}_K^T\mathbf{x}$ are the projection of $\mathbf{x}$ onto the range vectors of $\mathbf{E}$, and $\mathbf{G}_K^T\mathbf{n}$ is the same projection of the noise. We have agreed to regard the latter as zero, and we obtain

$$\mathbf{G}_K^T\mathbf{x} = \boldsymbol{\Lambda}_K^{-1}\mathbf{G}_K^T\mathbf{y}\,,$$

the dot products of the range of $\mathbf{E}$ with the solution. Hence, it must be true, because the range vectors are orthonormal, that

$$\tilde{\mathbf{x}} \equiv \mathbf{G}_K\mathbf{G}_K^T\mathbf{x} \equiv \mathbf{G}_K\boldsymbol{\Lambda}_K^{-1}\mathbf{G}_K^T\mathbf{y}\,, \tag{3.4.41}$$

$$\tilde{\mathbf{y}} = \mathbf{E}\tilde{\mathbf{x}} = \mathbf{G}_K\mathbf{G}_K^T\mathbf{y}\,, \tag{3.4.42}$$

which is identical to the particular solution (3.4.17). The residuals are

$$\tilde{\mathbf{n}} = \mathbf{y}-\tilde{\mathbf{y}} = \mathbf{y}-\mathbf{E}\tilde{\mathbf{x}} = \mathbf{y}-\mathbf{G}_K\mathbf{G}_K^T\mathbf{y} = (\mathbf{I}_L-\mathbf{G}_K\mathbf{G}_K^T)\mathbf{y} = \mathbf{Q}_G\mathbf{Q}_G^T\mathbf{y}\,. \tag{3.4.43}$$

One again has $\tilde{\mathbf{n}}^T\tilde{\mathbf{y}} = 0$.

Expression (3.4.43) shows that multiplication by $\mathbf{Q}_K\mathbf{Q}_K^T = \mathbf{I} - \mathbf{G}_K\mathbf{G}_K^T$ projects a vector onto the nullspace of $\mathbf{E}$, just as $\mathbf{G}_K\mathbf{G}_K^T$ projects onto its range. Such operators have an *idempotent* property,

$$(\mathbf{I} - \mathbf{G}_K\mathbf{G}_K^T)^n = (\mathbf{I} - \mathbf{G}_K\mathbf{G}_K^T),\ n = \text{integer}$$

–projection onto the nullspace is invariant. For future reference, notice that the reduced decomposition (3.4.38) permits writing,

$$\mathbf{E}^T(\mathbf{E}\mathbf{E}^T)^{-1}\mathbf{E} = \mathbf{E}(\mathbf{E}^T\mathbf{E})^{-1}\mathbf{E}^T = \mathbf{G}_K\mathbf{G}_K^T\,; \tag{3.4.44}$$

hence, (3.4.41) is

$$\tilde{\mathbf{x}} = \mathbf{E}^T(\mathbf{E}\mathbf{E}^T)^{-1}\mathbf{E}\mathbf{x} = \mathbf{E}(\mathbf{E}^T\mathbf{E})^{-1}\mathbf{E}^T\mathbf{x}\,, \tag{3.4.45}$$

and thus

$$\mathbf{Q}_G\mathbf{Q}_G^T = (\mathbf{I} - \mathbf{E}^T(\mathbf{E}\mathbf{E}^T)^{-1}\mathbf{E}) = (\mathbf{I} - \mathbf{E}(\mathbf{E}^T\mathbf{E})^{-1}\mathbf{E}^T)\,, \tag{3.4.46}$$

and the latter is also idempotent (see (3.3.11)).

The bias of the solution (3.4.17) or (3.4.41) is

$$< \tilde{\mathbf{x}} - \mathbf{x} > = \mathbf{G}_K\boldsymbol{\Lambda}_K^{-1}\mathbf{G}_K^T < \mathbf{y} > - \sum_{i=1}^{N} \alpha_i\mathbf{g}_i = -\mathbf{Q}_G\boldsymbol{\alpha}_G\,, \tag{3.4.47}$$

and so the solution is biased unless $\boldsymbol{\alpha}_G = 0$.

The uncertainty is

$$\begin{aligned}\mathbf{P} = D^2(\tilde{\mathbf{x}} - \mathbf{x}) &=< \mathbf{G}_K\boldsymbol{\Lambda}_K^{-1}\mathbf{G}_K^T(\mathbf{y}_0 + \mathbf{n} - \mathbf{y}_0)(\mathbf{y}_0 + \mathbf{n} - \mathbf{y}_0)^T\mathbf{G}_K\boldsymbol{\Lambda}_K^{-1}\mathbf{G}_K^T > \\ &\quad + < \mathbf{Q}_G\boldsymbol{\alpha}_G\boldsymbol{\alpha}_G^T\mathbf{Q}_G^T > \\ &= \mathbf{G}_K\boldsymbol{\Lambda}_K^{-1}\mathbf{G}_K^T < \mathbf{n}\mathbf{n}^T > \mathbf{G}_K\boldsymbol{\Lambda}_K^{-1}\mathbf{G}_K^T + \mathbf{Q}_G < \boldsymbol{\alpha}_G\boldsymbol{\alpha}_G^T > \mathbf{Q}_G^T \\ &= \mathbf{G}_K\boldsymbol{\Lambda}_K^{-1}\mathbf{G}_K^T\mathbf{R}_{nn}\mathbf{G}_K\boldsymbol{\Lambda}_K^{-1}\mathbf{G}_K^T + \mathbf{Q}_G\mathbf{R}_{\alpha\alpha}\mathbf{Q}_G^T \\ &= \mathbf{C}_{\tilde{x}\tilde{x}} + \mathbf{Q}_G\mathbf{R}_{\alpha\alpha}\mathbf{Q}_G^T\,,\end{aligned} \tag{3.4.48}$$

and $\mathbf{R}_{\alpha\alpha}$ are the second moments of the coefficients of the nullspace vectors.

Under the special circumstances that the residuals, $\mathbf{n}$, are white noise, with $\mathbf{R}_{nn} = \sigma_n^2\mathbf{I}$, (3.4.48) reduces to

$$\mathbf{P} = \sigma_n^2 \mathbf{G}_K \boldsymbol{\Lambda}_K^{-2} \mathbf{G}_K^T + \mathbf{Q}_G \mathbf{R}_{\alpha\alpha} \mathbf{Q}_G^T . \tag{3.4.49}$$

Either case shows that the uncertainty of the minimal solution is made up of two distinct parts. The first part, the solution covariance, $\mathbf{C}_{\tilde{x}\tilde{x}}$, arises owing to the noise present in the observations and generates uncertainty in the coefficients of the range vectors; the second contribution arises from the missing nullspace vector contribution. Either term can dominate. The magnitude of the noise term depends largely upon the ratio of the noise variance, σ_n^2, to the smallest nonzero singular value, λ_K^2. $\mathbf{R}_{\alpha\alpha}$ may be entirely unknown, or an estimate of its value might be available from prior information (e.g., on the basis of the difference between the expected variance of $\mathbf{x}$ and the estimated variance of $\tilde{\mathbf{x}}$).

3.4.2 The Singular Vector Expansion and Singular Value Decomposition

Instead of using the least-squares method already described to find solutions to sets of linear simultaneous equations, consider the possibility, suggested by the eigenvector method, of expanding the solution $\mathbf{x}$ in a set of orthonormal vectors. Equation (3.3.2) involves one vector, $\mathbf{x}$, of dimension N, and two vectors, $\mathbf{y}$, $\mathbf{n}$, of dimension M. We would like to use spanning orthonormal vectors but cannot expect, with two different vector dimensions involved, to use just one set: $\mathbf{x}$ can be expanded exactly in N, N-dimensional orthonormal vectors; and similarly, $\mathbf{y}$ and $\mathbf{n}$ can be exactly represented in M, M-dimensional orthonormal vectors. There are an infinite number of ways to select two such sets. But one particularly useful pair can be found, based upon the structure of $\mathbf{E}$.

The simple development leading to the discussion of the above solutions was based upon the theorem about the eigenvectors of a matrix $\mathbf{E}$ which was symmetric, so that they were guaranteed to be an orthonormal spanning set. Let us construct such a matrix out of an arbitrary $\mathbf{E}$. Put

$$\mathbf{B} = \begin{Bmatrix} \mathbf{0} & \mathbf{E}^T \\ \mathbf{E} & \mathbf{0} \end{Bmatrix}, \tag{3.4.50}$$

which by definition is not only square (dimension $M + N$ by $M + N$) but symmetric. Thus, $\mathbf{B}$ satisfies the theorem just alluded to, and the eigenvalue problem

$$\mathbf{B}\mathbf{q}_i = \lambda_i \mathbf{q}_i \tag{3.4.51}$$

will give rise to $M+N$ orthonormal eigenvectors $\mathbf{q}_i$ (an orthonormal spanning set) whether or not the λ_i are distinct or nonzero. Writing out (3.4.51),

$$\begin{Bmatrix} \mathbf{0} & \mathbf{E}^T \\ \mathbf{E} & \mathbf{0} \end{Bmatrix} \begin{bmatrix} q_{1i} \\ \cdot \\ q_{Ni} \\ q_{N+1,i} \\ \cdot \\ q_{N+M,i} \end{bmatrix} = \lambda_i \begin{bmatrix} q_{1i} \\ \cdot \\ q_{Ni} \\ q_{N+1,i} \\ \cdot \\ q_{N+M,i} \end{bmatrix}, \tag{3.4.52}$$

where q_{pi} is the p-th element of $\mathbf{q}_i$. Taking note of the zero matrices, (3.4.52) may be rewritten

$$\mathbf{E}^T \begin{bmatrix} q_{N+1,i} \\ \cdot \\ q_{N+M,i} \end{bmatrix} = \lambda_i \begin{bmatrix} q_{1i} \\ \cdot \\ q_{Ni} \end{bmatrix}, \tag{3.4.53}$$

$$\mathbf{E} \begin{bmatrix} q_{1i} \\ \cdot \\ q_{Ni} \end{bmatrix} = \lambda_i \begin{bmatrix} q_{N+1,i} \\ \cdot \\ q_{N+M,i} \end{bmatrix}. \tag{3.4.54}$$

Let

$$\mathbf{u}_i = \begin{bmatrix} q_{N+1,i} \\ \cdot \\ q_{N+M,i} \end{bmatrix}, \quad \mathbf{v}_i = \begin{bmatrix} q_{1i} \\ \cdot \\ q_{Ni} \end{bmatrix}, \quad \text{or, } \mathbf{q}_i = \begin{bmatrix} \mathbf{v}_i \\ \mathbf{u}_i \end{bmatrix}, \tag{3.4.55}$$

that is, defining the first N elements of $\mathbf{q}_i$ to be $\mathbf{v}_i$ and the last M to be $\mathbf{u}_i$. Then (3.4.53)–(3.4.54) are

$$\mathbf{E}\mathbf{v}_i = \lambda_i \mathbf{u}_i, \tag{3.4.56}$$

$$\mathbf{E}^T \mathbf{u}_i = \lambda_i \mathbf{v}_i. \tag{3.4.57}$$

If (3.4.56) is left multiplied by $\mathbf{E}^T$, and using (3.4.57), one has

$$\mathbf{E}^T \mathbf{E} \mathbf{v}_i = \lambda_i^2 \mathbf{v}_i. \tag{3.4.58}$$

Similarly, left multiplying (3.4.57) by $\mathbf{E}$ and using (3.4.56) produces

$$\mathbf{E}\mathbf{E}^T \mathbf{u}_i = \lambda_i^2 \mathbf{u}_i. \tag{3.4.59}$$

These last two equations show, surprisingly, that the $\mathbf{u}_i$, $\mathbf{v}_i$ each separately satisfy two independent eigenvector/eigenvalue problems of the square symmetric matrices $\mathbf{E}\mathbf{E}^T$, $\mathbf{E}^T\mathbf{E}$. If one of M, N is much smaller than the other, one need only solve the smaller of the two eigenvalues for either of $\mathbf{u}_i$, $\mathbf{v}_i$, with the other set calculated from (3.4.56) or (3.4.57).

The $\mathbf{u}_i$, $\mathbf{v}_i$ are called *singular vectors*, and the λ_i are the *singular values*. By convention, the λ_i are ordered in decreasing numerical value. Also by convention, they are all nonnegative (taking the negative values of λ_i produces singular vectors differing only by a sign from those corresponding to the positive roots, and thus they are not independent vectors). Equations (3.4.56)–(3.4.57) provide a relationship between each $\mathbf{u}_i$ and each $\mathbf{v}_i$. But because in general, $M \neq N$, there will be more of one set than another. The only way Equations (3.4.56)–(3.4.57) can be consistent is if $\lambda_i = 0$, $i > \min(M, N)$ [where $\min(M, N)$ is read as "the minimum of M and N"]. Suppose $M < N$. Then (3.4.59) is solved for $\mathbf{u}_i$, $1 \leq i \leq M$, and (3.4.57) is used to find the corresponding $\mathbf{v}_i$. There are $N - M$ $\mathbf{v}_i$ that are not generated this way but which can be found using the Gram-Schmidt method.

Let there be K nonzero λ_i; then

$$\mathbf{E}\mathbf{v}_i \neq 0\,, \quad 1 \leq i \leq K\,. \tag{3.4.60}$$

These $\mathbf{v}_i$ are known as the *range vectors of* $\mathbf{E}$ or the *solution range vectors.*" For the remaining N-K vectors $\mathbf{v}_i$,

$$\mathbf{E}\mathbf{v}_i = 0\,, \quad K + 1 \leq i \leq N\,, \tag{3.4.61}$$

known as the *nullspace vectors of* $\mathbf{E}$ or the *nullspace of the solution.* If $K < M$, there will be K of the $\mathbf{u}_i$ such that

$$\mathbf{E}^T\mathbf{u}_i = 0\,, \text{ or } \mathbf{u}_i^T\mathbf{E} \neq 0\,, \quad 1 \leq i \leq K\,, \tag{3.4.62}$$

which are the *range vectors of* $\mathbf{E}^T$ and M-K of the $\mathbf{u}_i$ such that

$$\mathbf{E}^T\mathbf{u}_i = 0\,, \text{ or } \mathbf{u}_i^T\mathbf{E} = 0\,, \quad K + 1 \leq i \leq M\,, \tag{3.4.63}$$

the *nullspace vectors of* $\mathbf{E}^T$ or the *data, or observation, nullspace vectors.* The nullspace of $\mathbf{E}$ is spanned by its nullspace vectors, the range of $\mathbf{E}$ is spanned by the range vectors, etc., in the sense, for example, that an arbitrary vector lying in the range is perfectly described by a sum of the range vectors.

Because the $\mathbf{u}_i$, $\mathbf{v}_i$ are complete orthonormal sets in their corresponding spaces, we can expand $\mathbf{x}$, $\mathbf{y}$, $\mathbf{n}$ without error:

$$\mathbf{x} = \sum_{i=1}^{N} \alpha_i \mathbf{v}_i\,, \quad \mathbf{y} = \sum_{j=1}^{M} \beta_i \mathbf{u}_i\,, \quad \mathbf{n} = \sum_{i=1}^{M} \gamma_i \mathbf{u}_i\,, \tag{3.4.64}$$

where $\mathbf{y}$ has been measured, so that we know $\beta_j = \mathbf{u}_j^T\mathbf{y}$. To find the solution, we need α_i, and to find the noise, we need the γ_i. Substitute (3.4.64) into the equations (3.3.2), and using (3.4.56)–(3.4.57),

$$\sum_{i=1}^{N} \alpha_i \mathbf{E}\mathbf{v}_i + \sum_{i=1}^{M} \gamma_i \mathbf{u}_i = \sum_{i=1}^{K} \alpha_i \lambda_i \mathbf{u}_i + \sum_{i=1}^{M} \gamma_i \mathbf{u}_i = \sum_{i=1}^{M} (\mathbf{u}_i^T\mathbf{y})\mathbf{u}_i\,. \tag{3.4.65}$$

Notice the differing upper limits on the summations. By the orthonormality of the singular vectors, (3.4.65) can be solved as

$$\alpha_i \lambda_i + \gamma_i = \mathbf{u}_i^T \mathbf{y}\,, \quad i = 1 \text{ to } M\,, \tag{3.4.66}$$

$$\alpha_i = (\mathbf{u}_i^T \mathbf{y} - \gamma_i)/\lambda_i\,, \quad \lambda_i \neq 0\,, \quad 1 \leq i \leq K\,. \tag{3.4.67}$$

In these equations, if $\lambda_i \neq 0$, nothing prevents setting $\gamma_i = 0$–that is,

$$\mathbf{u}_i^T \mathbf{n} = 0\,, \quad 1 \leq i \leq K \tag{3.4.68}$$

should we wish, which would have the effect of making the noise norm as small as possible. Then (3.4.67) produces

$$\alpha_i = \frac{\mathbf{u}_i^T \mathbf{y}}{\lambda_i}\,, \quad 1 \leq i \leq K\,. \tag{3.4.69}$$

But, because $\lambda_i = 0$, $i > K$, the only solution for these values of i to (3.4.66) is $\gamma_i = \mathbf{u}_i^T \mathbf{y}$, and α_i is indeterminate. These γ_i are nonzero, meaning that there is always a residual, except in the event (unlikely with real data) that

$$\mathbf{u}_i^T \mathbf{y} = 0\,, \quad K + 1 \leq i \leq N\,. \tag{3.4.70}$$

This last equation is called a *solvability condition* in direct analogy to (3.4.16).

The solution obtained in this manner now has the following form:

$$\tilde{\mathbf{x}} = \sum_{i=1}^{K} \frac{\mathbf{u}_i^T \mathbf{y}}{\lambda_i} \mathbf{v}_i + \sum_{i=K+1}^{N} \alpha_i \mathbf{v}_i\,, \tag{3.4.71}$$

$$\tilde{\mathbf{y}} = \mathbf{E}\tilde{\mathbf{x}} = \sum_{i=1}^{K} (\mathbf{u}_i^T \mathbf{y}) \mathbf{u}_i\,, \tag{3.4.72}$$

$$\tilde{\mathbf{n}} = \sum_{i=K+1}^{M} (\mathbf{u}_i^T \mathbf{y}) \mathbf{u}_i\,. \tag{3.4.73}$$

The coefficients of the last N-K of the $\mathbf{v}_i$ in Equation (3.4.71), the solution nullspace vectors, are arbitrary, representing structures in the solution about which the equations provide no information. A nullspace is always present unless $K = N$. The solution residuals are directly proportional to the nullspace vectors of $\mathbf{E}^T$ and will vanish only if $K = M$, or the solvability conditions are met

Just as in the square symmetric case, no choice of the coefficients of the solution nullspace vectors can have any effect on the size of the residuals. If we choose once again to exercise Occam's Razor, and regard the simplest solution as best, then setting the nullspace coefficients to zero,

$$\tilde{\mathbf{x}} = \sum_{i=1}^{K} \frac{\mathbf{u}_i^T \mathbf{y}}{\lambda_i} \mathbf{v}_i \,. \tag{3.4.74}$$

Along with (3.4.73), this is the *particular-SVD solution* (a terminology explained in the next subsection). It simultaneously minimizes the residuals and the solution norm. With $< \mathbf{n} > = 0$, the bias of (3.4.74) is

$$< \tilde{\mathbf{x}} - \mathbf{x} > = - \sum_{i=K+1}^{N} \alpha_i \mathbf{v}_i \,. \tag{3.4.75}$$

The solution uncertainty is

$$\mathbf{P} = \sum_{i=1}^{K} \sum_{j=1}^{K} \mathbf{v}_i \frac{\mathbf{u}_i^T < \mathbf{n}\mathbf{n}^T > \mathbf{u}_j}{\lambda_i \lambda_j} \mathbf{v}_j^T + \sum_{i=K+1}^{N} \sum_{j=K+1}^{N} \mathbf{v}_i < \alpha_i \alpha_j > \mathbf{v}_j^T \,. \tag{3.4.76}$$

If the noise is white with variance σ_n^2, or if a row-scaling matrix $\mathbf{W}^{-T/2}$ has been applied to make it so, then (3.4.76) becomes

$$\mathbf{P} = \sum_{i=1}^{K} \sigma_n^2 \frac{\mathbf{v}_i \mathbf{v}_i^T}{\lambda_i^2} + \sum_{i=K+1}^{N} < \alpha_i^2 > \mathbf{v}_i \mathbf{v}_i^T \tag{3.4.77}$$

where it was also assumed that $< \alpha_i \alpha_j > = < \alpha_i^2 > \delta_{ij}$ in the nullspace. The influence of very small singular values on the uncertainty is clear: In the solution (3.4.71) or (3.4.74), there are error terms $\mathbf{u}_i^T \mathbf{n}/\lambda_i$ that are greatly magnified by small or nearly vanishing singular values, introducing large terms proportional to σ_n^2/λ_i^2 into (3.4.77).

The decision to set to zero the projection of the noise onto the range of $\mathbf{E}^T$ as we did in Equations (3.4.68), (3.4.73) needs to be examined. Should we make some other choice, the solution norm would decrease, but the residual norm would increase. Determining the desirability of such a tradeoff requires understanding of the noise structure–in particular, (3.4.68) imposes rigid structures onto the residuals.

3.4.2.1 The Singular Value Decomposition

The singular vectors and values have been used to provide a convenient pair of orthonormal spanning sets to solve an arbitrary set of simultaneous equations. The vectors and values have another use, however, in providing a decomposition of $\mathbf{E}$.

Define $\boldsymbol{\Lambda}$ as the $M \times N$ matrix whose diagonal elements are the λ_i, in order of descending values in the same order, $\mathbf{U}$ as the $M \times M$ matrix whose columns are the $\mathbf{u}_i$, $\mathbf{V}$ as the $N \times N$ matrix whose columns are the $\mathbf{v}_i$ and whose other elements are 0. As an example, suppose $M = 3$, $N = 4$; then

$$\boldsymbol{\Lambda} = \begin{Bmatrix} \lambda_i & 0 & 0 & 0 \\ 0 & \lambda_2 & 0 & 0 \\ 0 & 0 & \lambda_3 & 0 \end{Bmatrix}.$$

Alternatively, if $M = 4$, $N = 3$

$$\begin{Bmatrix} \lambda_1 & 0 & 0 \\ 0 & \lambda_2 & 0 \\ 0 & 0 & \lambda_3 \\ 0 & 0 & 0 \end{Bmatrix},$$

therefore extending the definition of a diagonal matrix to nonsquare ones.

Precisely as with matrix $\mathbf{G}$ considered above, column orthonormality of $\mathbf{U}$, $\mathbf{V}$ implies that these matrices are orthogonal,

$$\mathbf{U}\mathbf{U}^T = \mathbf{I}_M\,, \tag{3.4.78}$$

$$\mathbf{U}^T\mathbf{U} = \mathbf{I}_M\,, \tag{3.4.79}$$

$$\mathbf{V}\mathbf{V}^T = \mathbf{I}_N\,, \tag{3.4.80}$$

$$\mathbf{V}^T\mathbf{V} = \mathbf{I}_N\,. \tag{3.4.81}$$

(It follows that $\mathbf{U}^{-1} = \mathbf{U}^T$, etc.) As with $\mathbf{G}$ in Section 3.4.1, should one or more columns of $\mathbf{U}$, $\mathbf{V}$ be deleted, the matrices will become semi-orthogonal.

The relations (3.4.56), (3.4.57) to (3.4.58), (3.4.59) can be written compactly as:

$$\mathbf{E}\mathbf{V} = \mathbf{U}\boldsymbol{\Lambda}\,, \tag{3.4.82}$$

$$\mathbf{E}^T\mathbf{U} = \mathbf{V}\boldsymbol{\Lambda}^T\,, \tag{3.4.83}$$

$$\mathbf{E}^T\mathbf{E}\mathbf{V} = \mathbf{V}\boldsymbol{\Lambda}^T\boldsymbol{\Lambda}\,, \tag{3.4.84}$$

$$\mathbf{E}\mathbf{E}^T\mathbf{U} = \mathbf{U}\boldsymbol{\Lambda}\boldsymbol{\Lambda}^T\,. \tag{3.4.85}$$

If we left multiply (3.4.82) by $\mathbf{U}^T$ and invoke (3.4.79), then

$$\mathbf{U}^T\mathbf{E}\mathbf{V} = \mathbf{U}^T\mathbf{U}\boldsymbol{\Lambda}\mathbf{U}\mathbf{V}^T = \boldsymbol{\Lambda}\,. \tag{3.4.86}$$

So $\mathbf{U}$, $\mathbf{V}$ diagonalize $\mathbf{E}$ (with *diagonal* having the extended meaning for a rectangular matrix as defined above.)

Right multiplying (3.4.82) by $\mathbf{V}^T$ produces

$$\mathbf{E} = \mathbf{U}\boldsymbol{\Lambda}\mathbf{V}^T\,. \tag{3.4.87}$$

This last equation represents a decomposition, called the *singular value decomposition* (SVD) of an arbitrary matrix, into two orthogonal matrices, $\mathbf{U}$, $\mathbf{V}$, and a usually nonsquare diagonal matrix $\boldsymbol{\Lambda}$.

There is one further step to take. Notice that for a rectangular $\boldsymbol{\Lambda}$, as in the examples above, one or more rows or columns must be all zero, depending upon the shape of the matrix. If any of the $\lambda_i = 0$, $i < \min(M, N)$, the corresponding rows or columns also will be all zeros. Let K be the number of nonvanishing singular values (the rank of $\mathbf{E}$). By inspection (multiplying it out), one finds that the last N-K columns of $\mathbf{V}$ and the last M-K columns of $\mathbf{U}$ are multiplied by zeros only. If these columns are dropped entirely from $\mathbf{U}$, $\mathbf{V}$ so that $\mathbf{U}$ becomes $M \times K$ and $\mathbf{V}$ becomes $N \times K$, and reducing $\boldsymbol{\Lambda}$ to a $K \times K$ square matrix, then the representation (3.4.87) remains exact, in the form

$$\mathbf{E} = \mathbf{U}_K \boldsymbol{\Lambda}_K \mathbf{V}_K^T, \tag{3.4.88}$$

the subscript indicating the number of columns, where $\mathbf{U}_K$, $\mathbf{V}_K$ are then only semi-orthogonal, and $\boldsymbol{\Lambda}_K$ is now square. Equation (3.4.88) should be compared to (3.4.38).

The singular value decomposition for arbitrary nonsquare matrices is apparently due to Carl Eckart (Eckart & Young, 1939; see the historical discussions in Haykin, 1986; Klema & Laub, 1980; or Stewart, 1993).[7] Derivations are given by Lanczos (1961), Noble and Daniel (1977), Strang (1986), and many other recent books on applied linear algebra. The crucial role it plays in inverse methods appears to have been first noticed by Wiggins (1972).

The SVD solution can be obtained by direct matrix manipulation rather than vector by vector. Consider once again finding the solution to the simultaneous equations (3.3.2), but first write $\mathbf{E}$ in its reduced SVD,

$$\mathbf{U}_K \boldsymbol{\Lambda}_K \mathbf{V}_K^T \mathbf{x} + \mathbf{n} = \mathbf{y}. \tag{3.4.89}$$

Left multiplying by $\mathbf{U}_K^T$ and invoking the semi-orthogonality of $\mathbf{U}_K$ produces

$$\boldsymbol{\Lambda}_K \mathbf{V}_K^T \mathbf{x} + \mathbf{U}_K^T \mathbf{n} = \mathbf{U}_K^T \mathbf{y}. \tag{3.4.90}$$

The inverse of $\boldsymbol{\Lambda}_K$ (square with all nonzero diagonal elements) is easily computed, and

$$\mathbf{V}_K^T \mathbf{x} + \boldsymbol{\Lambda}_K^{-1} \mathbf{U}_K^T \mathbf{n} = \boldsymbol{\Lambda}_K^{-1} \mathbf{U}_K^T \mathbf{y}. \tag{3.4.91}$$

But $\mathbf{V}_K^T \mathbf{x}$ is the dot product of the first K of the $\mathbf{v}_i$ with the unknown $\mathbf{x}$. Equation (3.4.91) thus represents statements about the relationship between

7 Eckart, a physicist turned oceanographer, had a somewhat controversial career. The SVD may turn out to have been his most important, if least credited, contribution.

dot products of the unknown vector, $\mathbf{x}$, with a set of orthonormal vectors, and therefore must represent the expansion coefficients of the solution in those vectors. If we set

$$\mathbf{U}_K^T \mathbf{n} = 0\,, \tag{3.4.92}$$

then

$$\mathbf{V}_K^T \mathbf{x} = \boldsymbol{\Lambda}_K^{-1} \mathbf{U}_K^T \mathbf{y}\,, \tag{3.4.93}$$

and hence

$$\tilde{\mathbf{x}} = \mathbf{V}_K \boldsymbol{\Lambda}_K^{-1} \mathbf{U}_K^T \mathbf{y}\,, \tag{3.4.94}$$

identical to the solution (3.4.74), which the reader is urged to confirm by writing it out explicitly. Substituting this solution into (3.4.89),

$$\mathbf{U}_K \boldsymbol{\Lambda}_K \mathbf{V}_K^T \mathbf{V}_K \boldsymbol{\Lambda}_K^{-1} \mathbf{U}_K^T \mathbf{y} + \mathbf{n} = \mathbf{U}_K \mathbf{U}_K^T \mathbf{y} + \mathbf{n} = \mathbf{y}$$

or

$$\tilde{\mathbf{n}} = (\mathbf{I} - \mathbf{U}_K \mathbf{U}_K^T)\mathbf{y}\,. \tag{3.4.95}$$

Let the full $\mathbf{U}$ and $\mathbf{V}$ matrices be rewritten as

$$\mathbf{U} = \{\mathbf{U}_K \quad \mathbf{Q}_u\}\,, \tag{3.4.96}$$

$$\mathbf{V} = \{\mathbf{V}_K \quad \mathbf{Q}_v\} \tag{3.4.97}$$

where $\mathbf{Q}_u$, $\mathbf{Q}_v$ contain the nullspace vectors. Then

$$\mathbf{E}\tilde{\mathbf{x}} + \tilde{\mathbf{n}} = \mathbf{y}\,, \quad \mathbf{E}\tilde{\mathbf{x}} = \tilde{\mathbf{y}}\,,$$

$$\tilde{\mathbf{y}} = \mathbf{U}_K \mathbf{U}_K^T \mathbf{y}\,, \quad \tilde{\mathbf{n}} = \mathbf{Q}_u \mathbf{Q}_u^T \mathbf{y} = \sum_{j=K+1}^{M} (\mathbf{u}_i^T \mathbf{y})\mathbf{u}_i\,, \tag{3.4.98}$$

which is identical to (3.4.72). Note that $\mathbf{Q}_u \mathbf{Q}_u^T = (\mathbf{I} - \mathbf{U}_K \mathbf{U}_K^T)$, $\mathbf{Q}_v \mathbf{Q}_v^T = (\mathbf{I} - \mathbf{V}_K \mathbf{V}_K^T)$, which are idempotent. The general solution is

$$\tilde{\mathbf{x}} = \mathbf{V}_K \boldsymbol{\Lambda}_K^{-1} \mathbf{U}_K^T \mathbf{y} + \mathbf{Q}_v \boldsymbol{\alpha}_v \tag{3.4.99}$$

where $\boldsymbol{\alpha}_v$ is now restricted to being the vector of coefficients of the nullspace vectors.

The solution uncertainty of (3.4.99) is

$$\begin{aligned}\mathbf{P} &= \mathbf{V}_K \boldsymbol{\Lambda}_K^{-1} \mathbf{U}_K^T < \mathbf{n}\mathbf{n}^T > \mathbf{U}_K \boldsymbol{\Lambda}_K^{-1} \mathbf{V}_K^T + \mathbf{Q}_v < \boldsymbol{\alpha}_v \boldsymbol{\alpha}_v^T > \mathbf{Q}_v^T \\ &= \mathbf{C}_{\tilde{x}\tilde{x}} + \mathbf{Q}_v < \boldsymbol{\alpha}_v \boldsymbol{\alpha}_v^T > \mathbf{Q}_v^T\end{aligned} \tag{3.4.100}$$

or

$$\mathbf{P} = \sigma_n^2 \mathbf{V}_K \boldsymbol{\Lambda}_K^{-2} \mathbf{V}_K^T + \mathbf{Q}_v \mathbf{R}_{\alpha\alpha} \mathbf{Q}_v^T \tag{3.4.101}$$

for white noise. The uncertainty of the residuals for white noise is

$$\mathbf{P}_{nn} = \sigma_n^2(\mathbf{I} - \mathbf{U}_K\mathbf{U}_K^T)\,. \tag{3.4.102}$$

Solution of simultaneous equations by SVD has several important advantages. Among other features, we can write down within one algebraic formulation the solution to systems of equations that can be under-, over-, or just-determined.[8] Unlike the eigenvalue/eigenvector solution for the square system, the singular values (eigenvalues) are always nonnegative and real, and the singular vectors (eigenvectors) can always be made a complete orthonormal set. Neither of these statements is true for the conventional eigenvector problem. Most important, however, the relations (3.4.56), (3.4.57) are a specific, quantitative statement of the connection between a set of orthonormal structures in the data and the corresponding presence of orthonormal structures in the solution. These relations provide a very powerful diagnostic method for understanding precisely why the solution takes on the form it does.

3.4.3 Some Simple Examples

The simplest underdetermined system is 1×2. Suppose $x_1 - 2x_2 = 3$ so that

$$\mathbf{E} = \{1 \quad -2\},\ \mathbf{U} = \{1\},\ \mathbf{V} = \left\{\begin{matrix} .447 & -.894 \\ -.894 & -.447 \end{matrix}\right\},\ \lambda_1 = 2.24$$

where the second column of $\mathbf{V}$ is in the nullspace of $\mathbf{E}$. The general solution is $\tilde{\mathbf{x}} = [.6 \quad -1.2]^T + \alpha_2\mathbf{v}_2$. Because $K = 1$ is the only possible choice here, it is readily confirmed that this solution satisfies the equation exactly, and a data nullspace is not possible.

The most elementary overdetermined problem is 2×1. Suppose

$$x_1 = 1\,,$$
$$x_1 = 3\,.$$

The appearance of two such equations is possible if there is noise in the observations, and they are written more properly as

$$x_1 + n_1 = 1\,,$$
$$x_1 + n_2 = 3\,.$$

[8] True, too, of the generalized least-squares formulation (3.3.66)–(3.3.68) or (3.3.38)–(3.3.40).

$\mathbf{E} = \{1 \quad 1\}^T$, $\mathbf{E}^T\mathbf{E}$ represents the eigenvalue problem of the smaller dimension, and

$$\mathbf{U} = \begin{Bmatrix} .707 & -.707 \\ .707 & .707 \end{Bmatrix}, \quad \mathbf{V} = \{1\}, \; \lambda_1 = \sqrt{2}$$

where the second column of $\mathbf{U}$ lies in the data nullspace, there being no solution nullspace. The general solution is $\mathbf{x} = x_1 = 2$, which if substituted back into the original equations produces

$$\mathbf{E}\tilde{\mathbf{x}} = \begin{bmatrix} 2 \\ 2 \end{bmatrix} = \tilde{\mathbf{y}}\,,$$

and hence there are residuals $\tilde{\mathbf{n}} = \tilde{\mathbf{y}} - \mathbf{y} = [1 \quad -1]^T$, which are necessarily proportional to $\mathbf{u}_2$. Evidently no other solution could produce a smaller l_2 norm residual than this one. The SVD produced a solution that compromised the contradiction between the two equations and is physically sensible.

The possibility of $K < M$, $K < N$ simultaneously is also easily seen. Consider the system:

$$\begin{Bmatrix} 1 & -2 & 1 \\ 3 & 2 & 1 \\ 4 & 0 & 2 \end{Bmatrix} \mathbf{x} = \begin{bmatrix} 1 \\ -1 \\ 2 \end{bmatrix},$$

which appears superficially just-determined. But the singular values are $\lambda_1 = 5.67$, $\lambda_2 = 2.80$, $\lambda_3 = 0$. The vanishing of the third singular values means that the row and column vectors are not linearly independent sets (not spanning sets)–indeed, the third row vector is just the sum of the first two (but the third element of $\mathbf{y}$ is not the sum of the first two, making the equations inconsistent). Thus, there are both solution and data nullspaces, which the reader might wish to find. With a vanishing singular value, $\mathbf{E}$ can be written exactly using only two columns of $\mathbf{U}$, $\mathbf{V}$ and the linear dependence is given explicitly as $\mathbf{u}_3^T\mathbf{E} = 0$.

Consider now the underdetermined system

$$x_1 + x_2 - 2x_3 = 1\,,$$
$$x_1 + x_2 - 2x_3 = 2\,,$$

which has no conventional solution at all, being a contradiction, and is thus simultaneously underdetermined and incompatible. If one of the coefficients is modified by a very small quantity, ϵ, to produce

$$x_1 + x_2 - (2+\epsilon)x_3 = 1\,,$$
$$x_1 + x_2 - 2x_3 = 2\,, \tag{3.4.103}$$

not only is there a solution, there are an infinite number of them, which the reader should confirm by computing the basic SVD solution and the nullspace. Thus, the slightest perturbation in the coefficients has made the system jump from having no solution to having an infinite number, an obviously disconcerting situation. Such a system is *ill-conditioned.* How would we know the system is ill-conditioned? There are several indicators. First, the ratio of the two singular values is determined by ϵ. In (3.4.103), if we take $\epsilon = 10^{-10}$, the two singular values are $\lambda_1 = 3.46$, $\lambda_2 = 4.1 \times 10^{-11}$, an immediate warning that the two equations are nearly linearly dependent. (In a mathematical problem, the nonvanishing of the second singular value is enough to assure a solution. As will be discussed later, the inevitable slight errors in $\mathbf{y}$ suggest small singular values are best regarded as actually being zero.)

A similar problem exists with the system:

$$x_1 + x_2 - 2x_3 = 1\,,$$
$$x_1 + x_2 - 2x_3 = 1\,,$$

which has an infinite number of solutions. But the change to

$$x_1 + x_2 - 2x_3 = 1\,,$$
$$x_1 + x_2 - 2x_3 = 1 + \epsilon$$

for arbitrarily small ϵ produces a system with no solutions in the conventional mathematical sense, although the SVD will handle the system without any difficulty.

Problems like these are simple examples of the practical issues that arise once one recognizes that unlike mathematical textbook problems, observational ones always contain inaccuracies; any discussion of how to handle data in the presence of mathematical relations must account for these inaccuracies as intrinsic, not as something to be regarded as an afterthought. But the SVD itself is sufficiently powerful that it always contains the information to warn of ill-conditioning, and by determination of K to cope with it, producing useful solutions.

3.4.3.1 The Neumann Problem

Consider the classical Neumann problem described in Chapter 1. The problem is to be solved on a 10×10 grid as stated in Equation (1.2.7), $\mathbf{A}_3\boldsymbol{\phi} = \mathbf{d}_3$. The singular values of $\mathbf{A}_3$ are plotted in Figure 3–10a; the largest one is $\lambda_1 = 7.8$, and the smallest nonzero one is $\lambda_{99} = 0.08$. As expected, $\lambda_{100} = 0$. The singular vector $\mathbf{v}_{100}$ corresponding to the zero singular value

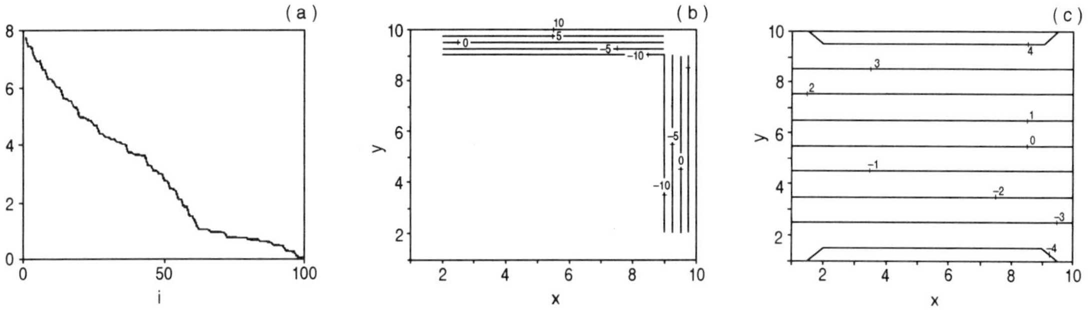

Figure 3–10. (a) Singular values of the coefficient matrix of the numerical Neumann problem. All λ_i are nonzero except the last one. (b) The 100th nullspace vector $\mathbf{u}_{100}$ of $\mathbf{A}_3$ such that $\mathbf{u}_{100}^T \mathbf{A}_3 = \mathbf{u}_{100}^T \mathbf{y} = 0$ defines the consistency or solvability conditions for the Neumann problem–that there be no net influx of source material from interior sources or across the boundaries. The form shown can be understood from the physical requirement that what enters across the left and bottom boundaries plus interior sources must balance what leaves across the top and right boundaries. The corresponding $\mathbf{v}_{100}$ is a constant. (c) Particular-SVD solution to the Neumann problem, with equal and opposite fluxes across the two horizontal walls, and vanishing flux across both vertical walls and in the interior. This specification satisfies the solvability conditions and leaves no residuals.

is not shown, because also as expected, it is a constant; $\mathbf{u}_{100}$, shown in Figure 3–10b, is not a constant and has considerable structure, which provides the solvability condition for the Neumann problem, $\mathbf{u}_{100}^T\mathbf{y} = 0$. The physical origin of the solvability condition is readily understood: The Neumann boundary conditions prescribe boundary flux rates, and the sum of the interior source strengths plus the boundary flux rates must equal zero; otherwise, no steady state is possible. If the boundary conditions are homogeneous, then no flow takes place through the boundary, and the interior sources must sum to zero. In particular, the value of $\mathbf{u}_{100}$ on the interior grid points is constant. The Neumann problem is thus a forward problem requiring one to deal with both a solution nullspace and a "data" solvability condition.

As an example of solution by the SVD, let there be unit positive flux into the box on the bottom boundary, unit positive flux out on the top and no interior sources. The resulting particular SVD solution is shown in Figure 3–10c. No residuals are left because the system was constructed as fully consistent, and there is an arbitrary constant that can be added ($\mathbf{v}_{100}$). The reader may wish to experiment with incompatible specifications for this problem. This is an example of a forward problem solved using an inverse method.

Related inverse problems are also easily formulated. The simplest of all would assert that ϕ is known and $\mathbf{d}_3$ is to be determined. In the present situation, one confirms that multiplication of the solution in Figure 3–10c by $\mathbf{A}_3$ produces $\mathbf{d}_3$. One can do a number of interesting experiments with the SVD. For example, if the equations imposing the boundary values are dropped, the resulting range vectors, $\mathbf{v}_i$, describe the particular solution of the partial differential equation, and the nullspace vectors describe the homogeneous one. In this way, one can pick apart the structure of the solution. A more interesting possibility is to withhold knowledge of the boundary conditions and ask for their determination, given the interior solution.

3.4.3.2 Relation of Least Squares to the SVD

What is the relationship of the SVD solution to the least-squares solutions? To some extent, the answer is already obvious from the orthonormality of the two sets of singular vectors. We begin by first asking when the simple least-squares solution will exist? Consider first the formally overdetermined problem, $M > N$. The solution (3.3.6) exists if and only if the matrix inverse exists. Substituting the SVD for $\mathbf{E}$, one finds

$$\left(\mathbf{E}^T\mathbf{E}\right)^{-1} = \left(\mathbf{V}_N\boldsymbol{\Lambda}_N^T\mathbf{U}_N^T\mathbf{U}_N\boldsymbol{\Lambda}_N\mathbf{V}_N^T\right)^{-1} = \left(\mathbf{V}_N\boldsymbol{\Lambda}_N^2\mathbf{V}_N^T\right)^{-1} \qquad (3.4.104)$$

where the semi-orthogonality of $\mathbf{U}_N$ has been used. Suppose that $K = N$, its maximum possible value; then $\boldsymbol{\Lambda}_N^2$ is $N \times N$ with *all nonzero diagonal elements* λ_i^2. The inverse in (3.4.104) may be found by inspection, using $\mathbf{V}_N\mathbf{V}_N^T = \mathbf{I}_N$,

$$\left(\mathbf{E}^T\mathbf{E}\right)^{-1} = \mathbf{V}_N\boldsymbol{\Lambda}_N^{-2}\mathbf{V}_N^T\,. \qquad (3.4.105)$$

Then the solution (3.3.6) becomes

$$\tilde{\mathbf{x}} = (\mathbf{V}_N\boldsymbol{\Lambda}_N^{-2}\mathbf{V}_N^T)\mathbf{V}_N\boldsymbol{\Lambda}_N\mathbf{U}_N^T = \mathbf{V}_N\boldsymbol{\Lambda}_N^{-1}\mathbf{U}_N^T\mathbf{y}\,, \qquad (3.4.106)$$

which is identical to the SVD solution (3.4.94). If $K < N$, $\boldsymbol{\Lambda}_N^2$ has at least one zero on the diagonal, no matrix inverse exists, and the conventional least-squares solution is not defined. The condition for its existence is thus $K = N$, the so-called *full rank overdetermined* case. The condition $K < N$ is called *rank deficient.* The dependence of the least-squares solution magnitude upon the possible presence of very small, but nonvanishing, singular values is obvious.

That the full-rank overdetermined case is unbiased, as previously asserted, can now be seen from

$$< \tilde{\mathbf{x}} - \mathbf{x} > = \sum_{i=1}^{N} \frac{(\mathbf{u}_i^T < \mathbf{y} >)}{\lambda_i}\mathbf{v}_i - \mathbf{x} = \sum_{i=1}^{N} \frac{\mathbf{u}_i^T\mathbf{y}_0}{\lambda_i}\mathbf{v}_i - \mathbf{x} = \mathbf{0}\,,$$

if $< \mathbf{n} > = \mathbf{0}$, assuming that the correct $\mathbf{E}$ is being used.

The identity of the SVD solution and the overdetermined full-rank solution (3.3.6) is also readily shown by directly substituting

$$\mathbf{x} = \sum_{i=1}^{N} \alpha_i \mathbf{v}_i \tag{3.4.107}$$

into the objective function (3.3.2), using the relation (3.4.56) and the orthogonality of $\mathbf{u}_i$. One finds the minimum at $\alpha_i = \mathbf{u}_i^T \mathbf{y}/\lambda_i$, $\lambda_i \neq 0$. If any singular value vanishes, the vector orthogonality proves that no other choice of α_i, $i \leq K$, can reduce J further, and so the particular-SVD solution produces the best possible minimum even when the system is rank deficient.

Now consider another least-squares problem, the one with the conventional purely underdetermined least-squares solution (3.3.53). When does that exist? Substituting the SVD into (3.3.53),

$$\begin{aligned}\tilde{\mathbf{x}} &= \mathbf{V}_M \boldsymbol{\Lambda}_M \mathbf{U}_M^T (\mathbf{U}_M \boldsymbol{\Lambda}_M \mathbf{V}_M^T \mathbf{V}_M \boldsymbol{\Lambda}_M^T \mathbf{U}_M^T)^{-1} \mathbf{y} \\ &= \mathbf{V}_M \boldsymbol{\Lambda}_M \mathbf{U}_M^T (\mathbf{U}_M \boldsymbol{\Lambda}_M^2 \mathbf{U}_M^T)^{-1} \mathbf{y}\,. \end{aligned} \tag{3.4.108}$$

Again, the matrix inverse exists if and only if $\boldsymbol{\Lambda}_M^2$ has all nonzero diagonal elements, which occurs only when $K = M$. Under that specific condition, the inverse is obtained by inspection, and

$$\tilde{\mathbf{x}} = \mathbf{V}_M \boldsymbol{\Lambda}_M \mathbf{U}_M^T (\mathbf{U}_M \boldsymbol{\Lambda}_M^{-2} \mathbf{U}_M^T) \mathbf{y} = \mathbf{V}_M \boldsymbol{\Lambda}_M^{-1} \mathbf{U}_M^T \mathbf{y} \tag{3.4.109}$$

$$\tilde{\mathbf{n}} = 0\,, \tag{3.4.110}$$

which is once again the particular-SVD solution (3.4.94)–with $K = M$ and the nullspace coefficients set to zero. This situation is usually referred to as the *full-rank underdetermined case.* Again, the possible influence of small singular values is apparent, and an arbitrary sum of nullspace vectors can be added to (3.4.109).

The bias of (3.4.108) is given by the nullspace elements, and its formal uncertainty is from the nullspace contribution. With $\tilde{\mathbf{n}} = \mathbf{0}$, the formal sample noise variance vanishes, and the particular-SVD solution covariance $\mathbf{C}_{\tilde{x}\tilde{x}}$ would be zero, if the sample variance is used. If the formally overdetermined problem is converted to an exact underdetermined one as in Equation (3.3.42), then the uncertainty is calculated solely from the nullspace contribution.

The particular-SVD solution thus coincides with the two simplest forms of least-squares solution and generalizes both of them to the case where the matrix inverses do not exist. *All of the structure imposed by the SVD, in particular the restriction on the residuals in* (3.4.68), *is present in the least-squares solutions.* If the system is not of full rank, then the simple least-squares solutions do not exist. The SVD generalizes these results by determining the elements of the solution lying in the range of $\mathbf{E}$ and giving an explicit structure for the resulting nullspace vectors.

The SVD has much flexibility. For example, it permits one to modify the simplest underdetermined solution to remove its greatest shortcoming, the necessity that $\tilde{\mathbf{n}} = \mathbf{0}$. One simply truncates the solution (3.4.74) at $K' < K < M$, thus assigning all vectors $\mathbf{v}_i$, $K' + 1 \leq i \leq K$, to an *effective nullspace* (or substitutes K' for K everywhere). The resulting residual is then

$$\tilde{\mathbf{n}} = \sum_{i=K'+1}^{K} (\mathbf{u}_i^T \mathbf{y}) \mathbf{u}_i \,, \tag{3.4.111}$$

with an uncertainty for $\tilde{\mathbf{x}}$, $\tilde{\mathbf{n}}$ given by (3.4.100)–(3.4.102), but with the upper limit being K' rather than K. Such truncation has the effect of reducing the solution covariance contribution to the uncertainty [recall (3.4.77)] but increasing the contribution owing to the nullspace (and increasing the potential bias). In the presence of singular values that are small compared to σ_n, the resulting overall reduction in uncertainty may be very great.

The solution now consists of three parts,

$$\tilde{\mathbf{x}} = \sum_{i=1}^{K'} \frac{\mathbf{u}_i^T \mathbf{y}}{\lambda_i} \mathbf{v}_i + \sum_{i=K'+1}^{K} \alpha_i \mathbf{v}_i + \sum_{i=K+1}^{N} \alpha_i \mathbf{v}_i \,, \tag{3.4.112}$$

where the middle sum contains the terms appearing with singular values too small to be employed for the given noise, and the third sum is the strict nullspace. Usually, one lumps the two nullspace sums together. The first sum, by itself, represents the particular-SVD solution in the presence of noise.

This consideration is extremely important: It says that despite the mathematical condition $\lambda_i \neq 0$, some structures in the solution cannot be estimated with sufficient reliability to be useful. The *effective rank* is then not the same as the mathematical rank.

Evidently, truncation of the SVD offers a simple method for controlling the ratio of solution and residual norms: As the nullspace grows by reducing K', it follows that the solution norm necessarily is reduced and that the residuals must grow, along with the size of the solution nullspace. The issue of how to choose K'–that is, *rank determination* in practice is an interesting one to which we will return.

The full-rank overdetermined least-squares solution leaves no solution nullspace but does produce a data nullspace (unless the special solvability conditions are met). In this case, we have the identity,

$$(\mathbf{I} - \mathbf{E}(\mathbf{E}^T\mathbf{E})^{-1}\mathbf{E}^T) = (\mathbf{I} - \mathbf{U}_M\mathbf{U}_M^T) = \mathbf{Q}_u\mathbf{Q}_u^T\,, \tag{3.4.113}$$

the idempotent projector of the data onto the nullspace of $\mathbf{E}^T$ (the matrix inverse is guaranteed to exist by the full-rank assumption). In the full-rank underdetermined case, there is no data nullspace, but there is a solution nullspace. In that situation, the relevant identity is

$$(\mathbf{I} - \mathbf{E}^T(\mathbf{E}\mathbf{E}^T)^{-1}\mathbf{E}) = (\mathbf{I} - \mathbf{V}_N\mathbf{V}_N^T) = \mathbf{Q}_v\mathbf{Q}_v^T\,, \tag{3.4.114}$$

the idempotent projector of $\mathbf{x}$ onto the solution nullspace. These identities follow immediately from introduction of the SVD and the definitions (3.4.96)–(3.4.97) and should be compared to the analogous result (3.4.46) for a square symmetric $\mathbf{E}$. The identities remain valid with both N, M replaced by the actual rank, K, for any $K \leq \min(N, M)$. Both identities prove useful in Chapter 6 for interpreting the Kalman filter and associated smoothers.

3.4.3.3 Row and Column Scaling

The effects on the least-squares solutions of the row and column scaling can now be understood. Suppose we have two equations

$$\begin{Bmatrix} 1 & 1 & 1 \\ 1 & 1.01 & 1 \end{Bmatrix} \begin{bmatrix} x_1 \\ x_2 \\ x_3 \end{bmatrix} + \begin{bmatrix} n_1 \\ n_2 \end{bmatrix} = \begin{bmatrix} y_1 \\ y_2 \end{bmatrix},$$

and there is no information about the noise covariance and no row scaling is reasonable, so $\mathbf{W} = \mathbf{I}$. The SVD of $\mathbf{E}$ is

$$\mathbf{U} = \begin{Bmatrix} 0.7059 & -0.7083 \\ 0.7083 & 0.7059 \end{Bmatrix}, \quad \mathbf{V} = \begin{Bmatrix} 0.5764 & -0.4096 & 0.7071 \\ 0.5793 & 0.8151 & 0.0000 \\ 0.5764 & -0.4096 & -0.7071 \end{Bmatrix},$$

$$\lambda_1 = 2.4536, \ \lambda_2 = .0058\,.$$

The SVD solutions, choosing ranks $K' = 1, 2$ in succession, are very nearly

$$\tilde{\mathbf{x}} \sim \frac{0.71(y_1 + y_2)}{2.45} \begin{bmatrix} .58 \\ .58 \\ .58 \end{bmatrix},$$

$$\sim \frac{0.71(y_1+y_2)}{2.45}\begin{bmatrix}.58\\.58\\.58\end{bmatrix}+\frac{0.71(y_1-y_2)}{.0058}\begin{bmatrix}-.41\\.82\\.41\end{bmatrix}, \tag{3.4.115}$$

respectively, so that the first term simply averages the two measurements, y_i, and the difference between them contributes with great uncertainty in the second term of the rank 2 solution owing to the very small singular value.

Now suppose that the covariance matrix of the noise is known to be

$$\mathbf{R}_{nn}=\begin{Bmatrix}1 & .999999\\ .999999 & 1\end{Bmatrix}$$

(an extreme case, chosen for illustrative purposes). Then put $\mathbf{W}=\mathbf{R}_{nn}$,

$$\mathbf{W}^{1/2}=\begin{Bmatrix}1.0000 & 1.0000\\ 0 & 0.0014\end{Bmatrix}, \quad \mathbf{W}^{-T/2}=\begin{Bmatrix}1.0000 & 0\\ -707.1063 & 707.1070\end{Bmatrix}.$$

The new system to be solved is

$$\begin{Bmatrix}1.0000 & 1.0000 & 1.0000\\ 0.0007 & 7.0718 & 0.0007\end{Bmatrix}\begin{bmatrix}x_1\\x_2\\x_3\end{bmatrix}=\begin{bmatrix}y_1\\ 707.1(-y_1+y_2)\end{bmatrix}$$

whose SVD is

$$\mathbf{U}=\begin{Bmatrix}0.1456 & 0.9893\\ 0.9893 & -0.1456\end{Bmatrix}, \quad \mathbf{V}=\begin{Bmatrix}0.0205 & 0.7068 & 0.7071\\ 0.9996 & -0.0290 & 0.0000\\ 0.0205 & 0.7068 & -0.7071\end{Bmatrix}$$

$$\lambda_1=7.1450\,,\ \lambda_2=1.3996\,.$$

The second singular value is now much larger relative to the first one, so that the two solutions are

$$\tilde{\mathbf{x}}\sim\frac{707(y_2-y_1)}{7.1}\begin{bmatrix}0\\1\\0\end{bmatrix},$$

$$\sim\frac{707(y_2-y_1)}{7.1}\begin{bmatrix}0\\1\\0\end{bmatrix}+\frac{y_1}{1.4}\begin{bmatrix}.71\\0\\.71\end{bmatrix}, \tag{3.4.116}$$

and the rank 1 solution is obtained from the difference of the observations, in contrast to the unscaled solution. The result is quite sensible; given the information that the noise in the two equations is nearly perfectly correlated, it can be removed by subtraction.

At full rank, that is, $K=2$, it can be confirmed that the solutions

(3.4.115) and (3.4.116) are identical, as they must be. It was previously asserted that in a full-rank formally underdetermined system, row scaling is irrelevant to $\tilde{\mathbf{x}}$, $\tilde{\mathbf{n}}$, as may be seen as follows,

$$\begin{aligned}\tilde{\mathbf{x}} &= \mathbf{E}'^T(\mathbf{E}'\mathbf{E}'^T)^{-1}\mathbf{y}' = \mathbf{E}^T\mathbf{W}^{-1/2}(\mathbf{W}^{-T/2}\mathbf{E}\mathbf{E}^T\mathbf{W}^{-1/2})^{-1}\mathbf{W}^{-T/2}\mathbf{y}\\ &= \mathbf{E}\mathbf{W}^{-1/2}\mathbf{W}^{1/2}(\mathbf{E}\mathbf{E}^T)^{-1}\mathbf{W}^{T/2}\mathbf{W}^{-T/2}\mathbf{y} = \mathbf{E}^T(\mathbf{E}\mathbf{E}^T)^{-1}\mathbf{y} \quad (3.4.117)\end{aligned}$$

where we used the result $(\mathbf{AB})^{-1} = \mathbf{B}^{-1}\mathbf{A}^{-1}$ and both inverses must exist, which is possible only in the full-rank situation. But the error covariance is quite different in the two cases:

$$\begin{aligned}(\mathbf{E}\mathbf{E}^T)^{-1} &= \begin{Bmatrix} 1.510 \times 10^4 & -1.505 \times 10^4 \\ -1.505 \times 10^4 & 1.505 \times 10^4 \end{Bmatrix}\\ (\mathbf{E}'\mathbf{E}'^T)^{-1} &= \begin{Bmatrix} 0.500 & -0.707 \\ -0.707 & 0.300 \end{Bmatrix}, \quad (3.4.118)\end{aligned}$$

which would give rise to very different error estimates (using a prior estimate σ_n^2 of the noise variance, because here the noise residuals vanish, a degenerate limit). In effect, the information provided in the row scaling with $\mathbf{R}_{nn}$ permits the SVD to nearly eliminate the noise at rank 1 by an effective subtraction, whereas without that information, the noise is reduced in the solution (3.4.115) at rank 1 only by direct averaging.

There is a subtlety in row weighting. Suppose we have two equations of form

$$\begin{aligned}10x_1 + 5x_2 + x_3 &= 1\,,\\ 100x_1 + 50x_2 + 10x_3 &= 2\,, \quad (3.4.119)\end{aligned}$$

after row scaling to make the expected noise variance in each the same. A rank 1 solution to these equations by SVD is $\tilde{\mathbf{x}} = [.0165 \quad .0083 \quad .0017]^T$, which produces residuals $\tilde{\mathbf{y}} - \mathbf{y} = [-0.79 \quad 0.079]^T$–much smaller in the second equation than in the first one.

Consider that the left side of the second equation is 10 times the first one; in effect we are saying that a measurement of 10 times the values of $10x_1 + 5x_2 + x_3$ has the same noise in it as a measurement of one times this same linear combination. The second equation clearly represents a much more accurate determination of this linear combination, and the equation should be given much more weight in determining the unknowns–and the SVD (and ordinary least squares) does precisely that. To the extent that one finds this result undesirable (one should be careful about why it is so found), there is an easy remedy–divide the equations by their row norms $\sum_j (E_{ij})^{1/2}$. But there may then be a contradiction if it was believed that the noise in all equations was the same to begin with.

An example of this situation is readily apparent in the box balances discussed in Chapter 2. Equations such as (2.4.2) for salt balance have row norms about 35 times larger than those (2.4.1) for the corresponding mass balance, simply because salinity is measured by convention on the Practical Salinity Scale, which produces ocean salinities near 35. Because there is nothing fundamental about the choice of salinity scale, it seems unreasonable to infer that the requirement of salt balance has an expected error 35 times smaller than for density. One usually proceeds in the obvious way by dividing the salt equations by their row norms as the first step. The second step is to ask whether anything further can be said about the relative errors of mass and salt balance, which would introduce a second, purely statistical row weight.

Consider two independent equations in two unknowns, for example,

$$x_1 + x_2 = 1\,,$$
$$2x_1 + x_2 = 2$$

with unique solution $x_1 = 1$, $x_2 = 0$. Now suppose that the right-hand side of the second equation is totally unknown. We have several possibilities for handling the situation. (1) Drop the second equation, and solve the first one as an underdetermined system, giving as the minimum norm solution $\tilde{x}_1 = 1/2$, $\tilde{x}_2 = 1/2$. (2) Downweight the second equation, multiplying it by some very small number. The minimum norm solution is the same as in (1). The advantage over (1) is that most software will compute the right-hand side of the original equations after the solution has been estimated, and the original set-up is unaltered. We thus find out that an estimate of the right-hand side from this solution is 1.5. The disadvantage is that we work with a 2×2 system rather than the 1×2 of (1). (3) Regard the right-hand side of the second equation as a new formal unknown, and rewrite the system as

$$\begin{aligned} x_1 + x_2 &= 1\,, \\ 2x_1 + x_2 - q &= 0\,. \end{aligned} \tag{3.4.120}$$

Solving for the minimum norm underdetermined solution now, we obtain $\tilde{x}_1 = 0$, $\tilde{x}_2 = 1$, $\tilde{q} = 1$, and the estimate of the right-hand side of equation two is 1. Why is this answer different from that in (1) and (2)? The reason is that the presence of the third unknown in the second equation in (3.4.120) provides the information that the unknown right-hand side of equation (2) is of the same magnitude as that of the unknowns x_1, x_2–information that is removed by downweighting or eliminating the equation altogether. The investigator must make his own choice of solution, dependent upon particular circumstances. But see the next section.

3.4.3.4 Column Scaling

In the least-squares problem, we formally introduced a column scaling matrix $\mathbf{S}$. Column scaling operates on the SVD solution exactly as it does in the least-squares solution, to which it reduces in the two special cases already described. That is, we should apply the SVD to sets of equations only where any knowledge of the solution element size has been removed first. If the SVD has been computed for such a column-scaled (and row-scaled) system, the solution is for the scaled unknown $\mathbf{x}'$, and the physical solution is

$$\tilde{\mathbf{x}} = \mathbf{S}^{T/2}\tilde{\mathbf{x}}' . \tag{3.4.121}$$

But there are occasions, with underdetermined systems, where a nonstatistical scaling may also be called for–the analogue to the situation considered above where a row scaling was introduced on the basis of possible nonstatistical considerations.

Example: Suppose we have one equation in two unknowns, the smallest example of an underdetermined system:

$$10x_1 + 1x_2 = 3 . \tag{3.4.122}$$

The particular-SVD solution produces $\tilde{\mathbf{x}} = [0.2970 \quad 0.0297]^T$ in which the magnitude of x_1 is much larger than that of x_2, and the result is readily understood: As we have seen, the SVD finds the exact solution, subject to making the solution norm as small as possible. Because the coefficient of x_1 in (3.4.122) is 10 times that of x_2, it is obviously more efficient in minimizing the norm to give x_1 a larger value than x_2.

Although we have demonstrated this dependence for a trivial example, similar behavior occurs for underdetermined systems in general. In many cases, this distribution of the elements of the solution vector $\mathbf{x}$ is desirable, the numerical value 10 appearing for good physical reasons. In other problems–and the geostrophic inversion problem is an example–the numerical values appearing in the coefficient matrix $\mathbf{E}$ are an accident (in the geostrophic problem, they are proportional to the distance steamed between hydrographic stations and the water depth). Unless one believed that velocities should be larger where the ship steamed further, or the water was deeper, then the solutions may behave unphysically. Indeed, in some situations the velocities are expected to be inverse to the water depth, and such a prior statistical hypothesis is best imposed after one has removed the structural accidents from the system. (The tendency for the solutions to be proportional to the column norms is not absolute. In particular, the equations themselves may actually preclude the proportionality.)

Take a positive-definite, diagonal matrix $\mathbf{S}$, and rewrite (3.3.2) as

$$\mathbf{E}\mathbf{S}^{T/2}\mathbf{S}^{-T/2}\mathbf{x} + \mathbf{n} = \mathbf{y}$$

Then,

$$\mathbf{E}'\mathbf{x}' + \mathbf{n} = \mathbf{y}\,.$$

Solving

$$\tilde{\mathbf{x}}' = \mathbf{E}'^T\left(\mathbf{E}'\mathbf{E}'^T\right)^{-1}\mathbf{y}\,,\ \tilde{\mathbf{x}} = \mathbf{S}^{-T/2}\tilde{\mathbf{x}}'\,. \tag{3.4.123}$$

How should $\mathbf{S}$ be chosen? Apply the recipe (3.4.123) for the simple one-equation example of (3.4.122):

$$\begin{aligned}
\mathbf{E}' = \{10/S_{11}^{1/2} \quad 1/S_{22}^{1/2}\},\ \mathbf{E}'\mathbf{E}'^T &= \frac{100}{S_{11}} + \frac{1}{S_{22}}\,, \\
\left(\mathbf{E}'\mathbf{E}'^T\right)^{-1} &= \left(\frac{S_{11}S_{22}}{100S_{22}+S_{11}}\right)\,, \\
\tilde{\mathbf{x}}' &= \begin{Bmatrix} 10/S_{11}^{1/2} \\ 1/S_{22}^{1/2} \end{Bmatrix} \left[\frac{S_{11}S_{22}}{100S_{22}+S_{11}}\right][3]\,, \\
\tilde{\mathbf{x}} = \mathbf{S}^{-T/2}\mathbf{x}' &= \begin{Bmatrix} 10/S_{11} \\ 1/S_{22} \end{Bmatrix} \left[\frac{S_{11}S_{22}}{100S_{22}+S_{11}}\right][3]\,.
\end{aligned} \tag{3.4.124}$$

The relative magnitudes of the elements of $\tilde{\mathbf{x}}$ are proportional to $10/S_{11}$, $1/S_{22}$. To make the numerical values of the elements of $\tilde{\mathbf{x}}$ the same, we should clearly choose $S_{11} = 10$, $S_{22} = 1$; that is, we should divide the elements of the first column of $\mathbf{E}$ by $\sqrt{10}$ and the second column by $\sqrt{1}$. The apparent rule (which is correct and general) is to divide each column of $\mathbf{E}$ by the square root of its length. The square root of the length may be surprising but arises because of the second multiplication by the elements of $\mathbf{S}^{-T/2}$ in (3.4.123). This form of column scaling should be regarded as nonstatistical in that it is based upon inferences from the numerical magnitudes of the columns of $\mathbf{E}$ and does not employ information about the statistics of the solution. Indeed, its purpose is to prevent the imposition of structure on the solution for which no statistical basis has been anticipated.

If the system is full-rank overdetermined, the column weights drop out, just as we claimed for least squares above. To see this, consider that in the full-rank case,

$$\begin{aligned}\tilde{\mathbf{x}}' &= \left(\mathbf{E}'^T\mathbf{E}'\right)^{-1}\mathbf{E}'^T\mathbf{y}\\ \tilde{\mathbf{x}} &= \mathbf{S}^{T/2}\left(\mathbf{S}^{1/2}\mathbf{E}^T\mathbf{E}\mathbf{S}^{T/2}\right)^{-1}\mathbf{S}^{1/2}\mathbf{E}^T\mathbf{y}\\ &= \mathbf{S}^{T/2}\mathbf{S}^{-T/2}\left(\mathbf{E}^T\mathbf{E}\right)^{-1}\mathbf{S}^{-1/2}\mathbf{S}^{1/2}\mathbf{E}^T\mathbf{y} = \left(\mathbf{E}^T\mathbf{E}\right)^{-1}\mathbf{E}^T\mathbf{y}\,. \end{aligned} \tag{3.4.125}$$

Note the importance of doing column scaling following the row scaling; otherwise, interpretation of the row norms becomes very difficult.

3.4.3.5 Solution and Observation Resolution

Typically, either or both of the set of vectors $\mathbf{v}_i$, $\mathbf{u}_i$ used to present $\mathbf{x}$, $\mathbf{y}$ will be deficient in the sense of the expansions in (3.4.2). Deficiency of one or the other or both is guaranteed if the effective system rank differs from one of M or N.

It follows immediately from Equations (3.4.3) that the particular-SVD solution is

$$\tilde{\mathbf{x}} = \mathbf{V}_K\mathbf{V}_K^T\mathbf{x} \tag{3.4.126}$$

and the data vector with which both it and the general solution are consistent is

$$\tilde{\mathbf{y}} = \mathbf{U}_K\mathbf{U}_K^T\mathbf{y}\,. \tag{3.4.127}$$

Define

$$\mathbf{T}_v = \mathbf{V}_K\mathbf{V}_K^T\,, \tag{3.4.128}$$

$$\mathbf{T}_u = \mathbf{U}_K\mathbf{U}_K^T\,, \tag{3.4.129}$$

the solution and observation (data) resolution matrices, respectively.

Interpretation of the data resolution matrix is slightly subtle. Suppose an element of $\mathbf{y}$ was fully resolved–that is, some row, j_0, of $\mathbf{U}_K\mathbf{U}_K^T$ were all zeros except for diagonal element j_0, which is one. Then a change of unity in y_{j_0} would produce a change in $\tilde{\mathbf{x}}$ that would leave unchanged all other elements of $\tilde{\mathbf{y}}$. If element j_0 is *not* fully resolved, then a change of unity in observation y_{j_0} produces a solution that leads to changes in other elements of $\tilde{\mathbf{y}}$. Stated slightly differently, if y_i is not fully resolved, the system lacks adequate information to distinguish equation i from a linear dependence on one or more other equations.

One can use these ideas to construct quantitative statements of which observations are the most important (data ranking). From Equation (3.4.5), $\text{trace}(\mathbf{T}_u) = K$, and the relative contribution to the solution of any particular constraint is given by the corresponding diagonal element of $\mathbf{T}_u$.

Consider the example (3.4.119) without row weighting. At rank 1,

$$\mathbf{T}_u = \begin{Bmatrix} 0.0099 & 0.099 \\ 0.099 & 0.9901 \end{Bmatrix},$$

showing that the second equation has played a much more important role in the solution than the first one, despite the fact that we asserted the expected noise in both to be the same. The reason is that described above; the second equation in effect asserts that the measurement is 10 times more accurate than in the first equation–and the data resolution matrix informs us of that explicitly. All of the statements made previously about resolution matrices now apply to $\mathbf{T}_u$, $\mathbf{T}_v$.

If row and column scaling have been applied to the equations prior to application of the SVD, the covariance, uncertainty, and resolution expressions apply in those new, scaled spaces. The resolution in the original spaces is

$$\mathbf{T}_v = \mathbf{S}^{T/2}\mathbf{T}_{v'}\mathbf{S}^{-T/2}, \tag{3.4.130}$$

$$\mathbf{T}_u = \mathbf{W}^{T/2}\mathbf{T}_{u'}\mathbf{W}^{-T/2}, \tag{3.4.131}$$

so that

$$\tilde{\mathbf{x}} = \mathbf{T}_v\mathbf{x}, \quad \tilde{\mathbf{y}} = \mathbf{T}_u\mathbf{y} \tag{3.4.132}$$

where $\mathbf{T}_{v'}$, $\mathbf{T}_{u'}$ are the expressions (3.4.128), (3.4.129) in the scaled space. The uncertainty in the new space is $\mathbf{P} = \mathbf{S}^{-T/2}\mathbf{P}'\mathbf{S}^{-1/2}$ where $\mathbf{P}'$ is the expression (3.4.100) or (3.4.101) in the scaled space.

We have seen an interpretation of three matrices obtained from the SVD: $\mathbf{V}\mathbf{V}^T$, $\mathbf{U}\mathbf{U}^T$, $\mathbf{V}\Lambda^{-2}\mathbf{V}^T$. The reader may well wonder, on the basis of the symmetries between solution and data spaces, whether there is an interpretation of the remaining matrix $\mathbf{U}\Lambda^{-2}\mathbf{U}^T$? Such an interpretation exists, but it will emerge most simply when we discuss constrained least squares and Lagrange multipliers.

3.4.3.6 Relation to Tapered and Weighted Least-Squares

In using least squares, a shift was made from the simple objective functions (3.3.4) and (3.3.47) to the more complicated (3.3.22) or (3.3.29). The change was made to permit a degree of control of the relative norms of $\tilde{\mathbf{x}}$, $\tilde{\mathbf{n}}$, and through the use of $\mathbf{W}$, $\mathbf{S}$ of the individual elements and the resulting uncertainties and covariances. Application of the weight matrices $\mathbf{W}$, $\mathbf{S}$ through their Cholesky decompositions to the equations prior to the use of the SVD is equally valid, thus providing the same amount of influence over the solution elements. The SVD provides its control over the solution norms, uncertainties, and covariances through choice of the effective rank

K'. This approach is different from the use of the extended objective functions (3.3.22), but the SVD is actually useful in understanding the effect of such functions.

Assume any necessary $\mathbf{W}$, $\mathbf{S}$ have been applied, but retain α^2 $(=1)$ as a marker. Then, the full SVD, including zero singular values and corresponding singular vectors, is substituted into (3.3.23),

$$\tilde{\mathbf{x}} = (\alpha^2\mathbf{I} + \mathbf{V}\Lambda^T\Lambda\mathbf{V}^T)^{-1}\mathbf{V}\Lambda^T\mathbf{U}^T\mathbf{y}\,,$$

and using the orthogonality of $\mathbf{V}$, we have

$$\tilde{\mathbf{x}} = \mathbf{V}(\Lambda^T\Lambda + \alpha^2\mathbf{I})^{-1}\Lambda^T\mathbf{U}^T\mathbf{y}\,. \tag{3.4.133}$$

The matrix to be inverted is diagonal and so

$$\tilde{\mathbf{x}} = \sum_{i=1}^{N} \frac{\lambda_i(\mathbf{u}_i^T\mathbf{y})}{\lambda_i^2 + \alpha^2}\mathbf{v}_i\,. \tag{3.4.134}$$

It is now apparent what the effect of tapering has done in least squares. The word refers to the tapering down of the coefficients of the $\mathbf{v}_i$ from the values they would have in the pure SVD. In particular, the guarantee that matrices like $(\mathbf{E}^T\mathbf{E} + \alpha^2\mathbf{I})$ would always have an inverse despite vanishing singular values is seen to follow because the inverse of the sum always exists, irrespective of the rank of $\mathbf{E}$. The simple addition of a positive constant to the diagonal of a singular matrix is a well-known method for making it have an inverse. Such methods are a form of what is usually known as *regularization* and are procedures for suppressing nullspaces.

The residuals of the tapered least-squares solution can be written in various forms. Equation (3.3.24) can be written (using the orthogonality of $\mathbf{U}$, $\mathbf{V}$),

$$\tilde{\mathbf{n}} = \alpha^2\mathbf{U}(\alpha^2\mathbf{I} + \Lambda\Lambda^T)^{-1}\mathbf{U}^T\mathbf{y} = \sum_{i=1}^{M} \frac{(\mathbf{u}_i^T\mathbf{y})\alpha^2}{\lambda_i^2 + \alpha^2}\mathbf{u}_i\,, \tag{3.4.135}$$

that is, the projection of the noise onto the range vectors $\mathbf{u}_i$ no longer vanishes. Some of the structure of the range of $\mathbf{E}^T$ is being attributed to noise, and it is no longer true that the residuals are subject to the rigid requirement (3.4.68) of having zero contribution from the range vectors. An increased noise norm is also deemed acceptable, as the price of keeping the solution norm small, by assuring that none of the coefficients in the sum (3.4.134) becomes overly large–values we can control by varying α^2; Wiggins (1972) discusses this form of solution. The covariance of this solution about its mean [Equation (3.3.25)] is readily rewritten as

$$\begin{aligned}
\mathbf{C}_{\tilde{x}\tilde{x}} &= \sum_{i=1}^{N}\sum_{j=1}^{N} \frac{\lambda_i\lambda_j \mathbf{u}_i \mathbf{R}_{nn} \mathbf{u}_j^T}{(\lambda_i^2+\alpha^2)(\lambda_j^2+\alpha^2)} \mathbf{v}_i \mathbf{v}_j^T \\
&= \sigma_n^2 \sum_{i=1}^{N} \frac{\lambda_i^2}{(\lambda_i^2+\alpha^2)^2} \mathbf{v}_i \mathbf{v}_i^T \\
&= \sigma_n^2 \mathbf{V}(\boldsymbol{\Lambda}^T\boldsymbol{\Lambda}+\alpha^2\mathbf{I}_N)^{-1}\boldsymbol{\Lambda}^T\boldsymbol{\Lambda}(\boldsymbol{\Lambda}^T\boldsymbol{\Lambda}+\alpha^2\mathbf{I}_N)^{-1}\mathbf{V}^T
\end{aligned} \tag{3.4.136}$$

where the second line is again the special case of white noise. The role of α^2 in controlling the solution variance, as well as the solution size, should be plain. The tapered least-squares solution is biased, but the presence of the bias can greatly reduce the solution variance. In agnostic situations where one has no real knowledge of any expected variation in the relative sizes of the individual elements of $\mathbf{x}$, $\mathbf{n}$, nor of any correlations amongst them, both $\mathbf{W}$, $\mathbf{S}$ are proportional to the identity. In this situation α^2 is often retained as a simple measure of the ratios of the diagonal elements of $\mathbf{W}$, $\mathbf{S}$ and used to control the relative norms of $\tilde{\mathbf{x}}$, $\tilde{\mathbf{n}}$. Study of the solution as a function of α^2 is known as *ridge regression* (Hoerl & Kennard, 1970a,b), but the interpretation of the results is clearer in the statistical methodology of Section 3.5. Elaborate techniques have been developed for determining the right value of α^2 (see Lawson & Hanson, 1974, or Hansen, 1992, for reviews).[9]

The uncertainty, $\mathbf{P}$, is readily found as

$$\begin{aligned}
\mathbf{P} &= \alpha^2 \sum_{i=1}^{N} \frac{\mathbf{v}_i\mathbf{v}_i^T}{(\alpha^2+\lambda_i^2)^2} + \sigma_n^2 \sum_{i=1}^{N} \frac{\lambda_i^2 \mathbf{v}_i\mathbf{v}_i^T}{(\lambda_i^2+\alpha^2)^2} \\
&= \alpha^2 \mathbf{V}(\boldsymbol{\Lambda}^T\boldsymbol{\Lambda}+\alpha^2\mathbf{I})^{-2}\mathbf{V}^T \\
&\quad + \sigma_n^2 \mathbf{V}(\boldsymbol{\Lambda}^T\boldsymbol{\Lambda}+\alpha^2\mathbf{I})^{-1}\boldsymbol{\Lambda}^T\boldsymbol{\Lambda}(\boldsymbol{\Lambda}^T\boldsymbol{\Lambda}+\alpha^2\mathbf{I})^{-1}\mathbf{V}^T
\end{aligned} \tag{3.4.137}$$

where one uses formally, $\mathbf{x} = \mathbf{V}\mathbf{V}^T\mathbf{x}$, $< \mathbf{x}\mathbf{x}^T > = \alpha^2\mathbf{I}$, and the contribution from the noise components is clearly separated.

The truncated SVD and the tapered SVD–tapered least-squares solutions produce the same qualitative effect: It is possible to increase the noise norm while decreasing the solution norm. Although the solutions differ somewhat, they both achieve a purpose stated above–to extend ordinary least squares in such a way that one can control the relative norms. The quantitative difference between them is readily stated: The truncated form makes a clear separation between range and nullspace in both solution and residual spaces;

[9] Hansen's (1992) discussion is particularly interesting because he exploits the generalized SVD, which is used to simultaneously diagonalize two matrices.

the particular SVD solution contains only range vectors and no nullspace vectors. The residual contains only nullspace vectors and no range vectors. The tapered form permits a merger of the two different sets of vectors with both solution and residuals containing some contribution from both formal range and nullspaces.

One advantage of the tapered form over the truncated SVD or simple least squares is worth noticing. A common empirical measure of a good least-squares fit is through the requirement that the residuals should be unstructured–that is, as nearly white noise as possible: $< \tilde{\mathbf{n}} > = 0$, $< \tilde{\mathbf{n}}\tilde{\mathbf{n}}^T > = \mathbf{I}$ as estimated by the sample averages. But if ordinary least squares (3.3.6) or the equivalent truncated SVD are used, the residuals cannot actually conform to this requirement because they lack the range vectors. That is,

$$< \left(\sum_{K+1}^{M} \beta_i \mathbf{u}_i\right)\left(\sum_{K+1}^{M} \beta_j \mathbf{u}_j\right)^T > \neq \mathbf{I}, \quad K > 0, \tag{3.4.138}$$

because any white-noise process must include contributions from the entire spanning set. If $K \ll M$, this problem may be undetectable. But if K approaches M, the possible structure in the residuals is so restricted by the few nullspace vectors available that it may produce highly non-random values. These considerations become paramount in Section 3.6.

3.4.3.7 Resolution of Tapered Solutions to Simultaneous Equations

The tapered least-squares solutions have an implicit nullspace, arising from the terms corresponding to zero singular values, or values small compared to α^2. Such solutions are often computed directly in the form (3.3.23)–(3.3.25) without ever bothering with the SVD–to save computing. But that solution form does a good job of hiding the existence of what should still be regarded as an effective nullspace.

To obtain a measure of solution resolution in the absence of the explicit $\mathbf{v}_i$ vectors, consider a situation in which the true solution were $\mathbf{x}_{j_0} \equiv \delta_{j,j_0}$– that is, unity in the j_0 element and zero elsewhere. Then, in the absence of noise (the resolution analysis applies to the noise-free situation), the correct value of $\mathbf{y}$ would be

$$\mathbf{E}\mathbf{x}_{j_0} = \mathbf{y}_{j_0}, \tag{3.4.139}$$

defining $\mathbf{y}_{j_0}$. If one actually knew (had measured) $\mathbf{y}_{j_0}$, what solution $\mathbf{x}_{j_0}$ would be obtained?

Tapered least squares produces [in the form (3.3.61)]

$$\tilde{\mathbf{x}}_{j_0} = \mathbf{E}^T(\mathbf{E}\mathbf{E}^T + \alpha^2\mathbf{I})^{-1}\mathbf{y}_{j_0} = \mathbf{E}^T(\mathbf{E}\mathbf{E}^T + \alpha^2\mathbf{I})^{-1}\mathbf{E}\mathbf{x}_{j_0}, \tag{3.4.140}$$

which is row (or column) j_0 of

$$\mathbf{T}_v = \mathbf{E}^T(\mathbf{E}\mathbf{E}^T + \alpha^2\mathbf{I})^{-1}\mathbf{E}. \tag{3.4.141}$$

Thus, we can interpret any row of $\mathbf{T}_v$ as the solution resolution for a Kronecker delta, correct solution, in that element. It is an easy matter, using the SVD of $\mathbf{E}$ and letting $\alpha^2 \to 0$ to show that (3.4.141) reduces to $\mathbf{V}\mathbf{V}^T$. These expressions apply in the row- and column-scaled space; Equations (3.4.130)–(3.4.131) are used to scale and rotate them into the original spaces.

An obvious variant of (3.4.141) follows from the alternative least-squares solution (3.3.23) and is

$$\mathbf{T}_v = (\mathbf{E}^T\mathbf{E} + \alpha^2\mathbf{I})^{-1}\mathbf{E}^T\mathbf{E}. \tag{3.4.142}$$

A solution resolution matrix is obtained similarly: Let $\mathbf{y}_{j_1}$ be zero, except for one in element j_1. Then (3.3.61) produces

$$\tilde{\mathbf{x}}_{j_1} = \mathbf{E}^T(\mathbf{E}\mathbf{E}^T + \alpha^2\mathbf{I})^{-1}\mathbf{y}_{j_1},$$

which if substituted into the original equations is

$$\mathbf{E}\tilde{\mathbf{x}}_{j_1} = \mathbf{E}\mathbf{E}^T(\mathbf{E}\mathbf{E}^T + \alpha^2\mathbf{I})^{-1}\mathbf{y}_{j_1} = \tilde{\mathbf{y}}_{j_1},$$

and thus

$$\mathbf{T}_u = \mathbf{E}\mathbf{E}^T(\mathbf{E}\mathbf{E}^T + \alpha^2\mathbf{I})^{-1}. \tag{3.4.143}$$

The alternate form from (3.3.23) is

$$\mathbf{T}_u = \mathbf{E}(\mathbf{E}^T\mathbf{E} + \alpha^2\mathbf{I})^{-1}\mathbf{E}^T, \tag{3.4.144}$$

which reduces to $\mathbf{U}\mathbf{U}^T$ as $\alpha^2 \to 0$. If row- and column-scaling matrices have been applied, the resolution matrices are modified in analogy to (3.4.130)–(3.4.132).

3.5 Using a Steady Model–Combined Least Squares and Adjoints

Consider now a modest generalization of the constrained problem Equation (3.3.2) in which the unknowns $\mathbf{x}$ are also meant to satisfy some constraints exactly, or nearly exactly, for example

$$\mathbf{A}\mathbf{x} = \mathbf{q}, \tag{3.5.1}$$

but to satisfy the observations (3.3.2) only approximately, in a least-squares sense. Equations like (3.5.1) will be referred to as the *model.* An example of a model occurs in acoustic tomography where we may have measurements of both density and velocity, and they are connected by the thermal wind equations (this case is written out by Munk & Wunsch, 1982). The distinction between the model (3.5.1) and the observations is usually an arbitrary one; $\mathbf{A}$ may well be some subset of the rows of $\mathbf{E}$, for which the corresponding error is believed negligible. What follows can in fact be obtained by imposing the zero-noise limit for some of the rows of $\mathbf{E}$ in the solutions already described. Furthermore, whether the model should be satisfied exactly, or should contain a noise element, too, is situation dependent. The thermal wind relationship is an approximation, and model error would normally be included in its enforcement, in which case the distinction between model and observations is purely conceptual. One should be wary of introducing exact equalities into estimation problems, because they carry the strong possibility of introducing small eigenvalues, or near singular relationships, into the solution, which may dominate the results.

Several approaches are now available to us. Consider for example, the objective function,

$$J = (\mathbf{Ex} - \mathbf{y})^T(\mathbf{Ex} - \mathbf{y}) + \alpha^2(\mathbf{Ax} - \mathbf{q})^T(\mathbf{Ax} - \mathbf{q}) \tag{3.5.2}$$

where $\mathbf{W}$, $\mathbf{S}$ have been applied if necessary and α^2 is retained as a tradeoff parameter. (This objective function corresponds to the requirement of a solution of the concatenated equation sets,

$$\begin{Bmatrix} \mathbf{E} \\ \mathbf{A} \end{Bmatrix} \mathbf{x} + \begin{bmatrix} \mathbf{n} \\ \mathbf{u} \end{bmatrix} = \begin{bmatrix} \mathbf{y} \\ \mathbf{q} \end{bmatrix} \tag{3.5.3}$$

in which $\mathbf{u}$ is the model noise, and the weight given to the model is $\alpha^2\mathbf{I}$.) By letting $\alpha^2 \to \infty$, the model can be forced to apply with arbitrary accuracy. For any finite α^2, the model is formally a soft constraint here because it is being applied only in a minimized sum of squares. The solution follows immediately from (3.3.6) with

$$\mathbf{E} \to \begin{Bmatrix} \mathbf{E} \\ \alpha\mathbf{A} \end{Bmatrix}, \quad \mathbf{y} \to \begin{Bmatrix} \mathbf{y} \\ \alpha\mathbf{q} \end{Bmatrix},$$

assuming the matrix inverse exists.

Alternatively, the model can be imposed as a perfect hard constraint with $\mathbf{u} = 0$. All prior covariances and scalings having been been applied, and Lagrange multipliers introduced, reduces the problem to one with an objective function

$$J = \mathbf{n}^T\mathbf{n} - 2\boldsymbol{\mu}^T(\mathbf{Ax} - \mathbf{q}) = (\mathbf{Ex} - \mathbf{y})^T(\mathbf{Ex} - \mathbf{y}) - 2\boldsymbol{\mu}^T(\mathbf{Ax} - \mathbf{q}), \quad (3.5.4)$$

which is just a variant of (3.3.47). To avoid confusion, it is important to realize that we have essentially interchanged the roles of the two terms in (3.3.47)–with the expression (3.5.1) to be exactly satisfied but the observations only approximately so.

Setting the derivatives of J with respect to $\mathbf{x}$, $\boldsymbol{\mu}$ to zero gives the normal equations

$$-\mathbf{E}^T\mathbf{y} + \mathbf{E}^T\mathbf{Ex} - \mathbf{A}^T\boldsymbol{\mu} = 0, \quad (3.5.5)$$

$$\mathbf{Ax} - \mathbf{q} = 0. \quad (3.5.6)$$

Equation (3.5.5) represents the adjoint, or *dual* model, for the adjoint or dual solution $\boldsymbol{\mu}$. We can distinguish two extreme cases, one in which $\mathbf{A}$ is square, $N \times N$, and of full rank, and one in which $\mathbf{E}$ has this property. In the first case,

$$\tilde{\mathbf{x}} = \mathbf{A}^{-1}\mathbf{q} \quad (3.5.7)$$

and from (3.5.5),

$$\mathbf{E}^T\mathbf{EA}^{-1}\mathbf{q} - \mathbf{E}^T\mathbf{y} = \mathbf{A}^T\boldsymbol{\mu} \quad (3.5.8)$$

or

$$\tilde{\boldsymbol{\mu}} = \mathbf{A}^{-T}(\mathbf{E}^T\mathbf{EA}^{-1}\mathbf{q} - \mathbf{E}^T\mathbf{y}). \quad (3.5.9)$$

Here, the values of $\mathbf{x}$ are completely determined by the full-rank, noiseless model, and the minimization of the deviation from the observations is passive. The Lagrange multipliers or adjoint solution, however, are useful, providing the sensitivity information, $\partial J/\partial\mathbf{q} = 2\boldsymbol{\mu}$. The uncertainty of this solution is zero because of the perfect model (3.5.6).

In the second case, from (3.5.5),

$$\tilde{\mathbf{x}} = (\mathbf{E}^T\mathbf{E})^{-1}[\mathbf{E}^T\mathbf{y} + \mathbf{A}^T\tilde{\boldsymbol{\mu}}] \equiv \tilde{\mathbf{x}}_u + (\mathbf{E}^T\mathbf{E})^{-1}\mathbf{A}^T\tilde{\boldsymbol{\mu}}$$

where $\tilde{\mathbf{x}}_u = (\mathbf{E}^T\mathbf{E})^{-1}\mathbf{E}^T\mathbf{y}$ is the ordinary, unconstrained least-squares solution. Substituting into (3.5.6) produces

$$\tilde{\boldsymbol{\mu}} = [\mathbf{A}(\mathbf{E}^T\mathbf{E})^{-1}\mathbf{A}^T]^{-1}(\mathbf{q} - \mathbf{A}\tilde{\mathbf{x}}_u) \quad (3.5.10)$$

and

$$\tilde{\mathbf{x}} = \tilde{\mathbf{x}}_u + (\mathbf{E}^T\mathbf{E})^{-1}\mathbf{A}^T[\mathbf{A}(\mathbf{E}^T\mathbf{E})^{-1}\mathbf{A}^T]^{-1}(\mathbf{q} - \mathbf{A}\tilde{\mathbf{x}}_u), \quad (3.5.11)$$

assuming $\mathbf{A}$ is full-rank underdetermined. The perfect model is underdetermined; its range is being fit perfectly, with its nullspace being employed to reduce the misfit to the data as far as possible. The uncertainty of this solution may be written (Seber, 1977)

$$\begin{aligned}\mathbf{P} &= D^2(\tilde{\mathbf{x}} - \mathbf{x}) \qquad (3.5.12)\\ &= \sigma^2 \left\{ (\mathbf{E}^T\mathbf{E})^{-1} - (\mathbf{E}^T\mathbf{E})^{-1}\mathbf{A}^T \left[\mathbf{A}(\mathbf{E}^T\mathbf{E})^{-1}\mathbf{A}^T\right]^{-1} \mathbf{A}(\mathbf{E}^T\mathbf{E})^{-1} \right\},\end{aligned}$$

which represents a reduction in the uncertainty of the ordinary least-squares solution (first term on the right) by the information in the perfectly known constraints. The presence in the inverse of terms involving $\mathbf{A}$ in these solutions is a manifestation of the warning about the possible introduction of components dependent upon small eigenvalues of $\mathbf{A}$.

Example: Consider the least-squares problem of solving

$$\begin{aligned} x_1 + n_1 &= 1 \\ x_2 + n_2 &= 1 \\ x_1 + x_2 + n_3 &= 3 \end{aligned}$$

with uniform, uncorrelated noise of variance 1 in each of the equations. The solution is then

$$\tilde{\mathbf{x}} = [1.3333 \quad 1.3333]^T$$

with uncertainty

$$\mathbf{P} = \begin{Bmatrix} 0.6667 & -0.333 \\ -0.333 & 0.6667 \end{Bmatrix}.$$

But suppose that it is known or desired that $x_1 - x_2 = 1$. Then (3.5.11) produces $\tilde{\mathbf{x}} = [1.8333 \quad 0.8333]^T$, $\boldsymbol{\mu} = 0.5$, $J = 0.8333$, with reduced uncertainty

$$\mathbf{P} = \begin{Bmatrix} 0.1667 & 0.1667 \\ 0.1667 & 0.1667 \end{Bmatrix}.$$

If the constraint is shifted to $x_1 - x_2 = 1.1$, the new solution is

$$\tilde{\mathbf{x}} = [1.8833 \quad 0.7833]^T$$

and the new objective function is $J = 0.9383$, a shift consistent with $\boldsymbol{\mu}$ and (3.3.56).

If neither $\mathbf{A}$ nor $\mathbf{E}$ is full rank, then the inverses appearing in equations (3.5.7), (3.5.11) will not exist. Either form can be used then, by replacing the inverses by, say, the particular SVD inverse. But in that case, the solution will not be unique if the combined observations and model leave a

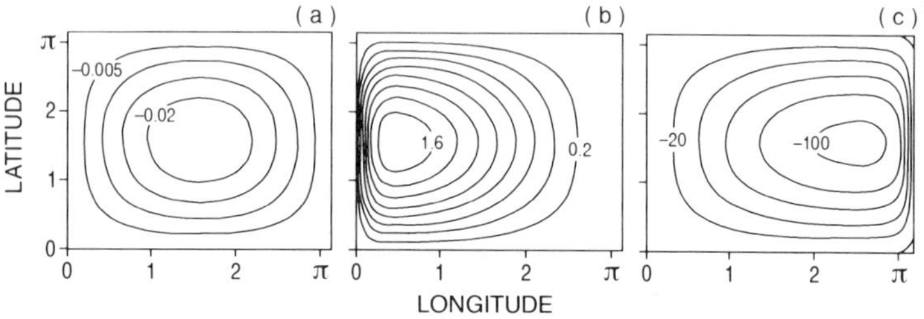

Figure 3–11. The Stommel Gulf Stream model solved using the adjoint: (a) depicts the windstress curl imposed, and (b) is the resulting transport streamfunction showing the expected westward intensification. The adjoint solution is shown in (c). Because it satisfies the adjoint equation, it shows an eastward intensification, consistent with a reversal of the sign of β.

nullspace in $\mathbf{x}$. The objective function (3.5.4) can be modified to have an extra term in $\mathbf{x}^T\mathbf{S}^{-1}\mathbf{x}$ if desired.

If the model has error terms, too, either in the forcing, $\mathbf{q}$, or in missing physics, it is modified to

$$\mathbf{Ax} + \mathbf{u} = \mathbf{q}\,. \tag{3.5.13}$$

A hard-constraint formulation can still be used, in which (3.5.13) is still to be exactly satisfied, imposed through an objective function of form,

$$J = (\mathbf{Ex} - \mathbf{y})^T(\mathbf{Ex} - \mathbf{y}) + \alpha^2\mathbf{u}^T\mathbf{u} - 2\boldsymbol{\mu}^T(\mathbf{Ax} + \mathbf{u} - \mathbf{q})\,. \tag{3.5.14}$$

It is again readily confirmed that the solutions using (3.5.2) or (3.5.14) are identical, and the hard/soft distinction is seen again to be artificial unless one truly has model equations with $\mathbf{u} = 0$. Equation (3.5.13) represents a model that is to be exactly satisfied; but it has an unknown *control* contribution, $\mathbf{u}$. Objective functions like (3.5.14) will be used extensively in Chapter 6. The most general form of objective function would be

$$J = \mathbf{n}^T\mathbf{R}^{-1}\mathbf{n} + \mathbf{x}^T\mathbf{S}^{-1}\mathbf{x} + \mathbf{u}^T\mathbf{Q}^{-1}\mathbf{u} - 2\boldsymbol{\mu}^T(\mathbf{Ax} + \mathbf{u} - \mathbf{q})\,. \tag{3.5.15}$$

If $\mathbf{A}$ is square and full rank, and $\mathbf{u} = 0$, one can readily confirm that $\mathbf{R}$ and $\mathbf{S}$ drop out of the solution.

Example: Let us apply these ideas to the Stommel Gulf Stream model. A code was written to solve by finite differences the nondimensional equation

$$\epsilon\nabla^2\phi + \frac{\partial\phi}{\partial x} = \hat{\mathbf{k}}\cdot\nabla\times\boldsymbol{\tau} \tag{3.5.16}$$

and is depicted in Figure 3–11 for the case $\epsilon = 0.05$ and $\phi = 0$ on the boundaries. The nondimensionalization and the basin dimension $0 \le x \le \pi$, $0 \le \phi \le \pi$ are those of Schröter and Wunsch (1986). The windstress curl was $\hat{\mathbf{k}}\cdot\nabla\times\boldsymbol{\tau} = -\sin x \sin y$.

The discretized form of the model is then the perfect $N \times N$ system

$$\mathbf{A}\mathbf{x} = \mathbf{q}\,, \quad \mathbf{x} = \{\phi_{ij}\}\,, \tag{3.5.17}$$

and $\mathbf{q}$ is the equivalently discretized windstress curl. The theory of partial differential equations shows that this system is full rank and generally well behaved. But let us ignore that knowledge and seek the values $\mathbf{x}$ that make the objective function (3.3.47)

$$J = \mathbf{x}^T\mathbf{x} - 2\boldsymbol{\mu}^T(\mathbf{A}\mathbf{x} - \mathbf{q}) \tag{3.5.18}$$

stationary with respect to $\mathbf{x}$, $\boldsymbol{\mu}$:

$$\mathbf{A}^T\boldsymbol{\mu} = \mathbf{x} \tag{3.5.19}$$

$$\mathbf{A}\mathbf{x} = \mathbf{q}\,. \tag{3.5.20}$$

$\mathbf{x}^T\mathbf{x}$ is readily identified with the solution potential energy. The solution $\boldsymbol{\mu}$, corresponding to the circulation of Figure 3–11b, is shown in Figure 3–11c. What is the interpretation? The Lagrange multipliers represent the sensitivity of the Stommel solution potential energy to perturbations in the windstress curl. We see that the sensitivity is greatest in the eastern half of the basin and indeed displays a boundary layer character. Schröter and Wunsch (1986) discuss this result in the context of the behavior of the Sverdrup interior of the Stommel model. A physical interpretation of the Lagrange multipliers can be inferred, given the simple structure of the governing equation (3.5.16) and the Dirichlet boundary conditions. This equation is not self-adjoint in the sense discussed by Morse and Feshbach (1953) or Lanczos (1961); the adjoint partial differential equation is of form

$$\epsilon\nabla^2\boldsymbol{\mu} - \frac{\partial\boldsymbol{\mu}}{\partial x} = \text{forcing}\ , \tag{3.5.21}$$

subject to mixed boundary conditions, and whose discrete form is (3.5.19), obtained by taking the transpose of the $\mathbf{A}$ matrix of the discretization. It is obvious from both (3.5.21) and Figure 3–11c that the adjoint solution represents flow streamlines in an ocean in which the sign of β has been reversed, resulting in an eastern boundary current, and the forcing is provided by the Stommel solution stream function. We can thus usefully think about the physics of an adjoint, or dual, ocean (or one might prefer the term *anti-ocean*) that governs the sensitivity of the real or "direct" ocean to parameter specifications. The structure of the $\boldsymbol{\mu}$ would change if J were changed.

The original objective function J is very closely analogous to the Lagrangian (not to be confused with the Lagrange multiplier) in classical mechanics. In mechanics, the gradients of the Lagrangian commonly are forces. The modified Lagrangian, J', is used in mechanics to impose various physical constraints, and the virtual force required to impose the constraints–for example, the demand that a particle follow a particular path–is the Lagrange multiplier. Lanczos (1970) has a good discussion. In an economics/ management context, the multipliers are usually called *shadow prices* because they are intimately related to the question of how much profit (the objective function) will change with a change in the availability or cost of a product ingredient.

More generally, there is a close connection between the stationarity requirements imposed upon various objective functions throughout this book and the mathematics of classical mechanics. In Chapter 6, this analogy will be exploited to introduce the Hamiltonian form of governing equations.

A conventional dynamical model is one that is properly posed, which can be interpreted here as meaning that $\mathbf{A}$ is full rank of dimension $M \times M$ but it need not actually be so. Let us examine the state vector $\mathbf{x}$, Lagrange multiplier $\boldsymbol{\mu}$, pair in a little more detail. As already noted if (3.5.19) is underdetermined, then (3.5.20) is overdetermined (and vice versa). Some insight is obtained if the pair is rewritten in the SVD form

$$\mathbf{V}\boldsymbol{\Lambda}\mathbf{U}^T\boldsymbol{\mu} = \mathbf{x}\,, \tag{3.5.22}$$

$$\mathbf{U}\boldsymbol{\Lambda}\mathbf{V}^T\mathbf{x} = \mathbf{q}\,. \tag{3.5.23}$$

Because of the structure of (3.5.22)–(3.5.23), the overdetermined system (whichever of the pair it is) will automatically satisfy the necessary solvability conditions, and an ill-posed model is readily handled.

Using the SVD inverse, (3.5.22)–(3.5.23) produce

$$\boldsymbol{\mu} = \mathbf{U}_K\boldsymbol{\Lambda}_K^{-2}\mathbf{U}_K^T\mathbf{q}\,. \tag{3.5.24}$$

We know that $\boldsymbol{\mu}$ are the sensitivity of J to perturbations in $\mathbf{q}$. Thus,

$$\frac{\partial J}{\partial \mathbf{q}} = 2\boldsymbol{\mu} = 2\mathbf{U}_K\boldsymbol{\Lambda}_K^{-2}\mathbf{U}_K^T\mathbf{q}\,, \tag{3.5.25}$$

where $\boldsymbol{\mu}$ contains the missing fourth matrix expression noticed in Section 3.4 and is the sensitivity of the objective function $\mathbf{x}^T\mathbf{x}$ to perturbations in the model elements $\mathbf{q}$. Taking the second derivative,

$$\frac{\partial^2 J}{\partial \mathbf{q}^2} = 2\mathbf{U}_K\boldsymbol{\Lambda}_K^{-2}\mathbf{U}_K^T\,, \tag{3.5.26}$$

is the Hessian of J. Evidently, if any of the λ_i become very small, the objective function will be extremely sensitive to small perturbations in the specification of $\mathbf{q}$, producing an effective nullspace of the problem. Equation (3.5.26) suggests that assertions that models are perfect can lead to difficulties. If the objective function (3.3.47) is used, Equation (3.5.26) represents the sensitivity to the data, $\mathbf{y}$.

3.5.1 Relation to Green's Functions

There is a close relationship between adjoint models and Green's functions. Consider any linear model, for example (1.2.4), the discrete Laplace equation with Dirichlet boundary conditions, which can be written as

$$\mathbf{Ax} = \rho\,. \tag{3.5.27}$$

To solve it, consider the collection of N adjoint problems

$$\mathbf{A}^T\mathbf{G}^T = \mathbf{I} \tag{3.5.28}$$

or

$$\mathbf{GA} = \mathbf{I}\,, \tag{3.5.29}$$

left multiplying (3.5.27) by $\mathbf{G}$, right multiplying (3.5.29) by $\mathbf{x}$, and subtracting,

$$\mathbf{G\ Ax} - \mathbf{G\ Ax} = \mathbf{G}\rho - \mathbf{x} \tag{3.5.30}$$

or

$$\mathbf{x} = \mathbf{G}\rho\,. \tag{3.5.31}$$

$\mathbf{G}$ is usually called the *Green's function*, which is seen to here satisfy a set of problems adjoint to the forward problem. The connection to the role of Green's functions in partial differential systems is laid out clearly in Morse and Feshbach (1953) and Lanczos (1961).

As in the general theory of Green's functions, there is an intimate relationship between the solution to (3.5.29) and the solution to the original problem, (3.5.27), for point disturbances. Consider the N separate problems

$$\mathbf{AX}_G = \mathbf{I} \tag{3.5.32}$$

where each column of $\mathbf{X}_G$ is the solution to a different right-hand side column of $\mathbf{I}_c$. As written, (3.5.32) is actually two different types of problem in a combined notation: In one type, a boundary value at one grid point is being set to 1, with zero everywhere else, and the interior sources, ρ, are all zero. In the other type, the boundary conditions are all zero, but one of the interior sources is being set to 1, all others being zero. Each separate problem in (3.5.32) gives rise to a solution vector $\mathbf{x}_G(j_0)$, where j_0 is the index of the nonzero boundary condition, or interior position source.

Now from (3.5.29), $\mathbf{X}_G = \mathbf{G} = \mathbf{A}^{-1}$–that is, each column of $\mathbf{G}$ corresponds to the solution to the forward problem for a unit disturbance at a particular boundary or grid point. Using the SVD for $\mathbf{A}$, it follows that

$$\mathbf{G} = \mathbf{U}\boldsymbol{\Lambda}^{-1}\mathbf{V}^T \tag{3.5.33}$$

(if $\mathbf{A}$ is not of full rank, reduced SVDs are used).

Now consider a different problem. We wish to solve (3.5.27) but *in addition* have the independent knowledge that at an interior point, i_0,

$$x_{j0} = \bar{\phi}_{j0}\,, \tag{3.5.34}$$

which is not the same as specifying a disturbance at this point–rather it is a piece of information (compare Lanczos, 1961, p. 207). With the addition of the original equations and boundary conditions, the problem is now formally overspecified. Unless the value of $\bar{\phi}_{j0}$ is chosen to be the specific value consistent with the solution to the original problem, or unless we are prepared to admit noise unknowns into the problem, the combination of (3.5.27) and (3.5.34) is a contradiction. Consider the consistency relationship that determines the unique value of $\bar{\phi}_{j0}$, which would permit a solution by forming the overdetermined system

$$\mathbf{A}_1\mathbf{x} = \boldsymbol{\rho}_1, \mathbf{A}_1 = \begin{Bmatrix} \mathbf{A} \\ \delta_{jj0} \end{Bmatrix},\ \boldsymbol{\rho}_1 = \begin{bmatrix} \boldsymbol{\rho} \\ \bar{\phi}_{j0} \end{bmatrix}. \tag{3.5.35}$$

The SVD of $\mathbf{A}_1$ will produce $N - \mathbf{v}_i$ and $N - \mathbf{u}_i$, corresponding to nonzero singular vectors. There will be one extra $\mathbf{u}_{N+1}$ in the nullspace. The solvability condition (3.4.70) is then

$$\mathbf{u}_{N+1}^T\boldsymbol{\rho}_1 = 0\,, \tag{3.5.36}$$

which is

$$\bar{\phi}_{j0} = -\frac{\mathbf{u}_{N+1}^{\prime T}(j_0)\boldsymbol{\rho}}{u_{N+1,N+1}(j_0)} \tag{3.5.37}$$

where $\mathbf{u}'_{N+1}$ is defined as the vector containing only the first N elements of $\mathbf{u}_{N+1}$ and j_0 is written as an argument in $\mathbf{u}'_{N+1}$ to show its dependence upon the particular location of the information; $u_{N+1,N+1}(j_0)$ is the $N+1$st element of the nullspace vector. The calculation (3.5.37) can be done for each interior point j_0 and can be done for all interior points simultaneously by appending a separate equation of form (3.5.34) to (3.5.27) for each j_0 as

$$\mathbf{A}\mathbf{x} = \boldsymbol{\rho}\,,$$

$$\mathbf{I}\mathbf{x} = \bar{\boldsymbol{\phi}} \tag{3.5.38}$$

where $\bar{\boldsymbol{\phi}} = [\bar{\phi}_{jo}]$ is the vector of data points, or

$$\mathbf{A}_1\mathbf{x} = \boldsymbol{\rho}_1, \mathbf{A}_1 = \begin{Bmatrix} \mathbf{A} \\ \mathbf{I} \end{Bmatrix}, \ \boldsymbol{\rho}_1 = \begin{bmatrix} \boldsymbol{\rho} \\ \bar{\boldsymbol{\phi}} \end{bmatrix}. \tag{3.5.39}$$

Let the nullspace of $\mathbf{A}_1^T$, $\mathbf{u}_i$, $N+1 \leq i \leq 2N$, form a matrix, $\mathbf{Q}_u$. The solvability conditions are

$$\mathbf{Q}_u^T \boldsymbol{\rho}_1 = 0 \tag{3.5.40}$$

or

$$\{\mathbf{u}_{iJ_2}\}^T \bar{\boldsymbol{\phi}} = -\{\mathbf{u}_{iJ_1}\}^T \rho$$

where $\{\mathbf{u}_{iJ_2}\}$ is the matrix composed of the elements of the nullspace $\mathbf{u}_i$ in positions $N+1 \leq j \leq 2N$, and $\{\mathbf{u}_{iJ_1}\}$ are the first N elements of these vectors. Thus,

$$\bar{\boldsymbol{\phi}} = -\{\mathbf{u}_{iJ_2}\}^{-T}\{\mathbf{u}_{iJ_1}\}^T \rho\,. \tag{3.5.41}$$

But $\boldsymbol{\phi}$ is then a solution to the original problem, and it must follow [Equation (3.5.31)] that

$$\mathbf{G} = -\{\mathbf{u}_{iJ_2}\}^{-T}\{\mathbf{u}_{iJ_1}\}^T\,. \tag{3.5.42}$$

If $\boldsymbol{\phi}$ are regarded as "data," then one can write instead,

$$\hat{\mathbf{x}} = \mathbf{G}\bar{\boldsymbol{\phi}}\,. \tag{3.5.43}$$

Thus, although $\mathbf{G}$ is the solution to the physical problem of a point disturbance at a boundary point, or an interior point (keeping in mind the distinction between the physics of these two cases), it also serves as a way of determining the consistency of a set of "data," (3.5.39), with the solution to the original model system. If $\bar{\boldsymbol{\phi}}$ are noisy, with white-noise contaminant, (3.5.43) will be a consistent solution to the equations agreeing as best as is possible with the observations.

3.6 Gauss-Markov Estimation, Mapmaking, and More Simultaneous Equations

The fundamental objective for least squares is minimization of the noise norm (3.3.4), although we complicated the discussion somewhat by introducing trade-offs against $\| \tilde{\mathbf{x}} \|$, various weights in the norms, and even the restriction that $\tilde{\mathbf{x}}$ should satisfy certain equations exactly. Least-squares methods, whether used directly as in (3.3.6) or indirectly through the vector representations of the SVD, are fundamentally deterministic–$\mathbf{W}$, $\mathbf{S}$, α^2 need not be given any statistical interpretation whatever–although sometimes one uses covariances for them. Statistics were used only to understand the sensitivity of the solutions to noise, and to obtain measures of the expected deviation of the solution from some supposed truth.

But there is another, radically different, approach to obtaining estimates of the solution to equation sets like (3.3.2), directed more clearly toward the physical goal: to find an estimate $\tilde{\mathbf{x}}$ which deviates as little as possible in the *mean square* from the true solution. That is, we wish to minimize the statistical quantities $< (\tilde{\mathbf{x}} - \mathbf{x})_i^2 >$. The next section is devoted to understanding how to find such an $\tilde{\mathbf{x}}$ (and the corresponding $\tilde{\mathbf{n}}$) through an excursion into statistical estimation theory. It is far from obvious that this $\tilde{\mathbf{x}}$ should bear any resemblance to one of the least-squares estimates, but as will be seen, under some circumstances the two are identical. Their possible identity is extremely useful but has apparently led many investigators to confuse the methodologies and therefore the interpretation of the result.

3.6.1 The Fundamental Result

Suppose we are interested in making an estimate of a physical variable $\mathbf{x}$, which might be a vector or a scalar and might be constant with space and time, or vary with either or both. To be definite, let $\mathbf{x}$ be a function of an independent variable $\mathbf{r}$, written discretely as $\mathbf{r}_j$ (it might be a vector of space coordinates, or a scalar time, or an accountant's label). Let us make some suppositions about what is usually called *prior information.* In particular, suppose we have an estimate of the low-order statistics describing $\mathbf{x}$–that is, we specify its mean and second moments:

$$< \mathbf{x} > = \mathbf{x}_0 \,, \quad < \mathbf{x}(\mathbf{r}_i)\mathbf{x}(\mathbf{r}_j)^T > = \mathbf{R}_{xx}(\mathbf{r}_i, \mathbf{r}_j) \,. \tag{3.6.1}$$

To have a concrete problem, one might think of $\mathbf{x}$ as being temperature at 700-m depth in the ocean (a scalar) and $\mathbf{r}_j$ as a vector of horizontal positions; $\mathbf{x}$ is the vector each of whose elements is the scalar value at a different position. Alternatively, in one dimension, the elements of $\mathbf{x}$ would be the salinity along a surface transect by a ship. Then r_j is the scalar of position, either time or distance, and $\mathbf{r}$ is the vector of all such positions. But if the field of interest is the velocity vector, then each element of $\mathbf{x}$ is itself a vector, and one can extend the notation in a straightforward fashion. To keep the notation a little cleaner, however, we will usually treat the elements of $\mathbf{x}$ as scalars.

Now suppose there exist observations, y_i, as a function of the same coordinate $\mathbf{r}_i$, with known second moments

$$\mathbf{R}_{yy} = <\mathbf{y}\mathbf{y}^T>, \quad \mathbf{R}_{xy}(\mathbf{r}_i, \mathbf{r}_j) = <\mathbf{x}(\mathbf{r}_i)\mathbf{y}(\mathbf{r}_j)^T>, \quad 1 \le i, j \le M \tag{3.6.2}$$

(the individual observation elements can also be vectors–for example, two or three components of velocity and a temperature at a point–but as with $\mathbf{x}$, the modifications required to treat this case are straightforward, and we will maintain the simplicity of assuming scalar observations). Could the measurements be used to make an estimate of $\mathbf{x}$ at a point $\tilde{\mathbf{r}}_\alpha$, which may not coincide with one of the places (labels) where an observation is available? The idea is to exploit the concept that finite covariances carry predictive capabilities from known variables to unknown ones. A specific example would be to suppose the measurements are of temperature $y(\mathbf{r}_j) = y_0(\mathbf{r}_j)+n(\mathbf{r}_j)$, where n is the noise, and we wish to estimate the temperature at different locations, perhaps on a regular grid $\tilde{\mathbf{r}}_\alpha$, $1 \le \alpha \le N$. This special problem is one of gridding or mapmaking (the tilde is placed on $\mathbf{r}_\alpha$ as a device to emphasize that this is a location where an estimate is sought; the numerical values of these places or labels are known). Alternatively, and somewhat more interesting, perhaps the measurements are more indirect, with $y(r_i)$ representing a velocity field component at depth in the ocean and believed connected through the thermal wind equation to the temperature field. We might want to estimate the temperature from measurements of the velocity.

Given the discussion immediately following Equation (3.2.30), we seek an estimate $\tilde{x}(\tilde{\mathbf{r}}_\alpha)$, whose dispersion about its true value, $x(\tilde{\mathbf{r}}_\alpha)$, is as small as possible–that is,

$$\mathbf{P}(\tilde{\mathbf{r}}_\alpha, \tilde{\mathbf{r}}_\alpha) = <(\tilde{x}(\tilde{\mathbf{r}}_\alpha) - x(\tilde{\mathbf{r}}_\alpha))(\tilde{x}(\tilde{\mathbf{r}}_\beta) - x(\tilde{\mathbf{r}}_\beta))> |_{\tilde{\mathbf{r}}_\alpha = \tilde{\mathbf{r}}_\beta}$$

is to be minimized (a minimum variance estimate). If we would like to answer the question for more than one point, and if we would like to understand the covariance of the errors of our estimates at various points $\tilde{\mathbf{r}}_\alpha$, then we can form a vector of values to be estimated, $\{\tilde{x}(\mathbf{r}_\alpha)\} \equiv \tilde{\mathbf{x}}$, and the uncertainty among them,

$$\begin{aligned}\mathbf{P}(\tilde{\mathbf{r}}_\alpha, \tilde{\mathbf{r}}_\beta) &= <(\tilde{x}(\tilde{\mathbf{r}}_\alpha) - x(\tilde{\mathbf{r}}_\alpha))(\tilde{x}(\tilde{\mathbf{r}}_\beta) - x(\tilde{\mathbf{r}}_\beta)) \\ &= <(\tilde{\mathbf{x}} - \mathbf{x})(\tilde{\mathbf{x}} - \mathbf{x})^T>, \qquad 1 \le \alpha \le N, \ 1 \le \beta \le N,\end{aligned} \tag{3.6.3}$$

where the *diagonal* elements are to be *individually* minimized.

What should the relationship be between data and estimate? At least initially, one might try a linear combination of data,

$$\tilde{x}(\tilde{\mathbf{r}}_\alpha) = \sum_{j=1}^{M} B(\tilde{\mathbf{r}}_\alpha, \mathbf{r}_j) y(\mathbf{r}_j) , \tag{3.6.4}$$

for all α, which makes the diagonal elements of $\mathbf{P}$ in (3.6.3) as small as possible. All the points $\tilde{\mathbf{r}}_\alpha$ can be treated simultaneously by letting $\mathbf{B}$ be an $M \times N$ matrix, and

$$\tilde{\mathbf{x}} = \mathbf{B}(\tilde{\mathbf{r}}_\alpha, \mathbf{r}_j)\mathbf{y} . \tag{3.6.5}$$

(This notation is mixed. Equation (3.6.5) is a shorthand for (3.6.4), in which the argument has been put into $\mathbf{B}$ explicitly as a reminder that there is a summation over all the data locations $\mathbf{r}_j$ for all mapping locations $\tilde{\mathbf{r}}_\alpha$.)

An important theorem, usually called the *Gauss-Markov theorem*, produces the values of $\mathbf{B}$ so as to minimize the diagonal elements of $\mathbf{P}$. The following heuristic derivation is based on that in Liebelt (1967): Substituting (3.6.5) into (3.6.3) and expanding,

$$\begin{aligned} \mathbf{P}(\tilde{\mathbf{r}}_\alpha, \tilde{\mathbf{r}}_\beta) &= < (\mathbf{B}(\tilde{\mathbf{r}}_\alpha, \mathbf{r}_j)\mathbf{y} - x(\tilde{\mathbf{r}}_\alpha))(\mathbf{B}(\tilde{\mathbf{r}}_\beta, \mathbf{r}_l)\mathbf{y} - x(\tilde{\mathbf{r}}_\beta))^T > \\ &\equiv < (\mathbf{B}\mathbf{y} - \mathbf{x})(\mathbf{B}\mathbf{y} - \mathbf{x})^T > |_{\alpha\beta} \\ &= \mathbf{B} < \mathbf{y}\mathbf{y}^T > \mathbf{B}^T - < \mathbf{x}\mathbf{y}^T > \mathbf{B}^T - \mathbf{B} < \mathbf{y}\mathbf{x}^T > + < \mathbf{x}\mathbf{x}^T > |_{\alpha\beta} \end{aligned} \tag{3.6.6}$$

(keep in mind that $\mathbf{y}$ is a function of the data positions $\mathbf{r}_j$, $\tilde{\mathbf{x}}$ is a function of the estimation positions $\tilde{\mathbf{r}}_\beta$, and $\mathbf{B}$ is a function of both). Using $\mathbf{R}_{xy} = \mathbf{R}_{yx}^T$, Equation (3.6.6) is

$$\mathbf{P} = \mathbf{B}\mathbf{R}_{yy}\mathbf{B}^T - \mathbf{R}_{xy}\mathbf{B}^T - \mathbf{B}\mathbf{R}_{xy}^T + \mathbf{R}_{xx} . \tag{3.6.7}$$

Notice that because $\mathbf{R}_{xx}$ represents the moments of $\mathbf{x}$ evaluated at the estimation positions, it is a function of $\tilde{\mathbf{r}}_\alpha$, $\tilde{\mathbf{r}}_\beta$, whereas $\mathbf{R}_{xy}$ involves covariances of $\mathbf{y}$ at the data positions with $\mathbf{x}$ at the estimation positions, and is consequently a function $\mathbf{R}_{xy}(\tilde{\mathbf{r}}_\alpha, \mathbf{r}_j)$.

Now, by completing the square (3.1.26), Equation (3.6.7) becomes

$$\mathbf{P} = (\mathbf{B} - \mathbf{R}_{xy}\mathbf{R}_{yy}^{-1})\mathbf{R}_{yy}(\mathbf{B} - \mathbf{R}_{xy}\mathbf{R}_{yy}^{-1})^T - \mathbf{R}_{xy}\mathbf{R}_{yy}^{-1}\mathbf{R}_{xy}^T + \mathbf{R}_{xx} . \tag{3.6.8}$$

The diagonal elements of (3.6.8) are the variances of the estimate at points $\tilde{\mathbf{r}}_\alpha$ about their true values. Because $\mathbf{R}_{xx}$ and $\mathbf{R}_{yy}$ are positive-definite, they and their inverses have positive diagonal elements (if they are only positive semidefinite, $\mathbf{R}_{yy}^{-1}$ has to be redefined, but we will ignore this pathology). By the symmetries present, then, the diagonal elements of all three terms in (3.6.8) are positive. Thus, minimization of any diagonal element of $\mathbf{P}$ is obtained by choosing $\mathbf{B}$ so that the first term vanishes, or

$$\mathbf{B}(\tilde{\mathbf{r}}_\alpha, \mathbf{r}_j) = \mathbf{R}_{xy}\mathbf{R}_{yy}^{-1} . \tag{3.6.9}$$

Then the minimum variance estimate is

$$\tilde{\mathbf{x}}(\tilde{\mathbf{r}}_\alpha) = \mathbf{R}_{xy}\mathbf{R}_{yy}^{-1}\mathbf{y} , \tag{3.6.10}$$

and the actual minimum value of the diagonal elements of $\mathbf{P}$ is found by substituting back into (3.6.7), producing

$$\mathbf{P}(\tilde{\mathbf{r}}_\alpha, \tilde{\mathbf{r}}_\beta) = \mathbf{R}_{xx}(\tilde{\mathbf{r}}_\alpha, \tilde{\mathbf{r}}_\beta) - \mathbf{R}_{xy}(\tilde{\mathbf{r}}_\alpha, \mathbf{r}_j)\mathbf{R}_{yy}^{-1}(\mathbf{r}_j, \mathbf{r}_k)\mathbf{R}_{xy}^{T}(\tilde{\mathbf{r}}_\beta, \mathbf{r}_k) . \tag{3.6.11}$$

The bias of (3.6.10) is

$$< \tilde{\mathbf{x}} - \mathbf{x} > = \mathbf{R}_{xy}\mathbf{R}_{yy}^{-1} < \mathbf{y} > - \mathbf{x} . \tag{3.6.12}$$

If $< \mathbf{y} > = \mathbf{x} = 0$, the estimator is unbiased and called a *best linear unbiased estimator*, or *BLUE*; otherwise, it is biased. It is not difficult to show that $\tilde{\mathbf{x}}$ is also the maximum likelihood estimate if the solution is jointly normal.

3.6.2 Linear Algebraic Equations

The result (3.6.9)–(3.6.11) is the abstract general case and is deceptively simple. Understanding it is far from trivial, and for many applications, some simplifications are very useful. Suppose the observations are related to the unknown vector $\mathbf{x}$ as in our canonical problem–that is, through a set of linear equations: $\mathbf{E}\mathbf{x} + \mathbf{n} = \mathbf{y}$. The measurement moments, $\mathbf{R}_{yy}$, can be computed directly:

$$\mathbf{R}_{yy} = < (\mathbf{E}\mathbf{x} + \mathbf{n})(\mathbf{E}\mathbf{x} + \mathbf{n})^T > = \mathbf{E}\mathbf{R}_{xx}\mathbf{E}^T + \mathbf{R}_{nn} \tag{3.6.13}$$

where the unnecessary but simplifying and often excellent assumption was made that the cross-terms of form

$$\mathbf{R}_{xn} = \mathbf{R}_{nx}^T = \mathbf{0} , \tag{3.6.14}$$

so that

$$\mathbf{R}_{xy} = < \mathbf{x}(\mathbf{E}\mathbf{x} + \mathbf{n})^T > = \mathbf{R}_{xx}\mathbf{E}^T , \tag{3.6.15}$$

that is, there is no correlation between the measurement noise and the actual state vector (e.g., that the noise in a temperature measurement does not depend upon whether the true value is 10° or 25°).

Under these circumstances, Equations (3.6.10), (3.6.11) take on the form:

$$\tilde{\mathbf{x}} = \mathbf{R}_{xx}\mathbf{E}^T(\mathbf{E}\mathbf{R}_{xx}\mathbf{E}^T + \mathbf{R}_{nn})^{-1}\mathbf{y}\,, \tag{3.6.16}$$

$$\tilde{\mathbf{n}} = \left\{\mathbf{I} - \mathbf{E}\mathbf{R}_{xx}\mathbf{E}^T(\mathbf{E}\mathbf{R}_{xx}\mathbf{E}^T + \mathbf{R}_{nn})^{-1}\right\}\mathbf{y}\,, \tag{3.6.17}$$

$$\mathbf{P} = \mathbf{R}_{xx} - \mathbf{R}_{xx}\mathbf{E}^T(\mathbf{E}\mathbf{R}_{xx}\mathbf{E}^T + \mathbf{R}_{nn})^{-1}\mathbf{E}\mathbf{R}_{xx}\,, \tag{3.6.18}$$

$$\begin{aligned}\mathbf{P}_{nn} = &\left\{\mathbf{I} - \mathbf{E}\mathbf{R}_{xx}\mathbf{E}^T(\mathbf{E}\mathbf{R}_{xx}\mathbf{E}^T + \mathbf{R}_{nn})^{-1}\right\} \times \\ &\mathbf{R}_{nn}\left\{\mathbf{I} - \mathbf{E}\mathbf{R}_{xx}\mathbf{E}^T(\mathbf{E}\mathbf{R}_{xx}\mathbf{E}^T + \mathbf{R}_{nn})^{-1}\right\}\,.\end{aligned} \tag{3.6.19}$$

These latter expressions are extremely important; they permit discussion of the solution to a set of linear algebraic equations in the presence of noise using information concerning the statistics of the noise and of the solution. Notice that they are *identical to the least-squares expression* (3.3.66)–(3.3.70) *if* $\mathbf{S} = \mathbf{R}_{xx}$, $\mathbf{W} = \mathbf{R}_{nn}$.

From the matrix inversion lemma, Equations (3.6.16)–(3.6.18) can be rewritten

$$\tilde{\mathbf{x}} = (\mathbf{R}_{xx}^{-1} + \mathbf{E}^T\mathbf{R}_{nn}^{-1}\mathbf{E})^{-1}\mathbf{E}^T\mathbf{R}_{nn}^{-1}\mathbf{y}\,, \tag{3.6.20}$$

$$\tilde{\mathbf{n}} = \left\{\mathbf{I} - \mathbf{E}(\mathbf{R}_{xx}^{-1} + \mathbf{E}^T\mathbf{R}_{nn}^{-1}\mathbf{E})^{-1}\mathbf{E}^T\mathbf{R}_{nn}^{-1}\right\}\mathbf{y}\,, \tag{3.6.21}$$

$$\mathbf{P} = (\mathbf{R}_{xx}^{-1} + \mathbf{E}^T\mathbf{R}_{nn}^{-1}\mathbf{E})^{-1}\,, \tag{3.6.22}$$

$$\begin{aligned}\mathbf{P}_{nn} = &\left\{\mathbf{I} - \mathbf{E}(\mathbf{R}_{xx}^{-1} + \mathbf{E}^T\mathbf{R}_{nn}^{-1}\mathbf{E})^{-1}\mathbf{E}^T\mathbf{R}_{nn}^{-1}\right\} \times \\ &\mathbf{R}_{nn}\left\{\mathbf{I} - \mathbf{E}(\mathbf{R}_{xx}^{-1} + \mathbf{E}^T\mathbf{R}_{nn}^{-1}\mathbf{E})^{-1}\mathbf{E}^T\mathbf{R}_{nn}^{-1}\right\}\,.\end{aligned} \tag{3.6.23}$$

Although these alternate forms are algebraically and numerically identical to (3.6.16)–(3.6.19), the size of the matrices to be inverted changes from $M \times M$ matrices to $N \times N$, where $\mathbf{E}$ is $M \times N$ (but note that $\mathbf{R}_{nn}$ is $M \times M$; the efficacy of this alternate form may depend upon whether the *inverse* of $\mathbf{R}_{nn}$ is known). Depending upon the relative magnitudes of M, N, one form may be much preferable to the other; finally, (3.6.22) has an important interpretation that will be discussed when we come to recursive methods. Recall, too, the options we had with the SVD of solving $M \times M$ or $N \times N$ problems. Equations (3.6.20)–(3.6.23) are identical to the alternative form least-squares solution (3.3.38)–(3.3.41) if $\mathbf{S} = \mathbf{R}_{xx}$, $\mathbf{W} = \mathbf{R}_{nn}$.

The solution (3.6.16)–(3.6.18) or (3.6.20)–(3.6.22) is an estimator; it was found by demanding a solution with the minimum dispersion about the true solution and is seen, surprisingly, to be identical with the tapered, weighted least-squares solution when the least-squares objective function weights are chosen to be the corresponding second-moment matrices of $\mathbf{x}$, $\mathbf{n}$. This correspondence of the two solutions often leads them to be seriously confused. It is essential to recognize that the logic of the derivations are quite distinct: We were free in the least-squares derivation to use weight matrices that were anything we wished–as long as appropriate inverses existed.

The correspondence of least squares with minimum variance estimation can be understood by recognizing that the Gauss-Markov estimator was derived by minimizing a quadratic objective function. The least-squares estimate was obtained from minimizing a summation that was a sample *estimate* of the Gauss-Markov objective function with $\mathbf{S}$, $\mathbf{W}$ properly chosen. The coincidence of the answers can be exploited in a number of ways. For example, we infer immediately that a resolution discussion directly analogous to Equations (3.4.142)–(3.4.144) for the least-squares solution is possible for the Gauss-Markov solution.

As with any statistical estimator, one must make *posterior* checks that the behavior of $\tilde{\mathbf{x}}$, $\tilde{\mathbf{n}}$ is consistent with the assumed prior statistics reflected in $\mathbf{R}_{xx}$, $\mathbf{R}_{nn}$, and any assumptions about their means or other properties. Such posterior checks are both essential and very demanding. One sometimes hears it said that estimation using Gauss-Markov and related methods is "pulling solutions out of the air" because the prior moment matrices $\mathbf{R}_{xx}$, $\mathbf{R}_{nn}$ often are only poorly known. But producing solutions that pass the test of consistency with the prior covariances can be very difficult. Solutions tend to be somewhat insensitive to the details of the prior statistics, and it is easy to become overly concerned with the detailed structure of $\mathbf{R}_{xx}$, $\mathbf{R}_{nn}$. As stated previously, it is also rare to be faced with a situation in which one is truly ignorant of the moments–*true ignorance* meaning that arbitrarily large or small numerical values of x_i, n_i would be acceptable. In the box inversions of Chapter 2 (to be revisited in Chapter 4), deep ocean velocity fields of order 1000 cm/s would be absurd, and their absurdity is readily asserted by choosing $\mathbf{R}_{xx} = \mathrm{diag}((10\ \mathrm{cm/s})^2)$, which reflects a mild belief that velocities are 0(10 cm/s) with no known correlations with each other. Testing of statistical estimates against prior hypotheses is a highly developed field in applied statistics, and we leave it to the references (e.g., Seber, 1977) for their discussion. Should such tests be failed, one must reject the solutions $\tilde{\mathbf{x}}$, $\tilde{\mathbf{n}}$ and ask why they failed–as it usually implies an incorrect model, $(\mathbf{E})$, or misunderstanding of the observational noise structure.

If the intention is that least-squares solutions are to be equivalent to the

Gauss-Markov ones, they must pass the same statistical tests. The simplest of these is that the weight matrices, thought to represent the covariances of $\mathbf{x}$, $\mathbf{n}$, should be shown after the fact to have been reasonable. Objective functions such as (3.3.36) must also have values consistent with the hypothesis. For example, substitution of the (scaled and weighted) solutions that are appropriate to (3.3.36) produce

$$< \tilde{J} > = < \mathbf{x}^T \mathbf{x} > \; + \; < \mathbf{n}^T \mathbf{n} > = N + M - K \,, \tag{3.6.24}$$

with an actual value consistent with a χ_ν^2 probability density, and the values of the individual terms of $\tilde{J}$ should, as previously discussed, prove consistent with a χ_1^2 distribution. Note that the number of degrees of freedom, ν, in $\tilde{J}$ would be approximately $N + M - K$, where K is the rank of $\mathbf{E}$. (Draper & Smith, 1982, Chapters 2 and 3 discuss such problems in detail.)

3.6.2.1 Use of Basis Functions

A superficially different way of dealing with prior statistical information is often commonly used. Suppose that the indices of x_i refer to a spatial or temporal position, call it r_i, so that $x_i = x(r_i)$. Then it is often sensible to consider expanding the unknown $\mathbf{x}$ in a set of basis functions, F_j–for example, sines and cosines, Chebyshev polynomials, ordinary polynomials, etc. One might write

$$x(r_i) = \sum_{j=1}^{L} \alpha_j F_j(r_i)$$

or

$$\mathbf{x} = \mathbf{F}\boldsymbol{\alpha} \,, \quad \mathbf{F} = \left\{ \begin{matrix} F_1(r_1) & F_2(r_1) & \cdots & F_L(r_1) \\ F_1(r_2) & F_2(r_2) & \cdots & F_L(r_2) \\ \cdot & \cdot & \cdot & \cdot \\ F_1(r_N) & F_2(r_N) & \cdots & F_L(r_N) \end{matrix} \right\} \,, \; \boldsymbol{\alpha}^T = [\alpha_1 \cdots \alpha_L]^T \,,$$

which, when substituted into (3.3.2), produces

$$\mathbf{L}\boldsymbol{\alpha} + \mathbf{n} = \mathbf{y} \,, \quad \mathbf{L} = \mathbf{E}\mathbf{F} \,. \tag{3.6.25}$$

If $L < M < N$, one can convert an underdetermined system into one that is formally overdetermined and, of course, the reverse is possible as well. More generally, one transfers the discussion of solution covariance, etc., to the expansion coefficients $\boldsymbol{\alpha}$. If there are special conditions applying to $\mathbf{x}$, such as boundary conditions at certain positions, r_B, a choice of basis function satisfying those conditions could be more convenient than appending them as additional equations.

It should be apparent, however, that the solution to (3.6.25) will have a covariance structure dictated in large part by that contained in the basis functions chosen; thus, there is no fundamental gain in employing basis functions although they may be convenient, numerically or otherwise. If $\mathbf{P}_{\alpha\alpha}$ denotes the uncertainty of $\boldsymbol{\alpha}$, then

$$\mathbf{P} = \mathbf{F}\mathbf{P}_{\alpha\alpha}\mathbf{F}^T \tag{3.6.26}$$

is the uncertainty of $\mathbf{x}$.

Example: The underdetermined system

$$\left\{\begin{matrix} 1 & 1 & 1 & 1 \\ 1 & -1 & -1 & 1 \end{matrix}\right\}\mathbf{x} + \mathbf{n} = \begin{bmatrix} 1 \\ -1 \end{bmatrix},$$

with noise variance $< \mathbf{n}\mathbf{n}^T > = .01\mathbf{I}$, has a solution, if $\mathbf{R}_{xx} = \mathbf{I}$, of

$$\tilde{\mathbf{x}} = \mathbf{E}^T(\mathbf{E}\mathbf{E}^T + .01\mathbf{I})^{-1}\mathbf{y} = [0 \pm 0.7 \quad 0.5 \pm 0.7 \quad 0.5 \pm 0.7 \quad 0 \pm 0.7]^T,$$
$$\tilde{\mathbf{n}} = [.0025 \pm 0.002 \quad -.0025 \pm 0.002]^T.$$

If the solution was thought to be large scale and smooth, one might use the covariance

$$\mathbf{R}_{xx} = \left\{\begin{matrix} 1 & .999 & .998 & .997 \\ .999 & 1 & .999 & .998 \\ .998 & .999 & 1 & .999 \\ .997 & .998 & .999 & 1 \end{matrix}\right\},$$

which produces a solution

$$\tilde{\mathbf{x}} = [0.18 \pm 0.05 \quad 0.32 \pm 0.04 \quad 0.32 \pm 0.04 \quad 0.18 \pm 0.05]^T,$$

$$\tilde{\mathbf{n}} = [4.6 \times 10^{-4} \pm 6.5 \times 10^{-5}, \ -0.71 \pm 0.07]^T,$$

which has a larger-scale property as desired and a smaller standard error.

If one attempts a solution as a first-order polynomial,

$$x_i = a + br_i, \ r_1 = 0, \ r_2 = 1, \ r_3 = 2, \ldots,$$

the system will become two equations in the two unknowns a, b:

$$\mathbf{EF}\begin{bmatrix} a \\ b \end{bmatrix} = \left\{\begin{matrix} 4 & 6 \\ -2 & -6 \end{matrix}\right\}\begin{bmatrix} a \\ b \end{bmatrix} + \mathbf{n} = \begin{bmatrix} 1 \\ -1 \end{bmatrix},$$

and if the covariance of a, b is the identity matrix,

$$[\tilde{a}, \tilde{b}] = [4 \times 10^{-4} \pm 0.07, \ 0.2 \pm 0.04],$$

$$\tilde{\mathbf{x}} = [4 \times 10^{-3} \pm 0.07 \quad 0.17 \pm 0.04 \quad 0.33 \pm 0.02 \quad 0.5 \pm 0.05]^T \,,$$

$$\tilde{\mathbf{n}} = [0.0002 \pm 0.0004 \,, \ -1.00 \pm 0.0005]^T \,,$$

which is also large scale and smooth but clearly different than that from the Gauss-Markov estimator. Although this latter solution has been obtained from a just-determined system, it is not clearly better. If a linear trend is expected in the solution, then the polynomial expansion is certainly convenient–although such a structure can be imposed through use of $\mathbf{R}_{xx}$ by specifying a growing variance with r_i.

3.6.3 Determining a Mean Value

Let the measurements of the physical quantity continue to be denoted y_i and suppose that each is made up of an unknown large-scale mean m, plus a deviation from that mean of θ_i. Then,

$$m + \theta_i = y_i \,, \quad 1 \le i \le M \tag{3.6.27}$$

or

$$\mathbf{D}m + \boldsymbol{\theta} = \mathbf{y} \,, \ \mathbf{D}^T = [1 \quad 1 \quad 1 \quad \cdots \quad 1]^T \,, \tag{3.6.28}$$

and we seek a best estimate, $\tilde{m}$, of m. In (3.6.27) or (3.6.28) the unknown $\mathbf{x}$ has become the scalar m, and the deviation of the field from its mean is the noise, that is, $\boldsymbol{\theta} \equiv \mathbf{n}$, whose true mean is zero. The problem is evidently a special case of the use of basis functions, in which only one function–a zeroth-order polynomial, m–is retained.

Set $\mathbf{R}_{nn} = \mathbf{C}_{nn} = < \boldsymbol{\theta}\boldsymbol{\theta}^T >$. If, for example, we were looking for a large-scale mean temperature in a field of oceanic mesoscale eddies, then $\mathbf{R}_{nn}$ is the sum of the covariance of the eddy field plus that of observational errors and any other fields contributing to the difference between y_i and the true mean m. To be general, suppose $\mathbf{R}_{xx} = < m^2 > = m_0^2$ and from (3.6.20),

$$\begin{aligned} \tilde{m} &= \left\{ \frac{1}{m_0^2} + \mathbf{D}^T \mathbf{R}_{nn}^{-1} \mathbf{D} \right\}^{-1} \mathbf{D}^T \mathbf{R}_{nn}^{-1} \mathbf{y} \\ &= \frac{1}{1/m_0^2 + \mathbf{D}^T \mathbf{R}_{nn}^{-1} \mathbf{D}} \mathbf{D}^T \mathbf{R}_{nn}^{-1} \mathbf{y} \end{aligned} \tag{3.6.29}$$

$$\tilde{\mathbf{n}} = \tilde{\boldsymbol{\theta}} = \mathbf{y} - \mathbf{D}\tilde{m} \tag{3.6.30}$$

($\mathbf{D}^T \mathbf{R}_{nn}^{-1} \mathbf{D}$ is a scalar). The expected uncertainty of this estimate is (3.6.22),

$$\mathbf{P} = \left\{ \frac{1}{m_0^2} + \mathbf{D}^T \mathbf{R}_{nn}^{-1} \mathbf{D} \right\}^{-1} = \frac{1}{1/m_0^2 + \mathbf{D}^T \mathbf{R}_{nn}^{-1} \mathbf{D}} \,. \tag{3.6.31}$$

As $m_0^2 \to \infty$, Equations (3.6.29)–(3.6.31) become the same expressions given by Bretherton, Davis, & Fandry (1976) for the mean of a field.

The estimates may appear somewhat unfamiliar; they reduce to more common expressions in certain limits. Let the θ_i be uncorrelated, with uniform variance σ^2; $\mathbf{R}_{nn}$ is then diagonal and (3.6.29) reduces to

$$\tilde{m} = \frac{1}{(1/m_0^2 + M/\sigma^2)\sigma^2} \sum_{i=1}^{M} y_i = \frac{m_0^2}{\sigma^2 + M m_0^2} \sum_{i=1}^{M} y_i \,, \tag{3.6.32}$$

where the relations $\mathbf{D}^T\mathbf{D} = M$, $\mathbf{D}^T\mathbf{y} = \sum_{i=1}^{M} y_i$ were used. The expected value of the estimate is

$$< \tilde{m} > = \frac{m_0^2}{\sigma^2 + M m_0^2} \sum_{i}^{M} < y_i > = \frac{m_0^2}{\sigma^2 + M m_0^2} M m \neq m \,, \tag{3.6.33}$$

that is, it is biased, as inferred above, unless $< y_i > = 0$, implying $m = 0$. $\mathbf{P}$ becomes

$$P = \frac{1}{1/m_0^2 + M/\sigma^2} = \frac{\sigma^2 m_0^2}{\sigma^2 + M m_0^2} \,. \tag{3.6.34}$$

Under the further assumption that $m_0^2 \to \infty$,

$$\tilde{m} = \frac{1}{M} \sum_{i=1}^{M} y_i \,, \tag{3.6.35}$$

$$P = \sigma^2 / M \,, \tag{3.6.36}$$

which are the ordinary average and its variance [the latter expression is the well-known square root of M rule for the standard deviation of an average–see Equation (3.5.27)]; $< \tilde{m} >$ in (3.6.35) is readily seen to be the true mean, but (3.6.29) is biased. However, the magnitude of (3.6.36) always exceeds that of (3.6.34)–acceptance of bias in the estimate (3.6.32) reduces the uncertainty of the result—a common trade-off in estimation problems.

Equations (3.6.29)–(3.6.31) are the general estimation rule–accounting through $\mathbf{R}_{nn}$ for correlations in the observations and their irregular distribution. Because many samples are not independent, (3.6.34) or (3.6.36) may be extremely optimistic. Equations (3.6.29)–(3.6.31) give the appropriate expression for the variance when the data are correlated (that is, when there are fewer degrees of freedom than the number of sample points). On the other hand, knowledge of the covariance structure of the noise can be exploited to reduce the uncertainty of the mean: Recall the reduced errors [Equation (3.4.118)] when the noise was known to be strongly positively correlated, thus permitting its reduction by subtraction.

The use of the prior estimate, m_0^2, is interesting. Letting m_0^2 go to infinity does not mean that an infinite mean is expected [(3.6.35) is finite]. This limit is merely a statement that there is no information whatever, before we start, as to the size of the true average–it could be arbitrarily large. Such a situation is, of course, unlikely, and even though we might choose not to use information concerning the probable size of the solution, we should remain aware that we could do so (the importance of the prior estimate diminishes as M grows–so that with an infinite amount of data it has no effect at all on the estimate).

It is very important not to be tempted into making a first estimate of m_0^2 by using (3.6.35), substituting into (3.6.32), thinking to reduce the error variance. For the Gauss-Markov theorem to be valid, the prior information must be truly independent of the data being used. If a prior estimate of m itself is available rather than just its mean square, the problem should be reformulated as one for the estimate of the perturbation about this value.

It is quite common in mapping and interpolation problems (taken up immediately below), to first estimate the mean of the field, to remove it, and then to map the residuals–here called θ. Such a procedure is a special case of a methodology often called (particularly in the geological literature) *kriging*,[10] which is discussed in Chapter 5.

3.6.4 Making a Map; Sampling, Interpolation, and Objective Mapping

A familiar oceanographic problem is to draw a set of contours from data that may have been observed irregularly in space. Such maps are usually the first step in understanding what one is measuring. A somewhat more sophisticated use of a map is to produce a regular grid of values to use in a numerical ocean model. An example with an irregular data distribution is shown in Figure 2–14. Ships observe the windfield wherever they happen to be, and the data are interpolated by investigators onto regular grids that are then used to drive model oceans.

Obtaining numbers on a regular grid from irregularly distributed observations is basically an interpolation or mapping problem–and as such may seem somewhat trivial. But it is far from trivial when one adds the requirement that the map should be accompanied by a useful estimate of the error of the values calculated at a grid point. For regions of a map surrounded by densely spaced data, the gridded value can be expected to be more accurate

[10] Pronounced with a soft "g."

than at a location effectively extrapolated from distant data points. (But what does one mean by "distant?") When using a map with a complex model, such inhomogeneities in the accuracy may be extremely important–as calculations could be in error because of large mapping errors in distant parts of the domain. Bretherton et al. (1976) introduced the subject of quantitative mapmaking into oceanography, and the book by Thièbaux and Pedder (1987) is devoted to the problem.

3.6.4.1 Sampling

The first question that must be addressed is whether the sampling of the field is adequate to make a useful map. This subject is a large and interesting one in its own right, and there are a number of useful references, including Bracewell (1978), Freeman (1965), Jerri (1977), or Butzer and Stens (1992), and we can only outline the basic ideas.

The simplest and most fundamental idea derives from consideration of a one-dimensional continuous function $f(q)$ where q is an arbitrary independent variable, usually either time or space, and $f(q)$ is supposed to be sampled uniformly at intervals Δq an infinite number of times (see Figure 3–12a) to produce the infinite set of sample values $\{f(n\Delta q)\}$, $-\infty \leq n \leq \infty$. The sampling theorem, or sometimes the Shannon-Whittaker Sampling Theorem[11] is a statement of necessary and sufficient conditions so that $f(q)$ can be perfectly reconstructable from the sample values. Let the Fourier transform of $f(q)$ be defined as

$$\hat{f}(r) = \int_{-\infty}^{\infty} f(q) e^{2i\pi rq} dq \,. \tag{3.6.37}$$

The sampling theorem asserts that a necessary and sufficient condition to perfectly reconstruct $f(q)$ from its samples is that

$$|\hat{f}(r)| = 0\,, \quad |r| \geq 1/(2\Delta q)\,. \tag{3.6.38}$$

The theorem produces an explicit formula for the reconstruction, the Shannon-Whittaker formula, which is

$$f(q) = \sum_{n=-\infty}^{\infty} f(n\Delta q) \frac{\sin[(2\pi/2\Delta q)(q - n\Delta q)]}{(2\pi/2\Delta q)(q - n\Delta q)} \,. \tag{3.6.39}$$

Mathematically, the Shannon-Whittaker theorem is surprising because it provides a condition under which a function at an uncountable infinity of points can be perfectly reconstructed from information only at a countable infinity of them. For present purposes, an intuitive interpretation is all

[11] In the Russian literature, Kotel'nikov's theorem.

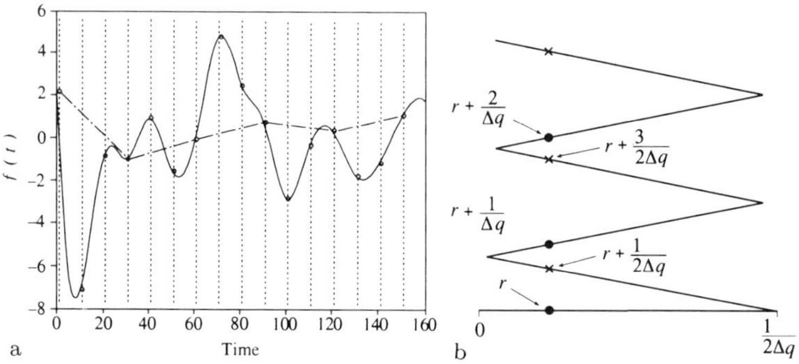

Figure 3–12. (a) Classical aliasing of a curve by undersampling. The original function is shown by the solid line. Samples every 10 time units (open circles) do an adequate job of capturing the variability in the function. But if the sampling interval is increased to 30 time units, the original function is grossly misrepresented (dashed line), and an attempt to calculate the derivative of the original curve from the coarse sampling interval would prove disastrously wrong. (b) Uniformly undersampled high frequencies or wavenumbers masquerade (alias) as lower frequencies or wavenumbers according to the Equation (3.6.40). The net result is a *folding* of the frequencies outside the baseband [Equation (3.6.41)] into apparent frequencies within the baseband. Here, a function $f(q)$ with Fourier transform $\hat{f}(r)$ is sampled at intervals Δq. The Fourier transform of the sampled function, $\hat{f}_s(r)$, sums contributions from the original Fourier transform as $\hat{f}_s(r) = \hat{f}(r) + \hat{f}(r \pm 1/\Delta q) + \hat{f}(r \pm 2/\Delta q) + \cdots$ (dotted points), which is potentially radically different from $\hat{f}(r)$. Points denoted "x" are aliased into a negative value $0 \geq r > -1/2\Delta q$–for example, $(r + 1/2\Delta q) - 1/\Delta q = r - 1/2\Delta q$.

we seek, and this is perhaps best done by considering a special case in which the conditions of the theorem are violated. Figure 3–12a displays an undersample curve. It is quite clear that there is at least one other curve, the one depicted with the broken line, which is completely consistent with all the sample points and which cannot be distinguished from it. If a pure sinusoid of frequency r_0 is sampled at intervals Δq, $\Delta q > 1/2r_0$, a little thought shows that the apparent frequency of this new sinusoid is

$$r_a = r_0 \pm \frac{n}{\Delta q} \tag{3.6.40}$$

such that

$$|r_a| \leq \frac{1}{2\Delta q}\,. \tag{3.6.41}$$

The samples cannot distinguish the true high-frequency sinusoid from a low frequency one, and the high frequency can be said to masquerade or *alias* as the lower-frequency one.[12] The Fourier transform of a sampled function is

[12] Aliasing is familiar as the stroboscope effect. Recall the appearance of the spokes of a wagon wheel in the movies. The spokes can appear to stand still, or move slowly forward or backward, depending upon the camera shutter speed relative to the true rate at which the spokes revolve.

easily seen to be periodic with period $1/\Delta q$ in the transform domain–that is, in the r space (Bracewell, 1978, and Hamming, 1973, have particularly clear discussions). Because of this periodicity, there is no point in computing its values for frequencies outside $|r| \leq 1/2\Delta q$ (we make the convention that this *baseband*, i.e., the fundamental interval for computation, is symmetric about $r = 0$, over a distance $1/2\Delta q$; see Figure 3–12b). Frequencies of absolute value larger than $1/2\Delta q$, the so-called Nyquist frequency, cannot be distinguished from those in the baseband, and they alias into it.

The consequences of aliasing range from the negligible to the disastrous. A simple, possibly trivial, example is that of the principal lunar tide, usually labeled M_2, with a period of 12.42 hours, $r = 1.932$ cycles/day. An observer measures the height of sea level at a fixed time, say 10 A.M. each day so that $\Delta q = 1$ day. Applying the formula (3.6.40), the apparent frequency of the tide will be .0676 cycles/day for a period of about 14.8 days. To the extent that the observer understands what is going on, he will not conclude that the principal lunar tide has a period of 14.8 days but will realize that he can compute the true period through (3.6.40) from the apparent one. But if he did not understand what was happening, he might produce some bizarre theory.[13]

Few situations are this simple. Consider Figure 2–2j, which shows a section of silicate across the North Atlantic Ocean. The observed variation includes all wavenumbers; an estimate of the wavenumber spectrum is displayed in Figure 3–13. (Spectra are discussed in many textbooks, e.g., Priestley, 1981. For present purposes it can be regarded as simply a numerical estimate of the wavenumber content of Figure 2–2i.) As one thins the sampling, the higher wavenumbers will be aliased into lower ones. Because the distribution is dominated by the low wavenumbers, the effects of this aliasing are perhaps not readily apparent to the eye. But for use in numerical models, one almost inevitably must differentiate observed fields one or more times. Consider what would happen if one tried to estimate the derivative of the curve in Figure 3–12a from the samples: The numerical values of the aliased data would be radically different from the correct values. It is this type of concern that leads one to single out the aliasing problem for special attention.

The reader may object that the Shannon-Whittaker theorem applies only to an infinite number of perfect samples and that one never has either per-

[13] There is a story, perhaps apocryphal, that one investigator was measuring the mass flux of the Gulf Stream at a fixed time each day. He was preparing to publish the exciting discovery that there was a strong 14-day periodicity to the flow, before someone pointed out to him that he was aliasing the tidal currents with a 12.42-hour period.

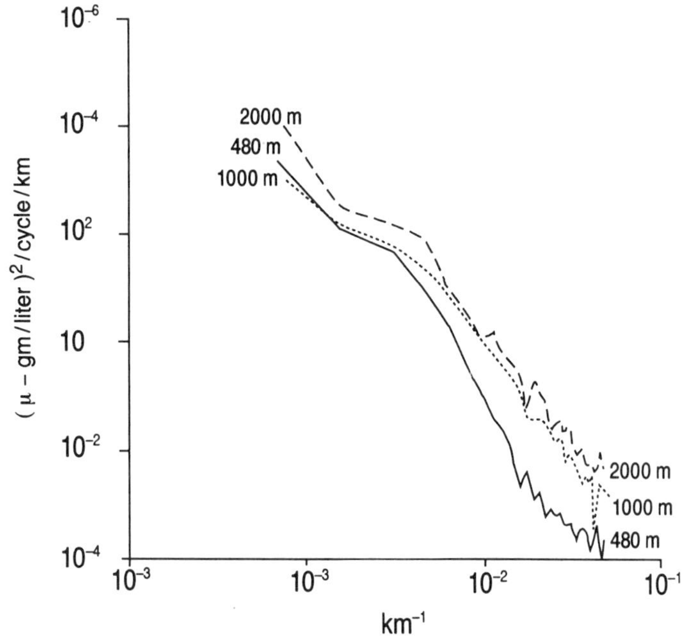

Figure 3–13. Estimated wavenumber spectrum of the silicate distribution along 25°N in the Atlantic, at three different depths. Such spectra are said to be *red*, because the energy increases as the wavelength increases (wavenumber decreases). It is the large-scale structure that is most visible in sections and maps. But if the field is differentiated, as it must be to use in conservation equations, the large-scale structures are suppressed, and the high wavenumbers–short scales–remain and are subject to aliases from inadequate sampling.

fect samples or an infinite number of them. In particular, it is true that if the duration of the data in the q domain is finite, then it is impossible for the Fourier transform to vanish over any finite interval (it follows from the so-called Paley-Wiener criterion and is usually stated in the form that "timelimited signals cannot be bandlimited"). Nonetheless, the rule of thumb that results from (3.6.39) has been found to be quite a good one. The deviations from the assumptions of the theorem are usually dealt with by asserting that sampling should be done so that

$$\Delta q \ll 1/2r_0 \,. \tag{3.6.42}$$

Many extensions and variations of the sampling theorem exist–taking account of the finite time duration (e.g., see Landau & Pollak, 1962), the use of burst-sampling and known-function derivatives, etc. (see Freeman, 1965; Jerri, 1977). Most of these variations are sensitive to noise. There are also extensions to multiple dimensions (e.g., Petersen & Middleton, 1962), which are required for mapmaking purposes. (An application, with discussion of the noise sensitivity, can be found in Wunsch, 1989.)

The subject will be left here for present purposes, with the comment that sampling theorems can be unforgiving–that is, once a function is undersampled, and unless the aliased signal is as simple as a known tidal contribution, it will be mappable, and differentiable, etc., but perhaps in a way that can be disastrously misleading. Consideration of sampling is critical to any discussion of field data.

3.6.5 One-Dimensional Interpolation

Supposing that the field has been adequately sampled, consider using only two observations $[y_1 \;\; y_2]^T = [x_1 + n_1 \;\; x_2 + n_2]^T$ located at positions $[r_1 \;\; r_2]^T$ where n_i are the observation noise. We require an estimate of $x(\tilde{r})$, where $r_1 < \tilde{r} < r_2$. The formula (3.6.39) is unusable; there are only two noisy observations, not an infinite number of perfect ones. We could try instead using linear interpolation:

$$\tilde{x}(\tilde{r}) = \frac{|r_2 - \tilde{r}|}{|r_2 - r_1|} y(r_1) + \frac{|r_1 - \tilde{r}|}{|r_2 - r_1|} y(r_2) \,. \tag{3.6.43}$$

If there are data points, r_i, $1 \le i \le M$, then another possibility is Aitken-Lagrange interpolation (Davis & Polonsky, 1965):

$$\tilde{x}(\tilde{r}) = \sum_{j=1}^{M} l_j(\tilde{r}) y_j \,, \tag{3.6.44}$$

$$l_j(\tilde{r}) = \frac{(\tilde{r} - r_1) \cdots (\tilde{r} - r_{j-1})(\tilde{r} - r_{j+1}) \cdots (\tilde{r} - r_M)}{(r_j - r_1) \cdots (r_j - r_{j-1})(r_j - r_{j+1}) \cdots (r_j - r_M)} \,. \tag{3.6.45}$$

Figure 3–14 shows these two examples.

Equation (3.6.43), (3.6.44)–(3.6.45) are only two of many possible interpolation formulas. When would one be better than the other? How good are the estimates? To answer these questions, let us take a different tack and employ the Gauss-Markov theorem, assuming we know something about the necessary covariances.

Suppose either $< x > = < n > = 0$ or that a known value has been removed from both (this just keeps our notation a bit simpler). Then,

$$\begin{aligned} \mathbf{R}_{xy}(\tilde{r}, r_j) \equiv & < x(\tilde{r}) y(r_j) > = < x(\tilde{r})(x(r_j) + n(r_j)) > \\ & = \mathbf{R}_{xx}(\tilde{r}, r_j) \end{aligned} \tag{3.6.46}$$

$$\begin{aligned} \mathbf{R}_{yy}(r_i, r_j) \equiv & < (x(r_i) + n(r_i))(x(r_j) + n(r_j)) > \\ & = \mathbf{R}_{xx}(r_i, r_j) + \mathbf{R}_{nn}(r_i, r_j) \,, \end{aligned} \tag{3.6.47}$$

where it has been assumed that $< x(r) n(q) > = 0$ for all r, q.

From (3.6.9), the best linear interpolator is

$$\tilde{\mathbf{x}} = \mathbf{B}\mathbf{y}, \quad \mathbf{B}(\tilde{r}, \tilde{r}_i) = \sum_{j=1}^{M} \mathbf{R}_{xx}(\tilde{r}, r_j) \left\{ \mathbf{R}_{xx} + \mathbf{R}_{nn} \right\}_{ji}^{-1} \tag{3.6.48}$$

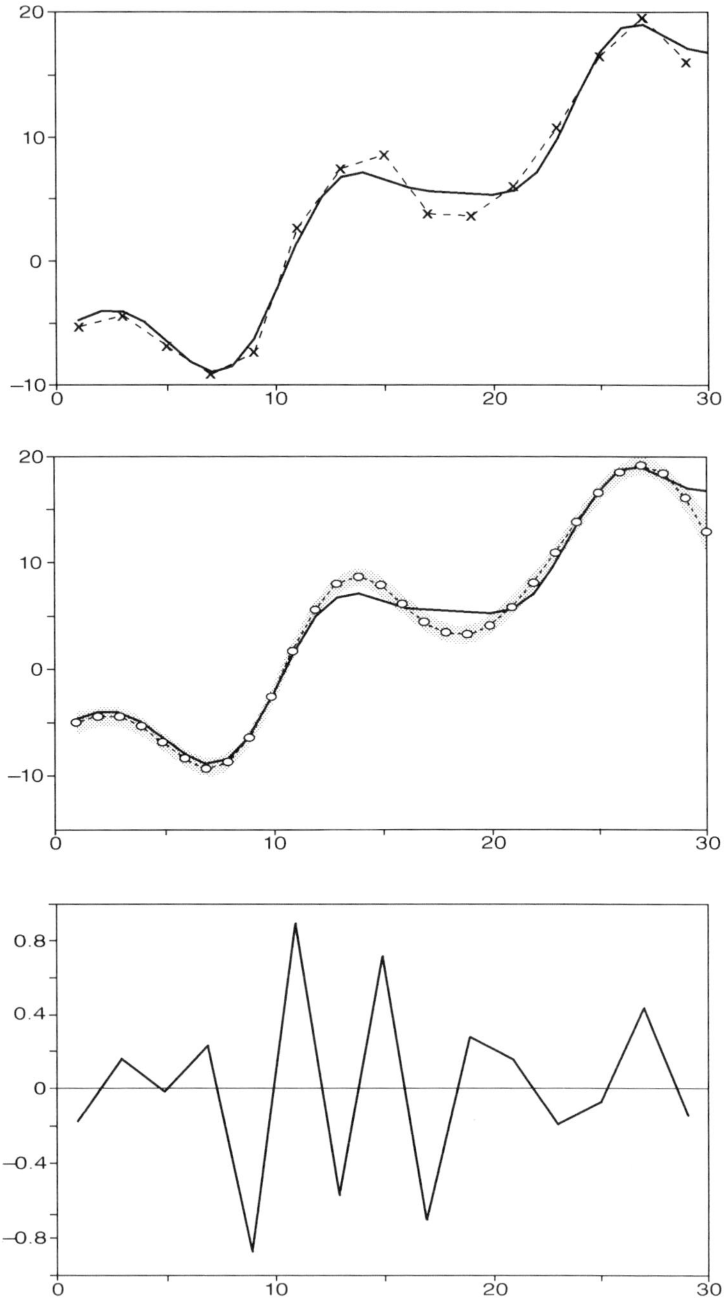

Figure 3–14. In the upper and middle panels, the solid curve represents the "true" values, generated as a function having covariance $S = 100\exp(-r^2/30)$. "Data" were then generated at every second point $(1, 3, \ldots)$ with pseudorandom white noise of variance 1. The upper panel shows linear interpolation of the data. Notice that the result interpolates literally, passing exactly through the point. The middle panel shows the result of using the Gauss-Markov estimate on the same pseudo-data. Estimated points do not agree exactly with the data, and an estimate of the expected one standard error is shown–perhaps the most important difference from the upper panel. Estimate and truth are generally consistent within two standard errors. The lowest panel displays the noise estimate at each data point from the Gauss-Markov estimate and appears, visually, suitably unstructured.

($\{\mathbf{R}_{xx} + \mathbf{R}_{nn}\}_{ji}^{-1}$ means the ji element of the inverse matrix), and the minimum possible error that results is

$$\mathbf{P}(\tilde{r}, \tilde{r}) = \mathbf{R}_{xx}(\tilde{r}, \tilde{r}) - \sum_{i}^{M} \sum_{j}^{M} \mathbf{R}_{xx}(\tilde{r}, r_j)\{\mathbf{R}_{xx} + \mathbf{R}_{nn}\}_{ji}^{-1} \mathbf{R}_{xx}(r_i, \tilde{r}) \quad (3.6.49)$$

[here $\mathbf{R}_{xx}$, $\mathbf{P}(\tilde{r}, \tilde{r})$ are both scalars], and $\tilde{\mathbf{n}} = \mathbf{y} - \tilde{\mathbf{x}}$.

Like the linear interpolation or the Aitken-Lagrange formula, or most other interpolation formulas, the optimal interpolator is simply a linear combination of the data. If any other set of weights $\mathbf{B}$ is chosen, then the interpolation is not as good in the mean-square error sense as it could be; the error of any such scheme can be obtained by substituting it into (3.6.7) and evaluating the result (the true covariances still need to be known.)

Looking back now at the two familiar formulas (3.6.43)–(3.6.45), it is clear what is happening: They represent a choice of $\mathbf{B}$. Unless the covariance is such as to produce one of the two sets of weights as the optimum choice, neither Aitken-Lagrange nor linear (nor any other common choice, like a spline) is the best one could do. Alternatively, if either of (3.6.43), (3.6.44)–(3.6.45) was thought to be the best one, they are equivalent to specifying the solution and noise covariances.

If interpolation is done for two points $\tilde{r}_\alpha$, $\tilde{r}_\beta$, the error of the two estimates will usually be correlated and represented by $\mathbf{P}(\tilde{r}_\alpha, \tilde{r}_\beta)$. Knowledge of the correlations between the errors in different interpolated points is often essential–for example, if one wishes to use uniformly spaced grid values so as to make estimates of derivatives of x. Such derivatives might be numerically meaningless if the mapping errors are small scale (relative to the grid spacing) and of large amplitude. But if the mapping errors are large scale compared to the grid, the derivatives may tend to remove the error and produce better estimates than for x itself.

Both linear and Aitken-Lagrange weights will produce estimates that are exactly equal to the observed values if $\tilde{r}_\alpha = r_i$–that is, on the data points themselves. Such a result is characteristic of *true interpolation.* In contrast, the Gauss-Markov estimate will differ from the data values at the data points, because the estimator attempts to reduce the noise in the data by averaging over all observations. The Gauss-Markov estimate is thus not a true interpolator; it is instead a *smoother* (smoothers will be encountered again in Chapter 6). One can recover true interpolation from the Gauss-Markov estimate if $\|\mathbf{R}_{nn}\| \to 0$, but being conscious that the matrix being inverted in (3.6.48) and (3.6.49) can become singular. If no noise is present, then the observed value is the correct one to use at a data point. The weights $\mathbf{B}$ can be complicated if there is any structure at all in either of $\mathbf{R}_{xx}$, $\mathbf{R}_{nn}$. The estimator takes explicit account of the expected spatial structure of both $\mathbf{x}$ and $\mathbf{n}$ to weight the data in such a way as to most effectively kill the noise relative to the signal. One is guaranteed that no other linear filter can do better.

If $\|\mathbf{R}_{nn}\| \gg \|\mathbf{R}_{xx}\|$, $\tilde{\mathbf{x}} \to 0$, manifesting the bias in the estimator–a bias introduced in the Gauss-Markov estimators so as to minimize the uncertainty (minimum variance about the true value). Thus, interpolated values tend toward zero, particularly far from the data points. For this reason, it is common to use expressions such as (3.6.29), (3.6.30) to first remove the mean, prior to mapping the residual, adding the estimated mean back in afterward. The interpolated values of the residuals are unbiased, because their true mean is nearly zero. Rigorous estimates of $\mathbf{P}$ for this approach require some care, as the mapped residuals contain variances owing to the uncertainty of the estimated mean (e.g., see Ripley, 1981, Section 5.2), but the corrections are commonly ignored.

The noise-free case would not normally be mapped with the Gauss-Markov estimator, and the presence of a realistic $\mathbf{R}_{nn}$ usually prevents singularity in $\mathbf{R}_{xx} + \mathbf{R}_{nn}$. Nonetheless, the general possibility of singularity should be examined and interpreted. This sum matrix is symmetric, and its SVD reduces to the symmetric form, (3.4.37),

$$\mathbf{R}_{xx} + \mathbf{R}_{nn} = \mathbf{U}\boldsymbol{\Lambda}\mathbf{U}^T . \tag{3.6.50}$$

If the sum covariance is positive-definite, $\boldsymbol{\Lambda}$ will be square with $K = M$, and the inverse will exist. If the sum covariance is not positive-definite but is only semidefinite, one or more of its singular values will vanish. The meaning is that there are *possible* structures in the data that have been assigned to neither the noise field nor the solution field. This situation is realistic if one is truly confident that $\mathbf{y}$ does not contain such structures. In that case, the solution

$$\tilde{\mathbf{x}} = \mathbf{R}_{xx}(\mathbf{R}_{xx} + \mathbf{R}_{nn})^{-1}\mathbf{y} = \mathbf{R}_{xx}(\mathbf{U}\boldsymbol{\Lambda}^{-1}\mathbf{U}^T)\mathbf{y} \tag{3.6.51}$$

will have components of the form $0/0$, the denominator corresponding to the zero singular values and the numerator to the absent, impossible, structures of $\mathbf{y}$. One can arrange that the ratio of these terms should be set to zero (e.g., by using the SVD). But such a delicate balance is not necessary. If one simply adds a small white-noise covariance to $\mathbf{R}_{xx}+\mathbf{R}_{nn} \to \mathbf{R}_{xx}+\mathbf{R}_{nn}+\epsilon^2\mathbf{I}$, one is assured by the discussion of tapering that $\mathbf{R}_{xx} + \mathbf{R}_{nn}$ is no longer singular–all structures in the field being assigned to either the noise or the solution (or in part to both).

Anyone using a Gauss-Markov estimator to make maps must do checks that the result is consistent with the prior estimates of $\mathbf{R}_{xx}$, $\mathbf{R}_{nn}$. Such checks include determining whether the difference between the mapped values at the data points and the observed values have numerical values consistent with the assumed noise variance; a further check involves the sample autocovariance of $\tilde{\mathbf{n}}$ and its test against $\mathbf{R}_{nn}$ (see books on regression for such tests). The mapped field should also have a variance and covariance consistent with the prior estimate $\mathbf{R}_{xx}$. If these tests are not passed, the entire result should be rejected.

A variant mapping problem is the construction of a streamfunction, $\Psi(\tilde{\mathbf{r}}_i)$, on a uniform grid, $\tilde{\mathbf{r}}_i$, from noisy measurements of a velocity field $[u(\mathbf{r}_j), v(\mathbf{r}_j)]$ at a collection of data points, $\mathbf{r}_j$. One has then a set of relations of the form

$$\Psi(\tilde{\mathbf{r}}_q) - \Psi(\tilde{\mathbf{r}}_{q'}) = \Delta y u(\mathbf{r}_j) + n(\mathbf{r}_j)$$
$$\Psi(\tilde{\mathbf{r}}_s) - \Psi(\tilde{\mathbf{r}}_{s'}) = -\Delta x v(\mathbf{r}_k) + n(\mathbf{r}_k)$$

where $\tilde{\mathbf{r}}_q$, $\tilde{\mathbf{r}}_{q'}$ are the grid points bracketing observation point $\mathbf{r}_j$ in the y–direction, over a distance Δy, and $\tilde{\mathbf{r}}_s$, $\tilde{\mathbf{r}}_{s'}$ bracket point $\mathbf{r}_k$ in the x–direction over a distance Δx, and are just another version of the problem of estimating the solution to a set of simultaneous equations.

3.6.6 Higher Dimensional Mapping

We can now immediately write down the optimal interpolation formulas for an arbitrary distribution of data in two or more dimensions. Let the positions where data are measured be the set $\mathbf{r}_j$ with measured value $\mathbf{y}(\mathbf{r}_j)$, containing noise $\mathbf{n}$. The mean value of the field is first estimated and subtracted from the measurements, and we proceed as though the true mean were zero. This problem was discussed by Bretherton et al. (1976); in meteorology, the method is associated with Gandin (1965). Fundamentally, it is nothing more than an application of the Gauss-Markov theorem in two (most commonly) dimensions. Fuller discussions may be found in Thièbaux and Pedder (1987) and Daley (1991).

One proceeds exactly as in the case where the positions are scalars, minimizing the expected mean-square difference between the estimated and the true field $\mathbf{x}(\tilde{\mathbf{r}}_\alpha)$. The result is (3.6.48), (3.6.49) except that now everything is a function of the vector positions. If the field being mapped is also a vector (e.g., two components of velocity) with known covariances between the two components, then the elements of $\mathbf{B}$ become matrices. The observations could also be vectors at each point.

An example of a two-dimensional map is shown in Figure 3–15a: The data points are the dots, while estimates of y on the uniform grid were wanted. The prior noise was described as $<\mathbf{n}> = 0$, $\mathbf{R}_{nn} = <n_i n_j> = \sigma_n^2 \delta_{ij}$, $\sigma_n^2 = 1$, and the true field covariance was $<\mathbf{x}> = 0$, $\mathbf{R}_{xx} = <\mathbf{x}(\mathbf{r}_i)\mathbf{x}(\mathbf{r}_j)> = P_0 \exp -|\mathbf{r}_i - \mathbf{r}_j|^2/L_2$, $P_0 = 25$, $L_2 = 9$. Figure 3–15b shows the estimated values and 3–15c the error variance estimate of the mapped values. Far from the data points, the estimated values are 0–that is, the mapped field goes asymptotically to the estimated true mean, and the error variance goes to the full value of 25, which cannot be exceeded. That is to say, when mapping far from any data point, the only real information available is provided by the prior statistics–the mean is 0, and the variance about that mean is 25. So the expected uncertainty of the mapped field in the absence of data cannot exceed the prior estimate of how far from the mean the true value is likely to be, with the best estimate being the mean itself.

The mapped field has a complex error structure even in the vicinity of the data points. Should a model be driven by this mapped field, one would need to make some provision in the model for accounting for the spatial change in the expected errors of this forcing.

In practice, most published objective mapping (often called *OI* for *objective interpolation*) has been based upon simple analytical statements of the covariances $\mathbf{R}_{xx}$, $\mathbf{R}_{nn}$ as used in the example; that is, they are commonly assumed to be spatially stationary and isotropic (depending on $|\mathbf{r}_i - \mathbf{r}_j|$ and not upon the two positions separately nor upon their orientation). The assumption is often qualitatively reasonable, but much of the ocean circulation is neither spatially stationary nor isotropic. Use of analytic forms removes the necessity for finding, storing, and computing with the potentially very large $M \times M$ data covariance matrices in which hypothetically every data or grid point has a different covariance with every other data or grid point. But the analytical convenience often distorts the solutions (see the discussion in Fukumori, Martel, & Wunsch, 1991).

3.6.7 Linear Combinations of Estimates and Mapping Derivatives

A common problem in setting up a general circulation model is to specify the fields of quantities like temperature, salinity, etc., on a regular model grid. One also often must specify the derivatives of these fields for use in equations like that of the advection-diffusion equation,

$$\frac{\partial C}{\partial t} + \mathbf{v} \cdot \nabla C = K \nabla^2 C \tag{3.6.52}$$

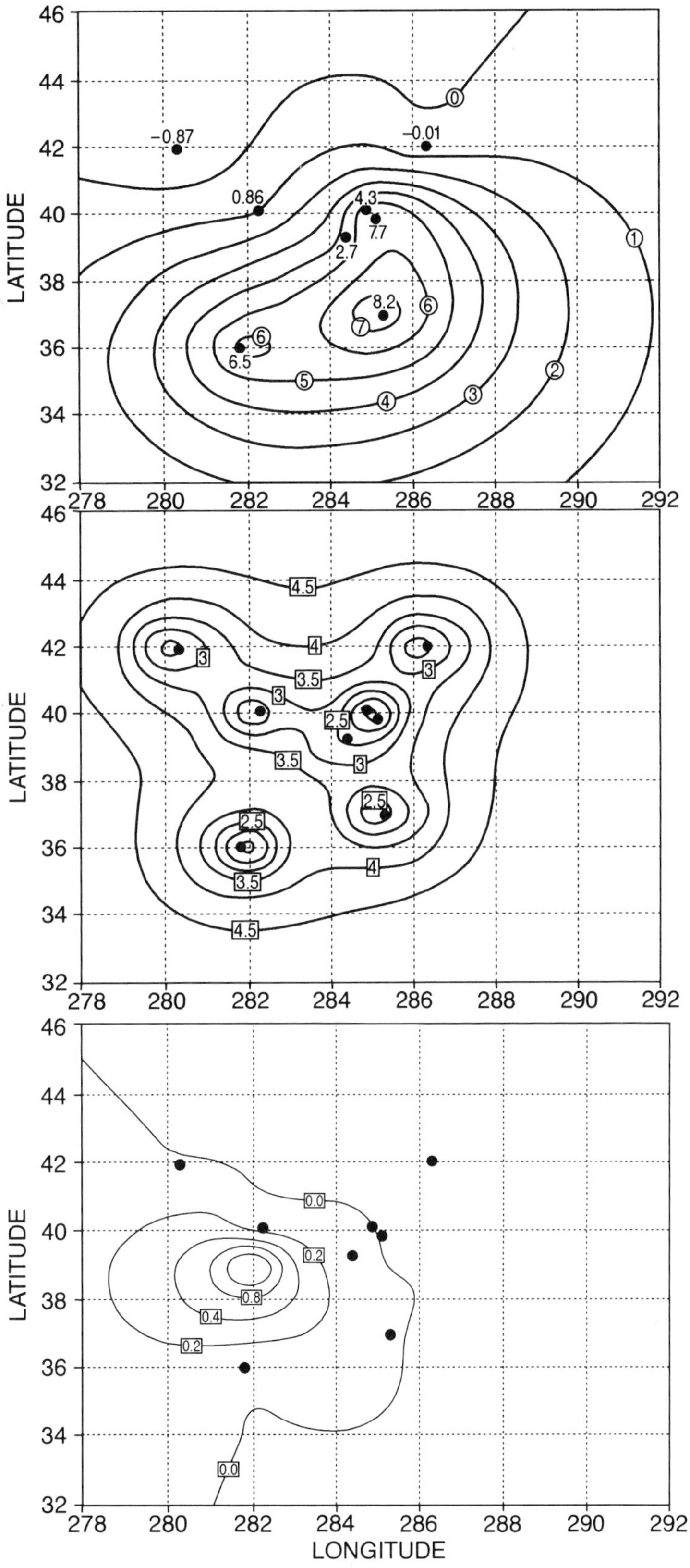

Figure 3–15. (a) Data points are assumed to be available at the locations marked with a solid dot and the values shown. The estimated field values on the regular grid (integer values of latitude and longitude) emulate what happens when data are interpolated for purposes of driving a model. The solution covariance was $R(|\mathbf{r}_i - \mathbf{r}_j|) = 25\exp(-|\mathbf{r}_i - \mathbf{r}_j|/9)$. The observation noise was assumed to be white of variance unity. Far from the data, the mapping tends toward the expected value (here, zero) because no other information about the correct value is available. (b) Standard error $\sqrt{P_{ii}}$ for the mapped field shown in (a). The values tend to 5–that is, $\sqrt{25}$–far from the data points, as the largest possible square error can never exceed the prior estimate of 25. In general, the expected errors are smallest near the data points. (c) One of the rows of $\mathbf{P}$, corresponding to the grid point on which the contours are centered (39°N, 282°E), displaying the correlations that occur in the expected errors of the mapped field at neighboring grid points. The variance was normalized to 1 for plotting.

where C is any scalar field of interest. Suppose one wished to estimate a derivative as a one-sided difference,

$$\frac{\partial C(\tilde{r}_1)}{\partial r} \sim \frac{C(\tilde{r}_1) - C(\tilde{r}_2)}{\tilde{r}_1 - \tilde{r}_2}. \tag{3.6.53}$$

Then one might think simply to subtract the two estimates made from Equation (3.6.48), producing

$$\Delta r \frac{\partial C(\tilde{r}_1)}{\partial r} \sim (\mathbf{R}_{xx}(\tilde{r}_1, r_j) - \mathbf{R}_{xx}(\tilde{r}_2, r_j))(\mathbf{R}_{xx} + \mathbf{R}_{nn})^{-1}_{jk}\mathbf{y}(r_k) \tag{3.6.54}$$

(a sum on j and k is implied).

Alternatively, suppose we tried to estimate $\partial C/\partial r$ directly from (3.6.5), using $\mathbf{x} = C(r_1) - C(r_2)$. $\mathbf{R}_{yy} = \mathbf{R}_{xx} + \mathbf{R}_{nn}$, which describes the data, does not change. $\mathbf{R}_{xy}$ does change:

$$\begin{aligned} \mathbf{R}_{xy} &= < (C(\tilde{r}_1) - C(\tilde{r}_2))(C(r_j) + n(r_j)) > \\ &= \mathbf{R}_{xx}(\tilde{r}_1, r_j) - \mathbf{R}_{xx}(\tilde{r}_2, r_j), \end{aligned} \tag{3.6.55}$$

which when substituted into (3.6.9) produces (3.6.54). *Thus, the optimal map of the finite difference field is simply the difference of the mapped values.* More generally, the optimally mapped value of any linear combination of the values is that linear combination of the maps (see Luenberger, 1969). Of particular importance is the estimate of an arbitrary linear combination of elements of $\tilde{\mathbf{x}}$, such as the finite difference derivative just considered, and the essential computation of their uncertainty. Consider any estimate $\tilde{\mathbf{x}}$, and a weighted sum

$$\tilde{H} = \mathbf{a}^T \tilde{\mathbf{x}} \tag{3.6.56}$$

where the constant vector $\mathbf{a}$ may be mostly zeros. The expected value of the sum is

$$< \tilde{H} > = \mathbf{a}^T < \tilde{\mathbf{x}} >, \tag{3.6.57}$$

whose bias depends directly on that of $\tilde{\mathbf{x}}$. If the uncertainty of $\tilde{\mathbf{x}}$ is $\mathbf{P}$, then one has immediately

$$< (\tilde{H} - H)^2 > = \mathbf{a}^T < (\tilde{\mathbf{x}} - \mathbf{x})(\tilde{\mathbf{x}} - \mathbf{x})^T > \mathbf{a} = \mathbf{a}^T \mathbf{P} \mathbf{a}. \tag{3.6.58}$$

3.7 Improving Solutions Recursively

An important idea in both least-squares approximation and estimation theory derives from the need to improve the result of an earlier computation with the arrival of some new data. In what follows, we initially will use the language of least squares, but because of the coincidence of the results for least squares with appropriate weight matrices, and minimum variance estimation, we will obtain the correct result for estimation problems, too.

Suppose we have solved the system (3.3.2), using any one of the procedures discussed above. Because we will add data, some extra notation is needed. Rewrite (3.3.2) as

$$\mathbf{E}(1)\mathbf{x}(1) + \mathbf{n}(1) = \mathbf{y}(1) \tag{3.7.1}$$

where the noise $\mathbf{n}(1)$ has zero mean and covariance matrix $\mathbf{R}_{nn}(1)$. Let the estimate of the solution to (3.7.1) from one of the estimators be written as $\tilde{\mathbf{x}}(1)$, with uncertainty $\mathbf{P}(1)$. As a specific example, suppose (3.7.1) is full-rank overdetermined and was solved using row-weighted least-squares solution as

$$\tilde{\mathbf{x}}(1) = \left(\mathbf{E}(1)^T \mathbf{R}_{nn}(1)^{-1} \mathbf{E}(1)\right)^{-1} \mathbf{E}(1)^T \mathbf{R}_{nn}(1)^{-1} \mathbf{y}(1) \tag{3.7.2}$$

with corresponding $\mathbf{P}(1)$ (no column weights are used because we know they are irrelevant for a full-rank overdetermined problem).

Some new observations, $\mathbf{y}(2)$, are obtained, with the error covariance of the new observations given by $\mathbf{R}_{nn}(2)$ so that the problem is now

$$\left\{\begin{matrix} \mathbf{E}(1) \\ \mathbf{E}(2) \end{matrix}\right\} \mathbf{x} + \begin{bmatrix} \mathbf{n}(1) \\ \mathbf{n}(2) \end{bmatrix} = \begin{bmatrix} \mathbf{y}(1) \\ \mathbf{y}(2) \end{bmatrix} \tag{3.7.3}$$

where $\mathbf{x}$ is the same unknown. We assume $< \mathbf{n}(2) > = \mathbf{0}$ and

$$< \mathbf{n}(1)\mathbf{n}(2)^T > = \mathbf{0}\,, \tag{3.7.4}$$

that is, no correlation of the old and new measurement error (this assumption is very important, and particular attention is called to it). A solution to (3.7.3) should give a better estimate of $\mathbf{x}$ than (3.7.1) alone because more observations are available. It is sensible to row weight the concatenated set to

$$\left\{\begin{matrix} \mathbf{R}_{nn}(1)^{-T/2}\mathbf{E}(1) \\ \mathbf{R}_{nn}(2)^{-T/2}\mathbf{E}(2) \end{matrix}\right\} \mathbf{x} + \begin{bmatrix} \mathbf{R}_{nn}(1)^{-T/2}\mathbf{n}(1) \\ \mathbf{R}_{nn}(2)^{-T/2}\mathbf{n}(2) \end{bmatrix} = \begin{bmatrix} \mathbf{R}_{nn}(1)^{-T/2}\mathbf{y}(1) \\ \mathbf{R}_{nn}(2)^{-T/2}\mathbf{y}(2) \end{bmatrix}. \tag{3.7.5}$$

Recursive weighted least squares seeks the solution to (3.7.5) without inverting the new, larger matrix by taking advantage of the existing knowledge of $\mathbf{x}$ already in hand from (3.7.2). Because of (3.7.4), the objective function corresponding to finding the minimum weighted error norm is

$$\begin{aligned} J = {} & [\mathbf{y}(1) - \mathbf{E}(1)\mathbf{x}]^T \mathbf{R}_{nn}(1)^{-1} [\mathbf{y}(1) - \mathbf{E}(1)\mathbf{x}] \\ & + [\mathbf{y}(2) - \mathbf{E}(2)\mathbf{x}]\, \mathbf{R}_{nn}(2)^{-1} [\mathbf{y}(2) - \mathbf{E}(2)\mathbf{x}]\,. \end{aligned} \tag{3.7.6}$$

Taking the derivatives with respect to $\mathbf{x}$, the least-squares solution is

$$\tilde{\mathbf{x}}(2) = \left\{\mathbf{E}(1)^T\mathbf{R}_{nn}(1)^{-1}\mathbf{E}(1) + \mathbf{E}(2)^T\mathbf{R}_{nn}(2)^{-1}\mathbf{E}(2)\right\}^{-1} \times \left\{\mathbf{E}(1)^T\mathbf{R}_{nn}(1)^{-1}\mathbf{y}(1) + \mathbf{E}(2)^T\mathbf{R}_{nn}(2)^{-1}\mathbf{y}(2)\right\}. \tag{3.7.7}$$

But one can manipulate (3.7.7) into (e.g., see Brogan, 1985, or Stengel, 1986)

$$\begin{aligned}\tilde{\mathbf{x}}(2) &= \tilde{\mathbf{x}}(1) + \mathbf{P}(1)\mathbf{E}(2)^T\left[\mathbf{E}(2)\mathbf{P}(1)\mathbf{E}(2)^T + \mathbf{R}_{nn}(2)\right]^{-1}\left[\mathbf{y}(2) - \mathbf{E}(2)\tilde{\mathbf{x}}(1)\right] \\ &= \tilde{\mathbf{x}}(1) + \mathbf{K}(2)\left[\mathbf{y}(2) - \mathbf{E}(2)\tilde{\mathbf{x}}(1)\right]\end{aligned} \tag{3.7.8}$$

where

$$\mathbf{K}(2) = \mathbf{P}(1)\mathbf{E}(2)^T\left[\mathbf{E}(2)\mathbf{P}(1)\mathbf{E}(2)^T + \mathbf{R}_{nn}(2)\right]^{-1} \tag{3.7.9}$$

$$\tilde{\mathbf{n}}(2) = \mathbf{y}(2) - \mathbf{E}(2)\tilde{\mathbf{x}}(2), \tag{3.7.10}$$

and this improved estimate has uncertainty

$$\mathbf{P}(2) = \mathbf{P}(1) - \mathbf{K}(2)\mathbf{E}(2)\mathbf{P}(1). \tag{3.7.11}$$

These last equations are algebraically identical to the alternate forms found from the matrix inversion lemma (3.1.25):

$$\begin{aligned}\tilde{\mathbf{x}}(2) = {} & \left(\mathbf{E}(2)^T\mathbf{R}_{nn}(2)^{-1}\mathbf{E}(2)\right)^{-1}\left\{\mathbf{P}(1) + \left[\mathbf{E}(2)^T\mathbf{R}_{nn}(2)^{-1}\mathbf{E}(2)\right]^{-1}\right\}^{-1}\tilde{\mathbf{x}}(1) \\ & + \mathbf{P}(1)\left\{\mathbf{P}(1) + \left[\mathbf{E}(2)^T\mathbf{R}_{nn}(2)^{-1}\mathbf{E}(2)\right]^{-1}\right\}^{-1}\left[\mathbf{E}(2)^T\mathbf{R}_{nn}(2)^{-1}\mathbf{E}(2)\right]^{-1} \times \\ & \mathbf{E}(2)\mathbf{R}_{nn}(2)^{-1}\mathbf{y}(2),\end{aligned} \tag{3.7.12}$$

$$\mathbf{P}(2) = \left\{\mathbf{P}(1)^{-1} + \mathbf{E}(2)^T\mathbf{R}_{nn}(2)^{-1}\mathbf{E}(2)\right\}^{-1}. \tag{3.7.13}$$

The two different boxed sets differ only in the matrix sizes to be inverted, and a choice between them is typically based upon computational loads (in some large problems, matrix inversion may prove less onerous than matrix multiplication).

The solution is just the least-squares solution to the full set but rearranged after a bit of algebra. The original data, $\mathbf{y}(1)$, and coefficient matrix, $\mathbf{E}(1)$, have disappeared, to be replaced by the first solution $\tilde{\mathbf{x}}(1)$ and its uncertainty $\mathbf{P}(1)$. That is to say, one need not retain the original data and $\mathbf{E}(1)$ for the new solution to be computed. Furthermore, because the new solution depends only upon $\tilde{\mathbf{x}}(1)$, $\mathbf{P}(1)$, the particular methodology originally employed for obtaining them is irrelevant (i.e., they might have actually been obtained from an educated guess or through some arbitrarily complex model computation). Finally, the structure of the improved solution (3.7.8) is interesting and suggestive. It is made up of two terms: the previous estimate plus a term proportional to the difference between the new observations $\mathbf{y}(2)$, and a prediction of what those observations should have been were the first estimate the wholly correct one and the new observations perfect. It thus has the form of a *predictor-corrector.*

The difference between the prediction and the forecast can be called the *prediction error*, but recall that there is observational noise in $\mathbf{y}(2)$. The new estimate is a weighted average of this difference and the prior estimate, with the weighting depending upon the details of the uncertainty of prior estimate and new data. The behavior of the updated estimate is worth understanding in various limits. For example, suppose the initial uncertainty estimate is diagonal, $\mathbf{P}(1) = \Delta^2\mathbf{I}$, or that one rotates $\mathbf{x}$ into a new space of uncorrelated uncertainty. Then

$$\mathbf{K}(2) = \mathbf{E}(2)^T\left(\mathbf{E}(2)\mathbf{E}(2)^T + \mathbf{R}_{nn}(2)/\Delta^2\right)^{-1}. \tag{3.7.14}$$

If the norm of $\mathbf{R}_{nn}(2)$ is small compared to that of $\Delta^2\mathbf{I}$ and if (to be specific only) the second set of observations is by itself full-rank underdetermined, then

$$\mathbf{K}(2) \to \mathbf{E}(2)^T\left(\mathbf{E}(2)\mathbf{E}(2)^T\right)^{-1}$$

and

$$\begin{aligned}\tilde{\mathbf{x}}(2) &= \tilde{\mathbf{x}}(1) + \mathbf{E}(2)^T\left(\mathbf{E}(2)\mathbf{E}(2)^T\right)^{-1}[\mathbf{y}(2) - \mathbf{E}(2)\tilde{\mathbf{x}}(1)] \\ &= \left[\mathbf{I} - \mathbf{E}(2)^T\left(\mathbf{E}(2)\mathbf{E}(2)^T\right)^{-1}\mathbf{E}(2)\right]\tilde{\mathbf{x}}(1) + \mathbf{E}(2)^T\left(\mathbf{E}(2)\mathbf{E}(2)^T\right)^{-1}\mathbf{y}(2)\end{aligned} \tag{3.7.15}$$

where $[\mathbf{I} - \mathbf{E}(2)^T(\mathbf{E}(2)\mathbf{E}(2)^T)\mathbf{E}(2)^{-1}]$ will be recognized as the nullspace projector (3.4.114) of $\mathbf{E}(2)$. The update is replacing the first estimate by the estimate from the second set of observations, which were deemed perfect, but keeping unchanged any components of $\tilde{\mathbf{x}}(1)$ in the nullspace of $\mathbf{E}(2)$ because no new information is available about them. Should the new observations be fully determined and perfect, then the previous estimate is wholly replaced by the estimate made from the new, low-noise observations.

At the opposite extreme, when the new observations are very noisy compared to the previous ones, $\mathbf{K}(2)$ will be comparatively small, and the previous estimate is left largely unchanged. The general case represents a weighted average of the previous and new data, the weighting depending both upon the relative noise in each and upon the structure of the observations relative to the structure of $\mathbf{x}$.

The matrix being inverted in (3.7.8)–(3.7.11) is the sum of the measurement error covariance $\mathbf{R}_{nn}(2)$, and the error covariance of the "forecast" $\mathbf{E}(2)\tilde{\mathbf{x}}(1)$. To see this, let $\boldsymbol{\gamma}(1)$ be the error component in $\tilde{\mathbf{x}}(1)$, which by definition has covariance $< \boldsymbol{\gamma}(1)\boldsymbol{\gamma}(1)^T > = \mathbf{P}(1)$. Then the expected covariance of the error of prediction is $< \mathbf{E}(2)\boldsymbol{\gamma}(1)\boldsymbol{\gamma}(1)^T\mathbf{E}(2)^T > = \mathbf{E}(2)\mathbf{P}(1)\mathbf{E}(2)^T$, which appears in $\mathbf{K}(2)$. Because of the assumptions (3.7.4) and $< \boldsymbol{\gamma}(1)x(1)^T > = \mathbf{0}$, it follows that

$$< \mathbf{y}(1)[\mathbf{y}(2) - \mathbf{E}(2)\tilde{\mathbf{x}}(1)] > = \mathbf{0}\,. \tag{3.7.16}$$

That is, the *innovation*, $\mathbf{y}(2) - \mathbf{E}(2)\tilde{\mathbf{x}}(1)$, is uncorrelated with the previous measurement.

It is useful to notice that Equations (3.5.11)–(3.5.12), the solution to the least-squares problem subject to certain perfect constraints imposed by a Lagrange multiplier, can be recovered from (3.7.8)–(3.7.13) by putting $\mathbf{E}(2) = \mathbf{A}$, $\mathbf{y}(2) = \mathbf{q}$, $\mathbf{P}(1) = (\mathbf{E}^T\mathbf{E})^{-1}$, $\mathbf{R}_{nn}(2) \to 0$. That is, this earlier solution can be conceived of as having been obtained by first solving the conventional least-squares problem and then being modified by the later information that $\mathbf{A}\mathbf{x} = \mathbf{q}$ with very high accuracy.

Finally, suppose given $\tilde{\mathbf{x}}(1)$, $\tilde{\mathbf{y}}(1)$ that we regard $\Delta\mathbf{y} = \mathbf{y}(2) - \mathbf{E}(2)\tilde{\mathbf{x}}(1)$ as the discrepancy between an initial estimate of $\mathbf{x}$ and what the new data suggest is correct. Putting $\Delta\mathbf{x} = \tilde{\mathbf{x}}(2) - \tilde{\mathbf{x}}(1)$, we have an ordinary estimation problem with $< \Delta\mathbf{x}\,\Delta\mathbf{x}^T > = \mathbf{P}(1)$,

$$\mathbf{E}(2)\Delta\mathbf{x} + \mathbf{n}(2) = \Delta\mathbf{y}\,.$$

The solution by the Gauss-Markov estimate (3.6.16)–(3.6.18) (or least squares) is

$$\Delta\tilde{\mathbf{x}} = \mathbf{P}(1)\mathbf{E}(2)^T(\mathbf{E}(2)\mathbf{P}(1)\mathbf{E}(2)^T + \mathbf{R}_{nn}(2))^{-1}\Delta\mathbf{y}\,,$$

which if added to $\tilde{\mathbf{x}}(1)$ produces (3.7.8).

The possibility of a recursion based on either of (3.7.8)–(3.7.11) or (3.7.12), (3.7.13) should now be obvious–all argument–1 variables being replaced by argument–2 variables, which in turn are replaced by argument–3 variables, etc. A practical example, as applied to altimetric data, may be seen in Wunsch (1991).

The computational load of the recursive solution needs to be addressed. If all of the constraints are available at once and used, the solution can be found as in any least-squares problem without ever computing the solution uncertainty (although its utility without the uncertainty may be doubted).

But if the constraints are divided and used in two or more groups, then the uncertainty must be computed one or more times to carry out the improvement. In general, owing to the need to compute the uncertainties, it is more efficient to use all of the constraints at once (if available, and if the computer can handle them) than it is to divide them into groups–unless special structures are present in the $\mathbf{E}(t)$. Oceanography has a particular need for recursive methods, however. The global-scale data flow is not well organized, and data tend to drift in to scientists over lengthy periods of time. It is a considerable advantage to be able to improve estimates when previously unavailable data finally appear.

The comparatively simple interpretation of the recursive, weighted least-squares problem will be used in Chapter 6 to derive the Kalman filter and suboptimal filters in a very simple form. It also becomes the key to understanding *assimilation* schemes such as *nudging*, *forcing to climatology*, and *robust diagnostic* methods.

If the initial set of equations (3.7.1) is actually underdetermined and should it have been solved using the SVD, one must be careful that $\mathbf{P}(1)$ includes the estimated error owing to the missing nullspace. Otherwise, these elements would be assigned zero error variance, and the new data could never affect them.

Consider another special case. Let there be a prior best estimate of the solution, which we will call $\tilde{\mathbf{x}}(0)$, and which is assumed to be zero. Specify an initial uncertainty (most often diagonal),

$$\mathbf{P}(0) = < \tilde{\mathbf{x}}(0)\tilde{\mathbf{x}}(0)^T > ,$$

and assuming a true mean of zero, treat this estimate and its uncertainty as the first set of data, replacing all the "1" estimates with those now relabeled "0":

$$\tilde{\mathbf{x}}(1) = \mathbf{0} + \mathbf{K}(1)\left[\mathbf{y}(1) - \mathbf{E}(1)\mathbf{0}\right] , \tag{3.7.17}$$

$$\mathbf{K}(1) = \mathbf{P}(0)\mathbf{E}(1)^T\left[\mathbf{E}(1)\mathbf{P}(0)\mathbf{E}(1)^T + \mathbf{R}_{nn}(1)\right]^{-1} ,$$

$$\mathbf{P}(1) = \mathbf{P}(0) - \mathbf{K}(1)\mathbf{E}(1)\mathbf{P}(0) . \tag{3.7.18}$$

The objective mapping estimate discussed in (3.6.16)–(3.6.18) is identical with (3.7.17)–(3.7.18). There, the prior estimate of the field at an interpolation point is 0; the prior uncertainty of the field corresponds to its estimated second moments, $\mathbf{R}_{xx}$, and the observation noise covariance is $\mathbf{R}_{nn}$. $\mathbf{K}$ interpolates from the data points $\mathbf{r}_i$ to the grid points $\tilde{\mathbf{r}}_\alpha$. Thus, as asserted, the Gauss-Markov mapping estimate coincides with the least-squares one, and we can regard objective mapping as a special case of recursive least squares.

Let us confirm more generally that the recursive least-squares result is identical to a recursive estimation procedure. Suppose there exist two estimates of an unknown vector $\mathbf{x}$, denoted $\tilde{\mathbf{x}}_a$, $\tilde{\mathbf{x}}_b$ with estimated uncertainties $\mathbf{P}_a$, $\mathbf{P}_b$, respectively. They are either unbiased or have the same bias–that is, $<\tilde{\mathbf{x}}_a> = <\tilde{\mathbf{x}}_b> = \mathbf{x}_B$. How should the two be combined to give a third estimate, $\tilde{\mathbf{x}}^+$, with minimum error? Try a linear combination

$$\tilde{\mathbf{x}}^+ = \mathbf{L}_a\tilde{\mathbf{x}}_a + \mathbf{L}_b\tilde{\mathbf{x}}_b\,. \tag{3.7.19}$$

If the new estimate is to be unbiased or is to retain the prior bias, it follows that

$$<\tilde{\mathbf{x}}^+> = \mathbf{L}_a <\tilde{\mathbf{x}}_a> + \mathbf{L}_b <\tilde{\mathbf{x}}_b> \tag{3.7.20}$$

or

$$\begin{aligned}\mathbf{x}_B &= \mathbf{L}_a\mathbf{x}_B + \mathbf{L}_b\mathbf{x}_B\\ \mathbf{L}_b &= \mathbf{I} - \mathbf{L}_a\,.\end{aligned} \tag{3.7.21}$$

Then the uncertainty is

$$\begin{aligned}<(\tilde{\mathbf{x}}^+ - \mathbf{x})(\tilde{\mathbf{x}}^+ - \mathbf{x})^T> &= <(\mathbf{L}_a\tilde{\mathbf{x}}_a + (\mathbf{I}-\mathbf{L}_a)\tilde{\mathbf{x}}_b - \mathbf{x})(\mathbf{L}_a\tilde{\mathbf{x}}_a + (\mathbf{I}-\mathbf{L}_a)\tilde{\mathbf{x}}_b - \mathbf{x})^T>\\ &= \mathbf{L}_a\mathbf{P}_a\mathbf{L}_a^T + (\mathbf{I}-\mathbf{L}_a)\mathbf{P}_b(\mathbf{I}-\mathbf{L}_a)^T\end{aligned} \tag{3.7.22}$$

where the assumption that the errors in $\mathbf{x}_a$, $\mathbf{x}_b$ are uncorrelated has been used. This expression is positive-definite; minimizing with respect to $\mathbf{L}_a$ yields immediately

$$\mathbf{L}_a = \mathbf{P}_b(\mathbf{P}_a + \mathbf{P}_b)^{-1}, \quad \mathbf{L}_b = \mathbf{P}_a(\mathbf{P}_a + \mathbf{P}_b)^{-1}\,.$$

The new combined estimate is then

$$\tilde{\mathbf{x}}^+ = \mathbf{P}_b(\mathbf{P}_a + \mathbf{P}_b)^{-1}\tilde{\mathbf{x}}_a + \mathbf{P}_a(\mathbf{P}_a + \mathbf{P}_b)^{-1}\tilde{\mathbf{x}}_b\,.$$

This last expression can be rewritten by adding and subtracting $\tilde{\mathbf{x}}_a$ as

$$\begin{aligned}\tilde{\mathbf{x}}^+ &= \tilde{\mathbf{x}}_a + \mathbf{P}_b(\mathbf{P}_a+\mathbf{P}_b)^{-1}\tilde{\mathbf{x}}_a + \mathbf{P}_a(\mathbf{P}_a+\mathbf{P}_b)^{-1}\tilde{\mathbf{x}}_b - (\mathbf{P}_a+\mathbf{P}_b)(\mathbf{P}_a+\mathbf{P}_b)^{-1}\tilde{\mathbf{x}}_a\\ &= \tilde{\mathbf{x}}_a + \mathbf{P}_a(\mathbf{P}_a + \mathbf{P}_b)^{-1}(\tilde{\mathbf{x}}_b - \tilde{\mathbf{x}}_a)\,.\end{aligned} \tag{3.7.23}$$

The uncertainty of the estimate (3.7.23) is easily evaluated as

$$\mathbf{P}^+ = (\mathbf{P}_a^{-1} + \mathbf{P}_b^{-1})^{-1}\,, \tag{3.7.24}$$

which, by straightforward application of the matrix inversion lemma (3.1.24), is

$$\mathbf{P}^{+} = \mathbf{P}_a - \mathbf{P}_a(\mathbf{P}_a + \mathbf{P}_b)^{-1}\mathbf{P}_a\,. \tag{3.7.25}$$

Equations (3.7.23)–(3.7.25) are the general rules for combining two estimates with uncorrelated errors.

Now suppose that $\tilde{\mathbf{x}}_a$ and its uncertainty are known and that there are measurements

$$\mathbf{E}(2)\mathbf{x} + \mathbf{n}(2) = \mathbf{y}(2) \tag{3.7.26}$$

with $<\mathbf{n}(2)> = 0$, $<\mathbf{n}(2)\mathbf{n}(2)^T> = \mathbf{R}_{nn}(2)$. From this second set of observations, we *estimate* the solution, using the Gauss-Markov estimator (3.6.20)–(3.6.22) with no prior estimate of the solution variance–that is, $\|\mathbf{R}_{xx}^{-1}\| \to 0$–so that

$$\tilde{\mathbf{x}}_b = \left(\mathbf{E}(2)^T\mathbf{R}_{nn}(2)^{-1}\mathbf{E}(2)\right)^{-1}\mathbf{E}(2)^T\mathbf{R}_{nn}(2)^{-1}\mathbf{y}(2) \tag{3.7.27}$$

$$\mathbf{P} = \left(\mathbf{E}(2)^T\mathbf{R}_{nn}(2)^{-1}\mathbf{E}(2)\right)^{-1}. \tag{3.7.28}$$

Subsituting (3.7.27), (3.7.28) into (3.7.23)–(3.7.25) and again using the matrix inversion lemma gives

$$\tilde{\mathbf{x}}^{+} = \tilde{\mathbf{x}}_a + \mathbf{P}_a\mathbf{E}(2)^T\left[\mathbf{E}(2)\mathbf{P}_a\mathbf{E}(2)^T + \mathbf{R}_{nn}(2)\right]^{-1}\left[\mathbf{y}(2) - \mathbf{E}(2)\tilde{\mathbf{x}}_a\right], \tag{3.7.29}$$

which is the same as (3.7.8); thus a recursive minimum variance estimate coincides with a corresponding weighted least-squares recursion. The covariance may also be confirmed to be (3.7.11). The alternate forms (3.7.12), (3.7.13) are also correct.

3.8 Estimation from Linear Constraints–A Summary

A number of different procedures for producing estimates of the solution to a set of noisy simultaneous equations of arbitrary dimension have been described here. The reader may wonder which of the variants makes the most sense to use in practice. There is no single best answer because in the presence of noise one is dealing with a statistical estimation problem, and one must be guided by model context and goals. A few general remarks might be helpful.

In any problem where data are to be used to make inferences about physical parameters, one typically needs some approximate idea of just how large the solution is likely to be and how large the residuals probably are. In this nearly agnostic case, where almost nothing else is known, and the problem is very large, the weighted, tapered least-squares solution (Section 3.3.2) is a good first choice—it is easily and efficiently computed and coincides with the Gauss-Markov and tapered SVD solutions for this situation if the weight matrices are the appropriate variances. Sparse matrix methods for its solution exist (e.g., Paige & Saunders, 1982) should that be necessary. Coincidence with the Gauss-Markov solution means that one can reinterpret it as a minimum-variance solution should one wish (and for Gaussian variables, it is also the maximum likelihood solution).

It is a comparatively easy matter to vary the trade-off parameter, α^2, to explore the consequences of any errors in specifying the noise and solution variances. Once a value for α^2 is known, the tapered SVD can be computed to understand the relationships between solution and data structures, their resolution, and their variance. For problems of small to moderate size (the meaning of *moderate* is constantly shifting, but it is difficult to examine and interpret matrices of more than order 1000×1000), the SVD, whether in the truncated or tapered forms is probably the method of choice–because it provides the fullest information about data and its relationship to the solution. Its only disadvantages are that one can be easily overwhelmed by the available information, particularly if a range of solutions must be examined, and it cannot take advantage of sparsity in large problems. The SVD has a flexibility beyond even what we have discussed should the investigator know enough to justify it. One could, for example, change the degree of tapering in each of the terms of (3.4.133)–(3.4.134) should there be reason to repartition the variance between solution and noise, or some terms could be dropped out of the truncated form at will.

The more general situation, in which structured solution and noise covariances are available, is then readily understood. These matrices are used to reduce the problem by coordinate transformation to ones in which the structure has been removed. At that point, the methods for unstructured problems are used, with the resulting solution, residuals, covariances, and resolution matrices being transformed back to the original physical spaces.

Both ordinary weighted least squares and the SVD applied to row- and column-weighted equations are best thought of as approximation, rather than estimation, methods and thus have a lot to recommend them. In particular, the truncated SVD does not produce a minimum variance estimate the way the tapered version can. On the other hand, the tapered SVD (along with the Gauss-Markov estimate, or the tapered least-squares solutions) produces the minimum variance property by tolerating a bias in the solution. Whether the bias is more desirable than a larger uncertainty is a decision that the user must make. But the reader is warned against the belief that there is any single best method whose determination should take precedence over understanding the problem physics.

A useful working definition now of an inverse method, distinguishing them from mere curve fitting, is that it quantifies the extent to which elements of a system have been determined by focusing on uncertainties in the solution. Different approaches have different desirable features, including (1) separation of nullspace uncertainties from those owing to observational noise, (2) ability to use prior statistical knowledge, (3) determination of orthogonal solution structures in terms of orthogonal data structures and of their relative importance (data ranking), and (4) ability to trade resolution against stability.

The statistical discussion here has been qualitative and intuitive with no claim to rigor. To some extent, the subject of inferring the ocean circulation from observations and dynamics has not yet evolved to the point where more than semiquantitative statistical tests seem warranted. One can expect that ultimately more refined tests leading to adoption or rejection of particular dynamical models will one day become necessary. The reader wishing a more careful account of the statistical underpinnings of the subject can make a beginning with Tarantola (1987) or Backus (1970a,b; 1988a) and the references there.

4
The Steady Ocean Circulation Inverse Problem

The purpose of this chapter is to bring to bear the mathematical machinery of Chapter 3 onto the problem of determining the oceanic general circulation. The focus is on the thermal wind equations, as outlined in Chapter 2, but the methods apply to a broad variety of problems throughout oceanography and science generally. The approach initially will be to work with a dataset that has toylike qualities–far smaller in size than one would use in practice and hence inaccurate as a representation of the ocean. Nonetheless, the data are taken from real observations, have the structure of a real situation, and raise many of the problems of practice. This approach permits the working through of a series of estimation problems in a form where complete results can be displayed readily. Realistic problems present the same issues, but if the uncertainty matrices are 1000×1000, there are problems of display and discussion. Following these simple examples, the results of a number of published computations addressing the ocean circulation will be described.

The central idea is very simple: The equations of motion (2.1.1)–(2.1.5) are to be combined with whatever observations are available so as to estimate the ocean circulation, along with an estimate of the errors. Recall from Chapter 2, Equation (2.5.1), that knowledge of the density field in a steady flow satisfying the thermocline equations (2.1.11)–(2.1.15) is formally adequate to determine completely the three components of the absolute velocity. But because the observed density field is corrupted owing to sampling errors, it cannot be assumed to be consistent with the equations of motion. The focus, therefore, of the present chapter is the problem of the inference of the flow field in the presence of noise in the observations under the assumption that a large-scale steady flow field exists. This is the classical

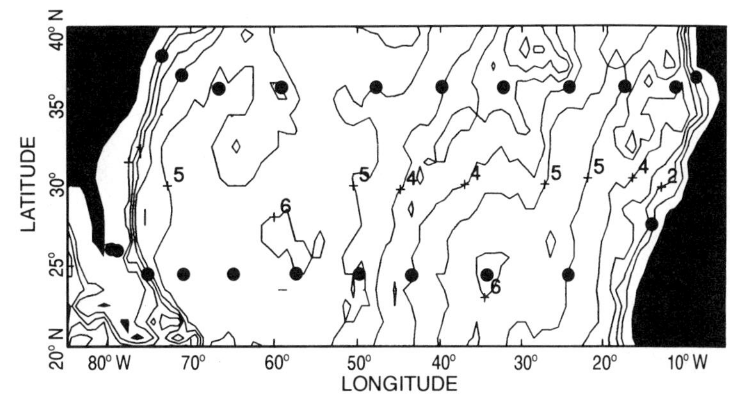

Figure 4–1. Positions of hydrographic stations used in example. The station data are real, but the horizontal spacing is much too large for accurate estimates of oceanic property fluxes. Depth contours are in kms.

problem of physical oceanography. The prototype problem is the linearized box balance equations of Chapter 2 which produce the canonical form

$$\mathbf{Ex} + \mathbf{n} = \mathbf{y}\,. \tag{4.0.1}$$

The specific problem supposes that there exist two hydrographic sections involving only 21 station pairs, as depicted in Figure 4–1, spanning the Atlantic Ocean, and where a full suite of measurements (temperature, salinity, oxygen, and nutrients) is available. The numbers we will use are quite realistic, because they have been obtained by subsampling the full Atlantic hydrographic section across 24°, 36°N displayed in Figure 2–2h,i. But the subsampling is so extreme that the reader is cautioned not to attempt to use the results as an estimate of the actual North Atlantic flow. In the vertical, the values will be used at the standard depths listed in Table 4–1 rather than at the original 2–decibar interval of the CTD measurements. Figure 4–2 displays the temperature of the three pseudosections we will use. The first section is in the Florida Straits where the Gulf Stream lies. It is separated by the Bahama Bank from the 24°N section. The 36°N section produces, with the other two sections, a closed volume typical of box model inversions. The topography shown is a correspondingly subsampled (not averaged) version of the true topography.

Table 4–1. *Standard depths (in decibars) used in example computation.*

0	50	100	150	200	250	300	400	500	600	700	800
900	1000	1100	1200	1300	1400	1500	1750	2000	2250	2500	2750
3000	3250	3500	3750	4000	4250	4500	4750	5000	5500	6000	7000

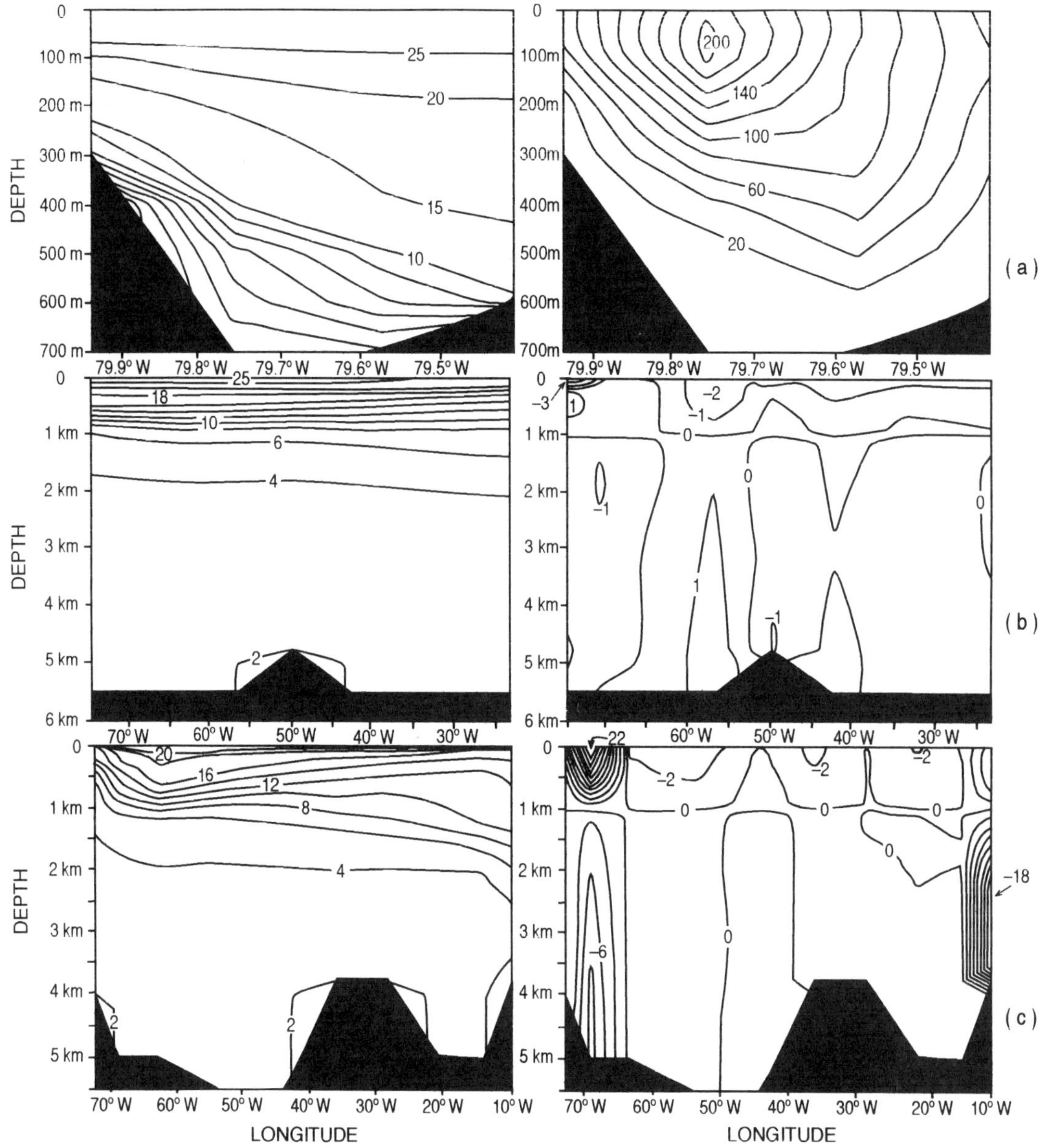

Figure 4–2. (a) Temperature (left panel) contours of the subsampled Florida Straits section. The contouring is nonuniform, with an interval of 1°C below 10°C and 5°C above 10°C. The right panel depicts the thermal wind relative to a bottom reference level. Depths are in meters. (b) Temperature (left panel) along nominal 24°N section. The contour interval is 2°C below 20°C and 5°C above 20°C. The right panel is the thermal wind relative to a 1000-decibar reference level. Longitude scales differ from those in the other two sections. (c) Temperature (left panel) in nominal 36°N section. The contour interval is 2°C. The right panel is the thermal wind relative to a 1000-decibar reference level.

4.1 Choosing a Model

Recall from Chapter 2 that in posing the problem of determining the circulation, a number of specific assumptions are often made, which can be summarized as:

1. geostrophic, hydrostatic balance as described by Equations (2.1.23)–(2.1.24);
2. conservation of mass [Equation (2.1.16)], and any other property for which conservation, or near-conservation, is appropriate or for which the sources and sinks are known, with Equation (2.4.4) being prototypical;
3. that the data describe either a near steady-state ocean, or the data are sufficiently synoptic that a dynamically and kinematically consistent ocean can be described;
4. that the mixing contribution to the conservation equations is negligible.

None of these assumptions is fundamental, and it is convenient to begin by broadening them. The assumption of a purely geostrophic flow is unnecessarily restrictive. Discussion of the time-dependent problem is postponed until Chapter 6, but within the steady-state context, the box balance equations readily accommodate further knowledge of the ocean circulation. In some regions, particularly at low latitudes (see Figure 2–9b,c), the mass and other property fluxes within the Ekman layer are very important. Let $F_E \pm \Delta F_E$ represent an estimate of the Ekman mass flux and its uncertainty. Then the conservation equation for mass (2.4.1) becomes

$$\sum_j \sum_q \rho_j(q)(v_{Rj}(q) + b_j)\delta_j \Delta a_j(q) + F_E \approx 0 \qquad (4.1.1)$$

or

$$\sum_j \sum_q \rho_j(q) b_j \delta_j \Delta a_j(q) + n = -\sum_j \sum_q \rho_j(q)\delta_j v_{Rj}(q)\Delta a_j(q) - F_E \quad (4.1.2)$$

where n includes the combined error of F_E and of the geostrophic elements; F_E itself will usually be computed from a sum over its spatially varying values. Similar terms are added to conservation equations in whatever layers are thought to carry the Ekman flux, both for mass and other tracer properties.

Determining where and how the ocean mixes is one of the central problems of physical oceanography and equations such as (2.1.18) and (4.1.1) are often employed with additional terms F_Q to represent mixing. Consider for example, an arbitrary property $C_j(q)$. Then

$$\sum_j \sum_q \rho_j(q) b_j \delta_j \Delta a_j(q) C_j(q) + F_Q + n$$
$$= -\sum_j \sum_q \rho_j(q) \delta_j v_{Rj}(q) C_j(q) \Delta a_j(q) - F_{EC} \quad (4.1.3)$$

where F_{EC} is the Ekman flux of C. If the F_Q are to be determined, then they can be introduced as new elements of $\mathbf{x}$. Commonly, in analogy to the eddy diffusion terms in (2.1.28)–(2.1.29), one writes a diffusive flux

$$F_Q = -\left\{ \frac{\partial}{\partial z}\left(K_{zz}\frac{\partial \rho C}{\partial z}\right) + \frac{\partial}{\partial x}\left(K_{xx}\frac{\partial C}{\partial x}\right) + \frac{\partial}{\partial y}\left(K_{yy}\frac{\partial C}{\partial y}\right) + \cdots \right\} \quad (4.1.4)$$

(the ellipsis indicates the possibility of mixed derivative terms), which appears in the integrated equation as

$$\sum_j \sum_q \rho_j(q) b_j \delta_j \Delta a_j(q) C_j(q) - K_{zz}(\partial \rho C/\partial z)_{q_u} a_{q_u}$$
$$- K_{zz}(\partial \rho C/\partial z)_{q_l} a_{q_l} - K_{xx}(\partial \rho C/\partial x) a_{q_x} + \cdots + n$$
$$= -\sum_j \sum_q \rho_j(q) \delta_j v_{Rj}(q) C_j(q) \Delta a_j(q) - F_{EC}\,. \quad (4.1.5)$$

The K_{ii} are treated as unknown, and the partial derivatives of the tracer C are known coefficients. Here, a_{q_u}, a_{q_l}, etc., are the horizontal areas of the upper and lower bounding surfaces of the volume, respectively, and the places where the partial derivatives are evaluated. The K_{ii} could be supposed spatially constant, or varying according to assumed functional forms, etc. If C is a biologically active tracer such as oxygen concentration, one may need to introduce source/sink terms, Q, as either new knowns or unknowns. A more likely situation is that there is some prior estimate $Q_0 \pm \Delta Q$. Such a situation can in turn be handled in two different ways: Q_0 can be made an element of $\mathbf{y}$, with the variance of ΔQ providing a noise contribution in the equation; alternatively, ΔQ can be regarded as an element of $\mathbf{x}$. These and other physical additions to the model can be introduced into the zero-order, purely geostrophic balance, while maintaining the linear form (4.0.1).

This discussion is far from exhaustive. Within the context of linear models, one can incorporate complex structures describing whatever physics, chemistry (or biology) that is reasonably believed to depict the observed fields. But if the model (4.0.1) fails to represent important governing physics, etc., one may be building in errors that no mathematical machinery can overcome.

To compute an estimate of the circulation and its resulting properties requires substantial further statements about the model. In particular, we recognize, and will regard as a *part of the model specification*, several further quantitative specifications:

1. choice of initial reference level;
2. choice of coordinate system (e.g., isopycnals, neutral surfaces, geopotentials, ...);
3. a statement of knowledge (or lack of it) concerning the statistics of $\mathbf{n}$, including the model error;
4. a statement of knowledge (or lack of it) concerning the statistics of $\mathbf{x}$;
5. other physics/chemistry.

4.2 The Initial Reference Level

How should the reference level for the thermal wind computation be chosen? The discussion in Chapter 2 has perhaps convinced the reader that there is no best choice. But there are rational and irrational ways to proceed.

If one knows nothing about the ocean circulation, then the choice of reference level is wholly arbitrary and could be any place between the surface and the bottom and might as well be simple–for example, a fixed pressure level. But much *is* known. Specifically, many users of hydrographic data have fairly firm ideas, based upon experience, about how the ocean behaves. We choose an initial reference level according to the principle that it should represent as best we are able our prior beliefs about the most reasonable ocean circulation. On that basis, some choices are clearly irrational: A surface reference level would lead in the region of Figure 2–2f, in Wüst's computation (Figure 2–11), or in the present sections, to a Gulf Stream that is strongest at the bottom and headed southward. Such a choice is unreasonable. Beyond that, the initial model is a test of the insights of the investigator. If one believes that the Gulf Stream reaches the bottom, then placing the reference level at the bottom would be a plausible starting point.

Alternatively, a reference level that lies between two dominant water masses moving in opposite directions (e.g., between the North Atlantic Deep Water and either the overlying Antarctic Intermediate Water or underlying Antarctic Bottom Water) would be plausible, too.

A fairly general principle can be enunciated:

> Estimates of the ocean circulation are made in the context of a long history of prior experience. The investigator, seeking to improve upon prior estimates, must regard as a possible outcome that the new data or ideas that have become available may prove inadequate to improve or modify the previous circulation scheme. Any model, including one whose specification requires a reference level, should be chosen so that if no improvement is possible, the investigator is satisfied that the initial model represents a description of the ocean circulation that is acceptable given previous data and ideas.

A reference level at the surface in the Gulf Stream region would fail to satisfy this principle. Sometimes the failings of such models become apparent only after solutions have been found, showing some unforeseen consequence, and there is no recourse but to go back and modify the model.

In the present case, there are a number of possible considerations. Measurements of the Florida Current (the name given to the Gulf Stream in the Florida Straits) suggest that the time-average mass flux is approximately 30 Sv to the north. Simple trial shows that any reference level with zero velocity within the water column produces mass fluxes that are far too small. If the bottom is chosen as a reference level with zero velocity, the mass flux found is about 24 Sv. But the uniform addition of about 20 cm/s in each station pair raises the flux to a value nearly equal to the prior best estimate of its value.

Specifying a reference-level velocity of zero is plainly a special case–in practice, one specifies the best estimate available of the velocity there, although zero is usually implied unless stated otherwise. Suppose it were thought that the actual flow at the reference level for some elements $j = j'$ was $x_{j'}^0 \pm \Delta x_{j'}^0$ cm/s (e.g., 20 ± 2 cm/s on the Florida Straits floor.

Then it is a simple matter to append to (4.0.1) further equations of the form

$$x_{j'} + n_{j'}^0 = x_{j'}^0 \tag{4.2.1}$$

with $n_{j'}^0 = \Delta x_{j'}^0$ being the noise estimate. Such equations increase the number of equations without increasing the number of statevector unknowns, x_j. Whether they increase the available information depends upon their relative noise levels. They do leave the structure of the model unchanged.

Yet another alternative but equivalent approach exists: One can set $x_{j'} =$

$x_{j'}^0$, and eliminate these unknowns from the problem, modifying $\mathbf{y}$ where appropriate, and increasing the estimated variance in the corresponding elements of $\mathbf{n}$. This method has the merit of decreasing the formal number of unknowns of the problem but with the loss of some flexibility.

Outside of the Florida Straits, the situation is less clear. Historically, the literature on the North Atlantic is divided between the older assumptions of comparatively shallow levels-of-no-motion and more recent use of deep ones. (As noted in Chapter 2, the shallow reference levels were more a consequence of the lack of deep data than any convincing indication of a zero velocity level.) Based upon "Wüstian" ideas, it is tempting to place a reference level between the large water masses observed in the North Atlantic, some emanating from the Arctic and some from the Antarctic. A view (slightly oversimplified) is that the southward moving North Atlantic Deep Water (NADW) between roughly 1300–3000 m in the western basin is sandwiched between Antarctic origin water masses above and below. Two choices would put an initial zero-level velocity at either the top or the bottom of the NADW, the upper one corresponding to the older shallow reference level.

In a definitive calculation, one tries to place the reference level as a function of position between the chosen water masses. Because the problem here is being simplified for illustration, we will use the simple choice of 3000 decibars (see the discussion in Section 2.1 of the meter/decibar correspondence) with the understanding that it does not fully represent all our prior knowledge and to that extent is an unwise choice. Where the bottom is shallower than 3000 decibars, including the entire Florida Straits, the bottom itself is used as the reference level.

With this choice, the geostrophic flows are readily computed. But doing this calculation involves some solution to the problem alluded to in Chapter 2 of the "bottom wedge" below the deepest common depth of a station pair. The various schemes for dealing with estimates of the flow in this wedge range from linear extrapolation (Wunsch & Grant, 1982) to extrapolation by objective mapping methods (Roemmich, 1983). One can extrapolate the temperature and salinity fields and compute the geostrophic shear from them, or extrapolate the geostrophic shear, etc. Here, we will use a straightforward method based upon the geometry depicted in Figure 4–3. The thermal wind requires knowledge of the horizontal gradient of the density, $\partial\rho/\partial x$ (in practice, one computes the dynamic topography from the temperature and salinity prior to taking the horizontal derivative, but the physics are identical). As shown in the figure, we can compute the difference along the sloping line between the shallower station and the deeper values at the adjoining station. These differences, along with the vertical differences

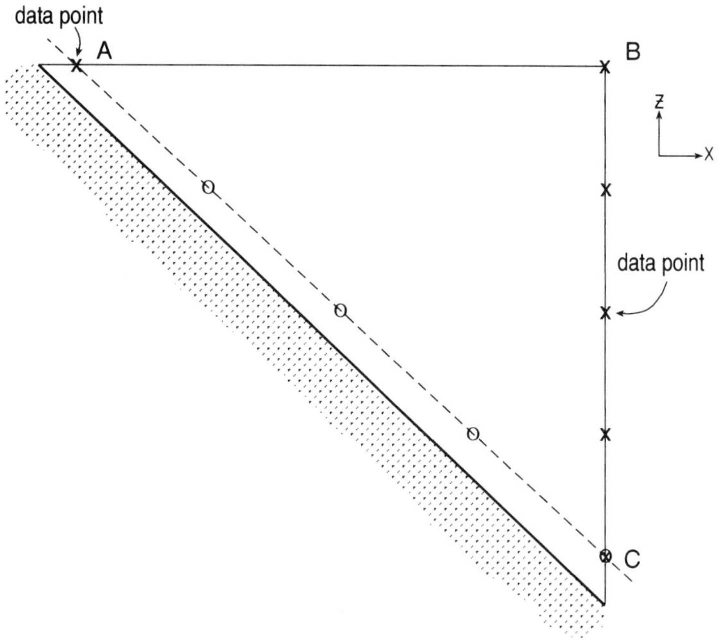

Figure 4–3. Geometry of the bottom wedge. Data points are assumed to be at the "x." A plane is fit to these values and used to interpolate to the open circle positions. The lateral gradients of density are then computed between the interpolated values and the data.

obtainable in the deeper station, contain information about the horizontal differences. This conclusion leads to the following simple procedure: fit by least squares a surface of form

$$\theta = a + bz + cx$$

to the temperature at the positions indicated (with a similar fit to the observed salinities). A minimum of three data points is required and is always available in the geometry. The resulting estimates of temperature and salinity along the sloping line are then used to calculate the dynamic topography and its horizontal derivatives, with due allowance made in transport calculations for the changing cross-sectional area. This method appears to be robust, and it is a local one in that unlike the objective mapping method of Roemmich (1983), it does not employ data from outside the particular bottom wedge. This localization seems desirable (although I would not defend it very vigorously), given the special physics applying over sloping topography.[1] If the vertical extent of the wedge is very great so that many standard depths are available in the deeper station, it may prove helpful to extend the plane to include quadratic terms or use other basis functions so as to account for any curvature that may be present in the deep property

[1] I am grateful to Prof. V. Kamenkovich for his suggestion that this method might be best.

fields. Here, we retain the linear rule. The resulting flows are depicted in Figure 4–2.

4.3 Simple Examples

4.3.1 Overall Mass Conservation and Fluxes Only

Consider the very simplest version of this problem: Only top-to-bottom mass conservation is sought, with no mixing considered. The top row of Table 4–2 lists all the elements of the **E** matrix, which then has this one row with 22 elements. The net geostrophic flow across 36°N is 31.5 Sv to the south, and the sum of the flows across the Florida Straits, and 24°N section is 11 Sv to the south. There is thus a net geostrophic convergence of about 20 Sv–that is, about 20 Sv more enters the box than leaves. (Recall that 1 Sv $= 10^6$ m^3/sec $\approx 10^9$ kg/s.) In addition, we expect a significant non-geostrophic Ekman flux across the bounding latitudes. Various estimates of these values exist, although their reliability is unclear. From zonal averages of the tables of Han and Lee (1981), we estimate the *annual average* zonal integral of the Ekman flux as 6.5 Sv (northward) at 24°N, and as -3 Sv (southward) at 36°N for a net Ekman convergence of 9.5 Sv. "Annual average" is emphasized because we are not sure this value is the most appropriate one to use with a synoptic station set. Observed hydrography is the result of wind and other forcing integrated over long periods of time, so that it is not obvious that the instantaneous windfield would be more appropriate.

We proceed with the present assumptions until some irreconcilable contradiction is found and therefore modify the constraint on the net geostrophic flux to reflect the requirement that the Ekman plus geostrophic mass fluxes must balance. Because the Ekman fluxes suggest a net inflow of 9.5 Sv (a convergence consistent with a subtropical gyre) that must be balanced geostrophically, and because the initial geostrophic flow also produces an inflow of 20 Sv, the geostrophic reference-level flow must provide a total box outflow of about $20 + 9.5 = 29.5$ Sv. This requirement becomes equation 1; its coefficients, $\mathbf{E}_{1j}$, are shown as line 1 in Table 4–2, with $\mathbf{y}_1$ also shown there. (Numbers in the text are commonly rounded for simplicity; those in the tables are more exact–e.g., the initial mass imbalance is closer to 29.3 than 29.5 Sv.)

For the second equation of the initial set, we use the results of a program monitoring the Florida Current (e.g., Molinari et al., 1985), which produced

Table 4–2. *Full* **E** *for the four-equation model of the ocean box of Figure 4–1.*[a]

	1	2	3	4	5	6	7	8	9	10	11	12	13	14	15	16	17	18	19	20	21
$\mathbf{E}_{1j}$	0.05	0.08	0.13	0.07	7.86	29.80	44.94	45.61	36.01	39.13	53.88	52.48	-6.33	-18.98	-31.15	-49.91	-33.62	-22.13	-23.22	-27.23	-25.08
$2j$	0.05	0.08	0.13	0.07	0.00	0.00	0.00	0.00	0.00	0.00	0.00	0.00	0.00	0.00	0.00	0.00	0.00	0.00	0.00	0.00	0.00
$3j$	0.05	0.08	0.13	0.07	7.86	29.80	44.94	45.61	36.01	39.13	53.88	52.48	0.00	0.00	0.00	0.00	0.00	0.00	0.00	0.00	0.00
$4j$	0.00	0.00	0.00	0.00	0.00	0.00	0.00	0.00	0.00	0.00	0.00	0.00	6.33	18.98	31.15	49.91	33.62	22.13	23.22	27.23	25.08
$\mathbf{y}^T$	-29.32	8.26	4.43	33.75																	
$\tilde{\mathbf{x}}^T$	0.37	0.68	1.02	0.55	0.00	0.01	0.01	0.01	0.01	0.01	0.02	0.02	0.03	0.09	0.14	0.23	0.15	0.10	0.11	0.12	0.11
$\tilde{\mathbf{n}}^T$	0.00	8.01	0.01	0.01																	
$\mathrm{diag}(\mathbf{P})^T$	1.00	0.99	0.98	1.00	1.00	0.93	0.85	0.85	0.90	0.89	0.79	0.80	0.99	0.95	0.87	0.66	0.85	0.93	0.93	0.90	0.92
$\sqrt{\mathrm{diag}(\mathbf{P})^T}$	1.00	1.00	0.99	1.00	1.00	0.97	0.92	0.92	0.95	0.94	0.89	0.89	1.00	0.98	0.93	0.81	0.92	0.97	0.96	0.95	0.96

[a] The first row of **E** is used to demand overall mass balance (geostrophic divergence equal and opposite to Ekman converge The second row imposes the observed Florida Straits mass flux, and the third and fourth rows impose mass conservation acros 24°N line (which for this purpose includes the Florida Straits) and the 36°N line, respectively (equal and opposite to the El flux, but carrying about 0.8 Sverdrups of inflow from the Arctic to the south as a required net flow). The first four columns the corresponding elements of x_i) control the Florida Straits, the next eight the 24°N section, and the remaining 10 (col 13–22), the 36°N section. In the first equation, the negative coefficients in the 36°N columns are required because a positive there contributes a negative flux to the box balance. The elements of **y** are: the required divergence of the reference-level vel the extra flow to the north in the Florida Straits required to reach about 30 Sv, and the net flows across the combined Fl Straits and 24°N lines, and the 36°N lines. $\tilde{\mathbf{x}}$ is the solution obtained from the Gauss-Markov estimate with these equations the prior statistical assumption that there is a uniform variance in the reference-level velocity. The large residual in the se equation shows that this solution must be rejected. The solution uncertainty and standard error are also listed but are irrele given the need to reject the solution.

a nominal annual average value of 31 ± 1 Sv. The reference-level calculation yields a transport of only 23 Sv. Here, the reference-level velocity must carry an extra 8 Sv–that is, $y(2) = 8$ Sv if we use 31 Sv as the best estimate of the Florida Straits transport.

The combined Florida Straits plus 24°N section also must compensate the 6.5 Sv of Ekman flux being blown to the north by the windfield. In addition, Coachman and Aagaard (1988) have estimated that 0.8 ± 0.2 Sv is transferred from the Pacific through the Bering Straits across the Arctic and into the Atlantic through the Fram Strait.[2] This water must flow across both the present zonal sections. So we write a third equation which asserts that the net geostrophic flux is $-6.5 - 0.8 = -7.3$ Sv (southward) across the combined 24°N section. The initial reference-level velocity produces -30 (to the south), and so $y(3) = +22.4$ Sv. Across 36°N, the initial geostrophic flow is 32 to the south, the Ekman flux requires 3 to the north, and there is 0.8 to the south, from the Fram Strait. Thus, the reference-level velocity needs to produce a net value of $y(4) = +35 + 3 - 0.8 = 34$ Sv (northward). The first equation, net mass balance, is the difference of equations 3 and 4, and we recognize that redundancy has been built into the constraints.

There are now four equations in 22 unknowns, of which only three equations will be independent. Estimates are required of the expected accuracies of these constraints.

4.3.1.1 Preliminary Variance Estimates

For illustrative purposes, suppose that the overall mass balance (equation 1) is believed accurate to about 1 Sv, that the Florida Straits transport is also known to this accuracy (consistent with the results of Molinari et al. and others), and that the two section equations involving the Ekman flux and the Fram Strait inflows are, because of the various difficulties involved, each accurate to ± 2 Sv. Therefore, put $\mathbf{R}_{nn} = \mathrm{diag}(1 \quad 1 \quad 2^2 \quad 2^2)$ (temporarily ignoring the fact that we know correlations are likely to exist in the errors occurring in these equations, if only because the same Ekman fluxes and station pair calculations occur in them).

An estimate of $\mathbf{R}_{xx}$ is also required. For illustration purposes, suppose for the moment that all reference-level velocities x_i are expected to have uniform variance of unity, $< x_i^2 > = 1$–that is, supposing ignorance of the fact that much larger values are anticipated for the floor of the Florida Straits, and much smaller ones for the deep interior flows. Thus, $\mathbf{R}_{xx} = \mathbf{I}_N$.

[2] The estimated range of the annual variation, 1946–1985, was 0.55–1.0 Sv and was based upon regressions with the windfield. The quoted uncertainty seems optimistic.

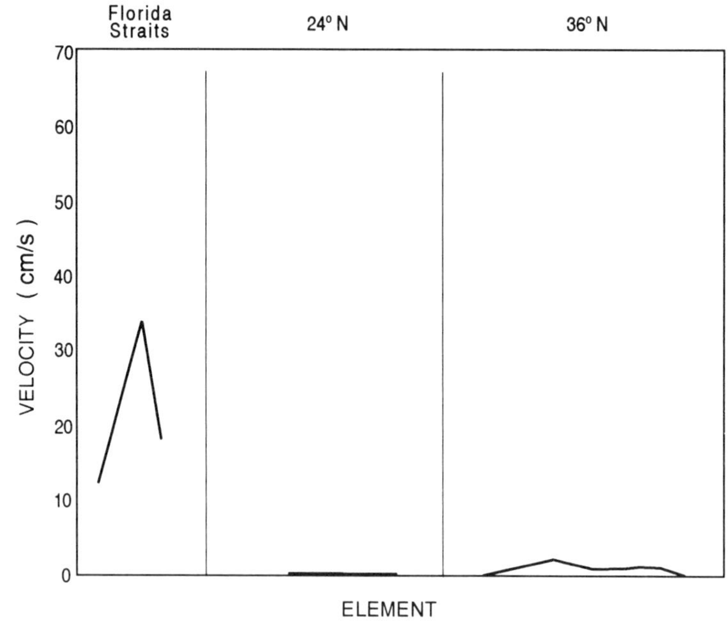

Figure 4–4. $\tilde{\mathrm{x}}$ from the Gauss-Markov estimate from the four-equation model, and using an appropriate variance for x_i. The values outside the Florida Straits have been multiplied by 10 for clarity.

4.3.2 Gauss-Markov Estimates

The advice provided at the end of Chapter 3 was that the Gauss-Markov estimate is a reasonable one to try first, so let us see what happens in the present case. Using Equation (3.6.16)–(3.6.18), the results are displayed in Table 4–2. Is this an acceptable solution? Although the elements of x_i are nonuniform, the result is only mildly in conflict with $\mathbf{R}_{xx}$. But examination of the residuals in Table 4–2 shows that there is a very large value (8 Sv) in the Florida Straits transport constraint. This residual is evidence that something is wrong with the model (the prior covariances are included in the definition of the model). Evidently there is a conflict between the requirement of solution elements of $\mathbf{O}(1)$, and the simultaneous demand that there be about 31 Sv passing through the Florida Straits, an unsurprising result. The important point, however, is that the solution provides ample warning to the investigator that something is going wrong. Here is a counterexample to the often-stated fear that in formally underdetermined problems, assumptions about covariances lead to "any solution one wants." One must reject this solution.

Let us now be a little more realistic in terms of what we already know, by giving the Florida Straits components of the reference velocity a prior variance of 1000 $\mathrm{cm}^2/\mathrm{s}^2$ and the deep water reference-level velocities a variance of 0.01 $\mathrm{cm}^2/\mathrm{s}^2$. The result is shown in Table 4–3 and Figure 4–4.

Table 4–3. $\tilde{\mathbf{x}}$, $\tilde{\mathbf{n}}$, *diag*($\mathbf{P}$), *the standard error* $[diag(\mathbf{P})^{1/2}]$, *and the full* $\mathbf{P}$ *for the Gauss-Markov solution where the prior variance reflects the expected variance of* $\mathbf{x}$ *in the different sections. The residuals,* $\tilde{\mathbf{n}}$, *are too small for statistical consistency, but in this case, they result from the artificial construction of the* y_i.

	1	2	3	4	5	6	7	8	9	10	11	12	13	14	15	16	17	18	19	20	21	22
$\tilde{\mathbf{x}}^T$	12.07	22.35	33.63	18.03	0.00	-0.01	-0.01	-0.01	-0.01	-0.01	-0.02	-0.02	0.03	0.08	0.14	0.22	0.15	0.10	0.10	0.12	0.11	0.02
$\tilde{\mathbf{n}}^T$	0.00	0.00	0.00	0.00																		
diag$(\mathbf{P})^T$	932.85	769.96	478.87	850.19	0.01	0.01	0.01	0.01	0.01	0.01	0.01	0.01	0.01	0.01	0.01	0.01	0.01	0.01	0.01	0.01	0.01	0.01
$\sqrt{\text{diag}(\mathbf{P})^T}$	30.54	27.75	21.88	29.16	0.10	0.10	0.09	0.09	0.10	0.09	0.09	0.09	0.10	0.10	0.09	0.08	0.09	0.10	0.10	0.10	0.10	0.10
full $\mathbf{P}$	932.85	-124.29	-187.07	-100.30	0.00	0.00	0.00	0.00	0.00	0.00	-0.01	-0.01	0.00	0.00	0.00	0.00	0.00	0.00	0.00	0.00	0.00	0.00
	-124.29	769.96	-346.24	-185.64	0.00	-0.01	-0.01	-0.01	-0.01	-0.01	-0.01	-0.01	0.00	0.00	0.00	0.00	0.00	0.00	0.00	0.00	0.00	0.00
	-187.07	-346.24	478.87	-279.41	0.00	-0.01	-0.01	-0.01	-0.01	-0.01	-0.02	-0.02	0.00	0.00	0.00	0.00	0.00	0.00	0.00	0.00	0.00	0.00
	-100.30	-185.64	-279.41	850.19	0.00	0.00	-0.01	-0.01	-0.01	-0.01	-0.01	-0.01	0.00	0.00	0.00	0.00	0.00	0.00	0.00	0.00	0.00	0.00
	0.00	0.00	0.00	0.00	0.01	0.00	0.00	0.00	0.00	0.00	0.00	0.00	0.00	0.00	0.00	0.00	0.00	0.00	0.00	0.00	0.00	0.00
	0.00	-0.01	-0.01	0.00	0.00	0.01	0.00	0.00	0.00	0.00	0.00	0.00	0.00	0.00	0.00	0.00	0.00	0.00	0.00	0.00	0.00	0.00
	0.00	-0.01	-0.01	-0.01	0.00	0.00	0.01	0.00	0.00	0.00	0.00	0.00	0.00	0.00	0.00	0.00	0.00	0.00	0.00	0.00	0.00	0.00
	0.00	-0.01	-0.01	-0.01	0.00	0.00	0.00	0.01	0.00	0.00	0.00	0.00	0.00	0.00	0.00	0.00	0.00	0.00	0.00	0.00	0.00	0.00
	0.00	-0.01	-0.01	-0.01	0.00	0.00	0.00	0.00	0.01	0.00	0.00	0.00	0.00	0.00	0.00	0.00	0.00	0.00	0.00	0.00	0.00	0.00
	0.00	-0.01	-0.01	-0.01	0.00	0.00	0.00	0.00	0.00	0.01	0.00	0.00	0.00	0.00	0.00	0.00	0.00	0.00	0.00	0.00	0.00	0.00
	-0.01	-0.01	-0.02	-0.01	0.00	0.00	0.00	0.00	0.00	0.00	0.01	0.00	0.00	0.00	0.00	0.00	0.00	0.00	0.00	0.00	0.00	0.00
	-0.01	-0.01	-0.02	-0.01	0.00	0.00	0.00	0.00	0.00	0.01	0.00	0.01	0.00	0.00	0.00	0.00	0.00	0.00	0.00	0.00	0.00	0.00
	0.00	0.00	0.00	0.00	0.00	0.00	0.00	0.00	0.00	0.00	0.00	0.00	0.01	0.00	0.00	0.00	0.00	0.00	0.00	0.00	0.00	0.00
	0.00	0.00	0.00	0.00	0.00	0.00	0.00	0.00	0.00	0.00	0.00	0.00	0.00	0.01	0.00	0.00	0.00	0.00	0.00	0.00	0.00	0.00
	0.00	0.00	0.00	0.00	0.00	0.00	0.00	0.00	0.00	0.00	0.00	0.00	0.00	0.00	0.01	0.00	0.00	0.00	0.00	0.00	0.00	0.00
	0.00	0.00	0.00	0.00	0.00	0.00	0.00	0.00	0.00	0.00	0.00	0.00	0.00	0.00	0.00	0.01	0.00	0.00	0.00	0.00	0.00	0.00
	0.00	0.00	0.00	0.00	0.01	0.00	0.00	0.00	0.00	0.00	0.00	0.00	0.00	0.00	0.00	0.00	0.01	0.00	0.00	0.00	0.00	0.00
	0.00	0.00	0.00	0.00	0.01	0.00	0.00	0.00	0.00	0.00	0.00	0.00	0.00	0.00	0.00	0.00	0.00	0.01	0.00	0.00	0.00	0.00
	0.00	0.00	0.00	0.00	0.01	0.00	0.00	0.00	0.00	0.00	0.00	0.00	0.00	0.00	0.00	0.00	0.00	0.00	0.01	0.00	0.00	0.00
	0.00	0.00	0.00	0.00	0.01	0.00	0.00	0.00	0.00	0.00	0.00	0.00	0.00	0.00	0.00	0.00	0.00	0.00	0.00	0.01	0.00	0.00
	0.00	0.00	0.00	0.00	0.01	0.00	0.00	0.00	0.00	0.00	0.00	0.00	0.00	0.00	0.00	0.00	0.00	0.00	0.00	0.00	0.01	0.00
	0.00	0.00	0.00	0.00	0.01	0.00	0.00	0.00	0.00	0.00	0.00	0.00	0.00	0.00	0.00	0.00	0.00	0.00	0.00	0.00	0.00	0.01

Now the Florida Straits velocities are much higher, and the residual in the Florida Straits mass flux equation is negligible. The residuals, n_i, in all the equations are vanishingly small–inconsistent with the prior estimate of $\mathbf{R}_{nn}$. But here the explanation is immediate. The y_i were constructed so that the equations are actually completely consistent, and the vanishing n_i need not lead to solution rejection in the present case. Table 4–3 lists the uncertainty and the standard error. Only a few of the x_i are marginally distinguishable from zero–not surprising when 22 state vector and four residual values are being determined from four dependent noisy equations.

In general, the averages of elements of the solution are considerably better determined than the individual elements. As an example, consider the mean velocity at the floor of the Florida Straits. Table 4–3 suggests that each of the four elements, x_i, $1 \le i \le 4$, have been determined with a standard error ranging from 22–31 cm/s. The simple average of the bottom velocity is 21.5 cm/s, with an error around ± 14 cm/s (as computed from the mean-square errors of each element).

Using expressions (3.6.29)–(3.6.31) for the uncertainty of a weighted sum, and with the weights vanishing except for $1 \le i \le 4$, the estimated mean is 24 cm/s with a standard error of 3.0 cm/s. Knowledge of the structure of the equations (now embodied in $\mathbf{P}$) produces an improved estimate, accounting properly for the strong correlations in the elements being averaged. Later, we will look at more interesting average and integral properties of the estimated flows.

Outside the Florida Straits, the adjustment to the initial thermal wind velocities is quite small. Figure 4–5 shows the flow in the 36°N section after inversion, which should be compared to the original flow in Figure 4–2c. Only a very slight adjustment has occurred. To the extent that the four equations represent the total of one's knowledge of the flow field, Figure 4–5 is the best estimate of the circulation, which suggests the importance of taking an initial state that represents as much as possible one's true knowledge of the circulation.

4.3.3 SVD and Tapered Estimates

For comparison with the Gauss-Markov estimate, we now solve the system using the SVD. A number of decisions are required. (1) The row norms of the system will be set to unity prior to imposition of the estimated errors because their sizes vary so greatly (the areal cross section of the Florida Straits is very small). (2) It is sensible to remove the accidental column norm dependence, recognizing that the velocities in the tightly spaced, shallow

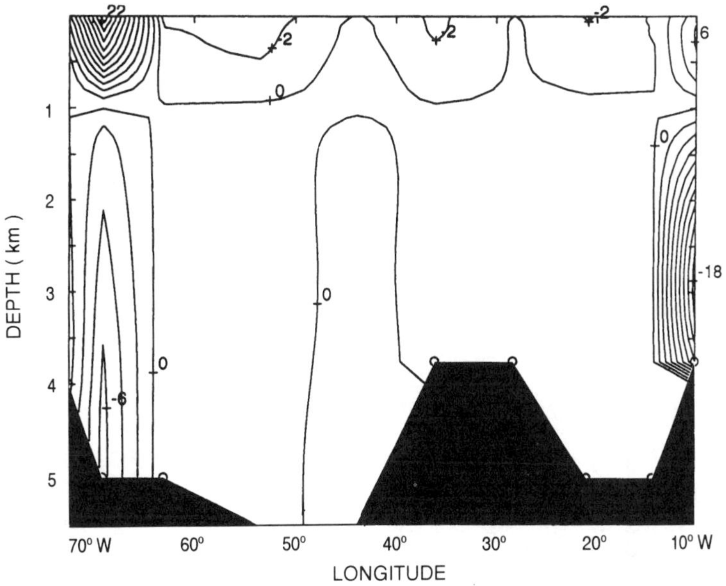

Figure 4–5. 36°N section velocity (cm/s) after inversion using the Gauss-Markov method on the four-equation model. Only slight changes in the 1000-decibar velocity are required, suggesting the importance of having the initial model reflect as much prior knowledge as possible.

water stations of the Florida Current are likely to be much higher than in the deep water of the widely spaced stations of the eastern Atlantic.

The prior noise covariances are imposed as a further row normalization through $\mathbf{R}_{nn}$, and a column normalization is then used to permit use of $\mathbf{R}_{xx}$ as a prior estimate of the solution elements. The resulting row- and column-scaled $\mathbf{E}$ produces singular values $\lambda_i = 43.33, \quad 0.20, \quad 0.09, \quad 0.00$, showing (as expected) that the maximum rank is $K = 3$. Table 4–4 lists the $\mathbf{u}_i$ and the first four (of 22) $\mathbf{v}_i$. Figure 4–6 shows the resulting solution–which is very similar to that obtained from the Gauss-Markov estimate, except that the solution elements within the subsections tend to be more uniform in the particular-SVD solution–as a result of the imposed column normalization. With $K = 3$, the estimated $\mathbf{n}$ is zero; there are no residuals from $\tilde{\mathbf{x}}$. As with the Gauss-Markov solution, the four equations are all perfectly consistent, and the solvability condition ($\mathbf{u}_4^T\mathbf{y} = 0$) is identically satisfied.

Examination of the $\mathbf{u}_i$, $\mathbf{v}_i$ is illuminating: $\mathbf{u}_1$ picks out the second equation, alone, imposing the Florida Straits transports. The corresponding $\mathbf{v}_1$ vanishes, except in the Florida Straits; $\mathbf{u}_2$ forms the sum of the first and third equations and takes the difference with the fourth, eliminating the Florida Straits velocities, leaving an equation for the divergence of the 24°N and 36°N mass flux. The corresponding $\mathbf{v}_2$ represents a positive inflow across 24°N and a negative flow (that is, into the box) across 36°N, the appropriate structure to balance a net convergence; $\mathbf{u}_4$, which lies in the

Table 4–4. *The fully scaled* $\mathbf{E}$ *to which the SVD is applied (row and column normalized, and then multiplied appropriately by Cholesky factors of* $\mathbf{R}_{xx}$, $\mathbf{R}_{nn}$*), the dimensional solution as a function of ranks* $K = 1$ *to* 3*, the rank* 3 *residual, the uncertainty* $\mathbf{P}_{nn}$ *(not including nullspace uncertainty) of the scaled and dimensional rank* 3 *solutions as well as the standard error of the rank* 3 *dimensional solution, again without the nullspace uncertainty. Also shown are the rank* 3 $\mathbf{U}$, $\mathbf{V}$, $\mathbf{T}_u$, $\mathbf{T}_v$ *(only the diagonal of the latter), which can be used to describe fully the solution structure.* $\mathbf{T}_v$ *is the data resolution matrix for the scaled equations, and the diag(*$\mathbf{T}_v$*) applies to both scaled and unscaled elements of* $\mathbf{x}$.

	1	2	3	4	5	6	7	8	9	10	11	12	13	14	15	16	17	18	19	20	21	22
$\mathbf{E}_{1j}$	0.02	0.03	0.03	0.02	0.02	0.04	0.05	0.05	0.05	0.05	0.06	0.06	-0.02	-0.03	-0.04	-0.05	-0.04	-0.03	-0.04	-0.04	-0.04	-0.02
$2j$	16.23	22.08	27.09	19.83	0.00	0.00	0.00	0.00	0.00	0.00	0.00	0.00	0.00	0.00	0.00	0.00	0.00	0.00	0.00	0.00	0.00	0.00
$3j$	0.01	0.02	0.02	0.01	0.01	0.03	0.03	0.03	0.03	0.03	0.03	0.03	0.00	0.00	0.00	0.00	0.00	0.00	0.00	0.00	0.00	0.00
$4j$	0.00	0.00	0.00	0.00	0.00	0.00	0.00	0.00	0.00	0.00	0.00	0.00	0.02	0.03	0.03	0.04	0.04	0.03	0.03	0.03	0.03	0.01
$\tilde{\mathbf{x}}^T$ for:																						
$K = 1$	25.32	25.32	25.32	25.32	0.00	0.00	0.00	0.00	0.00	0.00	0.00	0.00	0.00	0.00	0.00	0.00	0.00	0.00	0.00	0.00	0.00	0.00
$K = 2$	25.32	25.32	25.32	25.32	-0.07	-0.07	-0.07	-0.07	-0.07	-0.07	-0.07	-0.07	0.07	0.07	0.07	0.07	0.07	0.07	0.07	0.07	0.07	0.07
$K = 3$	25.32	25.32	25.32	25.32	-0.01	-0.01	-0.01	-0.01	-0.01	-0.01	-0.01	-0.01	0.14	0.14	0.14	0.14	0.14	0.14	0.14	0.14	0.14	0.14
$\tilde{n}$	0.00	0.00	0.00	0.00																		
$\mathrm{diag}(\mathbf{P}'_n)^T$	0.00	0.00	0.00	0.00	1.74	6.61	9.97	10.12	7.99	8.68	11.96	11.65	2.06	6.18	10.15	16.27	10.96	7.21	7.57	8.87	8.17	1.63
$\mathrm{diag}(\mathbf{P}_n)^T$	0.28	0.28	0.28	0.28	0.27	0.27	0.27	0.27	0.27	0.27	0.27	0.27	0.36	0.36	0.36	0.36	0.36	0.36	0.36	0.36	0.36	0.36
$\mathrm{diag}(\mathbf{P}_n)$	0.53	0.53	0.53	0.53	0.52	0.52	0.52	0.52	0.52	0.52	0.52	0.52	0.60	0.60	0.60	0.60	0.60	0.60	0.60	0.60	0.60	0.60
$\mathbf{U}$	0.00	0.89	-0.06	-0.45																		
	1.00	0.00	0.00	0.00																		
	0.00	0.32	-0.62	0.72																		
	0.00	-0.32	-0.78	-0.53																		
$\mathbf{v}_1^T$	0.37	0.51	0.63	0.46	0.00	0.00	0.00	0.00	0.00	0.00	0.00	0.00	0.00	0.00	0.00	0.00	0.00	0.00	0.00	0.00	0.00	0.00
$\mathbf{v}_2^T$	0.00	0.00	0.00	0.00	0.12	0.23	0.28	0.29	0.25	0.26	0.31	0.31	-0.11	-0.19	-0.24	-0.30	-0.25	-0.20	-0.21	-0.22	-0.21	-0.10
$\mathbf{v}_3^T$	0.00	0.00	0.00	0.00	-0.11	-0.21	-0.25	-0.26	-0.23	-0.24	-0.28	-0.28	-0.12	-0.21	-0.27	-0.34	-0.28	-0.22	-0.23	-0.25	-0.24	-0.11
$\mathbf{v}_4^T$	-0.55	0.46	-0.44	0.54	0.00	0.00	0.00	0.00	0.00	0.00	0.00	0.00	0.00	0.00	0.00	0.00	0.00	0.00	0.00	0.00	0.00	0.00
$\mathrm{diag}\left(\mathbf{T}_v(3)\right)^T$	0.14	0.26	0.39	0.21	0.03	0.10	0.15	0.15	0.12	0.13	0.17	0.17	0.03	0.08	0.13	0.21	0.14	0.09	0.10	0.11	0.10	0.02
$\mathbf{T}_u(3)$	0.80	0.00	0.32	-0.24																		
	0.00	1.00	0.00	0.00																		
	0.32	0.00	0.48	0.38																		
	-0.24	0.00	0.38	0.72																		

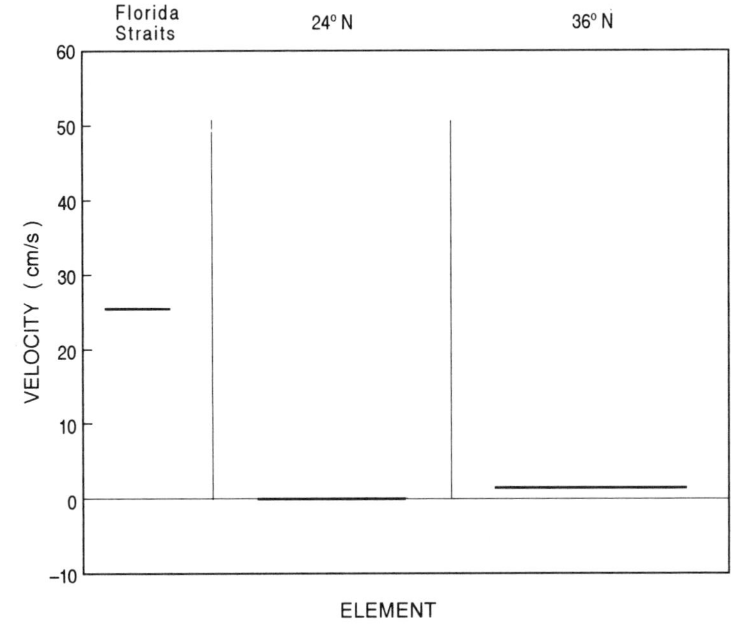

Figure 4–6. Rank 3 SVD solution from the four-equation model. The non-Florida Straits values are again multiplied by 10. The uniformity of the values in the different sections reflects the column normalization that was applied. The solution is qualitatively the same as for the Gauss-Markov estimate but differs quantitatively (the Gauss-Markov estimate is the minimum variance estimated, and this particular SVD solution is not).

data nullspace, produces the solvability condition $-.45y(1) + 0.72y(13) - 0.53y(4) = 0$, which is, in terms of the row- and column-scaled equations, a statement that the box convergence can be represented by the difference of the fluxes across the two separate bounding latitudes.

The uncertainty in the solution as listed in Table 4–4 is much less than for the Gauss-Markov estimate, because here it does not include the uncertainty owing to the nullspace–that is, what is tabulated is $\mathbf{C}_{\tilde{x}\tilde{x}}$ and not the full uncertainty. The diagonal of $\mathbf{T}_v$ is also shown; the general resolution is poor, but not surprising, given only three independent equations. The full $\mathbf{T}_u$ is shown, and the reader is urged to examine its structure.

4.3.4 Adding Temperature and Salt Equations

To make the situation slightly more interesting, we now include equations governing the movement of temperature and salt. Six equations are added, three for temperature across the three sections and three for the salt flux. Recall that salinity, S, is the salt/unit mass, and it is the total salt, ρS, that is conserved, rather than salinity. But ρ differs only slightly from 1.035×10^3 kg/m^3, and numerically one cannot distinguish salt conservation from salinity conservation. We do not expect that temperature is conserved overall, as the atmosphere vigorously exchanges heat with the ocean. The temperature equations will be used purely diagnostically–given no weight in the calculation–as a demonstration of the use of equations for purposes of determining property fluxes from the inferred circulation.

To use the salt conservation equations, an estimate must be made of the flux owing to the Ekman component. The mean salinity of the upper 50 meters at 36°N is about 36.2, and at 24°N it is about 36.9 (the reader is reminded that these numbers, while numerically realistic, are not expected to be accurate). We assign an Ekman salinity flux at 24°N to be this mean salinity, times the estimated Ekman mass flux there, or about +6.5 Sv×36.9. At 36°N, the value is −3 Sv × 36.2, that is, southward. (A more accurate value would be obtained if there were adequate data to integrate the Ekman flux as a function of position and depth times the local salt content.) These values must be balanced across both sections. In addition, the Arctic inflow, estimated as 0.8 Sv, will carry salt with it. What salinity should be assigned? Because we do not know the positions in the sections where the net mass flux is carried, we cannot accurately assign the salt content. For want of a better estimate, assume that the net southward flow carries the section mean salinity, which is about 35.2 at both 24°N and 36°N.

Because salinity differs only slightly from a value around 35, the conditioning of the system of equations is improved if one conserves instead of salt, the salt anomaly, $\rho\Delta S = \rho(S - S_0)$, where S_0 is a reference value, perhaps the local mean. The major difficulty here is in understanding the physical significance of the resulting conservation equation for ΔS, or more precisely, the extent to which exact conservation is justified. For the bounding volume, mass conservation is being written as

$$\sum_j E_{1j} x_j + n_1 = -F_E\,, \tag{4.3.1}$$

and the equation for salt conservation is

$$\sum_j E_{1j} \bar{S}_j x_j + n_2 = -\bar{S}_E F_E\,, \tag{4.3.2}$$

$\bar{S}_j$ being defined as the vertical mean salinity, and $\bar{S}_E$ as the mean salinity carried by the Ekman flux. It is the nearly constant value of $\bar{S}_j$ that suggests using the salinity anomaly. Multiplying (4.3.1) by S_0 and subtracting from (4.3.2) produces a new equation in $\Delta\bar{S}_j = \bar{S}_j - S_0$,

$$\sum_j E_{1j} \Delta\bar{S}_j x_j + n_2 - S_0 n_1 = -\bar{S}_E F_E + S_0 F_E\,. \tag{4.3.3}$$

Although the coefficients in (4.3.3), $E_{2j} = E_{1j}\Delta\bar{S}_j$, will have a greater tendency to independence of those in E_{1j} than do $E_{1j}\bar{S}_j$, the square residual to be left in (4.3.3) has grown from $< n_2^2 >$ to $< n_2^2 > \; + < S_0^2 n_1^2 >$

(assuming n_1 and n_2 are uncorrelated). The second term on the right is of order $35^2 \cdot 1^2$ Sv^2 if the rms value of n_1 is 1 Sverdrup. It is thus not clear that one has achieved very much–effectively doubling the expected mean-square residual in the salt conservation equation.

There are two circumstances in which using salinity anomaly conservation would clearly be beneficial: (1) mass conservation is anticipated to be satisfied exactly, so that $n_1 \equiv 0$; or (2) when n_1 or n_2 are strongly positively correlated, leading to cancellation of the two error terms in (4.3.3) rather than the simple addition of their variances. Whether either of these two cases can be assumed to be relevant is discussed below when we describe attempts to estimate the oceanic freshwater flux. For the time being, we will retain salt conservation as a constraint, because the physical basis is most clear, and the change from (4.3.2) to (4.3.3) may not be relevant if the accuracy of the calculation is adequate to remove concerns about numerical independence of the mass and salt equations. Specification of the behavior of the residuals will control the solution no matter how the equations are reorganized.

Even though they will be only diagnostic, similar numerical estimates must be made for the temperature flux equations. The mean temperature in the upper 50 m at 24°N is about 26.7°C, and at 36°N is about 21.3°C. The section mean temperature is 5.6°C at 24°N and 6.1°C at 36°N, and these values are used to assign the Ekman temperature transport as well as a nominal value for the net southward flow.

The uncertainty in the salinity of the mean southward flow necessarily introduces an uncertainty into the two individual flux equations. One way of removing the uncertainty is to work with the divergence of the salt flux, in which the mean salinity advection term would not appear to a first approximation (if the section mean salinities were very different, there would be a residual error). For present purposes, however, it is more interesting to recognize that we have introduced a correlated error into the two salt equations, and to use the information that the error is correlated rather than rewriting the equation set. The correlated error is estimated to have a magnitude of about 1 Sv times the mean salinity. $\mathbf{R}_{nn}$ is therefore modified from the previous diagonal form to that shown in Table 4–5.

The resulting Gauss-Markov estimate, in Table 4–5 and Figure 4–7, hardly differs from the one without salt conservation. The singular values of the scaled matrix are

$$\lambda_i = 43.34, \quad 0.20, \quad 0.09, \quad 8.3 \times 10^{-6};$$

Table 4–5. E *for 10-equation box constraints.* R_{nn} *includes very large values on the diagonal for those equations being used purely diagnostically, and a covariance (visible in the last two rows and columns) of the estimated errors (about 1 Sv × 35) is imposed to reflect the uncertainty in assigning the salinity of the Fram Strait inflow. Prior to use, the positive-definite nature of* R_{nn} *was confirmed. Also displayed are the diagonal of* R_{xx} *(no covariances were imposed), the solution elements, and the uncertainty and standard error of* $\tilde{\mathrm{x}}$.

1	2	3	4	5	6	7	8	9	10	11	12	13	14	15	16	17	18	19	20	21	22
											E										
0.05	0.08	0.13	0.07	7.86	29.80	44.94	45.61	36.01	39.13	53.88	52.48	-6.33	-18.98	-31.15	-49.91	-33.62	-22.13	-23.22	-27.23	-25.08	-5.01
0.05	0.08	0.13	0.07	0.00	0.00	0.00	0.00	0.00	0.00	0.00	0.00	0.00	0.00	0.00	0.00	0.00	0.00	0.00	0.00	0.00	0.00
0.05	0.08	0.13	0.07	7.86	29.80	44.94	45.61	36.01	39.13	53.88	52.48	0.00	0.00	0.00	0.00	0.00	0.00	0.00	0.00	0.00	0.00
0.00	0.00	0.00	0.00	0.00	0.00	0.00	0.00	0.00	0.00	0.00	0.00	6.33	18.98	31.15	49.91	33.62	22.13	23.22	27.23	25.08	5.01
0.86	1.39	2.10	1.37	78.98	157.38	226.85	227.45	195.05	219.88	279.09	295.49	0.00	0.00	0.00	0.00	0.00	0.00	0.00	0.00	0.00	0.00
0.86	1.39	2.10	1.37	78.98	157.38	226.85	227.45	195.05	219.88	279.09	295.49	0.00	0.00	0.00	0.00	0.00	0.00	0.00	0.00	0.00	0.00
0.00	0.00	0.00	0.00	0.00	0.00	0.00	0.00	0.00	0.00	0.00	0.00	33.78	103.31	198.75	282.60	187.94	156.79	162.22	158.56	152.31	48.59
1.64	3.03	4.58	2.48	0.00	0.00	0.00	0.00	0.00	0.00	0.00	0.00	0.00	0.00	0.00	0.00	0.00	0.00	0.00	0.00	0.00	0.00
1.64	3.03	4.58	2.48	279.72	1047.02	1578.11	1601.46	1266.14	1376.75	1894.01	1846.97	0.00	0.00	0.00	0.00	0.00	0.00	0.00	0.00	0.00	0.00
0.00	0.00	0.00	0.00	0.00	0.00	0.00	0.00	0.00	0.00	0.00	0.00	221.66	666.45	1097.41	1755.24	1181.98	781.13	819.96	959.69	885.73	178.88
											R_{nn}										
												1.00	0.00	0.00	0.00	0.00	0.00	0.00	0.00	0.00	0.00
												0.00	1.00	0.00	0.00	0.00	0.00	0.00	0.00	0.00	0.00
												0.00	0.00	4.00	0.00	0.00	0.00	0.00	0.00	0.00	0.00
												0.00	0.00	0.00	4.00	0.00	0.00	0.00	0.00	0.00	0.00
												0.00	0.00	0.00	0.00	10^{12}	0.00	0.00	0.00	0.00	0.00
												0.00	0.00	0.00	0.00	0.00	10^{12}	0.00	0.00	0.00	0.00
												0.00	0.00	0.00	0.00	0.00	0.00	10^{12}	0.00	0.00	0.00
												0.00	0.00	0.00	0.00	0.00	0.00	0.00	122×10^{15}	0.00	0.00
												0.00	0.00	0.00	0.00	0.00	0.00	0.00	0.00	4900.00	1225.00
												0.00	0.00	0.00	0.00	0.00	0.00	0.00	0.00	1225.00	4900.00
											$\mathrm{diag}(\mathrm{R}_{xx})^T$										
1000.00	1000.00	1000.00	1000.00	0.01	0.01	0.01	0.01	0.01	0.01	0.01	0.01	0.01	0.01	0.01	0.01	0.01	0.01	0.01	0.01	0.01	0.01
											$\tilde{\mathrm{x}}^T$										
12.07	22.35	33.64	18.04	0.00	-0.01	-0.01	-0.01	-0.01	-0.01	-0.02	-0.02	0.03	0.08	0.14	0.22	0.15	0.10	0.10	0.12	0.11	0.02
											$\tilde{\mathrm{n}}^T$										
												-0.16	0.29	0.49	0.65	-594.11	-298.15	-218.17	-1107.93	5.04	17.98
											$\mathrm{diag}(\mathrm{P})^T$										
932.85	769.94	478.87	850.04	0.01	0.01	0.01	0.01	0.01	0.01	0.01	0.01	0.01	0.01	0.01	0.01	0.01	0.01	0.01	0.01	0.01	0.01
											$\sqrt{\mathrm{diag}(\mathrm{P})^T}$										
30.54	27.75	21.88	29.16	0.10	0.10	0.09	0.09	0.10	0.09	0.09	0.09	0.10	0.10	0.09	0.08	0.09	0.10	0.10	0.09	0.10	0.10

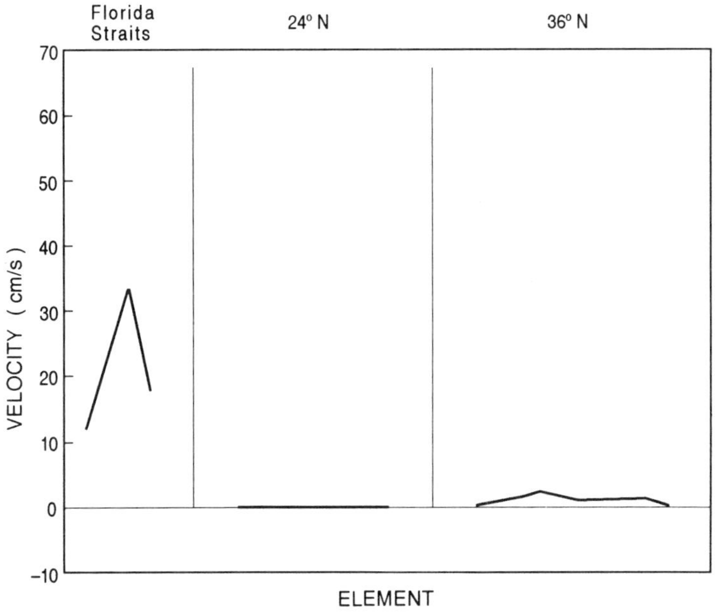

Figure 4–7. Gauss-Markov estimated **x** from the 10-constraint model. The non-Florida Straits values are multiplied by 10. The solution uncertainty is listed in Table 4–5.

that is, the two additional equations are not independent of the original ones, introducing a fourth nonzero singular value that is much smaller than the other three. Table 4–6 lists the particular-SVD solutions for $K = 3 - 5$, and the x_i for $K = 4$ and 5 are plotted in Figure 4–8. At rank 4, the solution in the Florida Straits deviates greatly from a constant (as the structure of $\mathbf{v}_4$ suggests it should), which is the system's way of balancing the combined mass and salt equations in the linear combination given by $\mathbf{u}_4$. But the structures introduced correspond to such a small singular value that the uncertainty, $\mathbf{C}_{\tilde{x}\tilde{x}}$, jumps by many orders of magnitude if one attempts to include this structure. The practical rank thus remains 3, and the SVD confirms the suspicion that the salt equations are so nearly linearly dependent upon the mass equations that little new information has been introduced into the system by the salt balance.

4.3.4.1 The Heat Budget

The various solutions, either the SVD or the Gauss-Markov, can be used with the temperature equations to diagnose the heat budget of the box. The temperature fluxes of the thermal wind with reference level at 1000 decibars were:

Florida Straits: 457°C–Sv; 24°N: -452°C–Sv; 36°N: 79°C–Sv.

The Ekman fluxes are estimated as:

Table 4–6. *Results of applying the SVD to the problem listed in Table 4–5.*

	1	2	3	4	5	6	7	8	9	10	11	12	13	14	15	16	17	18	19	20	21	22
$\tilde{\mathbf{x}}^T$ $K=2$	25.32	25.32	25.32	25.32	-0.07	-0.07	-0.07	-0.07	-0.07	-0.07	-0.07	-0.07	0.07	0.07	0.07	0.07	0.07	0.07	0.07	0.07	0.07	0.07
$K=3$	25.32	25.32	25.32	25.32	-0.01	-0.01	-0.01	-0.01	-0.01	-0.01	-0.01	-0.01	0.14	0.14	0.14	0.14	0.14	0.14	0.14	0.14	0.14	0.14
$K=4$	71.20	99.99	52.60	-148.83	-1.22	0.08	0.12	0.12	-0.01	-0.07	0.02	-0.09	-0.48	-0.17	0.15	-0.04	-0.05	0.33	0.38	0.18	0.42	1.48
$\tilde{n}^T$													-0.16	0.29	0.49	0.65	-594.11	-298.15	-218.17	-1107.93	5.04	17.98
$U(:,1-4)^T$													0.00	1.00	0.00	0.00	0.00	0.00	0.00	0.00	0.00	0.00
													0.89	0.00	0.32	-0.32	0.00	0.00	0.00	0.00	0.01	-0.01
													-0.06	0.00	-0.62	-0.78	0.00	0.00	0.00	0.00	-0.02	-0.02
													-0.01	0.00	0.00	0.01	0.21	0.00	0.00	0.00	0.65	-0.73
$V(:,1-4)^T$	0.37	0.51	0.63	0.46	0.00	0.00	0.00	0.00	0.00	0.00	0.00	0.00	0.00	0.00	0.00	0.00	0.00	0.00	0.00	0.00	0.00	0.00
	0.00	0.00	0.00	0.00	0.12	0.23	0.28	0.29	0.25	0.26	0.31	0.31	-0.11	-0.19	-0.24	-0.30	-0.25	-0.20	-0.21	-0.22	-0.21	-0.10
	0.00	0.00	0.00	0.00	-0.11	-0.21	-0.25	-0.26	-0.23	-0.24	-0.28	-0.28	-0.12	-0.21	-0.27	-0.34	-0.28	-0.22	-0.23	-0.25	-0.24	-0.11
	-0.11	-0.25	-0.11	0.51	0.45	-0.06	-0.12	-0.12	0.00	0.05	-0.03	0.08	0.22	0.19	-0.01	0.17	0.16	-0.13	-0.17	-0.03	-0.20	-0.42
$\text{diag}(\mathbf{C}_{\tilde{x}\tilde{x}})^T$ rank 3	0.00	0.00	0.00	0.00	1.74	6.61	9.97	10.12	7.99	8.68	11.95	11.64	2.06	6.18	10.15	16.26	10.95	7.21	7.56	8.87	8.17	1.63
	0.01	0.01	0.01	0.01	1.32	2.57	3.16	3.18	2.83	2.95	3.46	3.41	1.44	2.49	3.19	4.03	3.31	2.69	2.75	2.98	2.86	1.28
$\text{diag}(\mathbf{C}_{\tilde{x}\tilde{x}})^T/10^9$ rank 4	0.15	0.74	0.15	3.24	2.52	0.05	0.18	0.16	0.00	0.03	0.01	0.07	0.59	0.45	0.00	0.37	0.31	0.20	0.34	0.01	0.47	2.20
$\mathbf{T}_u(4)$													0.80	0.00	0.32	-0.24	0.00	0.00	0.00	0.00	0.00	0.00
													0.00	1.00	0.00	0.00	0.00	0.00	0.00	0.00	0.00	0.00
													0.32	0.00	0.48	0.38	0.00	0.00	0.00	0.00	0.01	0.01
													-0.24	0.00	0.38	0.72	0.00	0.00	0.00	0.00	0.01	0.01
													0.00	0.00	0.00	0.00	0.04	0.00	0.00	0.00	0.14	-0.15
													0.00	0.00	0.00	0.00	0.00	0.00	0.00	0.00	0.00	0.00
													0.00	0.00	0.00	0.00	0.00	0.00	0.00	0.00	0.00	0.00
													0.00	0.00	0.00	0.00	0.00	0.00	0.00	0.00	0.00	0.00
													0.00	0.00	0.01	0.01	0.14	0.00	0.00	0.00	0.43	-0.47
													0.00	0.00	0.01	0.01	-0.15	0.00	0.00	0.00	-0.47	0.53

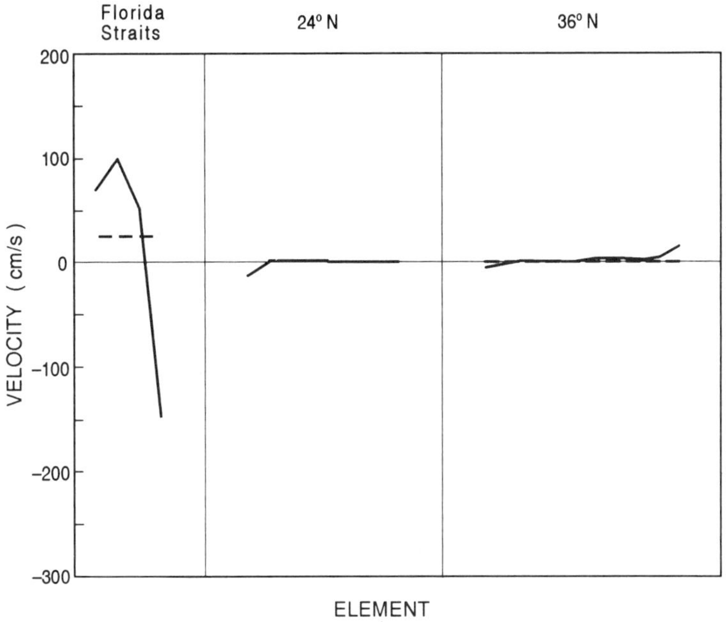

Figure 4–8. SVD solution at ranks 4 and 5 (solid curves) for the reduced box model, employing both mass and salt constraints. Introduction of the fifth singular value produces a marked increase in the Florida Straits reference-level magnitudes and an unacceptable strong flow to the south on the eastern edge, showing that the practical rank remains $K = 4$.

Florida Straits: assumed 0; 24°N: 173°C–Sv; 36°N: -64°C–Sv.

(The units are degrees Celsius–Sverdrups, °C–Sv.) From substitution of the Gauss-Markov solution into the original constraint equations, we deduce as temperature fluxes from the reference-level velocity $\tilde{\mathbf{x}}$:

Florida Straits: 137°C–Sv; 24°N: -21°C–Sv; 36°N: 198°C–Sv.

Adding all three contributions and summing the Florida Straits and 24°N values produces a net flux across the combined sections of 294°C–Sv, and for 36°N of 213°C–Sv. In the present solution, mass balance was achieved to a high degree of accuracy (the mass residual was 0.2 Sv), and we can treat the flows as being mass conserving, thus permitting the conversion of the temperature fluxes into heat fluxes. The heat capacity of seawater is very nearly $4 \times 10^{12} J/10^9$ kg/°C, and doing the multiplications produces a flux of 1.18×10^{15}W [1.2 petawatts (PW)] across the combined Florida Straits 24°N line, and of 0.85×10^{15}W across the 36°N line.

This calculation is comparatively straightforward; the real problem is estimating its accuracy. For example, one might be tempted to calculate the heat flux divergence between the two lines as $1.18 - .85 = 0.33$PW, an estimate of the heat flux divergence to the atmosphere, of special interest in studies of climate and climate change. But how much confidence should one place in the fluxes and in the divergence? There are three contributors to the heat or temperature flux across each line: (1) the thermal wind, (2) the Ekman flux, and (3) the reference-level velocity. The uncertainty of the flux owing to the reference-level velocity is easily found. The temperature flux estimate across any line owing to the reference-level velocity is the linear combination

$$\tilde{H}_x \equiv \sum_{j\epsilon J}\left[\sum_q \rho_j(q)\Delta a_j(q)\theta_j(q)\right]\tilde{x}_j = \sum_{j\epsilon J} h_j\,\tilde{x}_j = \mathbf{h}^T\tilde{\mathbf{x}}$$

$$\{h_j\} = \left\{\sum_q \rho_j(q)\Delta a_j(q)\theta_j(q)\right\}, \tag{4.3.4}$$

where $\theta_j(q)$ is the temperature in pair j at depth q, and $j\epsilon J$ is used to denote those station pair indices confined to the particular hydrographic line in question. In the present example, column j of the fourth row of $\mathbf{E}$ in Table 4–2 is just the combination, h_j, in the Florida Straits, and the fifth and sixth rows contain the corresponding elements for 24°N and 36°N, respectively. The uncertainty of the linear combination $\tilde{H}_x$ is (3.6.58),

$$<\left(\tilde{H}_x - H_x\right)^2> = \mathbf{h}^T\mathbf{P}_J\mathbf{h}, \tag{4.3.5}$$

where $\mathbf{P}_J$ denotes $i, j\epsilon J$–that is, the submatrix of $\mathbf{P}$ corresponding to the velocity elements involved in the temperature flux. For the combined 24°N line, we have a variance of $252(°\text{C–Sv})^2$, which produces a standard error of 0.06 PW, and for 36°N the variance is $67.4(°\text{C–Sv})^2$ for a standard error of 0.03 PW. These uncertainties are small compared to the estimated heat flux, despite the determination of the reference-level velocity from the equivalent of only three noisy equations. Their smallness reflects the well-determined nature of linear combinations of x_i, even when individual elements are poorly determined.

The uncertainty of the heat flux here is dominated by other contributions. Suppose there were a 10% error in the Ekman flux (an optimistic value). At 24°N, the heat flux uncertainty then has a value of 0.07 PW, and at 36°N it is 0.03 PW. What remains is the uncertainty owing to the thermal wind. Computing its uncertainty is very difficult and is postponed to later in the chapter.

4.3.5 More About Errors

The production of reasonable estimates of the errors to be expected in the various equations is essential to obtaining reliable, realistic solutions. It is in this area that the investigator's skill is crucial–to account for the details that will be different in every case. A few general remarks are in order.

Consider first the equation asserting that mass is conserved in a particular volume of ocean. How accurately should one expect this equation to be satisfied? By way of example, suppose the area of the box is (1000 km $\times$ 1000 km $= 10^{12}\text{m}^2$). We can distinguish several reasons why perfect mass conservation might not be a sensible constraint. One reason is the physical one, that the ocean is not strictly in a steady state and is capable of temporary regional storage of water. Should we tolerate as reasonable a 1 Sv imbalance in the flow-in minus the flow-out? One Sverdrup is $1 \times 10^6\text{m}^3/\text{s}$. Dividing this flux by the area produces a sealevel change of $10^6\text{m}^3/\text{sec}/10^{12}\text{m}^2 \approx 10^{-6}\text{m/sec}$, which if sustained for a month (about 2.6×10^6s) would lead to a sealevel change of several meters. Such changes are not observed; hence, we assert that imbalances greater than 1 Sv over such areas in the real ocean are implausible if sustained for more than a few days. Thus, model error owing to the steadiness assumption should be much less than 1 Sv.

Are we therefore able to assert that the sum in Equation (4.1.1) should vanish to within much less than 1 Sv? The answer is "no," not because the physics are not accurate to this level but because the individual terms of the equation cannot be estimated from the data to such accuracy.

Several separate sources of error must be considered. First, and probably least important, are the instrumental errors leading to finite precision in the determination of temperature and salinity, which coupled with uncertainties in the equation of state, generate errors in ρ, and hence its gradients and the computation of the thermal wind. A discussion of the accuracy of the equation of state may be found in Millero, Tung, Bradshaw, and Schleicher (1980). Oceanographers claim that the modern CTD system produces temperatures to an absolute accuracy of 0.005°C and that modern salinities are determinable to about 0.003 (e.g., Lewis & Perkin, 1978; Pond & Pickard, 1983). Then there are navigational errors–not only does no ship ever know its position perfectly, it takes several hours to do a hydrographic station, and during that time the ship may drift in strong currents or high winds and one usually assigns some nominal position to the station even though the top and bottom measurements may be several kilometers apart. There is internal wave noise. Then there are the errors involved in making the finite difference approximations contained in (2.4.1), (4.1.1) and equivalents. All these things are probably comparatively minor, because geostrophy is particularly forgiving when integrated over several stations, and many of the errors tend to cancel.

Consider that if a navigation error occurs and a station is misassigned, as depicted in Figure 4–9, the geostrophic mass transport assigned to the

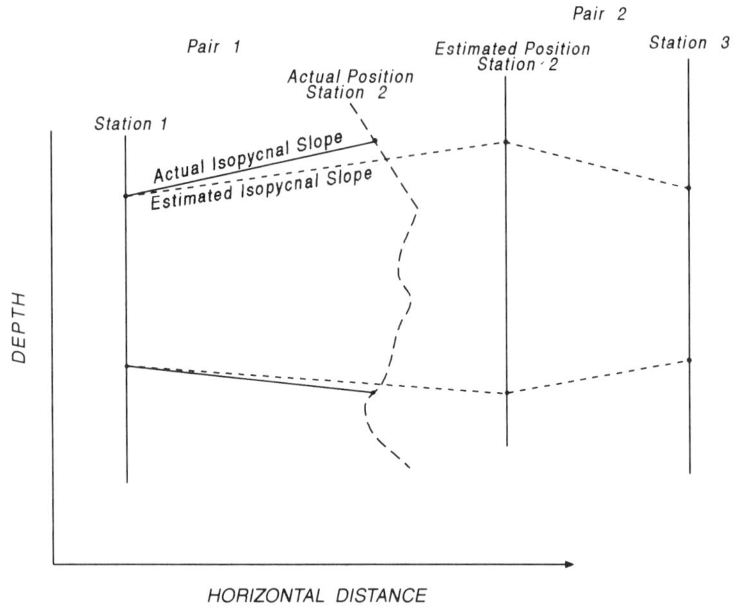

Figure 4–9. In regions where the isopycnals do not intersect the seafloor, geostrophy is forgiving of navigation errors when transports are computed. In the situation shown, the error in the position of station 2 causes an underestimate in the geostrophic velocity. But because the station pair area is erroneously overestimated, the transport remains accurate. Furthermore, in the second pair shown, the position error in station 2 causes the velocity to be overestimated, and the average velocity of pairs one and two remains accurate. If the navigational error is a function of depth, however, as would be true of deep-water stations made in strong surface velocity conditions, or if the bottom is encountered, full compensation no longer occurs.

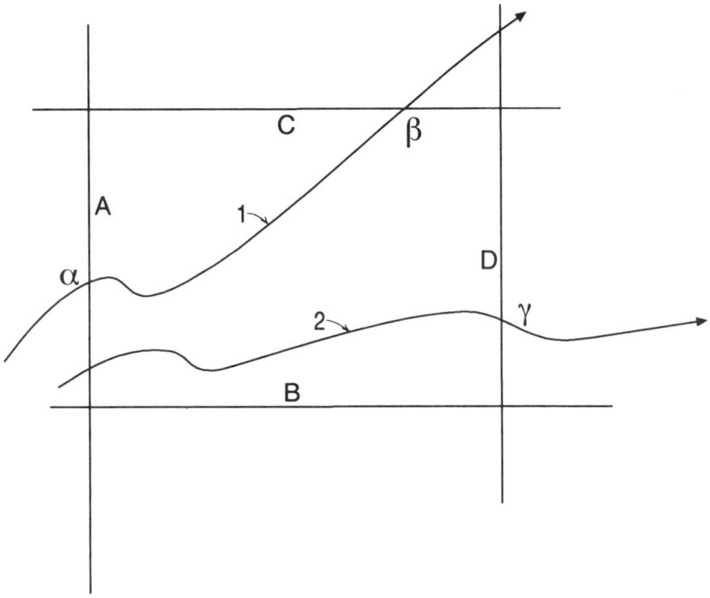

Figure 4–10. Aliasing errors can be caused by combining hydrographic sections that are not truly synoptic: Here, the Gulf Stream was observed to enter the box at point α when sections A, B, and D were measured, with the Stream leaving at point β along trajectory 1. But when section C was measured, the path had moved to trajectory 2, leaving the box at point γ. Thus, one has a data set in which the Stream appears to enter the box but not leave it. The expected error in the mass balance and other conservation equations is then very large.

first pair depicted will be erroneous but will tend to be compensated in the second pair (full cancellation, however, is dependent upon the bottom shape). The integral property of the fluxes, as computed by equations such as (4.3.4), leads to cancellation of many intermediate errors. A more serious error occurs over sloping topography, where none of the various methods for estimating the geostrophic flow has a rigorous basis.

The largest potential error is an alias arising from the usually necessary assumption that the data are synoptic. Consider the (somewhat extreme, but realistic) case depicted in Figure 4–10. The Gulf Stream was observed along a trajectory like that labeled "1," crossing section A, at the point marked α, when sections A and D were measured. It was in the position "2" when the sections B and C were measured and the outward crossing was at γ. Thus, the bounding box shown, of a type commonly used to generate constraints in inverse models, has a hydrography such that the Gulf Stream was observed to enter, but no hydrography depicts it leaving. This aliasing error can be handled in an inverse context by stipulating a large expected error in the box balance. In the case just described (similar to one encountered in Wunsch, 1978), the error is potentially equal to the full mass transport of the Gulf Stream, and the constraints in this particular box must be given such small weights that they are effectively useless in determining the solution. The problem occurs at a lesser level every time we bound an oceanic region, no matter how small, with nonsynoptic data, and is one of the primary reasons that box inversions have tended to focus on hydrographic sections terminating at land on both ends.

A related but different error occurs in combining the hydrography with the Ekman flux estimates. Even if the hydrographic data are synoptic, and the Ekman fluxes are regarded as accurate, it is far from clear that the Ekman flux appropriate to use in (4.1.1) is the climatological average, or the one appropriate to the days (or months or years) immediately surrounding the acquisition of the hydrographic data. The apparent interannual fluctuations of the Ekman fluxes have been discussed by Levitus (1988).

In Chapter 3, we saw that if the errors in the equations are correlated, we should use that knowledge. But there has been no systematic effort made to understand the structure of $\mathbf{R}_{nn}$, other than along its diagonal. If $\mathbf{R}_{nn}$ is used as though it were diagonally dominant, when there are known significant correlations in the equation errors, we are failing to use information that in principle can be used to improve the solution [recall (3.4.116)].

4.3.6 Changing the Reference Level

In many places, knowledge of the ocean circulation is so crude that there is no obvious best choice of a reference level. As already noted, in the North Atlantic, arguments have been made for both a shallow and a deep choice, above and below the North Atlantic Deep Water, respectively. It is worth remembering that if enough internally consistent constraints can be added, so that the state vector is fully resolved, the solution would be independent of the choice.

But in the common situation where there is little to choose between two or more possibilities, one may be driven simply to discuss the results of circulation estimates based upon different selections. If the results are very different but completely acceptable solutions, one has an important statement about the types of circulations consistent with present knowledge.

The particular-SVD solution shows the implications of a reference-level shift. Let

$$\tilde{\mathbf{x}}_1 = \mathbf{V}_K \boldsymbol{\Lambda}_K^{-1} \mathbf{U}_K^T \mathbf{y} = \mathbf{V}_K \mathbf{V}_K^T \mathbf{x}_1 \tag{4.3.6}$$

be the solution to (4.0.1) for a given reference level at rank K. A shift in level is equivalent to adding a constant (with depth) vector, $\mathbf{d}$, to the initial reference-level velocity, so that (4.3.6) is modified to

$$\mathbf{E}(\mathbf{x}_2 + \mathbf{d}) + \mathbf{n} = \mathbf{y}$$

or

$$\mathbf{E}\mathbf{x}_2 + \mathbf{n} = \mathbf{y} - \mathbf{E}\mathbf{d}\,. \tag{4.3.7}$$

The particular-SVD solution with the new reference level is

$$\begin{aligned}\tilde{\mathbf{x}}_2 &= \mathbf{V}_K \mathbf{V}_K^T \mathbf{x}_2 = \mathbf{V}_K \boldsymbol{\Lambda}_K^{-1} \mathbf{U}_K^T (\mathbf{y} - \mathbf{E}\mathbf{d}) = \mathbf{V}_K \boldsymbol{\Lambda}_K^{-1} \mathbf{U}_K^T \mathbf{y} - \mathbf{V}_K \mathbf{V}_K^T \mathbf{d} \\ &= \tilde{\mathbf{x}}_1 - \mathbf{V}_K \mathbf{V}_K^T \mathbf{d}\,. \end{aligned} \tag{4.3.8}$$

Then the estimated total velocity from the two different solutions is

$$\tilde{\mathbf{q}}_1 = \mathbf{q}_R + \tilde{\mathbf{x}}_1 \tag{4.3.9}$$

$$\tilde{\mathbf{q}}_2 = \mathbf{q}_R + \mathbf{d} + \tilde{\mathbf{x}}_2 \tag{4.3.10}$$

where $\mathbf{q}_R$ is the relative velocity from the original reference level. Projecting both (4.3.9), (4.3.10) onto the range vectors,

$$\mathbf{V}_K \mathbf{V}_K^T \tilde{\mathbf{q}}_1 = \mathbf{V}_K \mathbf{V}_K^T \mathbf{q}_R + \mathbf{V}_K \mathbf{V}_K^T \tilde{\mathbf{x}}_1\,, \tag{4.3.11}$$

$$\mathbf{V}_K \mathbf{V}_K^T \tilde{\mathbf{q}}_2 = \mathbf{V}_K \mathbf{V}_K^T \mathbf{q}_R + \mathbf{V}_K \mathbf{V}_K^T \mathbf{d} + \mathbf{V}_K \mathbf{V}_K^T \tilde{\mathbf{x}}_2 = \mathbf{V}_K \mathbf{V}_K^T \mathbf{q}_R + \mathbf{V}_K \mathbf{V}_K^T \tilde{\mathbf{x}}_1 \tag{4.3.12}$$

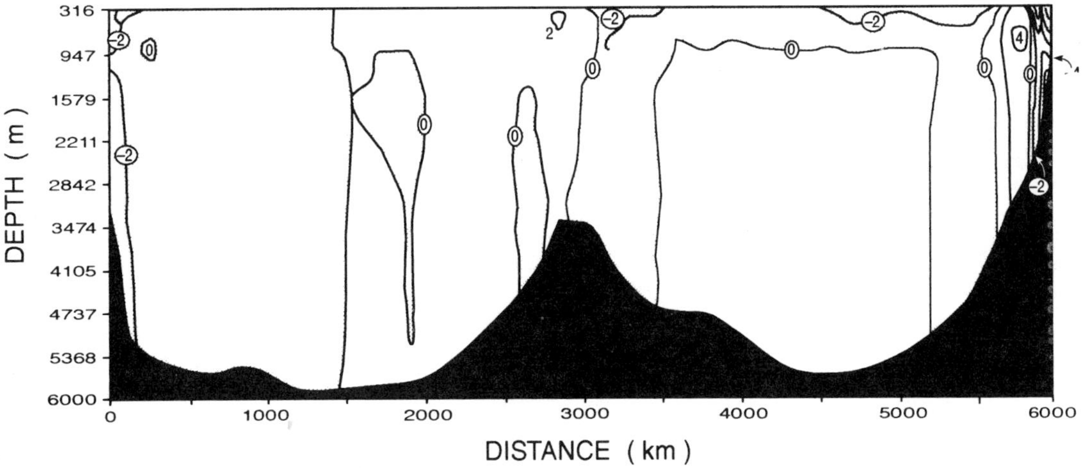

Figure 4–11. Smoothed flow field in the 24°N section in Figure 2–2h representing structures of the relative plus reference-level velocities that are present independent of the initial choice of reference level (from Wunsch & Grant, 1982). The contours are in cm/s.

by (4.3.8). That is, the total velocity, relative plus reference level, has a projection onto the range that is uniquely determined independent of the reference level. These projected total velocities were termed the *uniquely determined components* by Wunsch (1978) and the result is a smoothed version of the total velocity field (see Figure 4–11). Fu (1984) showed that this smoothed velocity field was identical to the reference-level velocity that would be determined by successively trying each of the standard depths in turn as a level-of-no-motion. A simple physical interpretation of the smoothed field is not otherwise readily apparent because the unique components do not by themselves necessarily satisfy any of the required constraints.[3] Wunsch (1978) suggested that there would be a tendency for these elements of the solution to be time invariant–they are often both large in scale and required for consistency with mass and other conservation requirements. But little has come of this suggestion, in part because it has not yet proved possible to separate time-dependent and time-invariant elements of the large-scale circulation.

The particular-SVD solution, in the context of the principle enunciated

[3] Spatially-averaged flows do not normally satisfy any constraints written in the original unaveraged physical space. The inability to write general equations governing flows described only in restricted wavenumber bands, even for dynamically linear flows, is a central difficulty in fluid mechanics in general and in oceanography in particular. A simple example of the difficulty is the description of a steady flow in a channel of changing cross-section. Even though the dynamics may be linear, all spatial scales in the fluid are required to satisfy the mass conservation constraint–thus coupling the scales.

above, finds the smallest deviation possible from the reference level that is consistent with the new information being provided by the constraints. Fiadeiro and Veronis (1982) proposed considering all possible purely horizontal reference levels, and choosing the one for which the resulting state-vector norm is smallest. That is, one successively tries all possible standard depths and then chooses as best the solution for which the state-vector norm is the smallest (the horizontal level of the smoothed solution with the smallest norm). In the absence of any better physical ideas, the choice of the minimal correction solution is reasonable. But it must be reckoned as itself an arbitrary way of selecting one solution out of an infinite number of possibilities–one might equally well decide that the smoothest state vector is the preferred one (e.g., Provost & Salmon, 1986).

4.4 Property Fluxes

One of the most important reasons for studying the ocean circulation is the need to compute the fluxes of properties, physical and chemical, which are central to the climate system. The movement by the ocean of heat, freshwater, carbon, etc., can dominate the global budgets of these quantities, and concerns about vulnerability to climate change lead one increasingly to determine what the ocean is doing and whether it is changing. Because the ocean is coupled to the atmosphere and exchanges these quantities with it, one must compute the *divergences* of the ocean circulation fluxes–as these are the rates at which the atmosphere must provide and absorb heat, etc., in different geographical areas. (A somewhat obvious point, which is nonetheless often overlooked, is that the numerical value of the flux of an oceanic property is of little interest, save as a secondary indicator of the vigor of the ocean circulation. Closed circuits of properties, e.g., of heat or freshwater, with no transfers to or from the atmosphere, even on a global scale, have no consequences. But the divergences of these properties are of first-order physical importance.)

It is comparatively easy to make estimates of oceanic property fluxes: One merely multiplies the water velocities by the local property concentrations and integrates (sums). What is much more difficult is producing useful estimates of the reliability of the results and of the consequent flux divergences.

The discussion of fluxes computed from the simplified models shows that property flux estimates will be no more reliable than the accuracy of the water movement estimates will permit. Consider any property, C, whose flux across a hydrographic section is required. In the context of the geostrophic physics we have been using, its flux across any ocean line, H_C, is made up of three parts:

$$H_C = H_E + H_R + H_x \tag{4.4.1}$$

where H_E, H_R, H_x are the Ekman, relative (thermal wind), and reference-level contributions, respectively. [In an ocean with important lateral mixing, one might wish to include its contribution as a fourth term in (4.4.1).]

The errors in the three terms are coupled by the equations for overall property distributions; for example, mass and other conservation constraints force a dependence of the reference-level velocity on the Ekman flux, and errors in the thermal wind mass flux are reflected in the inversion as well. A rigorous discussion would account for this coupling, but for present purposes, it appears adequate to discuss the contributions of the terms separately.

We have already seen in Equation (4.3.5) that the uncertainty of H_x is readily computed. The uncertainty owing to the Ekman component is somewhat more problematic. In principle, estimates of the uncertainty of the Ekman flow can be made from the twice-per-day surface stress analyses coming from meteorological agencies and models of the Ekman layer. If the Ekman velocity at any location $\mathbf{r}_j, z_j$, with local cross-section area Δa_j at time t, is $q_E(\mathbf{r}_j, z_j, t)$, and if the property concentration is $C(\mathbf{r}_j, z_j, t)$, the Ekman flux is

$$H_E = \sum_j C(\mathbf{r}_j, z_j, t)\rho(\mathbf{r}_j, z_j, t)\Delta a(\mathbf{r}_j, z_j, t) q_E(\mathbf{r}_j, z_j, t)\,. \tag{4.4.2}$$

In the computation done with the box model above, (4.4.2) was replaced by

$$\tilde{H}_E = 50 \text{ m} \cdot L \cdot \overline{\rho q_E}\,\overline{C} \tag{4.4.3}$$

where L is the section length, and the bar represents an average over the first 50 m of the water column along the entire section over some long time period. In Chapter 2, we discussed some of the difficulties that have plagued attempts to show the existence of Ekman layers at sea, and quantitatively, to relate the water movement to windstress. Because property flux estimates are often employed as though they represented climatological averages, one must include in the uncertainty of the Ekman component its poorly understood interannual variability as well as the suspected systematic errors in the windfield estimates described in Chapter 2. At the present time, it is virtually impossible to make a useful estimate of the discrepancy between H_E and $\tilde{H}_E$. Extant values (including some by the author) are little more than plausible guesses. A cautious investigator might choose to place uncertainties on $\tilde{H}_E$ approaching 100%.

In some circumstances, fortunately, the Ekman (or Ekman-like) flux is relatively weak. At high latitudes, the increase in f often makes the values negligible compared to that in the geostrophic flows. Some properties C–for example, the nutrients silicate and nitrate–are naturally depleted near the seasurface in many areas, and the directly wind-driven contribution there is minimal. At low latitudes, and especially for surface intensified fields such as temperature and oxygen, the Ekman contribution may well be the dominant one.

How reliable is $\tilde{H}_R$, owing to the relative velocity? Here, the situation is even murkier. Figure 4-12 shows an estimate of the relative velocity, $\tilde{q}_R$, for the 24°N section, using all the available stations (not the subset of the reduced box model) and here computed using a more reasonable 3000-decibar reference level (omitting two very shallow stations on the western side). The most conspicuous feature is the complex cellular structure owing to the presence of the mesoscale eddy field–a result that contrasts sharply with the classical picture of an ocean circulation taking place in horizontal, laminar layers.

$\tilde{H}_R$ can be written

$$\tilde{H}_R = \sum_j \sum_s C_j(s) T_j(s) = \text{vec}(\mathbf{C})^T \, \text{vec}(\mathbf{T}) \tag{4.4.4}$$

where $T_j(s)$ is the area and mass-weighted, reference-level velocity in pair j in depth range s and is thus the transport in each subarea of the section. Equation (4.4.4) is used to make the obvious point that in a turbulent flow field the transport by the relative velocity is proportional to the second-moment matrix of the two vectors, vec($\mathbf{C}$), vec($\mathbf{T}$). Figure 4–13 depicts these two vectors, with $\mathbf{C} = \boldsymbol{\theta}$–the temperature field, their product, and their cumulative sum for the 24°N section (values below the seafloor were not plotted in the individual vectors and were removed in the cumulative sum). Visually, at least, the vectors appear poorly correlated, and the accumulating sum has something of the character of a random walk (compare, for example, the examples in Wunsch, 1992, of purely random accumulating sums). If the goal is a climatologically significant value of $\tilde{H}_R$, we require temporal statistics. Unfortunately, when hydrographic sections have been measured only once, we are driven to using the spatial variability as a substitute for measurements of the time dependence in another ergodic hypothesis. But understanding the reliability of Equation (4.4.4) through

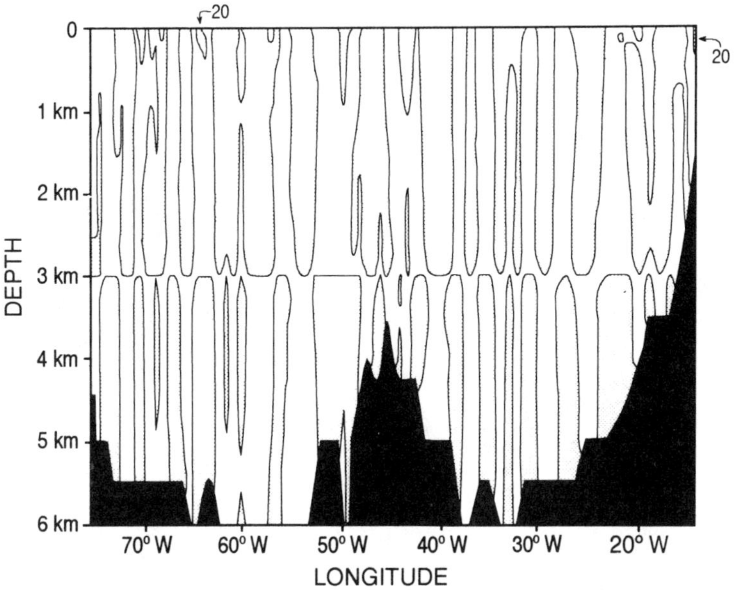

Figure 4–12. Relative vel0-city (thermal wind) in the 24°N section of Figure 2–2h, using all 90 station pairs (but two very shallow stations on the western end have been omitted from the plot). The most notable feature is the cellular structure, largely but not entirely owing to the presence of the mesoscale eddy field. The contours are in cm/s. The unlabeled contour is 0.

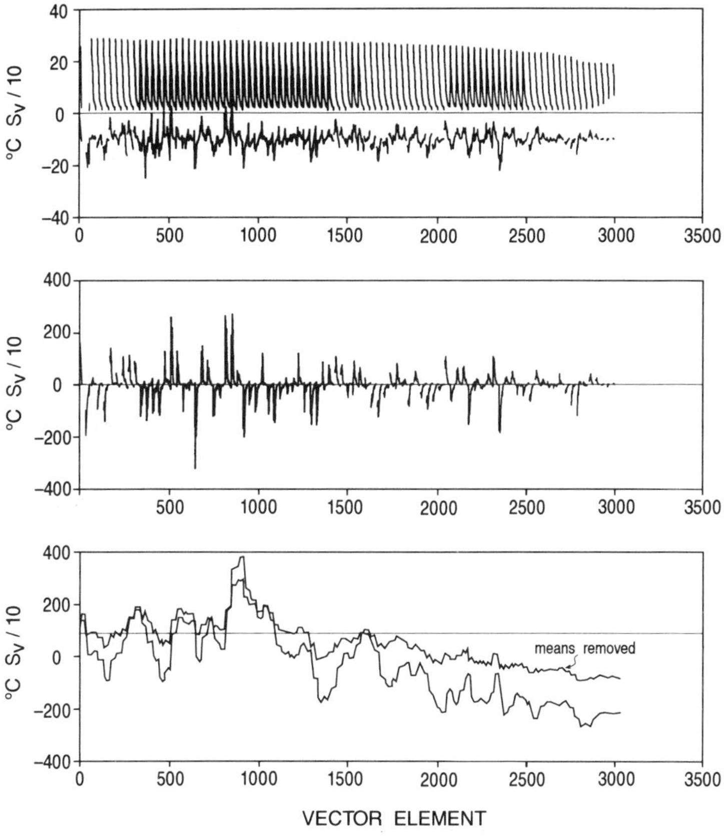

Figure 4–13. *Upper panel.* The upper curve is temperature, vec($\boldsymbol{\theta}$), in °C, and the lower curve is the thermal wind transport, vec($\mathbf{T}$), in tenths of Sverdrups. The dominant pattern in temperature is the sawtooth effect of the top-to-bottom temperature, which generates a large variance but also much of the correlation with the transport (velocity) field. Gaps occur where a standard depth lies within the bottom topography. *Middle panel.* The elements $\text{vec}(\boldsymbol{\theta})_i \text{vec}(\mathbf{T})_i$ making up the dot product of vec($\boldsymbol{\theta}$) and vec($\mathbf{T}$) and whose sum is the thermal wind contribution in Equation (4.4.1). The spatial means of both vectors are nonzero. *Lower panel.* The accumulating sum of $\text{vec}(\boldsymbol{\theta})_i\ \text{vec}(\mathbf{T})_i$ (in the middle panel), and the accumulating sum of $(\text{vec}(\boldsymbol{\theta})_i - \overline{\text{vec}(\boldsymbol{\theta})})\ (\text{vec}(\mathbf{T})_i - \overline{\text{vec}(\mathbf{T})})$ where the bar denotes the spatial aver age. Approximately one-half the value is contributed by the spatial means and one-half by the variance about the mean. The climatological representativeness of the flux calculations from the synoptic sections is determined in part by the stability of these accumulating sums.

substitution of spatial for temporal variability is rendered unsatisfactory because the fields involved are extremely inhomogeneous, both in the vertical and horizontal dimensions. Much of the variance in Equation (4.4.4) arises from the strong vertical gradients in temperature, but much of the correlation with velocity also lies in the vertical dimension.

To explore the issue qualitatively, however, let us ignore the spatial inhomogeneity in Equation (4.4.4). As with all second moments, we can separate the mean fields, $\overline{\text{vec}(\mathbf{C})}$, $\overline{\text{vec}(\mathbf{T})}$, from the fluctuating ones, $\text{vec}(\mathbf{C})' = \text{vec}(\mathbf{C}) - \overline{\text{vec}(\mathbf{C})}$, $\text{vec}(\mathbf{T})'$. The mean temperature is 8.6°C, and the mean transport is $-.008$ Sv; the temperature standard deviation is 7.2°C, and that of the transport is 0.2 Sv. The correlation coefficient between the fluctuating parts of the temperature and transport fields is -0.03. If the means are first removed from the two vectors, the cumulative sum changes by about a factor of 2 (Figure 4–13). That is to say, the numerical values of $\overline{\text{vec}(\boldsymbol{\theta})^T}\ \overline{\text{vec}(\mathbf{T})}$ and $\text{vec}(\boldsymbol{\theta})'^T\ \text{vec}(\mathbf{T})'$ are comparable. Thus, even in this quiet section, the numerical value of the flux owing to the relative velocity is directly dependent upon the reliability of the correlation between the area-weighted velocity and the temperature. Of particular concern is the very small correlation coefficient of -0.03. Although there were 2,775 elements in each of the vectors, the two fields do not contain so many degrees of freedom–each of the vectors is strongly autocorrelated.

An analysis of the reliability of the correlation coefficient can be developed along the general lines of Section 3.6.3 and is discussed at length in statistics texts (e.g., Kendall & Stuart, 1976). To keep the discussion simple, suppose that there are six degrees of freedom in each station pair (i.e., about six functions are required to describe the vertical structure of temperature and salinity–see Fukumori & Wunsch, 1991) and that the amplitudes of these functions are independent random variables with no correlation between pairs. The estimate of the correlation coefficient thus has been made with about 460 degrees of freedom. Is the correlation significantly different from zero? At 95% confidence we should reject the hypothesis that the correlation, r, is nonzero unless $z \equiv \sqrt{460-3}(0.5\cdot\log(1+r)/(1-r)) > 1.96$ (Anderson, 1984). The data produce $z \approx 0.64$, suggesting that the correlation is indistinguishable from zero, and the contribution of the fluctuation terms to the temperature flux is not statistically significant. Assuming that the mean terms are accurately estimated (which is by no means clear), the result of the present crude analysis suggests that the net temperature flux by the relative velocity is likely to contain errors in the vicinity of 100%, because the contribution from $\text{vec}(\boldsymbol{\theta}')^T\ \text{vec}(\mathbf{T}')$ may vary qualitatively and quantitatively from one realization to the next. To repeat, however, this calculation is of doubtful validity because of the spatial inhomogeneities in the data. Different ways of organizing the calculations can lead to radically different results: One can remove vertical and/or horizontal averages of velocity and temperature fields prior to computing correlations and variances (see Bennett, 1978, and Bryden, Roemmich, & Church, 1991, for examples) which–assuming, for example, that the average properties are exact and invariant–leads to much smaller formal uncertainties. In the presence of topography, however, the use of vertical or horizontal means introduces further inhomogeneities into the calculations.

If $\mathrm{vec}(\boldsymbol{\theta})'^T\ \mathrm{vec}(\mathbf{T})'$ is nonetheless regarded as the eddy heat flux, the failure to obtain a statistically significant value is consistent with moored array experiments in the open ocean (e.g., Fu, Keffer, Niiler, & Wunsch, 1982), assuming the present spatial average is equivalent through the ergodic assumption to time averages at a point. In other words, the eddy heat flux may be so small at any single location that one fails to find a statistically significant value of the correlation between the velocity and temperature fields. But even a very small correlation, when multiplied by the rms velocities and temperatures and integrated across an entire ocean basin, may produce physically important eddy flux totals. The argument is complicated by the presence of boundary currents that contribute indistinguishably from eddies but which are permanent features of the circulation. The 24°N section is a somewhat special one: The fields appear fairly homogeneous, with no obvious intense boundary currents (the Bahama Bank having permitted removal of the Gulf Stream from the discussion).

By way of contrast, consider the 36°N section (Figure 2–2j), in which there is no land mass to separate the boundary current region from the oceanic interior, making it more typical of open ocean sections used for property flux calculations. The estimated relative velocity in this section is shown in Figure 4–14. The intense velocities associated with the Gulf Stream (and an apparent meander or eddy associated with it) are the most conspicuous features, rendering it radically different from the 24°N section. In the interior, an intense eddy field again dominates any large-scale flow also possibly present.

For 36°N, the cumulating sums of $\mathrm{vec}(\boldsymbol{\theta})_i\,\mathrm{vec}(\mathbf{T})_i$, and $\mathrm{vec}(\boldsymbol{\theta})'_i\,\mathrm{vec}(\mathbf{T})'_i$–that is, with and without the vector means removed–are depicted in Figure 4–15. Here, the mean temperature is 8.3°C, the mean transport is 0.07 Sv, and the standard deviations are 5.6°C and 0.5 Sv, respectively. The correlation coefficient between the vectors is 0.024. Figure 4–15 suggests that the "eddy" contribution to the thermal wind-carried temperature flux is quantitatively important–shifting the net flux by more than a factor

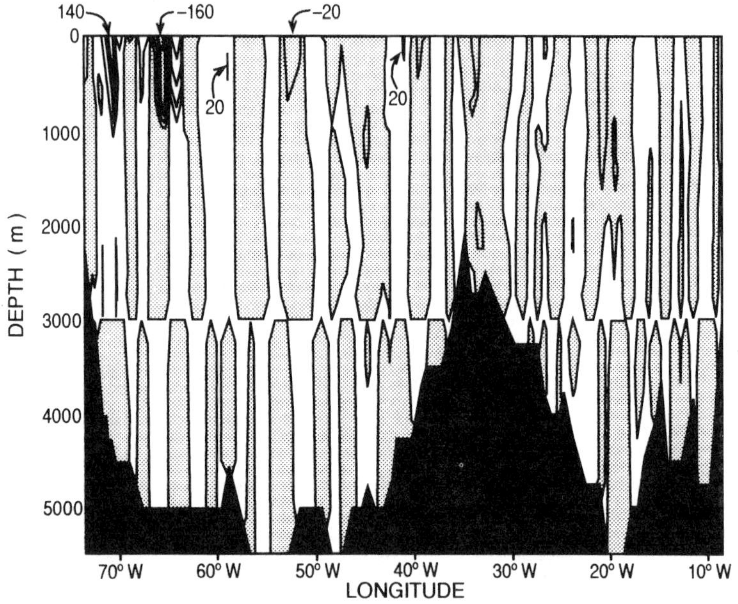

Figure 4–14. Relative velocity (cm/s) with a 3000-decibar reference level in the 36°N section using all 101 station pairs. In contrast with the 24°N section, no land mass separates the Gulf Stream from the section interior, and the extremely intense flows associated with the Stream, its meanders, and recirculation are evident on the western side of the section.

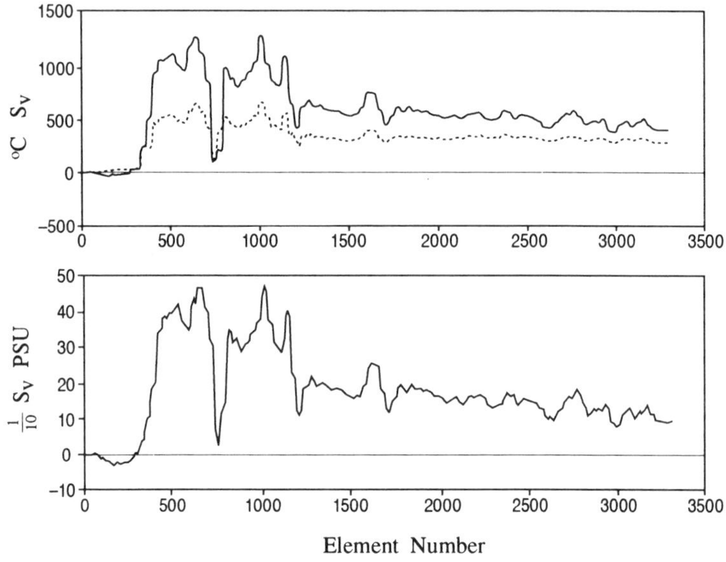

Figure 4–15. (Upper panel) Cumulating sum of $\text{vec}(\boldsymbol{\theta})_i \text{vec}(\mathbf{T})_i$ and $(\text{vec}(\boldsymbol{\theta})_i - \overline{\text{vec}(\boldsymbol{\theta})})(\text{vec}(\mathbf{T})_i - \overline{\text{vec}(\mathbf{T})})$ for the 36°N section (compare to panel of Figure 4–13). Here, the extremum associated with the Gulf Stream and the intense flows on its eastern edge are particularly apparent. Units are °C · Sv. (Lower panel) Cumulating sum of the salinity anomaly for the same section, showing a similar behavior to temperature.

of two if it is omitted. But again, the simple statistical test applied to determine the significance of the correlation coefficient in the 24°N line is even more dubious here, because Figures 4–14 and 4–15 show that the underlying supposition of homogeneous statistical behavior in the fields is inapplicable. For present purposes, we confine our conclusions to two points. (1) The property flux owing to the thermal wind cannot be computed even qualitatively by multiplying the zonal mean transport by the zonal mean property: One must account for the correlations between the velocity field and the property distribution. For some properties, the sign of the apparent flux is reversed if only the zonal means are employed. (2) The correlations of velocity and property are very weak and probably not statistically significant when computed from a single realization of the section. Nonetheless, the apparent correlations are large enough to compete with and overwhelm the products of the zonal averages. But because only slight modifications of the velocity and property fields can produce large changes in the apparent correlation function, the numerical values obtained from single realizations may be unreliable.

The situation is sufficiently complex that a definitive understanding of the stability of heat fluxes as climatological means will likely only emerge through repeated measurements of the fields so that the statistical basis does not have to depend upon employing spatial averages in place of the required temporal ones.

4.5 Application to Real Data Sets

4.5.1 Large Box Inversions

A considerable literature and body of experience has accumulated concerning the use of methods like the ones just outlined to determine elements of the ocean circulation. Most of the problems in dealing with inverse solutions are oceanographic rather than mathematical or statistical.

Consider as an illustration the study of Rintoul and Wunsch (1991) of the circulation in the North Atlantic Ocean, based upon the complete versions of the sections used in the reduced box example. Actual station positions may be seen in Figure 4–16. There are roughly 100 stations along each nominal latitude line–a number chosen to adequately sample the mesoscale eddy field, which is apparent in Figures 4–12 and 4–14. The constraints on the system included total mass and salt conservation plus zonal flux constraints on them as in the reduced example. In addition, requirements in individual layers of the form

$$\sum_{j=1}^{N} \delta_j \Delta r_j \int_{p_m}^{p_{m+1}} \rho C_j (v_{Rj} + b_j) dp + (w_m^* a_m C_m) - (w_{m+1}^* a_{m+1} C_{m+1}) + n_m = 0 \tag{4.5.1}$$

were imposed. Here, Δr_j are station spacings, and the integrations (done numerically) were carried out using the pressure p as the vertical coordinate,

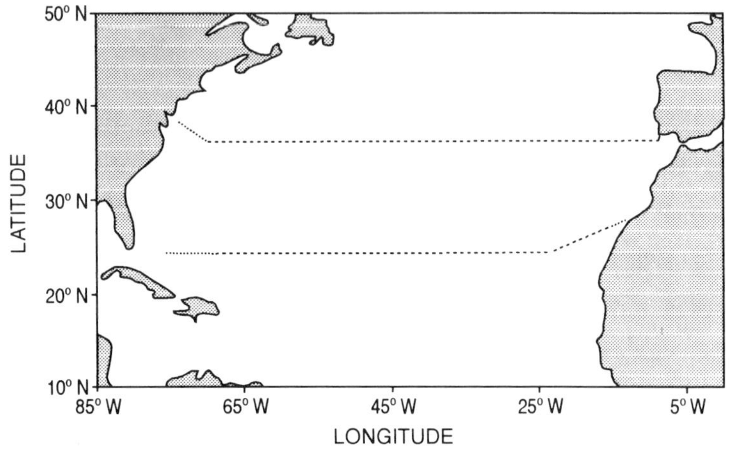

Figure 4–16. The positions of the hydrographic stations used by Rintoul and Wunsch (1991), corresponding to the values in Figures 2–2h-j and 4–12 to 4–15. The spatial sampling where the 36°N north section crosses the Gulf Stream, and near the continents on both ends of the sections, is much denser than elsewhere.

so that p_m is the pressure depth of isopycnal m; a_m is the horizontal areal cross section of the box in Figure 4–16. The w^* are a parameterization of all processes capable of carrying fluid across a surface of constant density (a representation of the poorly known mixing processes). Although McDougall (1987b) has criticized such parameterizations in inverse calculations, there is no reason to believe that they misrepresent the effect of mixing at the levels of accuracy permitted by real observations. The isopycnals used are listed in Table 4–7.[4]

Because the focus of the investigation was the property fluxes and divergences in the subtropical Atlantic, constraint equations were written for mass, heat, salt, oxygen, and the nutrients silicate and nitrate. Phosphate was combined with oxygen concentration to form a synthetic tracer called "PO" $= 135[PO_4] + [O_2]$ (see Broecker, 1974) where $[PO_4]$ denotes the phosphate concentration and $[O_2]$ that of oxygen. Under certain circumstances, "PO" can be treated as approximately conservative–that is, independent of biological activity, which is not true for either phosphate or oxygen alone (or any of the other nutrients).

The particular-SVD solution was employed with estimates of the expected errors in each of the equations and with column weighting by the column norms. No noise correlations were specified. The nutrient constraints were

[4] Several formulations of the net mixing process are possible. McDougall's (1987a,b) representation in terms of so-called neutral surfaces would undoubtedly be more accurate in an absolute sense. The main issue is whether the errors incurred by employing isopycnals rather than neutral surfaces are the dominant contribution to $\mathbf{n}$, or more likely to be swamped by other more significant problems such as the lack of simultaneity in the data. A coordinate system based upon the local vertical rather than the normal to isopycnals is also possible. One can then enforce structure on the mixing estimates by using the general tensor form, discussed by Redi (1982). Answers should not depend upon the coordinate system chosen–albeit, proper selection of the coordinates can vastly simplify solution of a problem.

Table 4–7. *Isopycnals used in the computations of Rintoul & Wunsch (1991) defining layer boundaries. Values are based upon potential density, but using three different reference pressures–0, 2000, 4000 decibars–to minimize the complications inherent in the nonlinear equation of state.*

Layer		
	Upper Boundary	Lower Boundary
1	Surface	$\sigma_\theta = 26.40$
2	$\sigma_\theta = 26.40$	26.80
3	26.80	27.10
4	27.10	27.30
5	27.30	27.50
6	27.50	27.70
7	27.70	$\sigma_2 = 36.82$
8	$\sigma_2 = 36.82$	36.89
9	36.89	36.94
10	36.94	36.98
11	36.98	37.02
12	37.02	$\sigma_4 = 45.81$
13	$\sigma_4 = 45.81$	45.85
14	45.85	45.87
15	45.87	45.89
16	45.89	45.91
17	45.91	45.93
18	45.93	Bottom

given such large uncertainties in the upper ocean that they could be regarded as purely diagnostic there. Two reference levels, 1300 and 3000 decibars, were explored. Although with mass constraints alone, the 1300-decibar level gave the smaller state-vector norm, the 3000-decibar level proved to be closer to the final best estimate when all constraints were used. Rintoul and Wunsch focus on a system with 217 unknowns (some of which are the w^*) in which the rank was estimated as $K = 20$. Figure 4–17 displays the resulting reference-level velocity and total flow–reference plus relative velocity–for the model with a 3000-decibar initial reference level. The complex cellular behavior is a real and important element of the ocean circulation.

The general noisiness of these sections led Rintoul and Wunsch (1991) to examine mainly the zonally integrated fluxes, which are both more reliably computed than the point velocities, and of more central importance in climate. The ergodic assumption that spatially averaged flows represent long-term time averages is plausible and useful, but it has never been proven valid.

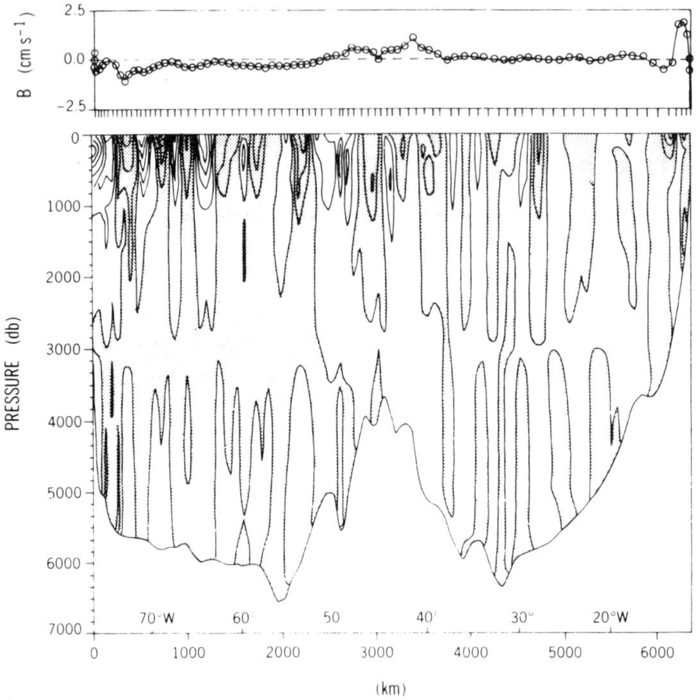

(a)

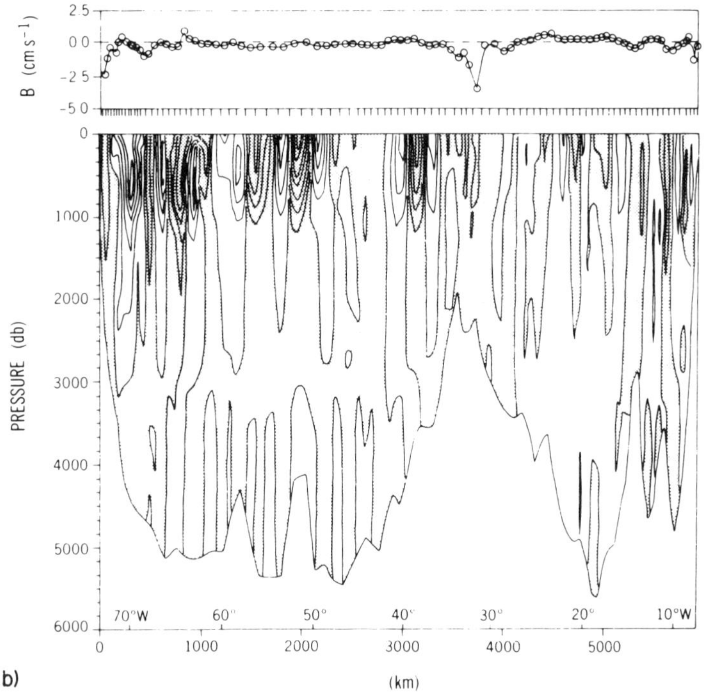

(b)

Figure 4–17. Rintoul and Wunsch (1991) reference-level velocity and total flow field, with rank $K = 20$ and a reference level at 3000 decibars for the 24°N (top), B, 36°N (lower) sections.

Figure 4–18 depicts the zonally integrated mass oxygen and nutrient fluxes across the two sections. In the mass flux, one sees the near-surface poleward moving water, much of which is bound up in the Gulf Stream circulation, a weaker poleward flux between about 600–1000 m, the southward flow, which is made up in part of NADW as well as Labrador Sea water, and a hint at 24°N at the very bottom of the weak flux of Antarctic Bottom Water. There are many details in these solutions, discussed in the original paper but omitted here. Note in particular, however, that the vertical structure of the oxygen flux is qualitatively different than the vertical structure of the mass flux. This result is an example of the point previously made–zonal mean property fluxes multiplied by zonal mean mass fluxes may be qualitatively incorrect estimates of the true values. One must adequately sample the horizontal (and vertical) structures in the fields and integrate them over the section area.

The Rintoul and Wunsch (1991) inversion is typical of a number of such calculations that have been published over the years, attempting to produce quantitative estimates of the flow fields and the property fluxes across regions of dimension ranging from entire ocean basins to smaller-scale boxes bounded by hydrography. A geographical survey of basin-scale inversions would include: (1) North Atlantic: Wunsch (1978), Wunsch and Grant (1982), Roemmich (1980), Roemmich and Wunsch (1985); (2) South Atlantic: Fu (1981), Rintoul (1991), Macdonald (1993); (3) North Pacific: Roemmich and McCallister (1989); (4) South Pacific: Wunsch, Hu, and Grant (1983); (5) Indian Ocean: Fu (1986), Macdonald (1993). Macdonald (1995) is apparently the first attempt at a global-scale inversion. Most of these results are based upon the SVD, or tapered SVD, with the major issues being the ones we have reviewed: the prior error estimates on the constraints, and the resulting choice of appropriate rank or tapering parameter, as the investigators sought to obtain solutions consistent with the prior statistics and to balance the competing demands of solution uncertainty against resolution.

Large regional inversions were undertaken by Thompson and Veronis (1980), the Tasman and Coral Seas; Tziperman and Hecht (1987), the eastern Mediterranean; Tziperman and Malanotte-Rizzoli (1991), the entire Mediterranean; and Roemmich (1981), the Caribbean Sea. Smaller regions were studied by Bingham and Talley (1991), the Kuroshio; and Joyce, Wunsch, and Pierce (1986) and Pierce and Joyce (1988), the Gulf Stream. All were again based upon SVD or tapered least-squares analyses.

The dominance of the eddy field in instantaneous flows led many of the authors listed above to focus on zonal or meridional averages of their results.

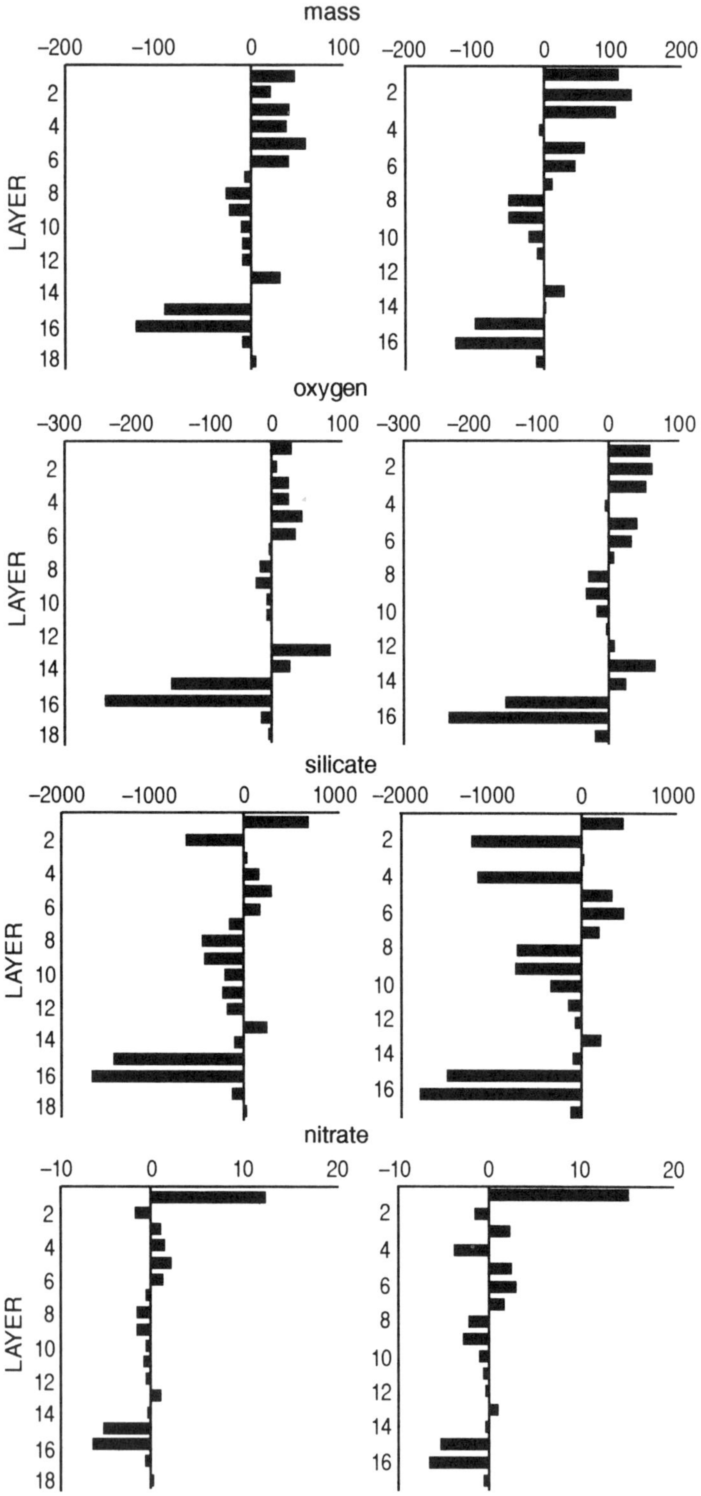

Figure 4–18. Zonally integrated meridional fluxes of properties from the box inversion of Rintoul and Wunsch (1991), with the 24°N section on the left and the 36°N section on the right.

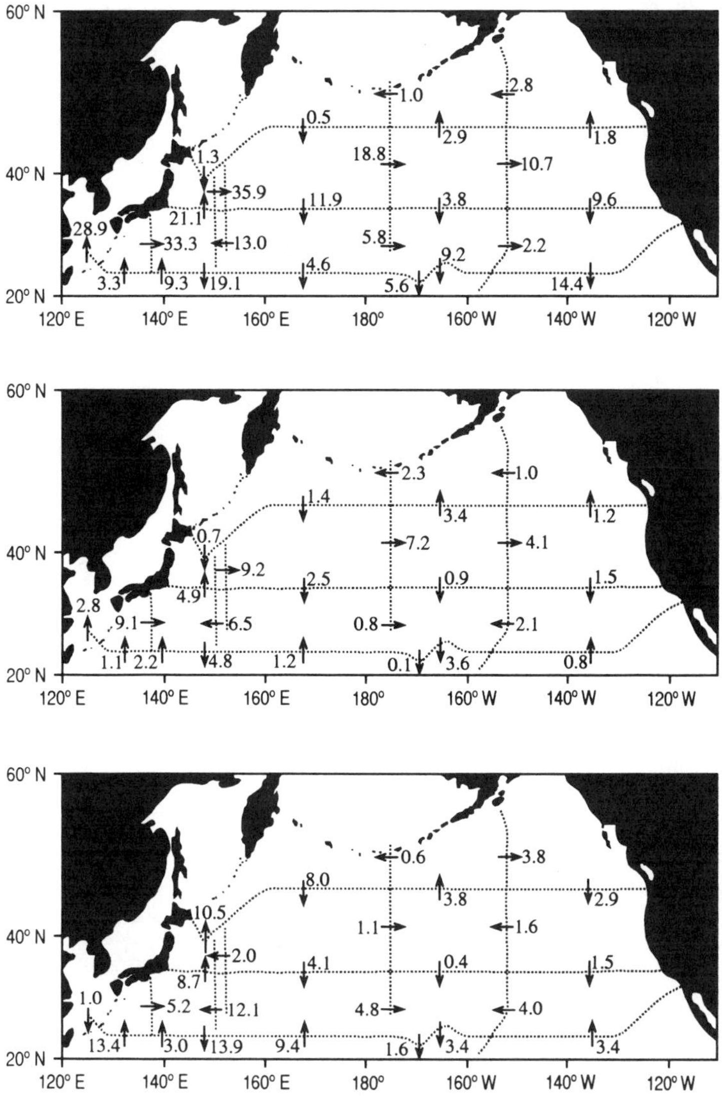

Figure 4–19. Mass fluxes Sv computed by Roemmich and McCallister (1989) along the segments of sections depicted. The three layers are defined to lie between the seasurface and $\sigma_\theta = 26.8$, 27.3, and $\sigma_2 = 36.96$.

There is no necessity of integrating entirely across ocean basins–any average tends to stabilize the flux results. Figure 4–19 is taken from Roemmich and McCallister (1989) and shows North Pacific results integrated along the segments depicted.

4.5.1.1 Eclectic Constraints

The linear constraint formalism permits one to write quantitative statements reflecting a wide variety of different observation types (called by Wunsch, 1984, *eclectic modeling*, reflecting the diversity of oceanographic data). For example, Joyce et al. (1986), Bingham and Talley (1991), and

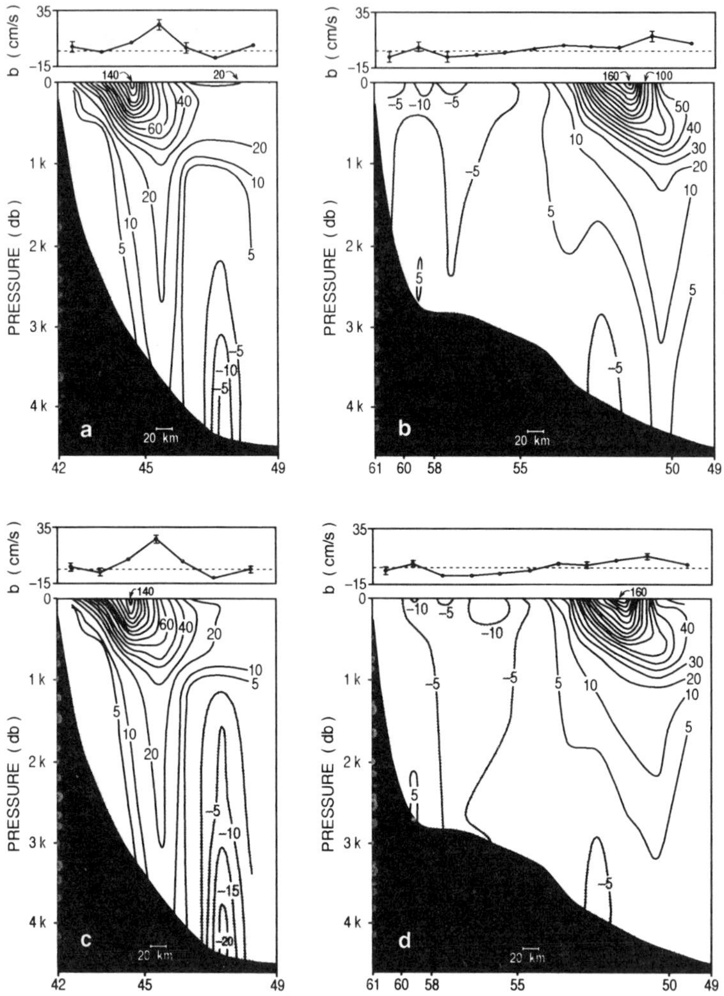

Figure 4–20. Velocities (cm/s) inferred in a region including the Gulf Stream by Joyce et al. (1986), who combined the box balances with constraints from so-called ADCP measurements. The estimated reference-level velocity is shown at the top of each section.

Bacon (1994) combined the hydrographic flux balances with direct velocity measurements from a shipboard device called an *acoustic Doppler current profiler* (ADCP; e.g., see Joyce, Bitterman, & Prada, 1982). The resulting fields from the Gulf Stream region, Figure 4–20, show a very strong Gulf Stream system embedded in strong reverse circulations on either side.

Pollard (1983) and Matear (1993) used moored current meter observations to add constraints to the hydrogaphic flux equations. Almost any measurement can be used (e.g., Martel & Wunsch, 1993b, used satellite altimeter data).

As previously stated, one simple way of studying the effects of various sums of nullspace vectors is to modify the initial reference level [recall Equation (4.3.7)]. Figure 4–21 displays two circulation pictures, as computed by

Wunsch and Grant (1982), for the North Atlantic circulation. One result was obtained using an initial zero level at 2000 decibars and the other at the bottom. At the time, there was little basis to choose between two such different hypotheses, and it is still not clear that one solution is more consistent with reality than the other. For this reason Wunsch and Grant labeled pictures such as these *cartoons* in the sense of a preliminary sketch done as preparation for a later, final painting; the final painting is still being constructed. (The more fundamental question, which is being side-stepped, is whether *any* steady circulation picture can possibly be an accurate representation of the ocean circulation. That is, motions such as those that dominate Figure 1–1 suggest that the movement of mass and properties are the result of integrations in space and time of rapidly varying turbulent flows. In that sense, representations like those seen in Figures 2–27, 2–29, or 4–21 are cartoons in another sense: They are highly schematized summaries of a complex world whose use is a trap for the unwary. An analogy would be the interpretation by a foreigner of the newspaper cartoon *Spiderman* as representing accurately life in the United States.)

When the first large-scale inverse results were published, the presence of strong columnar structure in the velocity field–apparent, for example, in Figures 4–5 and 4–17–led some to conclude that the result was wrong. The inference seems mainly to have been based upon the deep-seated Wüstian tradition that the ocean circulation moves in horizontal slabs. As the ubiquity and importance of the eddy field in the ocean has come to be accepted by most (but not all!) oceanographers, the expectation that synoptic flow fields should resemble a climatological average has gradually diminished. It is conceivable that a Wüstian picture would emerge from intervals over time of pictures such as Figure 4–17 (recall also Figure 2–27). Owing to the lack of data, no such computation has ever been attempted in the open ocean, and for the moment at least, the slablike movement of the ocean circulation must be regarded as counter to all the evidence.

4.5.1.2 Resolution

The small number of constraints in most large-scale applications of inverse methods, relative to the dimensions of the state vector, implies that the overall solution resolution tends to be poor. Figure 4–22b shows the resolution from a particular SVD solution (Wunsch, 1978) for the North Atlantic, in which elements of the reference-level velocity widely separated spatially are determined in linear combination. These linear combinations often correspond to water mass characteristics that are nearly identical despite the physical separation.

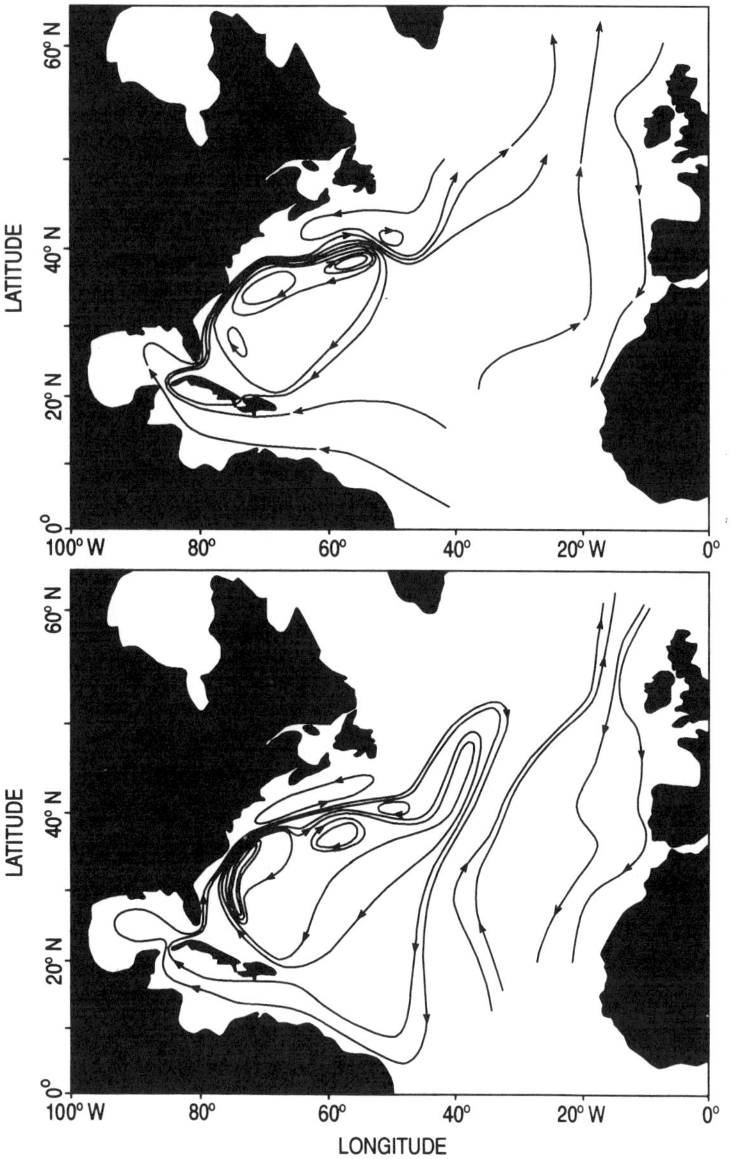

Figure 4–21. Transport cartoons drawn by Wunsch and Grant (1982) for the main thermocline (the density range $26.50 \leq \sigma_\theta \leq 27.50$) using the particular-SVD solution. The upper figure shows the solution with an initial reference level at 2000 decibars, and the lower one used the bottom. These solutions differ only in the nullspace of $\mathbf{E}$.

In contrast (Figure 4–22c,d), a number of published inversions (e.g., Roemmich, 1981; Rintoul, 1991) showed compact resolution–meaning, as discussed in Chapter 3, that the SVD solution is able to determine local averages of the reference-level velocity, which means that the resolution has a particularly simple interpretation. Such resolution is a reflection of spatially compact distributions of the properties being balanced in the constraint equations.

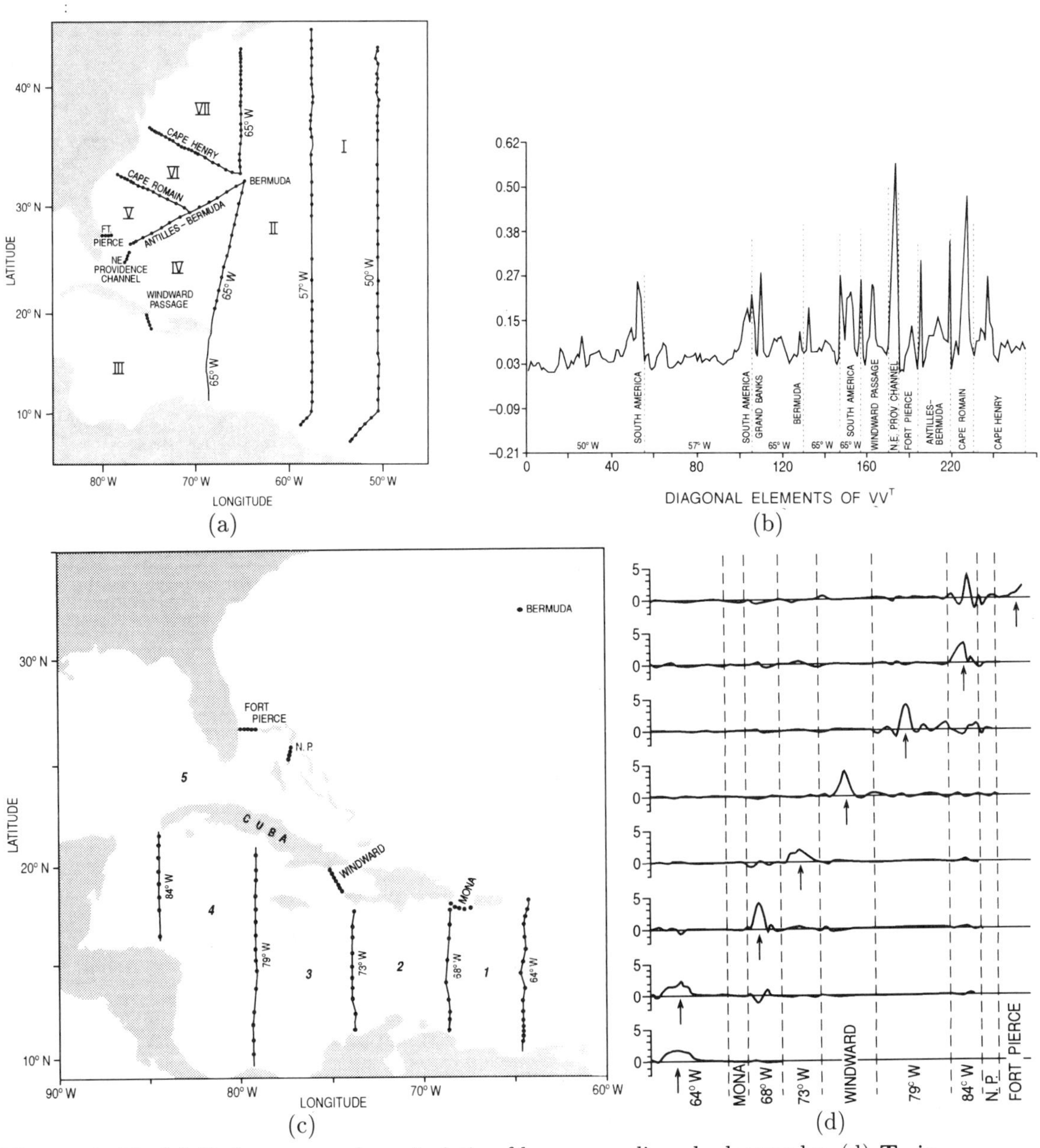

Figure 4–22. (a) Hydrography used by Wunsch (1978) to define conservation constraints; nominal geographical names are given to the sections. Roman numerals define box numbers. (b) Diagonal of $\mathbf{T}_v$ at rank 20, showing the highly variable resolution of the Wunsch (1978) solution, which is characteristic of large complicated inverse problems. The different elements of the solution may differ greatly in their reliability. (c) Hydrography used by Roemmich (1981) to study the Caribbean circulation. The lines are given names that are used as locators in Figure 4–22d. Five areas are defined by the hydrography. (d) $\mathbf{T}_v$ in Roemmich's (1981) rank $K = 27$ solution, showing generally compact resolution in contrast to Figure 4–22b. Arrows denote the station-pair position for which the resolution has been computed.

4.5.2 Heat Flux Calculations

The general problem of computing property fluxes across hydrographic lines has been discussed already in part. The most fundamental property is mass, because it is the movement of the water that carries the other properties with it, and the dynamical equations we use are written in terms of the velocity per unit mass (later we take up the possibility that there are significant lateral diffusive fluxes in the open ocean). Of the secondary properties, the heat (temperature) flux has a special importance because the flux divergence must be consistent with the atmospheric exchange of heat with the ocean, with all its implications for weather and climate.

Bryan (1962) reviewed early attempts to compute the oceanic heat flux, which until the work of Model (1950) and Jung (1952), had apparently been regarded as negligible compared to the atmospheric transport. Much of the present intense interest in the subject arose, however, when Vonder Haar and Oort (1973) suggested that in the northern hemisphere the meridional flux of heat by the ocean might dominate that of the atmosphere, reaching in their estimate, 74% of the total at 20°N. In a subsequent study, Oort and Vonder Haar (1976) examined the global budget and its annual cycle, apparently confirming the large oceanic contributions.

Such large values were surprising and stimulated efforts to better understand how much heat the ocean moves. Four methods have been used:

(1) *Atmospheric residuals.* The Vonder Haar and Oort (1973), Oort and Vonder Haar (1976), and Carissimo, Oort, and Vonder Haar (1985) estimates are computed as the meridional flux divergence of the ocean from a residual of the atmosphere. The heat flux divergence at the top of the atmosphere is known from space-borne measurements; estimates of the meridional flux within the atmosphere can be based upon conventional daily meteorological analyses. The difference between what is going in and out of the top of the atmosphere and what the atmosphere carries itself must be carried by the ocean.

This methodology involves finding small differences of very large numbers, whose value is taken to be the oceanic meridional heat flux, and the residual uncertainties are very large, consistent with zero oceanic flux.

(2) *Bulk formula estimates.* Here, one computes the estimates of thermodynamic exchanges, F, between ocean and atmosphere directly through various empirical formulas. There are several terms that must be accounted for, including those for direct solar heating, conduction, evaporative terms, etc. This difficult subject has been discussed in detail by Isemer and Hasse (1985), Bunker (1976), Large and Pond (1981), and others. A typical

expression, for the so-called sensible heat, is

$$F_{\text{sensible}} = \rho C_s h_p (\theta_s - \theta_{10}) U_{10}$$

where C_s is an empirical coefficient, h_p is the heat capacity of water, θ_s, θ_{10} are the seasurface temperature and the air temperature at 10 m, respectively, and U_{10} is the windspeed at 10 m. Obtaining accurate, simultaneous measurements of these quantities from ships has proved very difficult, and the value of C_s is itself uncertain and probably not constant.

A number of authors (Bunker, 1976; Talley, 1984; Hsiung, 1985; Isemer & Hasse, 1985) combined these expressions with compilations of millions of shipboard observations of the wind, relative humidity, cloud cover, etc., to produce maps of the air/sea transfers. Unfortunately, there are very large errors, in the range of 30–50+ W/m^2, both systematic and random, in these expressions (Talley, 1984; Hastenrath, 1984; Garrett, Outerbridge, & Thompson, 1993b; Large, McWilliams, & Doney, 1994). Unless there is unexpected progress in reducing the physical errors in the bulk formulas, it is likely that these methods will be superseded.

(3) *Atmospheric boundary layer analyses.* Estimates can be made from the daily atmospheric analyses of the flux divergence at the base of the atmosphere. That is, the ordinary meteorological forecast models require local sources and sinks of heat at the air/sea interface, and which, if correctly estimated, must be consistent with what the ocean is providing. This method has not been used often, but as meteorological models and analyses become more realistic, this method should become a competitive one.

(4) *The "direct method."* This approach is based upon the divergence of the direct estimates in the ocean of the type we have been considering. One calculates the heat carried by the circulation, with the temperature transports converted to heat fluxes.

In the previous discussion of temperature fluxes, a conversion was made to a heat flux, multiplying the temperature fluxes by the heat capacity of seawater, which is approximately $h_p = 4 \times 10^3$ J/kg/°C (4×10^{12} ergs/gm/°C) (Millero, Perron, & Desnoyers, 1973). A complicating factor is that the conversion is meaningless unless one is dealing with a closed, mass-conserving system (Montgomery, 1974). If mass is not conserved, a temperature flux in °C would be very different from one computed in kelvins. That is, temperature fluxes are dependent upon the temperature scale and can only be regarded as equivalent to heat fluxes if the system is closed. Write the temperature as $\theta_0 + \theta'$ where θ_0 is an arbitrary reference value and θ' is the deviation from it, used as the water temperature. Let $q_j(s)$ be the total water velocity. Then the temperature flux across any bounding fluid surface is

$$H_h = \sum_j \sum_s \rho_j(s) q_j(s) \Delta a_j(s) h_p \left[\theta_0 + \theta_j'(s)\right] \tag{4.5.2}$$
$$= \theta_0 h_p \sum_j \sum_s \rho_j(s) q_j(s) \Delta a_j(s) + h_p \sum_j \sum_s \rho_j(s) q_j(s) \Delta a_j(s) \theta_j'(s)\,,$$

which clearly depends upon θ_0 unless $\sum_j \sum_s \rho_j(s) q_j(s) \Delta a_j(s) = 0$–that is, mass is conserved.

In the one box example discussed above, the residual mass imbalance in the system after inversion was quite small (0.2 Sv), and one could convert quantities such as the Florida Current temperature flux into a heat flux because the temperature of the return flow was known. One sometimes sees discussions of the supposed heat flux of an element of an open system–for example, of the Florida Current or of the Ekman flow. It is asserted, for example, that because the average temperature of the Florida Current was observed to shift by $\Delta\theta$°C that therefore the heat flux must have shifted by $h_p \Delta\theta\ T_{\mathrm{FS}}$ where T_{FS} is the Florida Current mass flux. But if the return flow temperature (across the 24°N section) also increased by a like amount, then the heat flux across the zonal section did not change at all. The temperature flux of the Florida Current is well defined in terms of whatever temperature scale is being used; its heat flux, in isolation from the rest of the system is meaningless (consider using a temperature scale in which $\theta_0 = 100$°C; then the supposed "heat flux" of the Florida Current would be negative).

Using a mass-conserving system, Rintoul and Wunsch (1991) estimated the heat flux across the combined Florida Straits–24°N line as 1.3 ± 0.2 PW. The error estimate is reasonable for the instantaneous value, but in view of the comments above about the relative velocity/temperature correlation, the uncertainty estimate may be very optimistic if the number is to represent a climatological mean. This heat flux, if it is truly representative of the long-term average, must be transferred to the atmosphere; otherwise the mean temperature of the North Atlantic would change. [Whether there is any evidence of changes is the subject of intense interest. Inferred shifts (Roemmich & Wunsch, 1984; Parrilla, Lavin, Bryden, Garcia, & Millard, 1994) are too slight to affect the present arguments.] The area of the North Atlantic poleward of 24°N is roughly $50 \times 10^{12}\mathrm{m}^2$; we thus deduce that there must be a transfer to the atmosphere of about $(1.3 \pm 0.2\mathrm{PW})/5 \times 10^{12}\mathrm{m}^2 = 26 \pm 4\mathrm{W/m}^2$.

The quantity of physical interest is the heat flux *divergence*: All of the 24°N flux must be removed from the ocean before the northern boundary is reached, apart from any slight amount possibly carried into the Arctic. In principle, the flux can be regionalized by subdividing the area of the ocean to the north. Rintoul and Wunsch (1991) estimated the net flux across 36°N also as 1.3 ± 0.2PW. Subtracting the two fluxes and dividing by the area between the two sections produces a flux divergence of about $0 \pm 6\text{W/m}^2$ (treating the two uncertainty estimates as independent, which is not strictly correct). The need for much more accurate estimates drives much of the present interest in the ocean circulation. Evidently, the major sources of error in estimating the flux of heat by the ocean are tied directly to the uncertainty in the water movement; until the circulation estimates are improved, there will be no corresponding improvement in the heat fluxes. The formal errors are, however, already much smaller than those usually quoted for bulk formula and other methods but are subject to the concerns already expressed.

Beginning with Bennett (1978), Stommel (1980), Bryden and Hall (1980), Wunsch (1980), Roemmich (1980), and others tried to obtain improved estimates of the oceanic meridional heat fluxes and divergences from versions of the direct method. Some of these estimates arise from what are recognizably inverse method results and include: Wunsch (1980) for the subtropical North Atlantic; Roemmich (1980) for the tropical Atlantic; Wunsch et al. (1983) for the South Pacific; Fu (1981) and Rintoul (1991) for the South Atlantic; Fu (1986) for the Indian Ocean; and Macdonald (1993) for the entire southern hemisphere at 30°S. These estimates mostly (but not universally) include uncertainty values. The focus has been on meridional heat fluxes both because of the availability of zonal transocean sections, which lend themselves to the box inverse method, and because the dominant transfers of heat in the overall climate system are meridional–equator to pole. But there is nothing precluding estimates of zonal heat fluxes and corresponding divergences by the same methods.

A number of the estimates of the meridional flux of heat are based upon what might be thought of as hybrid methods; that is, they can all be included in the framework of Equations (4.1.1) but typically do not derive the reference-level velocities from a formal inverse solution. Rather, they make some plausible assumption about the reference-level velocity and produce uncertainty estimates of varying degrees of completeness, often just ignoring the nullspace components. For example, Jung (1952) used Riley's (1951) estimates of the North Atlantic circulation; Bryan (1962) and Bennett (1978) assumed the validity of Sverdrup balance and that the windfield estimates were very accurate. None of these authors discussed the impact of errors in the assumed model flow field.

The most influential of the estimates has been that of Bryden and Hall (1980) and Hall and Bryden (1982) for the North Atlantic near 24°N. For a number of reasons, the results are more accurate than almost anywhere else. It is useful to examine their calculations in the inverse context to understand why that might be.

The calculation was made using the 1957 hydrographic section from the Bahama Bank to Africa (Figure 2–2i), which is an earlier, lower spatial resolution version of the 1981 section at that latitude used above. In the Florida Straits, they used the various direct measurements of the Gulf Stream to calculate a temperature flux, which translates into 1.2 PW after the mass budget is closed (it includes a small Ekman flux).

In the 24°N section, they chose a reference level for the relative velocity, $q_{R_j}(s)$, in each station pair such that the vertical sum, on s, vanishes. The reference-level velocity, x_j, is then the vertical average velocity in pair j. The temperature flux across 24°N is computed as the sum of the three usual terms: Ekman, relative, and reference-level velocities. The Ekman and relative-velocity components are found as previously described. One must estimate the reference-level velocity. Hall and Bryden (1982) imposed the mass conservation constraint

$$\sum_j \sum_s x_j \rho_j(s) \Delta a_j(s) = -\text{Florida Straits mass flux} + \text{Ekman mass flux}$$
$$= -M_F$$

or

$$\sum_j T_j = -M_F \tag{4.5.3}$$

where $T_j = \sum_s x_j \rho_j(s) \Delta a_j(s)$ is now defined as the mass transport in pair j owing to the reference-level velocity and is equivalent to equation 4 of the reduced box model. The temperature transport of the reference-level velocity is

$$H_x = \sum_j T_j \overline{\theta}_j = \mathbf{T}^T \overline{\boldsymbol{\theta}}\,, \tag{4.5.4}$$

where $\overline{\theta}_j$ is the mean temperature of station pair j. Hall and Bryden (1982) plotted $\overline{\theta}_j$ as a function of j (Figure 4–23a), showing that it was nearly constant $\overline{\theta}_j = \theta_1$. Thus, the temperature transport of the reference-level velocity is simply

$$H_x = \theta_1 \sum_j T_j = -\theta_1 M_F\,, \tag{4.5.5}$$

and the problem is solved once the Ekman transport has been specified.

It is useful to analyze this computation in the inverse context. The constraints being imposed are: (1) a fixed mass flux in the Florida Straits, and (2) overall mass conservation. There are 38 unknown reference-level velocities remaining in the pairs east of the Florida Straits and only one constraint (4.5.4) to determine them. Notice that the Florida Straits velocities have been removed from the problem and only appear indirectly in M_F. If an SVD were used to solve the system, there would be one range and 37 nullspace velocity vectors. The uncertainty of the coefficients of the range components derives from that in the Florida Straits mass flux and in the Ekman flux. The replacement of the sum (4.5.3) by (4.5.5) is equivalent to arguing that all 37 nullspace vectors are orthogonal to the heat content vector, which is proportional to the mean temperature in Figure 4–23a. The accuracy of the computation therefore depends directly upon the assumption that there are no mass-conserving flows (nullspace vectors) that are large in the station pairs where Figure 4–23a displays significant deviations from a constant. Hall and Bryden (1982) cite inverse calculations and direct current meter records to propose that these flows are small or absent.

This discussion is easily rendered quantitative. Figure 4–23b (solid line) shows the mean vertical temperature in each of the 89 station pairs from the higher-resolution 1981 section along 24°N. Again, apart from the two boundaries and a slight rise over the Mid-Atlantic Ridge, the station-pair mean temperatures are remarkably constant. The single mass conservation constraint (4.5.3) now has one range vector and 88 nullspace vectors. Four arbitrarily chosen, but typical, nullspace vectors are displayed in Figure 4–23c. In general, they show a large mass flux in one pair, j, with a weak distributed mass-conserving return flow distributed nearly uniformly over the entire remainder of the section. Visually, such vectors are nearly orthogonal to the mean temperature vector $\overline{\boldsymbol{\theta}}$, depicted in Figure 4–23b, and thus even energetic flows lying in the nullspace will not generate significant temperature fluxes. (The possibility of strong currents on the eastern and western sides is clearly an exception to this conclusion, and they have to be excluded by appeal to independent evidence. It is a simple matter to bound the temperature fluxes in these two regions if prior velocity variances there are available. Wunsch et al. (1983) further discuss the projection of temperature content onto nullspaces, but there are some misprints in the formulas there. DeSzoeke and Levine, 1981, attempt to find a station distribution in which the vertically averaged temperature is as constant as possible.)

By way of contrast with the 24°N section, consider the same fields at 36°N, also shown in Figure 4–23b. There is more structure in the sta-

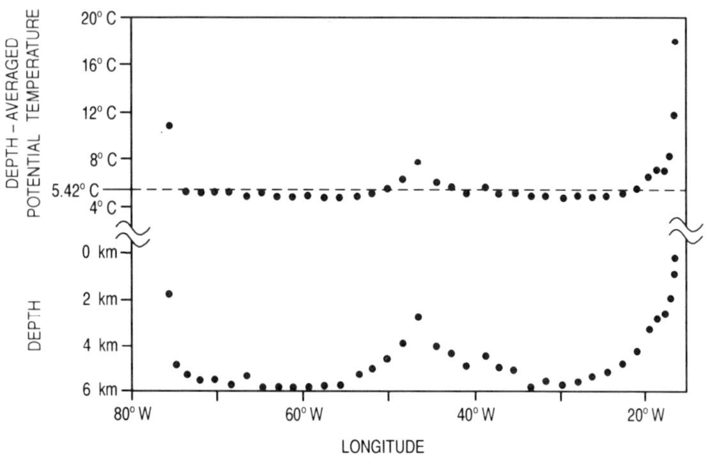

Figure 4–23a. Bryden and Hall (1980) estimate of the mean temperature in each station pair of the 24°N section obtained in 1957, as well as the station depth (lower curve). Station-pair average temperatures are remarkably uniform, apart from a few outliers associated with extreme topography.

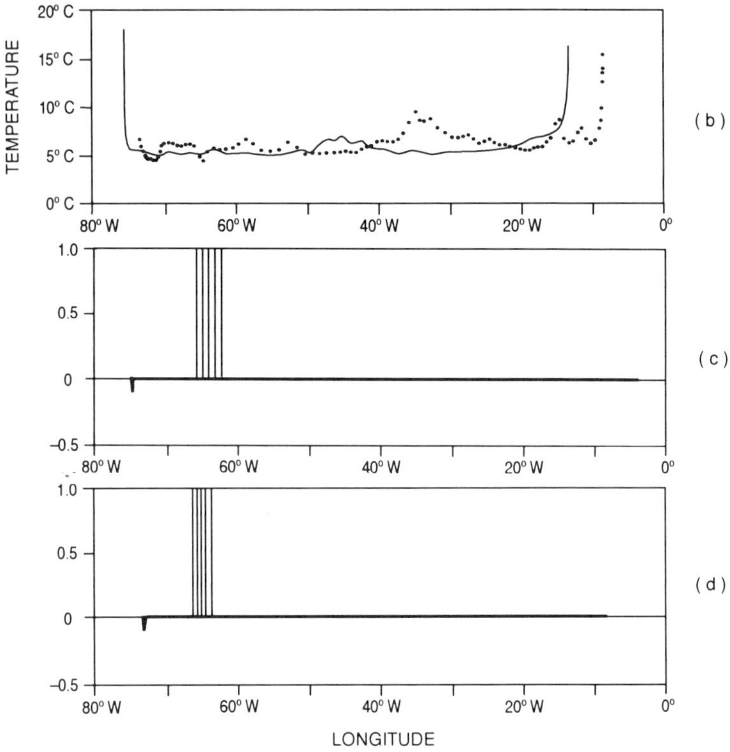

Figure 4–23b,c,d. (b) Same as Figure 4–23a, except computed for the higher-resolution 1981 section as well as the section along 36°N (dotted). Away from the two edges, the 24°N station-pair temperatures are remarkably uniform, but the 36°N section shows a greater baroclinic structure more typical of midlatitudes and includes the Gulf Stream region contributions. (c) Four nullspace vectors of the single mass-conservation constraint Equation (4.5.3) along 24°N. These are typically a large flow between a single pair (of either sign), and a return flow of opposite sign, conserving mass, nearly uniformly distributed in the rest of the section. Because of the near-uniformity of the station pair temperatures, such flows generate only slight heat flux contributions for even quite vigorous velocities. (d) Same as (c) except at 36°N, but now because of the much greater variability in station-pair mean temperatures, these mass-conserving flows are capable of producing significant heat fluxes. Because the nullspace contributions are indeterminate, one has a much greater formal uncertainty in the result.

tion average temperature, much of it on the west, associated with the Gulf Stream system, and the feature associated with the greater elevation of the Mid-Atlantic Ridge. There is also a warming trend toward the east, absent at 24°N and reflecting the large-scale baroclinic structure. A few typical nullspace vectors for the 36°N line, subject only to mass conservation such as (4.4.3), are depicted in Figure 4–23d. Mass-conserving flows are possible in which a peak transport can coincide with a local temperature maximum or minimum in Figure 4–23b, and in which the flux is returned, broadly, over the entire rest of the section, at temperatures 2–3°C and more different. If a 10 cm/s reference-level velocity is directed poleward in 4 km of water over a distance of 100 km, and is returned over the rest of the section at a mean temperature 2°C different, the corresponding heat flux is 0.3 PW. Hence, using only a single mass-conserving constraint renders the result quite uncertain–owing to the possibility of nullspace flows. This result contrasts with the 24°N section–with the Gulf Stream removed from the section–and with its notably flat interior isotherms. Although computations analogous to the Hall and Bryden (1982) one have been made elsewhere (e.g., Toole & Raymer, 1985, in the Indian Ocean, and Bryden et al., 1991, for the North Pacific), the results are not as accurate.

The remedy for this problem is to employ additional constraints that may require more or better observations, as in the calculations of Roemmich and Wunsch (1985) and Rintoul and Wunsch (1991), so as to reduce the number of nullspace vectors.

Once a mass-conserving flow has been found, one can decompose it into various spatial averages (zonal, vertical, etc; see Bennett, 1978, or Hall & Bryden, 1982) along with similar averages for the temperature field. The heat flux can be broken into subcomponents owing to the zonally averaged flow, or the vertically averaged one, etc. Insight into the circulation can be gained in this way, although such decompositions are not unique, and if the flows are not mass-conserving, they are meaningless.

4.5.3 Property Fluxes and the Spatial Resolution Problem

The ocean circulation moves properties through covariances between the flow field and the property distribution. Quantitative estimates of the property fluxes thus demand adequate sampling of these fields. Property fluxes have often been computed using representations of the ocean having extremely coarse resolution, whether from hydrographic data or in numerical models. The unpalatable question facing anyone computing property fluxes is whether an undersampled version of the velocity fields can produce quantitatively useful values.

To suggest the magnitude of the problem, consider Figure 2-19b,c; the coldest temperatures apparently nearly coincide with the maximum southward-moving velocity in the deep western boundary current. Figure 4–24 shows an expanded view of the 36°N section temperatures and relative velocities crossing the Gulf Stream. The so-called Warm Core of the Gulf Stream coincides with the highest velocities observed. In both these examples, the velocity temperature products $q_i(s)\,\theta_i(s)$ will be very large and of uniform sign. If the ocean is not modeled in such a way as to capture these velocity and temperature extremes, the property flux will be underestimated–that is, biased toward zero. Almost all numerical models of the ocean fail to produce sufficiently large meridional fluxes of heat, at least in part owing to inadequate representation of the property-velocity extremes.

Some of the boundary current problem is evident if the heat flux computed across 36°N from the reduced box model, 0.85 PW, is compared to that of Rintoul and Wunsch (1991), who estimated it as 1.3 ± 0.2 PW by using the full-station resolution. In contrast, the estimated flux across 24°N in the reduced box model (1.2 PW) is much closer to the Rintoul and Wunsch (1991) estimate of 1.3 ± 0.2 PW. The resolution requirement in the quiescent 24°N section, with no Gulf Stream present, and the nearly featureless interior isotherms render the section much less sensitive to sampling errors. (Both 24°N calculations set the Florida Current temperature transports to essentially the same values.)

Notice that in these examples, the problem is not that of resolving the interior eddy field; it is the failure to resolve the boundary currents. Thus, even schemes that seem to greatly improve the way in which circulation models calculate property fluxes (Gent & McWilliams, 1990; Danabagoglu, McWilliams, & Gent, 1994) cannot describe realistically much of the most important flow elements.

The role of the eddy field itself remains somewhat enigmatic because it has proved remarkably difficult to obtain any statistically significant eddy temperature fluxes in the oceanic interior (e.g., Fu et al., 1982). There are a few isolated, specific regions where significant correlations of the velocity and temperature have been found from moored arrays–near the Gulf Stream itself (Brown, Owens, & Bryden, 1986), in the Drake Passage (Bryden, 1979), and where the Circumpolar Current crosses steep topography (Bryden & Heath, 1985). But these are special regimes; in the interior,

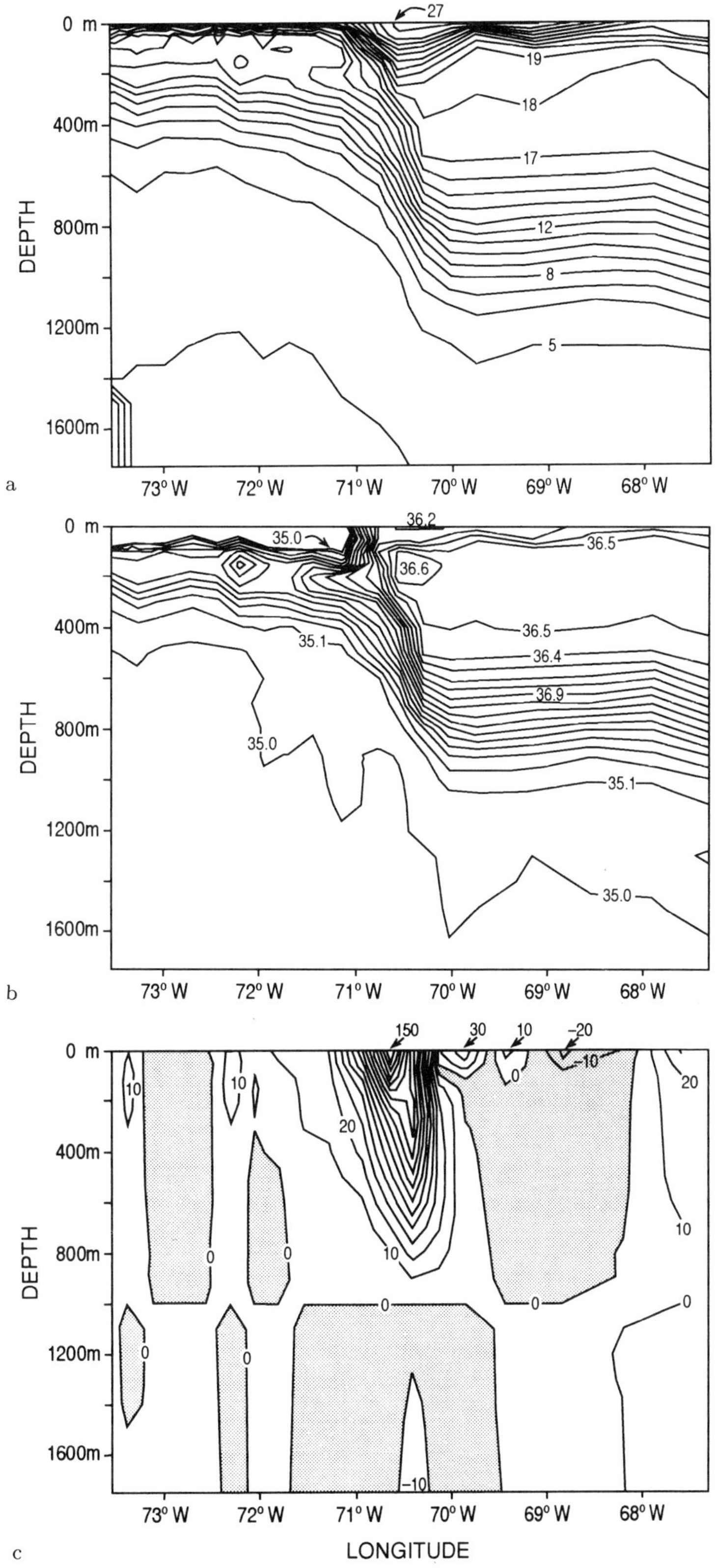

Figure 4–24. Expanded view of the temperature field (°C) in the vicinity of the Gulf Stream (a), as well as the salinity (b), and relative velocity (cm/sec) (c), with a 3000-decibar reference level for the 36°N section. The velocity extremes tend to coincide with the property extremes (one sees the *warm core* of the Gulf Stream). Because the property flux is the mean second-order product of the transport and temperature fields, accurate results require adequate sampling of the extreme values.

as previously described, the eddy fluxes appear small over most of the ocean. The difficulty is that as we have seen, even small fluxes, when multiplied by the enormous areas of an oceanic section, can translate into physically important contributions.

4.5.4 Nutrient and Related Fluxes

Although we have used the temperature flux as a prototypical property, there are many other fields whose flux divergences are of interest. Among them are the nutrients such as phosphate and silicate, as well as oxygen, carbon, etc., which are intimately bound up with the oceanic biological fields and ultimately with climate. Oxygen behaves somewhat analogously to temperature, as it is readily exchanged with the atmosphere, whereas the nutrients, like salt, do not have significant atmospheric pathways. Unlike temperature and salt, however, none of these other fields contributes significantly to the water density; hence, they can be expected to show somewhat different correlations with velocity than do fields that are dynamically active.

In principle, biochemical knowledge of the ocean permits one to obtain independent estimates of the interior and boundary sources and sinks of the nutrients and oxygen. With such estimates, equations such as (4.1.5) with known sources, Q, become useful constraints for determining the circulation. In practice, this independent knowledge is so imprecise that most large-scale estimates of oxygen and nutrient consumption and production are based instead on physical knowledge of the circulation, with the flux divergences computed from equations such as (4.1.5) used diagnostically (zero-weighted) with Q treated as another unknown.

The computations are the same as discussed for temperature, with one computing the covariance of the estimated flow field (Ekman plus relative geostrophic plus reference-level velocities) with C. All the evidence (e.g., Figure 2–2j) shows that eddy noise and boundary current extremes are present here, too, and the same high spatial resolution is required for accuracy. Indeed, the spatial resolution problem is perhaps most extreme for fields such as silicate, which are bottom intensified. For example, the complete 36°N section produces a silicate flux, owing to the reference-level velocity (3000-decibar reference level), of 170 moles/kg/s Sv, whereas the subsampled reduced section with the same reference level produces the qualitatively different 708 μmoles/kg/s Sv. The differences arise from the failure in the latter to capture the extreme values of the fluxes near the Mid-Atlantic Ridge and other topography.

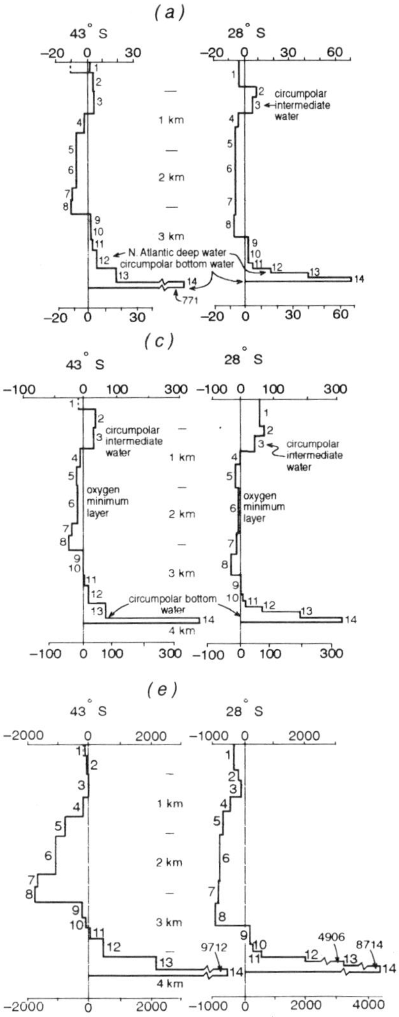

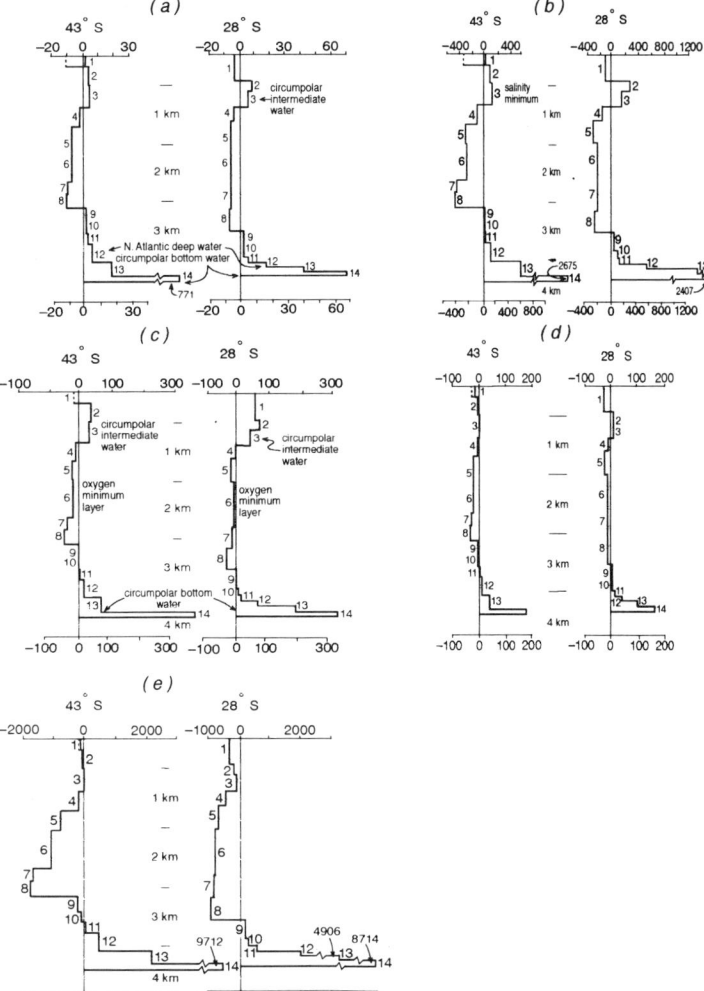

Figure 4–25. Zonally integrated estimates of property flux densities in the Pacific Ocean at 43°S and 28°S (Wunsch et al., 1983). These can be contrasted with the North Atlantic results shown in Figure 4–18. The properties are: (a) mass (10^9kg/s/km), (b) salt (10^6kg/s/km), (c) oxygen (ml/l × 10^9kg/s/km), (d) phosphate (10^3 – moles/kg/km), and (e) silicate (10^3 – moles/kg/km).

Figure 4–25 shows the estimated silicate flux across 43°S and 28°S in the South Pacific Ocean as estimated by Wunsch et al. (1983) using the sections of Stommel et al. (1973)–which, however, are still undersampled. North Atlantic estimates for silicate, oxygen, and the other nutrients have been published by Rintoul and Wunsch (1991). Martel and Wunsch (1993a) calculated the carbon flux across 25°N in the Atlantic from the finite-difference model described in Section 4.5.6, but they concluded that the 1° resolution of the model was too coarse to distinguish the flux from zero. Similarly, the estimates by Brewer, Goyet, and Dyrssen (1989) for nutrient and carbon flux are probably not reliable owing to an extremely coarse grid.

4.5.5 Freshwater Fluxes

The movement of freshwater is one of the key elements of the climate system (water vapor is the most important of the greenhouse gases). A knowledge of where the ocean transfers moisture to and from the atmosphere is critical to understanding the global water cycle and climate. The difficulty here is that the ocean carries laterally enormous masses of fluid–and it is the minute divergence of these masses that is the water flux divergence to the atmosphere. So for example, if the Gulf Stream should carry 100 Sv of freshwater poleward, the interior Atlantic might return only 99.99 Sv to the south–the remaining 0.01 Sv being given up to the atmosphere (100 Sv of freshwater corresponds to about 103.5×10^9 kg/s of mass flux, because seawater is about 3.5% salt by weight). How does one find such small residuals when the underlying numbers are uncertain by at least 1 Sv?

Let $\mathbf{T}$ again denote the matrix of transport elements in station pairs j in depth intervals s, but now representing the total velocity field, whatever its physics (i.e., including the Ekman component where appropriate), and let its uncertainty be $\mathbf{P}_T$. Then mass conservation across a transoceanic section can be written

$$\mathrm{vec}(\mathbf{T})^T \mathbf{1} + n_1 = -F\,, \tag{4.5.6}$$

where $\mathbf{1}$ is a column vector containing 1 in each element and F is the excess of precipitation over evaporation in the volume bounded by the section. Conservation of salt is

$$\mathrm{vec}(\mathbf{T})^T \ \mathrm{vec}(\mathbf{S}) + n_2 = 0 \tag{4.5.7}$$

[another version of Equation (4.3.2)]. Equations (4.5.6)–(4.5.7) might have been used in estimating $\mathbf{T}$, but not necessarily so. The term n_1 is typically an order of magnitude larger than any plausible estimate of the size of F, and the ability to make any useful estimate of F relies upon the comparatively slight variations in *salinity* across the section. If $S_0 = \overline{\mathrm{vec}(\mathbf{S})}$ (the section mean salinity), then $\mathbf{S}' = \mathbf{S} - S_0\mathbf{1}$ and (4.5.7) is

$$\mathrm{vec}(\mathbf{T})^T \left[\mathrm{vec}(\mathbf{S}') + S_0\,\mathrm{vec}(\mathbf{1})\right] + n_2 = 0$$

or

$$\mathrm{vec}(\mathbf{T})^T \ \mathrm{vec}(\mathbf{S}') + S_0(-F - n_1) + n_2 = 0\,. \tag{4.5.8}$$

Thus,

$$F = \left(\mathrm{vec}(\mathbf{T})^T \ \mathrm{vec}(\mathbf{S}') + n_2\right)/S_0 - n_1\,. \tag{4.5.9}$$

Under the assumption that $< n_1 > = < n_2 > = 0$, the best estimate of F is then

$$\tilde{F} = \text{vec}(\mathbf{T})^T \ \text{vec}(\mathbf{S}')/S_0 \tag{4.5.10}$$

with an uncertainty directly computable as

$$\mathbf{P}_s = \text{vec}(\mathbf{S}')^T \mathbf{P}_T \ \text{vec}(\mathbf{S}')/S_0^2 \,, \tag{4.5.11}$$

which will typically be attractively small because numerically $\| \ \text{vec}(\mathbf{S}') \ \|$ is quite small. It is this variance that is usually quoted in the literature as the uncertainty of oceanic water budget estimates. Unfortunately, there are a number of problems. Exactly as in the discussion of the accuracy of Equation (4.3.3), the true variance of (4.5.10) includes contributions from n_1 and n_2/S_0. Should these two errors be uncorrelated, $\mathbf{P}_s$ will increase to $< n_1^2 > + < (n_2/S_0)^2 >$, anticipated to be about $(1\text{–}2 \ \text{Sv})^2$, thus swamping any reasonable value for $\tilde{F}$. If the errors in the mass balance equation (4.5.6) and in the salt balance equation (4.5.7) are strongly and positively correlated, then $n_1 - n_2/S_0 \approx 0$, with the errors subtracting. Perfect applicability of this latter assumption appears to be the (implicit) assumption made in water budget computations, so that the variance is (4.5.11).

Although a plausible idea, the accuracy to which the two comparatively large numbers actually cancel is unknown. The cancellation would be inhibited by any approximations involving assumed temperature-salinity relationships, extrapolations in the bottom-wedges, and systematic errors in the salinity determination. No quantitative study of the resulting errors appears ever to have been made, and until such a study is carried out, one should remain cautious in interpreting the published estimates of freshwater divergences in the ocean circulation.

Furthermore, if the freshwater flux and its divergence as estimated from hydrographic sections are to be interpreted as representing *climatological* values, rather than estimates at a single moment in time, one has the same difficulty as appeared in the use of the temperature flux (4.5.4). That is, the mass flow and salt anomaly fields are dominated by the eddy-scale fluctuations, with the resulting relative velocity transport contributing to the salt anomaly flux through a very weak correlation field. Figure 4–15 shows the result for the 36°N section. As with the temperature flux, the contribution of the weak correlations becomes significant when integrated across the ocean basin. Again, Figure 4–15 exhibits visually something of the nature of a random walk pattern (including the gross inhomogeneities of the Gulf Stream and its recirculation regime). The stability of this contribution, in time, which is a major fraction of the total flux, remains unknown.

Knowledge of the error budgets of oceanic property fluxes is generally in an unsatisfactory state. Subject, therefore, to the warning that the published freshwater flux divergences may be unreliable estimates of the climatological means, the reader is referred to Stommel (1980), Hall and Bryden (1982), and Wijffels, Schmitt, Bryden, and Stigebrandt (1992) for calculations of freshwater fluxes and divergences in the ocean.

Like heat, water flux divergences to and from the atmosphere can in principle be estimated by independent methods. Direct estimates are hindered by the very poor database for over-ocean precipitation. Atmospheric model estimates of water flux divergence are also not very trustworthy (e.g., Peixoto & Oort, 1983). In the long term, the direct oceanic calculations, whatever their present difficulties, appear to be the most promising method by which large-area estimates of atmosphere/ocean water transfers will be made.

4.5.6 Finite-Difference Models

Geostrophic box inverse models have usually been written for large volumes of ocean because the data came from a few long transoceanic hydrographic lines. Consequently, the budget constraints of mass, salt, etc., have been used to describe very large oceanic regions. An advantage of these large volumes is that hydrography and flows in the ocean are much more stable on the large scales than they are for the mesoscale. (There is not much evidence for large-scale shifts in oceanic hydrography even over decades; see the discussion in Roemmich & Wunsch, 1984; Levitus, 1989; Wunsch, 1992; Bindoff & Wunsch, 1992; Bindoff & Church, 1992; Parrilla et al., 1994.)

By way of contrast, modern oceanic general circulation models (GCMs) (e.g., Figure 1–1a) attempt to resolve and describe regions of the ocean on grids written at a fraction of a degree of latitude and longitude (at present at $1/6°$ or less). There has thus been a large gap in scale between the ability to describe the ocean directly with observations, in comparison with the ability of a numerical model to produce numbers on arbitrarily defined grids.

The discrepancy has nothing to do with the inverse methodologies and everything to do with the distribution of observations. Consider that a conventional approach to deriving the partial differential equations of fluid dynamics is first to write conservation laws for mass, momentum, etc., in small control volumes of fluid. Permitting the volumes to become differentially small produces, in the limit, the corresponding differential equations. If the partial differential equations are discretized for use in a numerical model, one is simply reversing the limit and recognizing that a numerical ocean model is nothing but a collection of simultaneous equations for conservation in a large number of fluid volumes.

A few attempts have been made to apply directly inverse methods to models formulated in finite-difference forms, including Tziperman and Hecht (1987), Mercier, Ollitrault, and Le Traon (1993), Matear (1993), Schlitzer (1993), and Martel and Wunsch (1993a). To explore the application of inverse methods to models approaching the resolution of GCMs, the latter authors considered the North Atlantic in the region shown in Figure 4–26 and discretized the generic steady conservation equation (2.1.6) with Laplacian diffusion as

$$\frac{u(i+1,j,k+1/2)\rho C(i+1,j,k+1/2)-u(i,j,k+1/2)\rho C(i,j,k+1/2)}{\Delta x}+\cdots+$$

$$\frac{w(i+1/2,j+1/2,k+1)\rho C(i+1/2,j+1/2,k+1)-w(i+1/2,j+1/2,k)\rho C(i+1/2,j+1/2,k)}{\Delta z}+\cdots$$

$$-K_{xx}(i+1,j,k+1/2)\frac{\left(\rho C(i+1,j,k+1/2)-2\rho C(i,j,k+1/2)+\rho C(i-1,j,k+1/2)\right)}{(\Delta x)^2}+\cdots$$

$$=0\,, \tag{4.5.12}$$

where i, j are horizontal indices, k is a vertical one, and ρC is being written as one variable (x here is a coordinate, not to be confused with the state vector, $\mathbf{x}$). A discretized form of the mass-conservation equation (2.1.14) was also written for each unit cell (Figure 4–26b). If $C(i,j,k+1/2)$ is known at each grid point, the collection of (4.5.11) plus mass conservation is a set of simultaneous equations for $u(i,j,k+1/2)$ and K_{xx}, etc. In analogy to the process used for the large box inversions, Martel and Wunsch (1993a) wrote $u(i,j,k+1/2) = u_R(i,j,k+1/2)+c(i,j)$, $v(i,j,k+1/2) = v_R(i,j,k+1/2)+b(i,j)$ where u_R, v_R are calculated from the thermal wind equations on the grid using an initial reference level, and b, c are the reference-level velocities. The state vector unknowns of the problem are $b, c, K_{xx}, \ldots$, and the equations are in the standard form (3.3.2). [In conventional forward modeling, one would specify the u, v, K on the grid and solve the linear equation set (4.5.12) for C on the grid. Here, the problem is inverse because C is known and elements (b, c, K) are the unknowns.]

At this stage, Martel and Wunsch (1993a) rearranged the problem somewhat to acknowledge that a perfect solution to (4.5.12) would be undesirable. The mixing terms in (4.5.12) require numerical second derivatives of the mapped fields, and it is preferable to reduce the differentiation of data insofar as is practical (recall the sampling discussion of Section 3.6). The

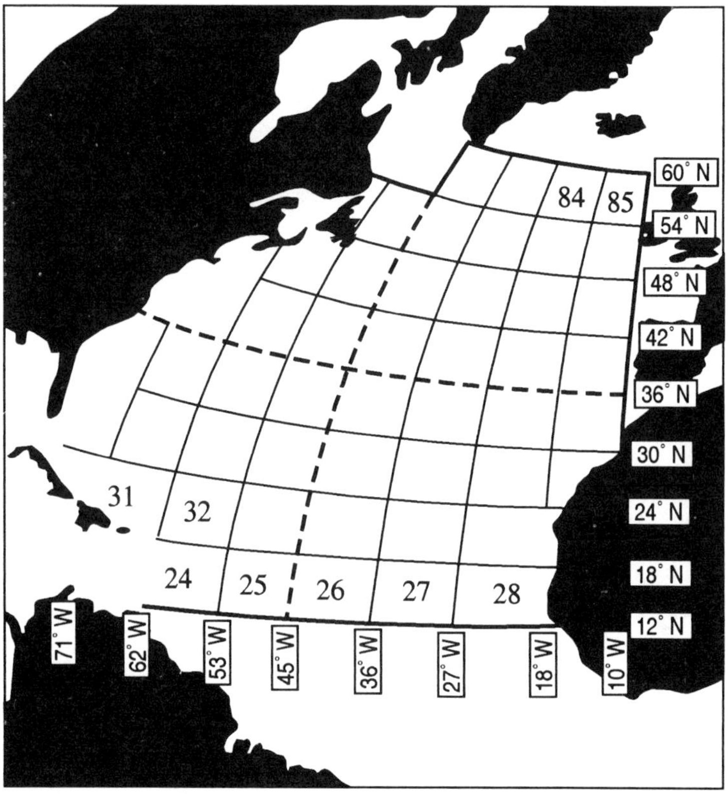

Figure 4–26. Domain of the model of Martel and Wunsch (1993a,b), in which objective mapping was used to produce hydrographic estimates on a uniform grid. A nested box model was used, in which geostrophic flows were required nearly to balance in the large region bounded by the heavy solid line and the continents. This region was then subdivided into four roughly equal regions, for which near-balance was required. Further subdivision occurred until a box size, labeled with the numbers, was reached.

finite difference form was thus integrated over a grid cell to produce a locally integrated balance:

$$\sum_{i,j,k\in\partial D} u(i+1,j,k+1/2)\rho C(i+1,j,k+1/2)a^x(i+1,j,k+1/2)$$
$$- u(i,j,k+1/2)\rho C(i,j,k+1/2)a^x(i,j,k+1/2) + \cdots$$
$$-K_{xx}(i,j,k+1/2)\frac{\left(\rho C(i,j,k+1/2) - \rho C(i-1,j,k+1/2)\right)}{\Delta x}a^x(i,j,k+1/2) + \cdots$$
$$+n(i,j,k) = 0\,, \tag{4.5.13}$$

which involves only first derivatives of the data. The notation, $i,j,k \in \partial D$ denotes the indices defining the boundary of differential cell ∂D, and $a^x(i,j,k)$ is the area of the interface normal to the x component of velocity $u(i,j,k+1/2)$, etc.; $n(i,j,k)$ has been introduced to represent the noise in the equations. Constraints such as the linear vorticity balance are consequences of the equations used and are thus implicit.

As the ocean is noisiest on the mesoscale, and apparently most stable on the largest scales, the numerical grid was made a nested one: the horizontal integration over the grid cell in (4.5.13) was taken over the entire North Atlantic, as depicted in Figure 4–26, as though the whole ocean were a single cell (vertical integration was, however, taken over each layer separately). But the flow in and out of this cell is written at a horizontal resolution of 1° along the zonal boundaries. This largest cell was then subdivided into four subcells, as depicted, and the integration taken over each of them (at this stage there are five basic balances written, including one redundant balance, for each layer). The subdivision then continues until one reaches the maximum resolution of 1°, the computer capacity is exceeded, or the data noise overwhelms the added information.

Martel and Wunsch (1993a) ended with the equivalent of about 9000 equations in 29,000 unknowns (the horizontal mixing unknowns were suppressed by giving them zero column weights and are not counted). They solved the system with the tapered least-squares solution, but it was obtained owing to the system size by a sparse conjugate gradient search method (Paige & Saunders, 1982). Their estimated absolute flow field at the seasurface is shown in Figure 4–27. The major limitation on this calculation arose from the spatial resolution: Even a 1° grid is much too coarse to capture the narrow boundary currents (recall Figures 2–19b,c and 4–24) with their associated property extremes, which appear necessary for accurate property flux estimates. Features such as those shown in the figures suggest that a horizontal spatial resolution near $1/10^\circ$ at midlatitudes is required at least near the boundaries. Except for the inversions based upon the original hydrographic sections (ungridded), no one has yet attempted a large-scale inversion on such a dense grid.

In the most ambitious GCM models, one uses equations permitting computation of the velocity, density, and tracer fields–that is, the velocity field is not specified independently, but all of (2.1.1)–(2.1.5) are used simultaneously. To the extent that the equations of motion carry information about the behavior of the density and velocity fields, one expects that the very sparse data distributions might in part be compensated by the ability to extrapolate them, using known dynamics, into regions of poor data coverage. Such ideas are the focus of a number of attempts to use inverse methods on GCM-like scales (e.g., Mercier et al., 1993, or Marotzke & Wunsch, 1993; Wunsch, 1994); we must, however, postpone the discussion until Chapter 5, because the resulting equations are fully nonlinear.

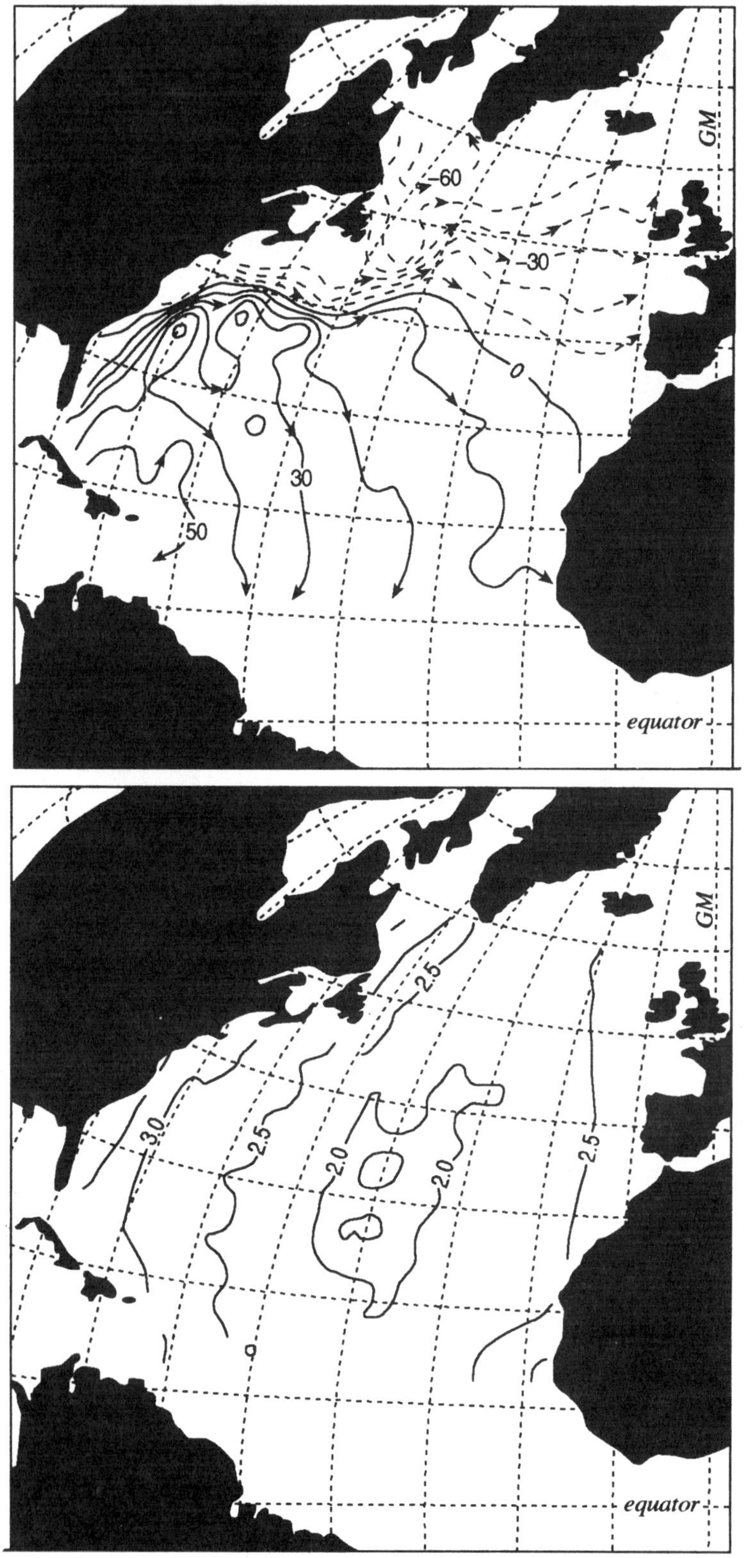

Figure 4–27. Absolute seasurface topography in centimeters (top) and its uncertainty (bottom) from the model in Figure 4–26. Only the standard error part of the uncertainty is shown.

4.5.7 Mixing Coefficients and Other Residuals

One of the central problems of physical oceanography is the determination and understanding of mixing processes. In the finite difference approach just discussed, the mixing tensor, **K**, occurred as part of the problem unknowns although it could not be distinguished from zero. Oceanic mixing processes are interesting and complicated (see, for example, Munk, 1966, 1981; Gregg, 1987; Thorpe, 1987; Gargett, 1989; Garrett, MacCready, & Rhines, 1993a). The great property tongues visible in the sections displayed in Chapter 2 have led many investigators to describe them as static balances described by Equation (2.1.17).

If a feature like the Mediterranean salt tongue is represented as a two-dimensional horizontal feature, (2.1.17) becomes

$$u\frac{\partial C}{\partial x} + v\frac{\partial C}{\partial y} = K_{xx}\frac{\partial^2 C}{\partial x^2} + K_{yy}\frac{\partial^2 C}{\partial y^2} \tag{4.5.14}$$

where Laplacian mixing has again been assumed, and for notational simplicity, the mixing tensor has been taken as diagonal–that is, $K_{xy} = 0$, and spatially slowly varying. The horizontal velocities in (4.5.13) are supposed to be geostrophic ones. In a number of exercises in the literature, lines of constant property are least-squares fit for K_{xx}, etc., given assumptions about the behavior of u, v and the relative importance of vertical processes (Richardson & Mooney, 1975; Needler & Heath, 1975; Tziperman, 1987; Hogg, 1987; and Fukumori, 1991, have all discussed the Mediterranean salt tongue).

A lot of care is required in interpreting the results. First, the representation of mixing through Laplacian terms is justified only by analogy with molecular processes and, while simple, can hardly be regarded as a rule based upon any fundamental principles. One might argue (see, for example, Salmon, 1986) that the even more simple representation using Newtonian-type relaxation such that

$$u\frac{\partial C}{\partial x} + v\frac{\partial C}{\partial y} = -\tau_C C \tag{4.5.15}$$

where τ_C is a constant, is all that is justified.

Furthermore, the numerical values of the eddy coefficients will depend upon the spatial scales over which the geostrophic flows are permitted to vary. Wunsch (1985) considered the finite-difference formulation (restricted to two dimensions with a constant, isotropic eddy coefficient, K) on the grid shown in Figure 4–28,

$$
\begin{aligned}
&(C_{i+1,j} + C_{i,j})u_{i,j} - (C_{i,j} + C_{i-1,j})u_{i-1,j} \\
&\qquad + \delta[(C_{i,j+1} + C_{i,j})v_{i,j} - (C_{i,j} + C_{i,j-1})v_{i,j-1}] \\
&\quad = 2K/\Delta x[C_{i+1,j} + C_{i-1,j} - 2C_{i,j} + \delta^2(C_{i,j+1} + C_{i,j-1} - 2C_{i,j})] \,,
\end{aligned}
\tag{4.5.16}
$$

subject to the continuity relation

$$
u_{i,j} - u_{i-1,j} + \delta(v_{i,j} - v_{i,j-1}) = 0 \,, \qquad \delta = \Delta y/\Delta x \,. \tag{4.5.17}
$$

In this set of equations, an attempt to estimate K will obviously depend upon what is asserted about u_{ij}, v_{ij} and vice versa. The problem worked out by Wunsch (1985) addressed the determination of u, v given K, but a more likely problem is the simultaneous determination of all three from a given tracer distribution. A realistic problem would assert some prior knowledge with stated uncertainty of $\tilde{u}$, $\tilde{v}$, $\tilde{K}$ and require a new best estimate. Hogg (1987) produced estimates of K using a streamfunction version of (4.5.16)–(4.5.17). The reader will appreciate, however, that should u, v be permitted to vary arbitrarily rapidly over the domain, $\tilde{K} = 0$ may be an adequate solution. On the other hand, if u, v are taken to be unvarying spatial constants, then numerically large values might have to be permitted in K (the equations need to be modified if K is rapidly varying to take into account the derivatives of the eddy coefficient). The structure of $\tilde{K}$ will depend upon the spatial structures permitted in the flow-field model, and the resulting numerical values can scarcely be regarded as fundamental constants of the ocean; they may be useful as summary descriptions (see also the discussion in Davis, 1994).

Equations (4.5.16)–(4.5.17) are homogeneous in u_{ij}, v_{ij}, K; in the absence of constraints setting an overall rate, a good solution would be $u_{ij} = v_{ij} = K = 0$. This outcome is the clock problem for steady tracer distributions as described in Chapter 2–a standing crop of stable, conservative tracers by itself requires no flow at all. There are several possibilities for setting the clock: (1) Thermal-wind balance coupled with mass conservation is the most powerful of the known rate-setting constraints; (2) boundary conditions on u_{ij}, v_{ij} would set a rate; and (3) a tracer with a known source or sink, but in presumed steady state, would also suffice.

The most influential and interesting of the curve-fitting computations was

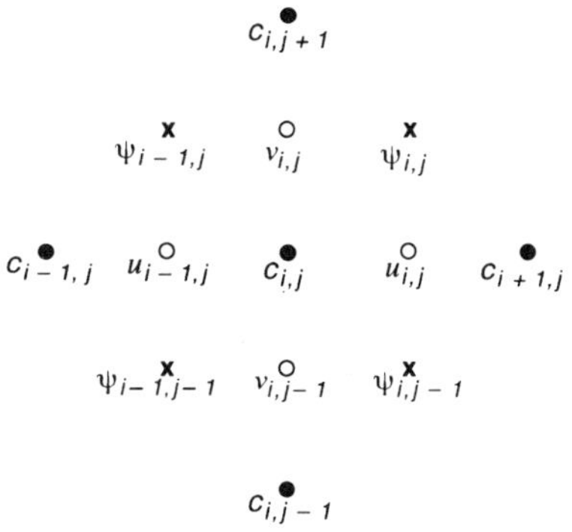

Figure 4–28. Numerical grid used by Wunsch (1985) to study the inference of a flow field from a steady tracer distribution. The tracer $\mathbf{C}_{ij}$ is defined on the solid circles, and the velocity field components on the open circles. When a streamfunction was employed, it was defined on the "x" positions. In all cases, the result of the numerical scheme is a set of noisy linear simultaneous equations.

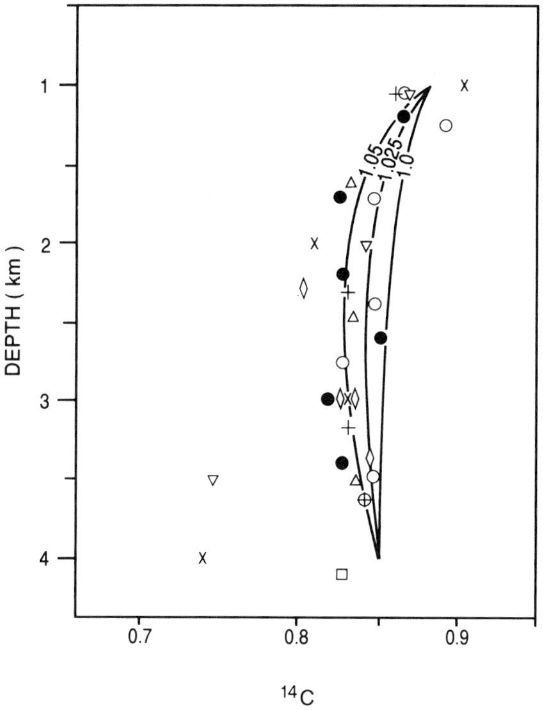

Figure 4–29. Radiocarbon data from the central Pacific Ocean, used by Munk (1966) to infer apparent vertical mixing and velocity rates. The curves are different fits of exponentials, involving the parameter $\lambda_C^2 = 1 + 4 (K\mu_C/w^2)$, where μ_C is the decay constant for radiocarbon, w is the vertical velocity, and K is the vertical diffusivity of Equation (4.5.19). The great scatter in the data will be apparent. Different symbols denote various measurement sets.

that of Munk (1966),[5] who examined the vertical balance

$$w\frac{\partial C}{\partial z} = \frac{\partial}{\partial z} K_{zz} \frac{\partial C}{\partial z}, \tag{4.5.18}$$

previously mentioned in Section 3.3.3. To preserve the discrete spirit of the methods used in this book, rewrite (4.5.18) in the form

$$(w_{i+1}C_{i+1} - w_i C_i)\Delta z_i - \{K_{i+1}(C_{i+1} - C_i) - K_i(C_i - C_{i-1})\} = 0 \tag{4.5.19}$$

where K_i is a discrete value of K_{zz}. Equation (4.5.19) is a collection of homogeneous simultaneous equations in w_i and K_i, which is inadequate to

[5] Munk called his exercise a recipe to make it clear that the results were not based upon fundamental physics. This distinction was not always appreciated by subsequent readers.

estimate w_i and K_i separately–that is, they are not resolved. The inadequacy of Equation (4.5.19) is a mathematical representation of the clock problem: A stable, conservative tracer cannot provide an overall rate of movement, and the homogeneous solution, $w_i = K_i = 0$, is a consistent result. We could solve (4.5.19) by the particular-SVD, leaving a nullspace, but Munk (1966), realizing he needed more constraints to separately resolve w_i and K_i, invoked the measured profiles of radiocarbon, R, displayed in Figure 4–29. Because radiocarbon is unstable, its governing differential equation is inhomogeneous of the form

$$w\frac{\partial R}{\partial z} = \frac{\partial}{\partial z}K_{zz}\frac{\partial R}{\partial z} - \lambda R \qquad (4.5.20)$$

where λ is a radioactive decay constant (not a singular value), which sets the clock rate. (Craig, 1969, showed that the radiocarbon measurements had been somewhat misinterpreted, but the changes required are not significant for our present discussion.) We can discretize (4.5.20) to the form

$$\begin{aligned}(w_{i+1}R_{i+1} - w_iR_i)\Delta z_i - \{K_{i+1}(R_{i+1} - R_i) - K_i(R_i - R_{i-1})\}\\ = -\lambda_i R_i \Delta z_i^2\,, \qquad (4.5.21)\end{aligned}$$

which with (4.5.19) are formal enough equations to produce estimates of w and K with no nullspace.

Munk (1966) further assumed that w and K were constants with depth; that is, he chose the solution to (4.5.19) and (4.5.21) to be $w_i = w_0$, $K_i = K_0$ and solved for the best-fitting values. This procedure may be viewed as either the application of the simplest form of the expansion in basis functions, or as an extreme version of a smoothness assumption, with the solution correlation being unity. The result of the calculation was the famous *canonical* value $K \approx 1\text{cm}^2/\text{s} = 10^{-4}\ \text{m}^2/\text{s}$, with $w \approx 10^{-7}$ m/s. Much later, Olbers and Wenzel (1989) reexamined Munk's numerical calculation and concluded that his equations were numerically somewhat ill conditioned and that the canonical value was an accident (no uncertainty estimates had been provided).

Munk's results were enthusiastically seized upon by a generation of chemists who were impressed by the beautiful simplicity of the resulting ocean circulation picture (one-dimensional and steady). Unfortunately, subsequent direct measurements of K_{zz}, pioneered by C. S. Cox (i.e., Osborne & Cox, 1972; Gregg, 1987; or Gargett, 1989; Toole, Polzin, & Schmitt, 1994, Ledwell, Watson, & Law, 1993, for recent work), has failed to substantiate nonequatorial values of K_{zz} higher than about $0.1\text{cm}^2/\text{s} = 10^{-5}\text{m}^2/\text{s}$. It is also very difficult, upon examination of any of the flow patterns depicted

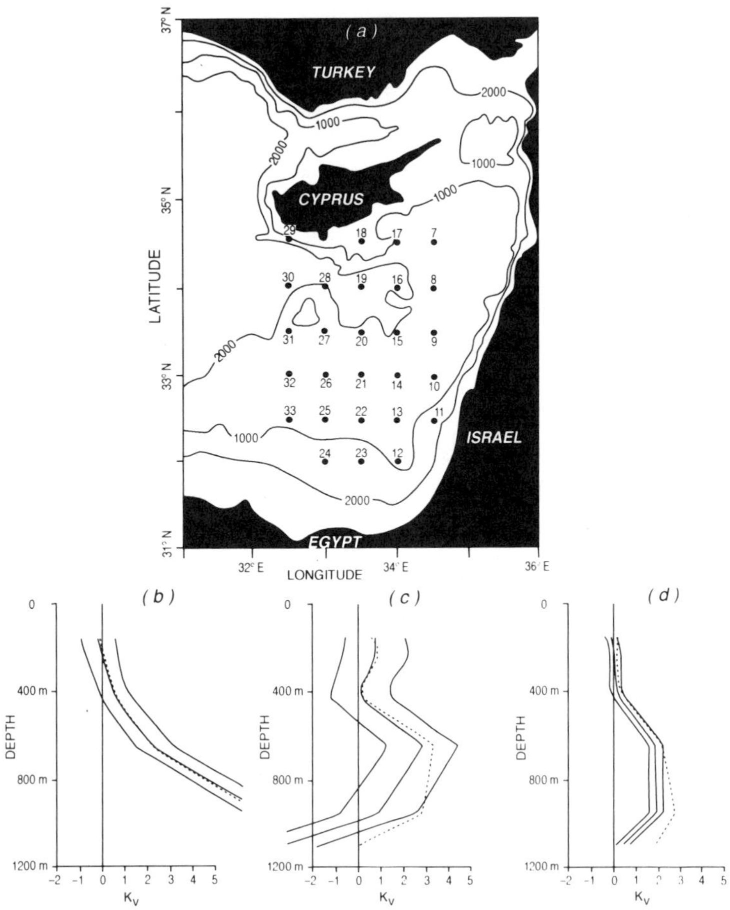

Figure 4–30. (a) Grid of stations in the eastern Mediterranean on which a hydrographic survey had been repeated several times, permitting averaging of the hydrography before inversion, or averaging flow characteristics obtained from several inversions (from Tziperman & Hecht, 1987). (b,c,d) Estimates of the eddy coefficients (Tziperman, 1988) obtained from three different inversions, differing only in the way in which the observations were averaged. Dashed lines resulted from an inversion with inequality constraints, which are described in Chapter 5.

in this book or in the wider literature, to be convinced that the ocean is even crudely described as having a one-dimensional structure and circulation. The true uncertainty of K_{zz} would be dominated by the model error and not the observational noise.

Tziperman (1988) studied the behavior of vertical eddy coefficient fits in finite-difference models, with the added advantage of having a set of repeated hydrographic surveys on a grid (Figure 4–30a). He showed that the numerical values for the eddy coefficients were sensitive to the precise form of data analysis. In particular, if the density data from a set of repeated surveys were averaged prior to use in equations like (4.5.16), quite different values of the eddy coefficients were obtained than if the estimates were made by using Equation (4.5.16) for each of the surveys separately, with the eddy coefficients then obtained by averaging each of the separate estimates. The geostrophic velocities were much less sensitive to the methodology than were the eddy coefficients (Figure 4–30b,c,d).

The central message is that oceanic mixing in such estimates is being deduced from the small residuals left as imbalances of the large advection into and out of oceanic volumes by the geostrophic flows. Numerical values of the residuals, as we have seen, are very sensitive to the presence of small singular values and the precise manner in which they are handled. But there is a more fundamental problem. The coefficient matrix, **E**, is constructed from observations of the density field and as such inevitably contains errors. It can be shown (e.g., Wunsch, 1994) that comparatively slight changes in the elements of **E** are capable of producing qualitative changes in the solution residuals. Estimation of the uncertainty of the resulting mixing coefficients requires the methods of nonlinear regression (e.g., Seber & Wild, 1989), which are briefly discussed in Chapter 5.

4.6 Climatologies and Box Models

So-called box models are much used as a representation of the ocean circulation, particularly in the chemical oceanographic community. They are best thought of as extremely coarse parameterizations of the finite-difference models, in which the differential volumes are permitted to grow to large, even ocean-basin scales, and in which the exchanges between the resulting boxes can only take place on the same spatial scale. (This last remark is made to draw a distinction with the model of Martel & Wunsch, 1993a, in which some of the box dimensions coincided with the entire North Atlantic, but in which the flows in and out of the box were permitted to vary every 1° along the box boundaries.)

In a typical box model, the physicochemical properties of the region–for example, its temperature, or oxygen, or other scalar property–are represented as a single value, supposed to be representative of the box, as though it were well mixed. If $\overline{C}_i$ is the value assigned to a property in box i, then the transfer of this property to a neighboring box, j, is often written as $J_{ij}\overline{C}_i$ where the single parameter J_{ij} encompasses the transfers between the boxes, which depending upon region, might lump together western boundary currents, interior flows, Ekman fluxes, and eddy fluxes.

There are numerous ways to represent property transfers between boxes–the single reservoir value times a transfer coefficient being just one; see Keeling and Bolin (1967). Variations include defining the transfers by specifying $\overline{C}_{bi}$ on each of its boundaries, b, with the transfers between boxes given by $J_{ij}\overline{C}_{bi}$ on their common boundary times. All of the various schemes in use are special cases of finite-difference techniques, including centered differences, up-stream differencing, etc. (see Roache, 1976).

There is thus a continuum of model representations of the ocean ranging from true differential volumes (in the latest general circulation models of the ocean, the differentials are still 10 km on a lateral side) on up to models in which entire oceans or even the global ocean is represented by 10 or 12 boxes (see Garçon & Minster, 1988; Bolin, Björkström, & Moore, 1989), with hybrids like that of Martel and Wunsch (1993a) in which the interbox fluxes can occur on much smaller scales than the defining boxes themselves.

Once the box model is set up, any of the linear inverse methods can be used to obtain estimates for J_{ij}, K, etc., and to study their uncertainty (Wunsch & Minster, 1982, provide a general discussion), although most authors have unfortunately contented themselves with finding *the* solution and give no estimate of its uncertainty in any form. (Exceptions include Bolin, Björkström, Holmen, & Moore, 1983, and Garçon & Minster, 1988.)

There is no doubt that box models provide useful qualitative estimates of the gross behavior of oceanic properties. The major issue is not their formal uncertainty but rather the systematic or model error that inevitably arises when property fluxes in turbulent fluids, controlled by large numbers of dynamical elements, are replaced in the model by truly drastic reduction in degrees of freedom. That is, one has once again the issues of spatial resolution and dimensionality. If property C is actually transported as $\overline{qC}$, where q is the velocity, and the bar represents a spatial average over the full turbulent flux (e.g., Figure 4–17) and is replaced by $\overline{q}\,\overline{C}$, is there any expectation of quantitatively useful results? The answer to this question can be found only on a case-by-case basis and only if it is actually examined by the box model user. Often it is possible to find numerical values of $\overline{q}$ so that tracer balance is achieved in the model. But the physical interpretation of the value is that of a summary parameterization rather than anything with predictive power should any element of the flow field change. (In other words, perturbations to the flow fields that are either highly structured spatially and/or turbulent produce physical consequences that may be completely different from those in a simplified laminar flow tuned to have the same tracer flux.)

Similarly, the use of one- or two-dimensional physical representations of a three-dimensional ocean always runs the serious risk that good fits, with apparently well-resolved parameters, actually misrepresent the true physics and chemistry. As a working rule, it is better to carry three-dimensional models with degrees of freedom adequate for representing the expected oceanic structure, even if the resulting situation is formally underdetermined. One can understand the nullspaces, describe their implications, and include them in the uncertainty discussion. Arbitrary elimination of nullspaces by dimension reduction runs a high risk of producing results that are misleadingly definite.

4.6.1 Climatologies

There is a very close connection between the questions raised by low-resolution box models and the use of so-called hydrographic climatologies. Levitus (1982) produced an analysis of the historical hydrography by averaging it in space and time over 2° squares and about 70 years, producing a climatology on a uniform horizontal grid at a set of vertical standard depths. This climatology has proved so convenient to use with numerical models that it has clearly come to represent, in the minds of some model users, a form of truth.[6] Apart from serious concerns over the very large inhomogeneity of the database that went into forming the climatology and that leads to serious variations with position and depth in its reliability as an average, the effect of a multi-year average over a spatial scale large compared to the Rossby radius-of-deformation on the resulting model flux properties is potentially extremely serious. One reduces the property and velocity extremes that are so important for flux computations, and entire water masses disappear.

Memery and Wunsch (1990) reported that in a box inversion carried out with the Levitus (1982) gridded values, the resulting heat transport by the relative velocity was a fraction of that obtained by using the raw station data (a result consistent with the reduction in value obtained above when the 36°N section was subsampled–roughly equivalent to spatially averaging a section). Schlitzer (1988, 1989), who also used the Levitus (1982) climatology, was forced to introduce horizontal eddy mixing terms into his geostrophic box balances because the resulting highly smoothed geostrophic flows were not energetic enough to provide property balance by themselves. As noted above, the eddy coefficients that emerge from such studies depend directly upon the degree of smoothing that is imposed. Unless one is convinced that the physics giving rise to the eddy coefficients are well founded and understood, the resulting numerical values are best thought of as a kind of engineering parameterization–useful for summarizing the present numerical values of the required fluxes, but probably not useful if one asks what would happen if the entire system were perturbed, for example, under a climate shift.

Climatologies raise other issues as well. The governing equations of mo-

[6] To the point where many of them refer to the gridded values as data–an unfortunate usage, because the climatology is remote from any actual observation.

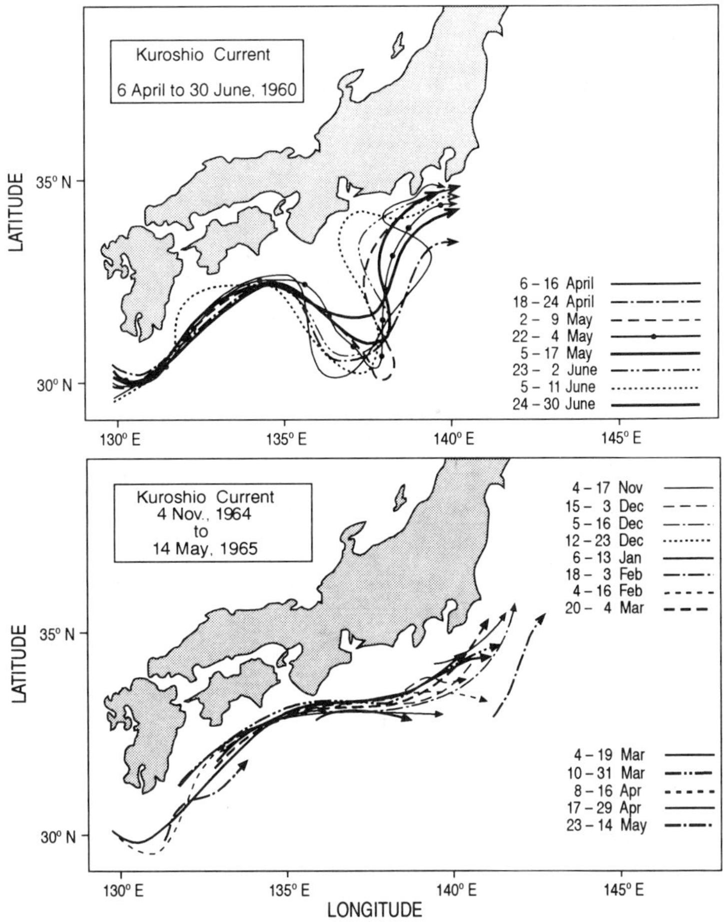

Figure 4–31. Two sets of paths of the Kuroshio showing two apparently stable states. The transition period (which may be seen as part of Shoji's 1972 complete sequence) is very short compared to the time that the Current spends at either extreme. Thus, the mean path is almost never occupied.

tion are nonlinear dynamically and kinematically, and it is far from clear whether an inferred time-average circulation is possible. Consider, for example, the case of the path of the Kuroshio near Japan. It is known (Shoji, 1972; see Figure 4–31) that the current has two stable paths in which it spends most of its time, a southern one which dominates, and a northern one. At rather rare intervals, the path switches from one state to the other, the transition times being of the order of months. The time-mean Kuroshio, computed possibly by averaging the hydrography over many years, would lie somewhere in between the two states–at a position where it rarely is.

Although the case of the Kuroshio is perhaps the most obvious, the same situation occurs with any attempt to average hydrography so as to produce time-average circulations. Currents will appear broadened and in unusual positions. Much time and effort have gone into producing time-average pictures, and important insights have emerged, but it is clear that the procedure is ultimately indefensible if it is intended to lead to a *quantitative* picture of the circulation. In other words, the elimination of time as an independent variable can distort one's conclusions in just the same way that reduction of a spatial domain from three- to two- or one-space dimensions can grossly bias the result. For this reason, Chapter 6 is devoted to a discussion of the special methods that can be used for studying time-dependent flows and property distributions.

4.7 The β–Spiral and Variant Methods

4.7.1 The β–Spiral

The requirement that the flow field satisfy the Cartesian coordinate equations of motion (2.1.11)–(2.1.18) in a minimum mean-square sense is the basis of the estimates made so far. But because the equations can be manipulated in various ways, eliminating some unknowns, or simply through rearrangement, the particular formulation used in the large-scale geostrophic box balances is not unique. Rather, it was dictated more by the wish to use data distributed in widely separated transoceanic sections than by any fundamental consideration. It will be no surprise to find that if different data distributions are considered, different forms of the equations prove useful.

The so-called β–spiral method of Stommel and Schott (1977) and Schott and Stommel (1978) is an important rearrangement of the equations used for pointwise balance. There are several ways to derive it, producing equivalent requirements.

Here, ignore for the time being the distinction in Equations (2.1.11)–(2.1.18) between the potential density, which is strictly used in (2.1.15)–(2.1.18), and the in situ density, which is required to compute the hydrostatic pressure and the thermal wind (the difference is negligible in the upper ocean). The box balances are most convenient for data distributed so as to bound volumes of ocean. But the differential equations represent pointwise balances, and in a mathematical sense, one can construct estimation methods that should work at a single point (although the distinction between point and areal balances is not, in practice, very meaningful).

If the density equation (2.1.15) is solved for w, and then differentiated by z,

$$\left(\frac{\partial\rho}{\partial z}\right)^2 \frac{\partial w}{\partial z} = -\left(\frac{\partial\rho}{\partial z}\right)\left(\frac{\partial u}{\partial z}\frac{\partial\rho}{\partial x} + u\frac{\partial^2\rho}{\partial x\partial z} + \frac{\partial v}{\partial t}\frac{\partial\rho}{\partial y} + v\frac{\partial^2\rho}{\partial z\partial y}\right) + \left(u\frac{\partial\rho}{\partial x} + v\frac{\partial\rho}{\partial y}\right)\frac{\partial^2\rho}{\partial z^2}, \tag{4.7.1}$$

the linear vorticity balance (2.1.20) permits eliminating w. The thermal wind equations (2.1.23)–(2.1.24) can be used to write $u(x,y,z) = u_R(x,y,z) + c(x,y)$, $v(x,y,z) = v_R(x,y,z) + b(x,y)$ where u_R, v_R are known from the density field. Substituting into (4.7.1) produces a partial differential equation in ρ involving the integration constants c, b,

$$(u_R + c)\left[\frac{\partial \rho}{\partial x} - \left(\frac{\partial \rho}{\partial z}\right)\frac{\partial^2 \rho}{\partial x \partial z}\right] + (v_R + b)\left[\frac{\partial \rho}{\partial y} - \left(\frac{\partial \rho}{\partial z}\right)\frac{\partial^2 \rho}{\partial z \partial y} - \frac{\beta}{f}\left(\frac{\partial^2 \rho}{\partial z^2}\right)^2\right] = 0 . \tag{4.7.2}$$

If b, c were known, (4.7.2) would be an equation to be solved for ρ subject to boundary conditions (the forward problem). As it is, one treats ρ as known and attempts to find b, c such that (4.7.2) is satisfied at each point x, y, z as best possible. If discretized, the result is a set of simultaneous equations (2.4.5) for $b(x,y)$, $c(x,y)$ that can be solved by any one of the methods available to us. Suppose that $(x,y) = (x_1, y_1)$ are fixed and that (4.7.2) is applied at a series of depths z_i, $1 \le i \le M$. There are then M equations in the two unknown $b(x_i, y_i)$, $c(x_i, y_i)$, and a solution and its uncertainty can be found.

In the original Stommel and Schott (1977) version of the β-spiral, and in most subsequent use, the balance was formulated in terms of density coordinates rather than z so as to avoid the problem of having to distinguish the in situ and potential densities. Potential density is used to define the isopycnals, and in situ density to compute the thermal wind. Following Stommel and Schott (1977) and Olbers et al. (1985), take (2.1.15) and write it as

$$w = -u\frac{\partial \rho}{\partial x}/\frac{\partial \rho}{\partial z} - v\frac{\partial \rho}{\partial y}/\frac{\partial \rho}{\partial z} = u\frac{\partial h}{\partial x} + v\frac{\partial h}{\partial y} \tag{4.7.3}$$

where

$$\frac{\partial h}{\partial x} = -\frac{\partial \rho}{\partial x}/\frac{\partial \rho}{\partial z}, \quad \frac{\partial h}{\partial y} = -\frac{\partial \rho}{\partial y}/\frac{\partial \rho}{\partial z} \tag{4.7.4}$$

and where $z = h(x, y, \rho)$ defines the depth of a surface of constant density (an isopycnal). With this coordinate change, the thermal wind equations are

$$\frac{\partial u}{\partial z} = \frac{g}{f}\frac{\partial \rho}{\partial y} = \frac{g}{f}\frac{\partial \rho}{\partial z}\frac{\partial h}{\partial y}, \tag{4.7.5}$$

$$\frac{\partial v}{\partial z} = \frac{g}{f}\frac{\partial \rho}{\partial x} = -\frac{g}{f}\frac{\partial \rho}{\partial z}\frac{\partial h}{\partial x}. \tag{4.7.6}$$

Differentiating (4.7.3) and substituting the thermal wind values produces

$$\frac{\partial w}{\partial z} = u\frac{\partial^2 h}{\partial x \partial z} + v\frac{\partial^2 h}{\partial y \partial z}\,. \tag{4.7.7}$$

But using the linear vorticity equation, (2.1.20), w can be eliminated, producing

$$u\frac{\partial^2 h}{\partial x \partial z} + v\left(\frac{\partial^2 h}{\partial y \partial z} - \frac{\beta}{f}\right) = 0\,. \tag{4.7.8}$$

Making the separation of u, v into relative and reference-level components, we have another version of the β–spiral equation that represents requirements leading to the determination of c, b. The reader will recognize the structural identity of (4.7.2) and (4.7.8).[7]

To see that a spiral is implied, we follow Bryden (1977). From (2.1.21)–(2.1.22), Equation (2.1.15) is

$$w\frac{\partial \rho}{\partial z} = -\frac{uf}{g}\frac{\partial v}{\partial z} + \frac{vf}{g}\frac{\partial u}{\partial z}\,. \tag{4.7.9}$$

Write $(u, v) = V(z)(\cos\theta(z),\, \sin\phi(z))$ and, substituting into (4.7.9),

$$\frac{d\theta(z)}{dz} = -\frac{g}{f}\frac{w\partial\rho/\partial z}{V^2}\,. \tag{4.7.10}$$

The linear vorticity equation shows that if $v \neq 0$–that is, if the flow is not purely zonal–then there must be a nonvanishing w. With $\partial\rho/\partial z < 0$, and $w > 0$, θ would increase upward (spiral counterclockwise upward). If β vanishes, there is no spiral, and the system reduces to a statement of parallel fields (see Defant, 1961, pp. 476–82). Stommel and Schott (1977) noted that the sign of the spirals could be changed with a time-dependent density term, for example, if the ocean were cooling.

In practice, the partial derivatives of the density field must be estimated

[7] If Equation (4.7.2) is written in spherical coordinates and using the pressure, $P = p/\rho_0$, rather than ρ, the equation is readily found to be

$$\frac{\partial P}{\partial \phi}\left(\frac{\partial^3 P}{\partial z^3}\frac{\partial^2 P}{\partial \lambda \partial z} - \frac{\partial^2 P}{\partial z^2}\frac{\partial^3 P}{\partial \lambda \partial z^2}\right) + \frac{\partial P}{\partial \lambda}\left(\frac{\partial^2 P}{\partial z^2}\frac{\partial^3 P}{\partial \phi \partial z^2} - \frac{\partial^3}{\partial z^3}\frac{\partial^2 P}{\partial \phi \partial z} + \cot\phi\left(\frac{\partial^2 P}{\partial z^2}\right)^2\right) = 0\,,$$

the so-called P–equation of Needler (1967). Here, $P = g\int_{z_0}^{z}\rho dz + P_0(x, y)$; P_0, which is independent of z, takes the role of the reference-level velocities, b, c. The P–equation (discussed recently by Wunsch, 1994) is identical to the β–spiral equation (4.7.2) except that the use of the pressure rather than the density raises the order of derivatives in z by 1.

from observations, and the system will be noisy, with rank deficiency issues. Schott and Stommel (1978) produced several examples of the use of these equations (see Figures 4–32b,c). It is instructive to consider how this procedure differs from that of the box models discussed above. A first obvious difference is that we must be able to form both x and y derivatives of the density field, requiring at a minimum two crossing lines of hydrography (Figure 4–32a). A box geometry such as that considered in Figure 4–16 would not do. On the other hand, if only a set of crossing lines is available, then the integral constraints for box balance would not be useful. So one major consideration in making a choice of form is the geometry of the available data.

As applied in practice by Schott and Stommel (1978), the derivatives in (4.7.8) were estimated by fitting straight lines to the hydrography over considerable distances, as depicted in Figures 4–32a. More sophisticated fits (e.g., Armi & Stommel, 1983, see Figure 4–32c) are obviously possible. But in general we can recognize from the mapping and basis function-fitting discussion of Chapter 3 that assumptions are being made about the spatial correlations of the flow fields so that derivatives and resulting flows are actually intended to apply over large areas (Davis, 1978a, discusses this spatial scale assumption explicitly) and not at a single point.

4.7.2 Variations on the β–Spiral

A number of variations on the β–spiral have been studied in the published literature. Stommel (1956) and Killworth (1983a) combined the hydrostatic, geostrophic equations (3.1.11)–(3.1.16) with an assumed Ekman pumping velocity to determine the absolute meridional velocity. Killworth (1983a), in particular, discusses the combination as a test of the consistency of the dynamics, but not as a practical method of velocity determination. But if estimates of the uncertainty of the Ekman pumping are used, there is no reason why it could not become a practical estimation method; indeed, given realistic estimates of the uncertainties, one should use *all* available information.

Just as the box balances can be extended to carry mixing terms, so can the β–spiral balances (Schott & Zantopp, 1979; Olbers & Willebrand, 1984), and in general, there are many variations possible, including nonlinear and time-dependent terms. Once the model equations are written down and error estimates made, one can proceed to use the inverse machinery to obtain solutions and their uncertainty.

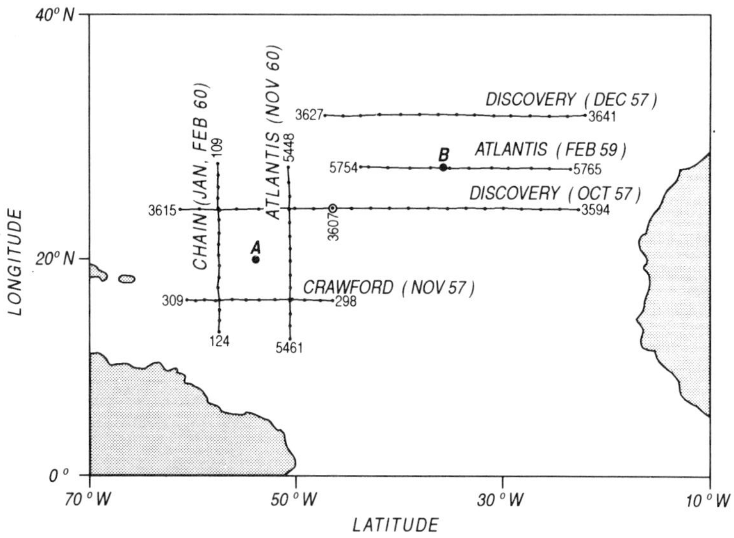

Figure 4–32a. Hydrographic sections used by Schott and Stommel (1978) to infer absolute currents at points A, B using the β–spiral constraints. Ship names, station numbers, and dates are indicated. The sections are used to estimate the isopycnal slopes at the two points.

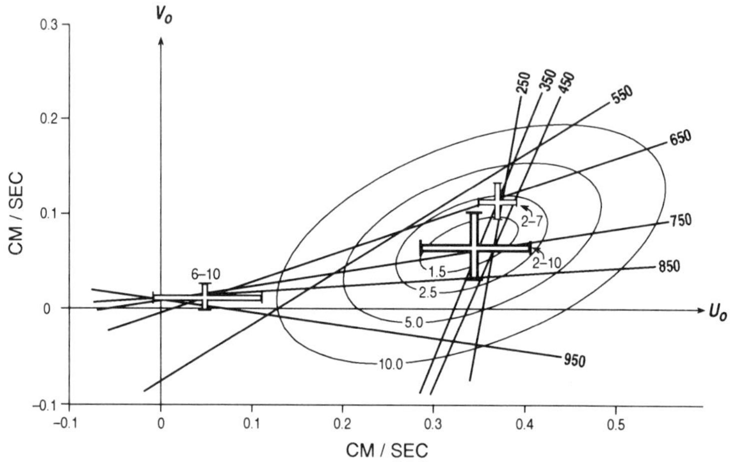

Figure 4–32b. β–spiral constraint equations at position B in Figure 4–32a as a function of depth. In a perfect, noise-free situation, all the lines should intersect at one point. Heavy crosses show minimum noise norm solution points using the constraints over differing depth ranges, along with the corresponding error bars based upon $\sqrt{P_{ii}}$; (u_0, v_0) correspond to (b, c) of the text. Labels on the equation lines denote depth of the β–spiral balance.

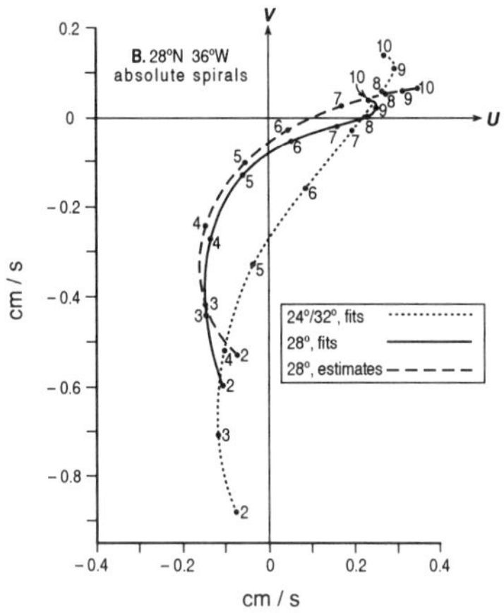

Figure 4–32c. Velocity spirals at position B in Figure 4–32a deduced from different assumptions about how to estimate slopes, and which depth ranges to use. Depths (in hundreds of meters) are shown on each curve. Dashed curve labeled "estimates" is from a visual determination of the isopycnal slopes. As with all estimation methods, there is not a unique best answer here.

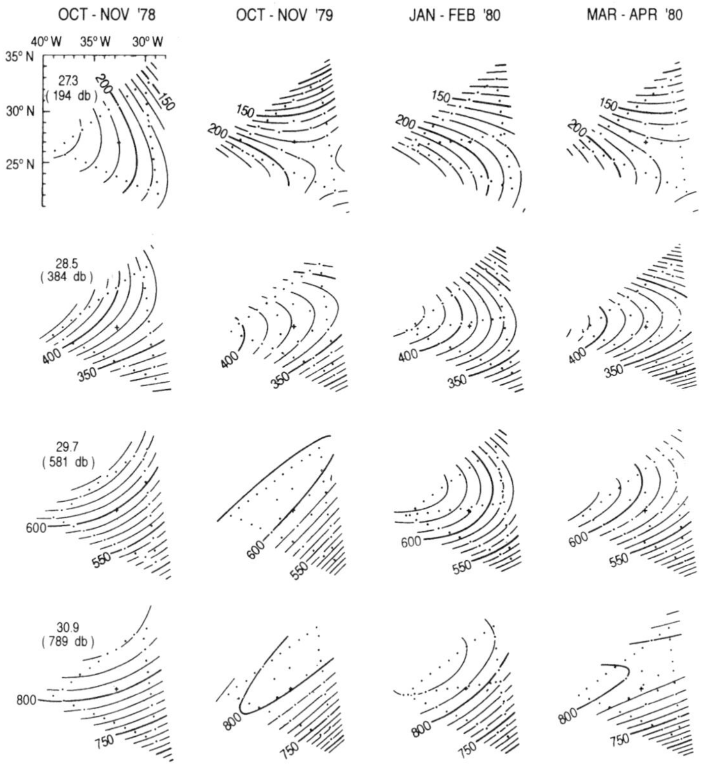

Figure 4–32d. Data distribution used by Armi and Stommel (1983) to perform a set of β–spiral calculations. The triangular observation grid is shown in latitude and longitude for each of four different time spans and for four different potential densities ($\sigma_p = 27.3, 28.5, 29.7, 30.9$), each referred to a reference pressure, p, shown in parentheses. The mapped depth of each isopycnal is shown in decibars for each time. A quadratic surface was fit to the data at different density levels and the resulting surface used to define the derivatives of the β–spiral equations. The procedure is a form of kriging, or expansion in basis functions, and imposes a basic scale on the results. Time variations in the field are apparent.

The broadest use of the β–spiral balances is in the work of Olbers et al. (1985) and Olbers and Wenzel (1989) for the North Atlantic and Southern Oceans. Figure 4–33 shows the horizontal velocity estimated by Olbers et al. (1985) in the North Atlantic. Knowledge of the density field and of the absolute horizontal velocity at a point render it comparatively simple to then estimate the vertical velocity at each point as well. Figure 2–23a shows the Olbers *et al.* (1985) estimate of vertical velocity in the North Atlantic. Because their model equations also included vertical mixing terms, they obtained estimates of those as well. But as with similar estimates from the box inversions, small errors in the coefficient matrix, **E**, can easily produce spurious apparent mixing. The remedy is to account properly for those errors in estimating the uncertainty of the estimated values–an accounting readily available through the nonlinear regression methods already mentioned.

One drawback of the use of nominal pointwise balances is that one does not enforce constraints that more properly apply over regions. So, for example, the Olbers et al. (1985) flows in Figure 2–23a do not satisfy the mass conservation conditions in which the horizontal and vertical flows at neigh-

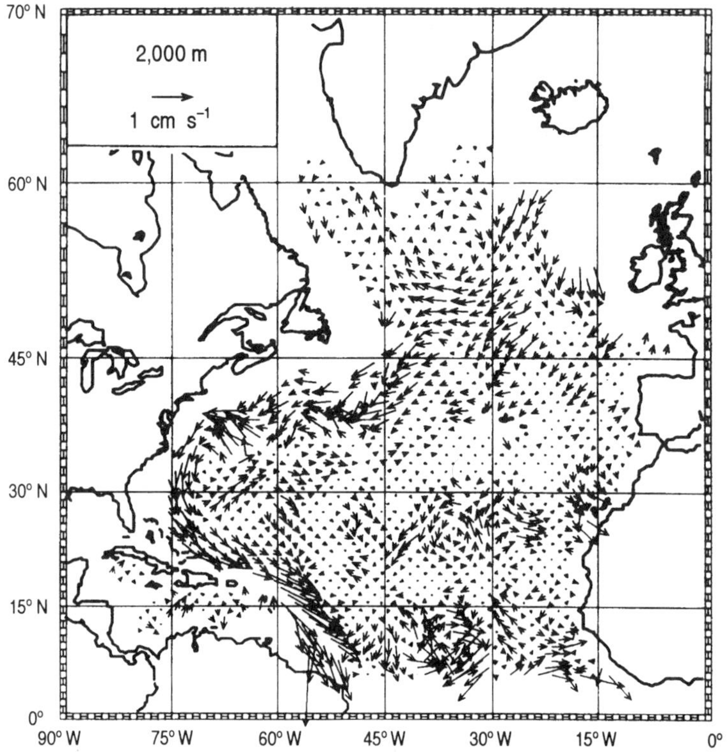

Figure 4–33. β–spiral estimate of absolute horizontal velocity at 2000 m. From Olbers, Wenzel, and Willebrand (1985).

boring grid points need to add to zero over finite volumes defined by several grid points. This problem was readily overcome by Wenzel (1986) and Fukumori (1991), who simply added such box constraints to the point balances of the β–spiral, showing the truism that the apparent distinctions amongst these methods are mainly ones of scale selection and rearrangement.

4.7.2.1 Some Summary Comments

This chapter has explored some of the practical issues of estimating the general circulation using the estimation methods outlined in Chapter 3. The original, slightly naive, motivation for the introduction of the methods was the so-called level-of-no-motion problem described in Chapter 2. But the perspective is now rather different. As we have seen, there is no level-of-no-motion problem in the mathematical sense–the density field alone, when combined with equations (2.1.1)–(2.1.5), produces the absolute flow field in (2.5.1). The problem of inferring the flow field from density is one of estimation–how to find a best solution, given noisy, irregularly distributed data, and to understand the reliability of the solution and any properties computed from it. [We have not considered dynamical regimes in which a different approximating set of equations would be required (e.g., strong nonlinearity, small Coriolis parameter, or time dependence). Although the time-dependent problem is taken up in Chapter 6, each such dynamical situation needs to be analyzed separately. The traditional oceanographic problem, which has been our focus, assumes the validity of Equations (2.1.1)–(2.1.5). If a different model is required, the investigator must provide it.]

Statistical estimation theory is a very large and complex subject, and only a fraction of the possible material relating to it has been alluded to here. In particular, we have focused on so-called 2–norm measures of fit. The choice of the 2–norm to measure distance, as well as the the solution that minimizes it, remains somewhat arbitrary although intimately connected to Gaussian statistics. But the notion that there is some unique right solution is a mirage. There is always observational noise, and any fluid flow formally has an infinite number of degrees of freedom–something is always subgridscale.

The minimum 2–norm solution is recognized as a useful way to define one best solution. Other investigators prefer to select smoothest solutions, as measured by some other partly arbitrary measure (e.g., Provost & Salmon, 1986; Wahba, 1990; McIntosh & Veronis, 1993), and there are other possibilities. The selection of a solution is less important than the production of a statement of which elements of the solution are well determined and are present or absent only because of the objective function used to select the solution.

4.7.2.2 Diagnostic Methods

Discussion of so-called diagnostic methods, of which many published versions exist, has been omitted here. Examples of such calculations are Sarkisiyan and Pastukhov (1970), Mellor, Mechoso, and Keto (1982), and Bogden, Davis, and Salmon (1993). These results, in which dynamical and kinematic constraints are employed to make inferences from observations, may appear much like those from inverse calculations. But they differ fundamentally from inverse methods in their omission of the analysis framework that identifies a true inverse method–the production of the uncertainty and resolution estimates–which would permit one to understand the significance of a particular circulation calculation. Some of the diagnostic methods could be extended readily to provide the necessary quantitative estimates, and that direction is surely a fruitful one for future effort.

4.7.2.3 Alleged Failure of Inverse Methods

There is a small but persistent literature (e.g., Bigg, 1986; Grose, Bigg, & Johnson, 1994) in which various authors claim to show the failure of various inverse methods applied to the ocean circulation. These papers usually start with a known answer generated from a model. Commonly, the inverse model differs in some significant way from the forward model used to generate the pseudodata (e.g., a highly viscous or time-dependent forward calculation is used to generate "data" that the inverse model treats as purely geostrophic; or the Ekman flux is omitted from the inverse model). They then proceed to demonstrate that numbers derived from some particular procedure fail to reproduce the known values to some greater or lesser extent. Although such papers claim to be using an inverse method, even where the forward and inverse models are physically consistent, these studies typically ignore the required uncertainty and resolution analyses and the posterior comparisons of solutions with prior statistics. Deviation of an estimate from a known answer is only meaningful if one can demonstrate that the discrepancy lies outside a properly computed uncertainty limit under conditions in which all posterior statistical checks have also been passed.

As we have been at pains to demonstrate, inverse procedures are variants of general statistical estimation methods; their failure can only occur if one employs an erroneous physical model or misrepresents the observation or model error structure, and even those problems are detectable by sensible posterior tests (recall the discussion of the problem of fitting a straight line to data). Any powerful tool can be dangerous to an inexperienced or naive user, but there is no evidence that the problem of determining the ocean circulation is so unusual that it leads to the failure of conventional methods of statistical inference.

5
Additional Useful Methods

This chapter describes a potpourri of ideas, techniques, and applications not covered in Chapter 3. A number of issues raised in Chapter 4–for example, how to impose inequality constraints–are discussed. We take up the problem of understanding the errors in the coefficient matrices **E**, which have not hitherto been much addressed. Further application of the SVD to a variety of problems, including how to best make maps of the ocean properties, is described. Few of the results are in any sense complete (many books exist devoted to linear programming alone, for example), but they are intended to indicate some of the ideas and methods that can be employed for quantitative description of the ocean circulation, which remain for future work.

The problems discussed in the last chapter raise a number of issues about the models that are difficult to address with the mathematical machinery already available. Among those which seem most important to deal with are the nonlinearities, which are in turn of at least two types, and the grossly underdetermined nature of the conventional large-scale box inversions.

5.1 Inequality Constraints; Nonnegative Least Squares

There are a number of aspects of the estimation problem of the general circulation that suggest the usefulness of being able to impose inequality constraints upon the solutions. Problems involving tracer concentrations usually demand that they remain positive; eddy coefficients are sometimes regarded as acceptable only when positive;[1] in some problems we may wish to impose directions, but not magnitudes, upon the flow fields.

Some of the figures in Chapter 2 (e.g., Figure 2–2) show huge tongues of

[1] The physical basis for this requirement is, however, flimsy; see Starr (1968).

properties extending for thousands of kilometers throughout the ocean. The relative salinity maximum of the Mediterranean water in Figure 2–2b can be traced all the way into the North Pacific Ocean. These property tongues have been accepted since the earliest days of oceanography as compelling evidence that water masses labeled by property extremes thereby move from the apparent source regions of the most extreme property values along the tongues into the open sea. The point was made in Chapter 2, however, that the property tongues represent a "standing crop" and do not necessarily represent either present-day water movements or even net mass fluxes. But if one is convinced that Mediterranean water of locally high salt content is moving past the Straits of Gibraltar at an average rate of about 1 Sv, then it becomes possible to argue qualitatively that the tongue represents some average direction of motion for the outflow waters.

There is no difficulty with these ideas except when they are naively combined with synoptic hydrographic sections. Confusion reigns when climatological features like the property tongues are simply assumed to be necessarily consistent with mass fluxes appearing in a short duration set of observations. It is apparent (e.g., Figure 2–21) that even such powerful flow features as the deep western boundary currents reverse their flow for extended periods. Simple calculation suggests that no contradiction arises if the flow along the Mediterranean salt tongue should reverse for many weeks or months–as long as it did not reverse for years at a time. In the presence of the eddy field, so apparent in Figure 1–1, one cannot sensibly require the flow at any given point to be in a direction required by a tracer feature.

On the other hand, with suitable care, the imposition of flow directions can be a potentially useful addition to the arsenal of information available about the circulation. Although one cannot reasonably require at any given point in an instantaneous picture that the flow has its climatological properties, it is reasonable to hypothesize that when integrated over large distances, climatology might be appropriate. Consider a band of latitudes $\phi_0 \leq \phi \leq \phi_1$ spanning the Mediterranean outflow. Then the zonal flow $u(\phi)$ can be in either direction with widely varying amplitudes at any given moment. But the integral over the whole band

$$\int_{\phi_0}^{\phi_1} u(\phi) d\phi < 0 \,, \tag{5.1.1}$$

that is, away from the Mediterranean. This *assumption* is a kind of ergodic hypothesis–that the spatial average in (5.1.1) would produce a value equivalent to the hypothetical long-time average at a point. Without necessarily accepting the validity of the assumption, one would like to be able to experiment with such constraints, and along with the positivity constraints on eddy coefficients and the like, it leads us to consider problems of the form

$$\mathbf{Ex} + \mathbf{n} = \mathbf{y} \tag{5.1.2}$$

$$\mathbf{Gx} \geq \mathbf{h}\,, \tag{5.1.3}$$

where the use of a greater-than inequality to represent the general case is purely arbitrary, multiplication by minus 1 readily reversing it; $\mathbf{G}$ is of dimension $M_2 \times N$.

Several cases need to be distinguished. (A) Suppose $\mathbf{E}$ is full rank and fully determined, the SVD solution to (5.1.2) by itself is $\tilde{\mathbf{x}}$, $\tilde{\mathbf{n}}$, and there is no solution nullspace. Substitution of the solution into (5.1.3) shows that the inequalities are either satisfied or that some are violated. In the first instance, we are finished, and the inequalities bring no new information. In the second case, the solution must be modified, and necessarily $\|\tilde{\mathbf{n}}\|$ will increase, given the noise-minimizing nature of the SVD solution by itself. It is possible that the inequalities are contradictory, in which case there is no solution at all.

(B) Now suppose that $\mathbf{E}$ is formally underdetermined so that there is a solution nullspace. If the particular-SVD solution violates one or more of the inequalities and requires modification, we can distinguish two subcases. (1) Addition of one or more nullspace vectors permits the inequalities to be satisfied. Then the solution residual norm will be unaffected, but $\|\tilde{\mathbf{x}}\|$ will increase. (2) The nullspace vectors by themselves are unable to satisfy the inequality constraints, and one or more range vectors are required to do so. Then both $\|\tilde{\mathbf{x}}\|$, $\|\tilde{\mathbf{n}}\|$ will increase.

Case (A) is the conventional one and is described by Lawson and Hanson (1974) and Strang (1986); the standard full treatment is Fiacco and McCormick (1968). The so-called Kuhn-Tucker-Karush theorem is a requirement for a solution $\tilde{\mathbf{x}}$ to exist. Its gist is as follows: Let $M \geq N$ and $\mathbf{E}$ be full rank; there are no $\mathbf{v}_i$ in the solution nullspace. If there is a solution, there must exist a vector, $\mathbf{q}$, of dimension M_2 such that

$$\mathbf{E}^T(\mathbf{E}\tilde{\mathbf{x}} - \mathbf{y}) = \mathbf{G}^T\mathbf{q}\,. \tag{5.1.4}$$

Now define

$$\mathbf{Gx} - \mathbf{h} = \mathbf{r}\,, \tag{5.1.5}$$

where the M_2 elements of $\mathbf{q}$ are divided into two groups. For group 1, of dimension m_1,

$$r_i = 0, \quad q_i \geq 0\,, \tag{5.1.6}$$

and for group 2, of dimension $m_2 = M_2 - m_1$,

$$r_i > 0, \quad q_i = 0\,. \tag{5.1.7}$$

To understand this theorem, recall that in the solution to the ordinary overdetermined least-squares problem, the left-hand side of (5.1.4) vanishes identically [(3.3.6) and (3.4.111)], being the projection of the residuals onto the range vectors $\mathbf{u}_i$ of $\mathbf{E}^T$. If this solution violates one or more of the inequality constraints, one must introduce into it structures that produce increased residuals.

Because there are no nullspace $\mathbf{v}_i$, the rows of $\mathbf{G}$ may each be expressed exactly by an expansion in the range vectors. In the second group of indices, the corresponding inequality constraints are already satisfied by the ordinary least-squares solution, and no modification of the structure proportional to $\mathbf{v}_i$ is required. In the first group of indices, the inequality constraints are marginally satisfied, at equality, only by permitting violation of the demand (3.4.111) that the residuals should be orthogonal to the range vectors of $\mathbf{E}$. If the ordinary least-squares solution violates the inequality, the minimum modification required to it pushes the solution to the edge of the acceptable bound but at the price of increasing the residuals proportional to the corresponding $\mathbf{u}_i$. The algorithm consists of finding the two sets of indices and then the smallest coefficients of the $\mathbf{v}_i$ corresponding to the group 1 indices required to just satisfy any initially violated inequality constraints. The version of the algorithm described by Lawson and Hanson (1974) is based upon the solution to the special case $\mathbf{G} = \mathbf{I}$, $\mathbf{h} = \mathbf{0}$–called "nonnegative least squares"–plus a transformation for the more general case to this one.

Fu (1981) extended the algorithm to the underdetermined/rank-deficient case in which the addition, to the original basic SVD solution, of appropriate amounts of the nullspace of $\mathbf{v}_i$ is capable of satisfying any violated inequality constraints. One simply chooses the smallest mean-square coefficients necessary to push the solution to the edge of the acceptable inequalities, producing the smallest norm solution. The residuals of the original problem do not increase because only nullspace vectors are being used. $\mathbf{G}$ must have a special structure for this to be possible, but Fu (1981) found that the method worked for his particular situation.

Tziperman and Hecht (1987) completed the algorithm by considering the general case of rank-deficiency/underdeterminism where the nullspace vectors by themselves are inadequate to produce a solution satisfying the inequalities. In effect, any inequalities "left over" are satisfied by invoking the smallest perturbations necessary to the coefficients of the range vectors

$\mathbf{v}_i$. Figure 4–30b shows an example of the shift in the estimated value of a vertical mixing coefficient originally estimated from a particular-SVD solution, and then when a positivity constraint was added, where the minimum required modification just touches zero at one point.

The nonnegative least-squares special case, $\mathbf{x} \geq 0$, is essential in many problems involving tracer concentrations, which are neccessarily positive.

Discussion in this section has focused on least squares and the SVD. The combination of inequality constraints with statistically based methods like the Gauss-Markov estimate seems not to have been developed.

5.2 Linear Programming and Eclectic Models

In a number of important oceanographic problems, the objective functions are linear rather than quadratic functions of the state vector $\mathbf{x}$. Property fluxes such as heat–for example (4.4.1)–(4.4.4)–are linear functions of $\mathbf{x}$. If one sought the extreme values of a property, C, carried by the ocean circulation, it would require finding the extremal values of the corresponding linear function. Least squares does not produce useful answers in such problems because linear objective functions achieve their minima or maxima only at plus or minus infinity unless the elements of $\mathbf{x}$ are bounded. The methods of *linear programming* are generally directed at finding extremal properties of linear objective functions subject to bounding constraints. In general terms, such problems can be written as

$$\text{minimize: } J = \mathbf{c}^T\mathbf{x}\,,$$

$$\mathbf{E}_1\mathbf{x} = \mathbf{y}_1 \tag{5.2.1}$$

$$\mathbf{E}_2\mathbf{x} \geq \mathbf{y}_2 \tag{5.2.2}$$

$$\mathbf{E}_3\mathbf{x} \leq \mathbf{y}_3\,, \tag{5.2.3}$$

$$\mathbf{a} \leq \mathbf{x} \leq \mathbf{b}\,, \tag{5.2.4}$$

that is, as a collection of equality and inequality constraints of both greater than or less than form, plus bounds on the individual elements of $\mathbf{x}$. In distinction to the least squares and minimum variance equations we have hitherto been discussing, these are hard constraints; they cannot be violated at all in an acceptable solution.

Linear programming problems are normally reduced to what is referred to as a *canonical form*, although different authors use different definitions of what it is. But all such problems are readily reduced to a form equivalent to

$$\text{minimize: } J = \mathbf{c}^T \mathbf{x}\,, \tag{5.2.5}$$

$$\mathbf{E}\mathbf{x} \le \mathbf{y} \tag{5.2.6}$$

$$\mathbf{x} \ge \mathbf{0}\,. \tag{5.2.7}$$

The use of a minimum rather than a maximum is readily reversed by introducing a minus sign, and the inequality is similarly readily reversed. The last relationship, requiring purely positive elements in $\mathbf{x}$, is obtained without difficulty by simple translation.

Linear programming problems are widespread in many fields, including especially financial and industrial management where they are used to maximize profits, or minimize costs, in, say, a manufacturing process. Necessarily then, the amount of a product of each type is positive, and the inequalities reflect such things as the need to consume no more than the available amounts of raw materials. General methodologies were first developed during the World War II in what became known as *operations research* (*operational research* in the U.K.) by Dantzig (1963), although special cases were known much earlier. Since then, given the economic stake in practical use of linear programming, immense effort has been devoted both to textbook discussion and efficient, easy-to-use software. (See, for example, Luenberger, 1984; Bradley, Hax, & Magnanti, 1977; and many others.) Given this highly accessible literature and software, we will not actually describe the methodologies of solution but merely make a few general points.

The solution algorithm that Dantzig used is usually known as the *simplex method* (a simplex is a convex geometric shape). It is a highly efficient search method conducted along the bounding constraints of the problem. In general, it is possible to show that the outcome of a linear programming problem falls into several distinct categories: (1) The system is *infeasible*, meaning that it is contradictory and there is no solution; (2) the system is unbounded, meaning that the minimum lies at negative infinity; (3) there is a unique minimizing solution; and (4) there is a unique finite minimum, but it is achieved by an infinite number of solutions $\mathbf{x}$.

The last situation is equivalent to observing that if there are two minimizing solutions, there must be an infinite number of them because then any linear combination of the two solutions is also a solution. Alternatively, if one makes up a matrix from the coefficients of $\mathbf{x}$ in Equations (5.2.5)–(5.2.7), one can ask whether it has a nullspace. If one or more such vectors exists, it is also orthogonal to the objective function, and it can be assigned an arbitrary amplitude without changing J. One distinguishes between *feasible solutions*, meaning those that satisfy the inequality and equality constraints but which are not minimizing, and *optimal solutions*, which are both feasible and minimize the objective function.

An interesting and useful feature of a linear programming problem is that equations (5.2.5)–(5.2.7) have a *dual*:

$$\text{maximize: } J_2 = \mathbf{y}^T \boldsymbol{\mu}\,, \tag{5.2.8}$$

$$\mathbf{E}^T \boldsymbol{\mu} \geq \mathbf{c} \tag{5.2.9}$$

$$\boldsymbol{\mu} \geq \mathbf{0}\,. \tag{5.2.10}$$

It is possible to show that the minimum of J must equal the maximum of J_2. The reader may want to compare the structure of the original (the *primal*) and dual equations with those relating the Lagrange multipliers to $\mathbf{x}$ discussed in Chapter 3. In the present case, the important relationship is

$$\frac{\partial J}{\partial y_i} = \mu_i\,. \tag{5.2.11}$$

That is, in a linear program, the dual solution provides the sensitivity of the objective function to perturbations in the constraint parameters $\mathbf{y}$. (Duality theory pervades optimization problems, and the relationship to Lagrange multipliers is no accident. See Strang, 1986; Luenberger, 1969; Cacuci, 1981; Hall & Cacuci, 1984; Rockafellor, 1993.) Some simplex algorithms, called the *dual simplex*, take advantage of the different dimensions of the primal and dual problems to accelerate solution.

In recent years much attention has focused upon a new, nonsimplex method of solution[2] known as the *Karmackar method*, which is meant to be vastly more efficient, although some controversy over its efficiency in practice remains. Stone and Tovey (1991) produce a unification of the simplex and Karmackar algorithms by reducing both to a form of least squares.

Wunsch (1984) wrote a system of equations like those used in earlier chapters for the geostrophic box inversions but for multiple boxes spanning the North Atlantic. A form suitable for linear programming was used: The soft constraints for mass, salt, etc., conservation were replaced by hard inequalities representing absolute maximum and minimum bounds. Individual hard bounds were set on the reference-level velocities. The hard bounds did not have any simple statistical interpretation–unlike the soft

[2] One of a few mathematical algorithms ever to be written up on the front page of *The New York Times* (19 November 1984, story by J. Gleick)–a reflection of the huge economic importance of linear programs in industry.

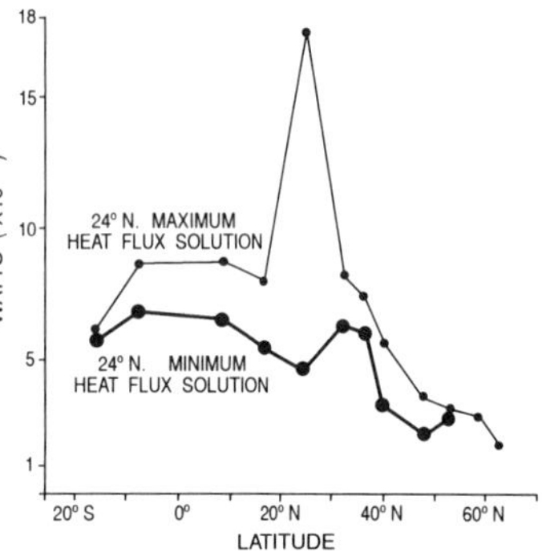

Figure 5–1. Heat fluxes as a function of latitude in the North Atlantic for solutions which maximize, and minimize, the flux across the 24°N line (after Wunsch, 1984). The values at other latitudes are not necessarily maxima or minima, merely values which are consistent with the solution. These results are extreme possibilities with many of the constraints up against their hard limits.

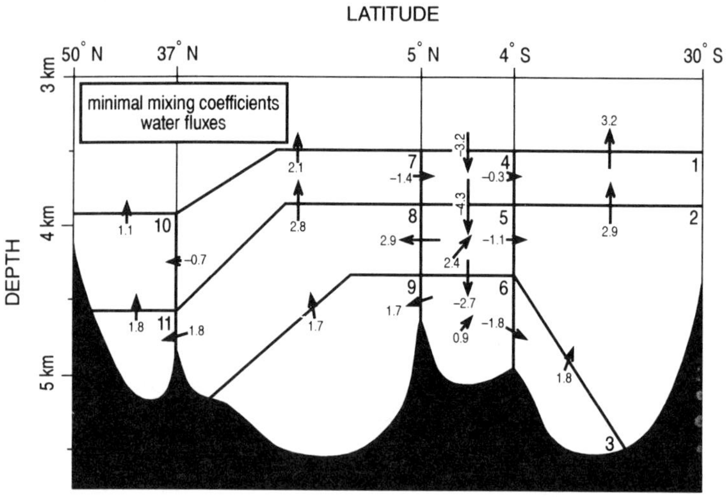

Figure 5–2. Linear programming solution (Schlitzer, 1988) showing the mass fluxes (Sv.) into various boxes in the eastern North Atlantic Ocean in a model that employed radiocarbon as well as hydrographic data. The objective function minimized a sum of positive mixing coefficients.

bounds of Gauss-Markov estimation or least squares. Rather, they represented partially subjective views, based upon experience, of what would be extreme acceptable values. The objective functions consisted of the heat fluxes across each of the ocean-spanning sections in the form (4.3.4). As the bounds on x_i are not in canonical form, permitting both positive and negative values, one introduces new variables, $\mathbf{x}^+$, $\mathbf{x}^-$, which are positive, and defines $\mathbf{x} = \mathbf{x}^+ - \mathbf{x}^-$.

The problem was solved by a simplex method. The bounding heat fluxes were sought in order to understand their range of uncertainty. There were approximately 500 constraints of the form (5.2.1)–(5.2.4) in approximately 900 unknown parameters, each of which had a set of upper and lower bounds. The resulting bounding heat fluxes are shown in Figure 5–1.

Such methods, in which bounds are sought, are especially useful in problems that from the point of view of the basic SVD are grossly underdetermined, and for which determining "*the*" value of quantities such as the heat flux is less important than understanding their possible range. If the resulting range is sufficiently small, the remaining uncertainty may be unimportant. For anyone interested in the large-scale properties of the circulation–for example, its chemical and biological flux properties–detailed determination of the flow at any given point may be beyond reach, and not very important, whereas the integrated property extrema may be well and readily determined. Wunsch (1984) called this type of model and solution *eclectic modeling* because inequality constraints are particularly flexible in accounting for the wide variety and inhomogeneous distribution of most oceanographic observations. These methods (Wunsch & Minster, 1982) are very powerful and, because of the ease with which positivity constraints are imposed, are the natural mathematical tool for handling tracers (see Wunsch, 1986). Schlitzer (1988, 1989) used the methodology to explore the carbon budget and biological productivity of the eastern North Atlantic (see Figure 5–2).

Hard-constraint optimizing solutions are rigid and unforgiving; the system will try as hard as it possibly can to force the objective function toward its extremes by driving critical inequalities right to their bounds, and by introducing structures into the solution that maximize (or minimize) the correlation between $\mathbf{c}$ and $\mathbf{x}$. Such solutions may appear esthetically distasteful, but one gets exactly what one has asked for. (Shepard, 1980, explored a method for compromising between the yielding soft constraints of least squares and the unyielding, hard constraints of linear programming by writing objective functions involving the logarithms of the elements of $\mathbf{x}$. Such objective functions cannot be handled by linear methods but require the nonlinear search procedures of Section 5.5.)

Linear programming is also valuable for solving estimation or approximation problems in which norms other than the 2-norms, which have been the focus of this book, are used. For example, suppose that one sought the solution to the constraints (3.3.2), $M > N$, but subject not to $J = \sum_i n_i^2$ being minimum, but seeking the minimum of $J = \sum_i |n_i|$ (a 1-norm). Such norms are less sensitive to outliers than are the 2-norms and are said to be *robust.* The maximum likelihood idea connects 2-norms to Gaussian statistics, and similarly, 1-norms are related to maximum likelihood in exponential statistics (see Arthnari & Dodge, 1981). The reduction of such problems to linear programming is carried out by setting $n_i = n_i^+ - n_i^-$, $n_i^+ \geq 0$, $n_i^- \geq 0$, and the objective function is

$$\text{min:}\ J = \sum_i (n_i^+ + n_i^-) \tag{5.2.12}$$

(see for example, Wagner, 1969). Other norms, the most important of which is the so-called infinity norm, which minimizes the maximum element of an objective function, are also reducible to linear programming (Arthnari & Dodge, 1981). The somewhat intricate relationship between the imposition of hard constraints on parameters and the corresponding use of softened ones is discussed by Backus (1988b).

5.3 Quantifying Water Mass Properties; Empirical Orthogonal Functions

Physical oceanography has retained some of its antique charm in the widespread use of beautiful hand-colored atlases of hydrographic properties, usually contour maps on geographical axes (e.g., some of Figure 2–2). In this way, the subject has held its connection to geography, from which it came, hence its name, and the tenacity with which some romantics have clung to the story-telling inspired by pictures. More recently, color computer displays have tended to displace the bound books of Wüst (1935), Fuglister (1960), etc., but the use of visualization has remained constant, and it will always be an important preliminary step in the use of hydrographic and all other types of data.

Ultimately, however, as suggested in Chapter 2, one recognizes that seawater is tagged by a point in an n-dimensional space of properties, writable as an n-tuple (temperature, depth, latitude, salinity, longitude, silicate, velocity,...), and one's eye has a very difficult time integrating the very large number of possible two-dimensional projections onto planes cutting through the n-space. Because the end use of the great investment in obtaining shipboard data is meant to be the extraction of numerical values for estimating physical properties, there have been a few attempts directed at representations and syntheses more convenient and powerful than property-property diagrams or sections. If one is attempting to understand the degree to which a model is consistent with the real ocean, a measure of distance in one particular two- or three-dimensional subspace of the n-space may give a very distorted picture of the true distance.

Some of the motivation for these studies has been the recognition that many of the properties that oceanographers measure are nearly redundant, appearing sometimes in geographical space as coincident tongues of, say, salinity and silica, or in property-property space as linear or other simple functional relationships. A quantitative description of the relationship becomes useful in several ways: finding an explanation; reducing the dimension of the n-dimensional space; using the relationships to produce estimates of properties of a poorly measured variable from measurements of a better measured one; reducing costs by substituting a cheap-to-measure variable for a functionally connected expensive-to-measure one.

In the comparatively small literature on this subject in the context of water masses, one might take note of Mackas et al. (1987), Fukumori and Wunsch (1991), Hamann and Swift (1991), Tomczak and Large (1989), and the summary in Bennett (1992).

In the wider context, however, one is dealing with the problem of efficient relationships amongst variables known or suspected to carry mutual correlations. Because of its widespread use, this subject is another one plagued by multiple discovery and thus multiple jargon. In different contexts, the problem is known as that of *principal components* (e.g., Jolliffe, 1986; Preisendorfer, 1988; Jackson, 1991), *empirical orthogonal functions* (EOFs), the Karhunen-Loève expansion (in mathematics and electrical engineering, e.g., Davenport & Root, 1958; Wahba, 1990), *proper orthogonal decomposition* (e.g., Berkooz, Holmes, & Lumley, 1993), etc.

Consider in the hydrographic context the method of empirical orthogonal functions as employed by Fukumori and Wunsch (1991). This method is particularly easy to explain, because as used by these authors, it is based directly upon the SVD. They define a hydrographic station as a column vector of all the properties (p-properties) measured at a station in sequence. That is, let $\boldsymbol{\theta}_i$ be the vector of potential temperatures at station i; let $\mathbf{S}_i$, $\mathbf{N}_i$, etc., be the corresponding $n \times 1$ vector of salinities, nitrate, etc., at that station. Form an extended column vector,

$$\mathbf{s}_i = \left[\boldsymbol{\theta}_i^T, \mathbf{S}_i^T, \mathbf{N}_i^T, \ldots\right]^T$$

(i.e., first n elements are temperatures, second n are salinities, etc.), and make up a matrix of all available hydrographic stations:

$$\mathbf{M}_2 = \{\mathbf{s}_1, \mathbf{s}_2, \ldots, \mathbf{s}_N\}, \tag{5.3.1}$$

which is a projection of the $N \times n \times p$–dimensional space onto two dimensions. If the SVD is applied to $\mathbf{M}_2$, then precisely as discussed in Chapter 3, the singular values and $\mathbf{u}_i$, $\mathbf{v}_i$ produce the most efficient possible representation of the matrix. In this context, then either the $\mathbf{u}_i$ or $\mathbf{v}_i$ are empirical orthogonal functions, or principle components. Fukumori and Wunsch (1991) called these the *form-2 modes.* If the matrix is ordered instead as

$$\mathbf{M}_1 = \{\{\boldsymbol{\theta}_i\}, \{\mathbf{S}_i\}, \{\mathbf{N}_i\} \ldots\}, \tag{5.3.2}$$

they called the result *form-1 modes.* There are advantages to each form; the representation is not unique–it cannot be because a high-dimensional space is still being projected onto a two-dimensional one. More generally, the column and row indices can be time, space, or any ordering or bookkeeping variable.

The mathematical underpinning of this methodology is the SVD in a form sometimes known as the *Eckart-Young-Mirsky theorem* (see Van Huffel & Vandewalle, 1991). This theorem states that the most efficient representation of a matrix $\mathbf{A}$ in the form

$$\mathbf{A} \approx \sum_{i}^{K} \mathbf{u}_i \lambda_i \mathbf{v}_i^T \tag{5.3.3}$$

where the $\mathbf{u}_i$, $\mathbf{v}_i$ are orthonormal is achieved by choosing the vectors to be the singular vectors. Efficiency is measured as either of the 2-norms of the difference between $\mathbf{A}$ and its representation for fixed K.

The connection to the subject of regression analysis is readily made by noticing that the sets of singular vectors are the eigenvectors of the two matrices $\mathbf{M}\mathbf{M}^T$, $\mathbf{M}^T\mathbf{M}$ (3.4.58)–(3.4.59). If each row of $\mathbf{M}$ is regarded as a set of observations at a fixed coordinate, then $\mathbf{M}\mathbf{M}^T$ is just proportional to the sample second-moment matrix of all the observations, and its eigenvectors, $\mathbf{u}_i$, are the EOFs. Alternatively, if each column is regarded as the observation set for a fixed coordinate, then $\mathbf{M}^T\mathbf{M}$ is the corresponding sample second-moment matrix, and the $\mathbf{v}_i$ are the EOFs.

The large literature already cited provides various statistical rules for use of EOFs. For example, the rank determination in the SVD becomes a test of the statistical significance of the contribution of singular vectors to the structure of $\mathbf{M}$. The use of EOFs, with various normalizations, scalings, and in various row/column physical spaces, is widespread–for example, Wallace and Dickinson (1972), Wallace (1972), Davis (1978b), and many others.

As an example, consider Figure 5–3a taken from Mercier et al. (1993), who show the vertical EOFs of the North Atlantic density field. That is, $\mathbf{M}_2$ has columns consisting only of the observed density, and the EOFs correspond to the $\mathbf{u}_i$ singular vectors. Figure 5–3b shows the percentage of the variance in the density field captured as a function of the number of singular vectors retained. A considerable advantage of such methods is that they permit a straightfoward extension of the two-dimensional objective mapping methods of Chapter 3 to three dimensions, without the need to manipulate three-dimensional covariances and data sets. The vertical covariances are captured in the EOFs ($\mathbf{u}_i$), and the coefficients of each $\mathbf{u}_i$ are mapped horizontally (these coefficients are $\lambda_i \mathbf{v}_i$). The best estimate at any location, using both vertical and horizontal covariance information, is just the sum of each vertical mode times its mapped coefficient at that

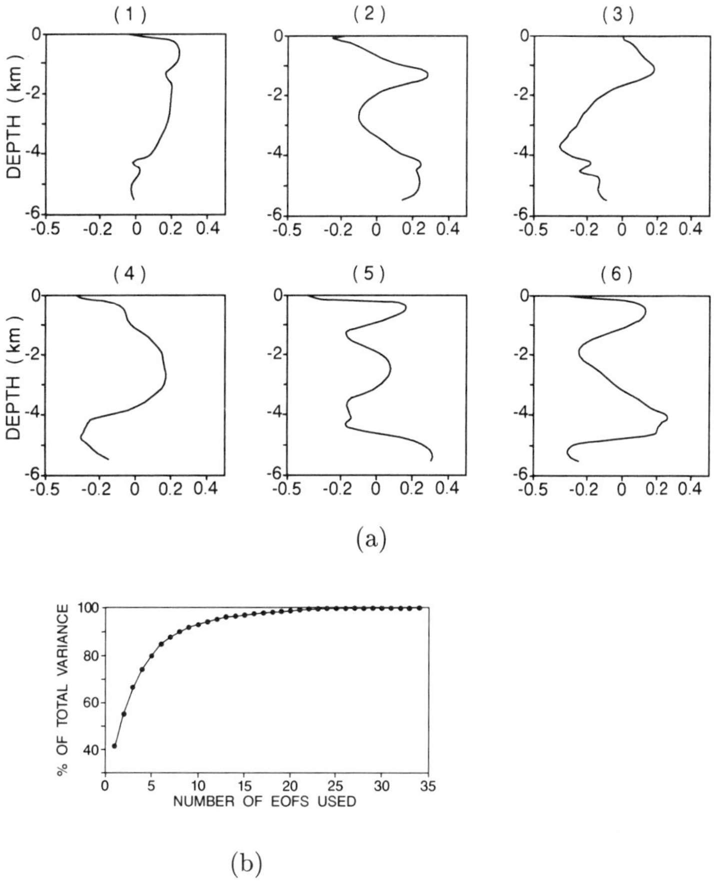

Figure 5–3. (a) The first six empirical orthogonal functions (EOFs, or $\mathbf{u}_i$ singular vectors) of the North Atlantic density field (Mercier, Ollitrault, & Le Traon, 1993). (b) Percentage variance in the density field deviation from a global mean profile, captured by the EOFs in Figure 5–3a. With 35 standard depths, all the variance is accounted for at rank 35.

point. Fukumori and Wunsch (1991) display numerous examples from the situation where multiple hydrographic properties are used in this manner, thus further exploiting covariances between properties in three dimensions as well as the purely geographic covariances for any single property.

[Businger and Golub (1969) showed that the SVD was readily extended to arbitrary complex matrices, the singular values remaining real and positive, and the two sets of singular vectors becoming complex. This application of the SVD has proven most useful in the context of arrays of time series–for example, the measured temperatures z_i, $1 \leq i \leq N$, as a function of time on a spatially distributed set of instruments. Although the complex SVD could be applied directly to the matrix of Fourier transforms, such use is usually not very interesting if the time series contain stochastic elements. Instead, one uses the cross spectra between pairs, i, j, of instruments to define an $N \times N$ Hermitian matrix $\boldsymbol{\Phi}(\omega)$, which is a function of frequency, ω (see, for example, Priestley, 1981). For any value of ω, the SVD of $\boldsymbol{\Phi}$ captures the structure of the amplitudes and phases among the various time series $z_i(t)$ and can, for example, be used to sort out propagating and traveling motions, among others.

As a simple example, consider the 3×3 matrix,

$$\mathbf{A} = \begin{Bmatrix} 2.52 & 2.05 - 0.06i & 1.51 - 0.24i \\ 2.05 + 0.06i & 2.45 & 2.00 + 0.24i \\ 1.51 + 0.24i & 2.00 - 0.24i & 1.94 \end{Bmatrix}.$$

Its SVD is

$$\mathbf{U} = \mathbf{V} = \begin{Bmatrix} 0.58 & -0.73 & 0.35 \\ 0.62 + 0.06i & 0.23 + 0.25i & -0.57 + 0.42i \\ 0.52 + 0.03i & 0.54 - 0.24i & 0.26 - 0.56i \end{Bmatrix}$$

$$\boldsymbol{\Lambda} = \{\mathrm{diag}(6.05 \quad 0.84 \quad 0.02)\}.$$

The Hermitian symmetry requires **U**, **V** to be identical. The magnitude and phase of these vectors represent, in the most efficient way, the coherence structure of the time (or space) series. This subject properly lies in the general area of time series analysis and is beyond our scope. For applications, see Wallace (1972) or Hogg (1981).]

Returning to the specifically hydrographic context, Mackas et al. (1987), working solely within the physical property space (i.e., the geographical coordinates are specifically ignored), attempted to calculate the mixing required to form a particular water type in terms of a number of originating, extreme properties. For example, let the water at some particular place be defined by an n-tuple of properties like temperature, salt, oxygen, etc. (θ, S, $O, \ldots$). It is supposed that this particular set of water properties is formed by mixing a number, N, of parent water masses, for example, "pure" Labrador Sea Water, Red Sea Intermediate Sea Water, etc.,[3] each defined as $[\theta_i, S_i, O_i, \ldots]$. Mathematically,

$$\begin{aligned} \theta &= m_1\theta_1 + m_2\theta_2 + \cdots + m_N\theta_N \\ S &= m_1 S_1 + m_2 S_2 + \cdots + m_N S_N \\ O &= m_1 O_1 + m_2 O_2 + \cdots + m_N O_N \\ &\cdots \\ m_1 &+ m_2 + m_3 + \cdots + m_N = 1, \end{aligned}$$

and one seeks the fractions m_i of each water type making up the local water mass. Because the fractions to be interpreted physically must be positive,

[3] Attaching pet names to water types is not necessary, but it is a tradition and gives many oceanographers a sense of intimacy with their subject, similar to the way physicists define quarks as having charm, and it does provide some geographic context.

$m_i \geq 0$, the resulting system must be solved either with a linear programming algorithm or one for nonnegative least squares. The equations can as always be rotated, scaled, etc., to reflect prior statistical hypotheses about noise, solution magnitude, etc. Mackas et al. (1987) and Bennett (1992) discuss this problem in detail.

5.4 Kriging and Other Variants of Gauss-Markov Estimation

A variant of the Gauss-Markov mapping estimators, often known as *kriging* (named for David Krige, a mining geologist), addresses the problem of a spatially varying mean field. Accounts of this methodology, which is a generalization of the ordinary Gauss-Markov estimator, may be found in Armstrong (1989), David (1988), Ripley (1981), and elsewhere. There are a number of ways to describe this method; here we follow Ripley (1981).

Consider the discussion in Section 3.6 of the fitting of a set of functions $f_i(\mathbf{r})$ to an observed field $y(\mathbf{r}_j)$. That is, we put

$$y(\mathbf{r}_j) = \mathbf{F}\boldsymbol{\alpha} + q(\mathbf{r}_j) \tag{5.4.1}$$

where $\mathbf{F}(\mathbf{r}) = \{f_i(\mathbf{r})\}$ is a set of basis functions and one seeks the expansion coefficients, $\boldsymbol{\alpha}$, and q such that the data, y, are *interpolated* (meaning reproduced exactly) at the observation points, although there is nothing to prevent further breaking up q into signal and noise components. If there is only one basis function–for example a constant–one is doing *kriging*, which is the determination of the mean prior to objective mapping of q, as discussed in Chapter 3. If several basis functions are being used, one has universal kriging, which is clearly related to the fitting of such functions in Equation (3.6.25). The main issue concerns the production of an adequate statement of the expected error, given that the q are computed from a preliminary regression to determine the $\boldsymbol{\alpha}$ (Ripley, 1981). The method is often used in situations where large-scale trends are expected in the data, and where one wishes to estimate and remove them before analyzing and mapping the q.

Because the covariances employed in objective mapping are simple to use and interpret only when the field is spatially stationary, much of the discussion of kriging uses instead what is called the *variogram*, defined as $V = <(y(\mathbf{r}_i)-y(\mathbf{r}_j))(y(\mathbf{r}_i)-y(\mathbf{r}_j))>$, which is related to the covariance and which is often encountered in turbulence theory as the *structure function.* Kriging is popular in geology and hydrology but deserves more oceanographic attention; oceanographic applications may be seen in Davis (1985) or Festa and Molinari (1992).

The geodetic community (e.g., Vaniček & Krakiwsky, 1986; Moritz, 1978) employs a highly developed form of Gauss-Markov estimation but under the label of *collocation.* Spline fitting (e.g., Wahba & Wendelberger, 1980; Wahba, 1990; Bennett, 1992) is also closely allied to Gauss-Markov methods where the norms involve derivatives of the field (called *semi-norms*). The first two references contain an extended discussion of cross-validation, a statistical method (omitted here) for testing adequacy of a set of fitted functions–for example, splines or another kriging basis.

5.5 Nonlinear Problems

The least-squares solutions examined thus far treat the coefficient matrix $\mathbf{E}$ as known. But in many of the cases encountered in practice, the elements of $\mathbf{E}$ are computed from data and are imperfectly specified. It is well known in the regression literature that treating $\mathbf{E}$ as known, even if $\mathbf{n}$ is increased beyond the errors contained in $\mathbf{y}$, can lead to significant bias errors in the least-squares and related solutions, particularly if $\mathbf{E}$ is nearly singular (e.g., Seber, 1977). The problem is known as that of *errors in regressors* or *errors in variables* (EIV); it manifests itself in the classical simple least-squares problem (3.3.3) where a straight line is being fit to data of the form $y_i = at_i + b$ but where the measurement positions t_i are partly uncertain rather than perfect.

In general terms, when $\mathbf{E}$ has errors, the model statement becomes

$$(\tilde{\mathbf{E}} + \Delta\tilde{\mathbf{E}})\tilde{\mathbf{x}} = \tilde{\mathbf{y}} + \Delta\tilde{\mathbf{y}} \tag{5.5.1}$$

where one seeks estimates, $\tilde{\mathbf{x}}$, $\Delta\tilde{\mathbf{E}}$, $\Delta\tilde{\mathbf{y}}$ where the old $\mathbf{n}$ is now broken into two parts: $\Delta\tilde{\mathbf{E}}\tilde{\mathbf{x}}$ and $-\Delta\tilde{\mathbf{y}}$. If such estimates can be made, the result can be used to rewrite (5.5.1) as

$$\tilde{\mathbf{E}}\tilde{\mathbf{x}} = \tilde{\mathbf{y}} \tag{5.5.2}$$

where the relation is to be exact.

That is, one seeks to modify the elements of $\mathbf{E}$ such that the observational noise in it is reduced to zero. In the β-spiral and box inverse problems, one would be seeking to modify the density estimates entering the problem so that the resulting density field and related velocities satisfy the equations exactly rather than approximately, as in the solutions we have discussed so far.

5.5.1 Total Least Squares

For some problems, although not directly relevant to the oceanographic one, the method of total least squares (TLS) is a powerful and interesting method. It is worth examining briefly to understand why it is not immediately useful and to motivate a different approach. Extensive discussions of the method may be found in Golub and van Loan (1980, 1989) and Van Huffel and Vandewalle (1991).

The SVD plays a crucial role in TLS. Consider, for example, that in Equation (3.1.11) we wrote the vector $\mathbf{y}$ as a sum of the column vectors of $\mathbf{E}$; to the extent that the column space does not fully describe $\mathbf{y}$, a residual must be left by the solution $\tilde{\mathbf{x}}$, and ordinary least squares can be regarded as producing a solution in which a new estimate, $\tilde{\mathbf{y}} \equiv \mathbf{E}\tilde{\mathbf{x}}$, of $\mathbf{y}$ is made; $\mathbf{y}$ is changed, but the elements of $\mathbf{E}$ are untouched. But suppose it were possible to introduce small changes in both the column vectors of $\mathbf{E}$ as well as in $\mathbf{y}$ such that the column vectors of the modified $\mathbf{E} + \Delta\mathbf{E}$ produced a spanning vector space for $\mathbf{y} + \Delta\mathbf{y}$; then the problem as stated would be solved.

The simplest problem to analyze is the full-rank, formally overdetermined one. Let $M \geq N = K$. Then, if we form the $M \times (N+1)$ augmented matrix

$$\mathbf{E}_a = \{\mathbf{E} \quad \mathbf{y}\}\,,$$

the solution we seek is such that

$$\{\tilde{\mathbf{E}} \quad \tilde{\mathbf{y}}\}\begin{bmatrix} \tilde{\mathbf{x}} \\ -1 \end{bmatrix} = \mathbf{0} \tag{5.5.3}$$

(exactly). If this solution is to exist, $[\tilde{\mathbf{x}}, -1]^T$ must lie in the nullspace of $\{\tilde{\mathbf{E}} \quad \tilde{\mathbf{y}}\}$. A solution is thus ensured by forming the SVD of $\{\mathbf{E} \quad \mathbf{y}\}$, setting $\lambda_{N+1} = 0$, and forming $\{\tilde{\mathbf{E}} \quad \tilde{\mathbf{y}}\}$ out of the remaining singular vectors and values. Then $[\tilde{\mathbf{x}} \quad -1]^T$ lies in the nullspace of the modified augmented matrix and must therefore be proportional to the nullspace vector $\mathbf{v}_{N+1}$, and

$$\{\Delta\tilde{\mathbf{E}} \quad \Delta\tilde{\mathbf{y}}\} = -\mathbf{u}_{N+1}\lambda_{N+1}\mathbf{v}_{N+1}^T\,. \tag{5.5.4}$$

Various complications can be considered, for example, if the last element of $\mathbf{v}_{N+1} = 0$; this and other special cases are considered by Van Huffel and Vandewalle (1991). Cases of nonuniqueness are treated by selecting the solution of minimum norm. A simple generalization applies to the underdetermined case: If the rank of the augmented matrix is p, one reduces the rank by one to $p - 1$. This same reference also discusses estimates of the solution error.

The TLS solution just summarized applies only to the case in which the errors in the elements of $\mathbf{E}$ and $\mathbf{y}$ are uncorrelated and of equal variance and in which there are no required structures–for example, where certain elements of $\mathbf{E}$ must always vanish (in the box inversions, zero elements often correspond to places where there are no physical flow pathways). More generally, changes in some elements of $\mathbf{E}$ require, for reasons of physics, specific corresponding changes in other elements of $\mathbf{E}$ and in $\mathbf{y}$, and vice versa (a perturbation in the density of a station pair implies a change in the transport in various model layers). Although Van Huffel and Vandewalle (1991) discuss generalizations of the TLS to cases in which structural constraints can be imposed upon the solution, the analysis becomes complex and has apparently not been used on the ocean circulation problem. The fundamental difficulty is that the model (5.5.1) presents a nonlinear estimation problem with correlated variables, and its solution requires modification of the linear procedures we have been using.

5.5.2 Method of Total Inversion

The simplest form of TLS does not readily permit the use of correlations and prior variances in the parameters appearing in the coefficient matrix and does not provide any way of maintaining the zero structure there. Tarantola and Valette (1982) and Tarantola (1987) introduced a generalization of the objective functions that we have been using to permit a general accounting for prior knowledge of covariances. Suppose we have a set of nonlinear constraints in a vector of unknowns $\mathbf{x}$,

$$\mathbf{g}(\mathbf{x}) + \mathbf{u} = \mathbf{q}\,. \tag{5.5.5}$$

This set of equations is the generalization of the linear models hitherto used [e.g., (3.5.13)]; $\mathbf{u}$ again represents any expected error in the specification of the model. An example of a scalar nonlinear model is

$$8x_1^2 + x_2^2 + u = 4\,.$$

In general, there will be some expectations about the behavior of $\mathbf{u}$. Without loss of generality, we take its expected value to be zero, and its covariance is $\mathbf{Q} = <\mathbf{u}\mathbf{u}^T>$. There is nothing to prevent us from combining $\mathbf{x}$, $\mathbf{u}$ into one single set of unknowns $\boldsymbol{\xi}$, and indeed if the model has some unknown parameters, $\boldsymbol{\xi}$ might as well include those as well. So (5.5.5) can be written

$$\mathcal{L}(\boldsymbol{\xi}) + \mathbf{q} = \mathbf{0}\,. \tag{5.5.6}$$

In addition, it is supposed that a reasonable initial estimate $\tilde{\boldsymbol{\xi}}(0)$ is available, with uncertainty $\mathbf{P}(0) \equiv <(\boldsymbol{\xi} - \tilde{\boldsymbol{\xi}}(0))(\boldsymbol{\xi} - \tilde{\boldsymbol{\xi}}(0))^T>$ (or the covariances of the $\mathbf{u}$, $\mathbf{x}$ could be specified separately if their uncertainties are not correlated). An objective function is written

$$J = (\mathcal{L}(\boldsymbol{\xi}) - \mathbf{q})^T \mathbf{Q}^{-1}(\mathcal{L}(\boldsymbol{\xi}) - \mathbf{q}) + (\boldsymbol{\xi} - \tilde{\boldsymbol{\xi}}(0))\mathbf{P}(0)^{-1}(\boldsymbol{\xi} - \tilde{\boldsymbol{\xi}}(0)), \quad (5.5.7)$$

whose minimum is sought. The presence of the weight matrices $\mathbf{Q}$, $\mathbf{P}(0)$ permits control of the elements most likely to change, which should not change at all [e.g., by introducing real zeros into $\mathbf{P}(0)$], as well as the stipulation of covariances. Tarantola and Valette (1982) labeled the use of similar objective functions and the determination of the minimum as the *method of total inversion* (although they considered only the case of perfect model constraints). We can regard it as a generalization of the process of minimizing objective functions, which led us to least squares in previous chapters.

In the context of the geostrophic inverse problem, we are interested in the solution of simultaneous equations of the form

$$\mathbf{E}\mathbf{x} + \mathbf{n} = \mathbf{y} \quad (5.5.8)$$

where errors, $\Delta\mathbf{E}$, exist in $\mathbf{E}$ as well as in $\mathbf{y}$, so that the actual problem is given by (5.5.1) where the $\Delta\mathbf{E}$ are correlated with the $\Delta\mathbf{y}$.

Consider an example for the two simultaneous equations

$$2x_1 + x_2 + n_1 = 1 \quad (5.5.9)$$

$$0 + 3x_2 + n_2 = 2 \quad (5.5.10)$$

where all the numerical values except the zero are now regarded as in error to some degree, corresponding to the errors in the hydrography used to calculate equations like (2.4.2). One way to proceed is to write the coefficients of $\mathbf{E}$ in the specific perturbation form (5.5.1). For example, we might write $E_{11} = 2 + \Delta E_{11}$, and define the unknowns $\boldsymbol{\xi}$ in terms of the ΔE_{ij}. Let us for illustration retain the full nonlinear form by setting

$$\xi_1 = E_{11},\ \xi_2 = E_{12},\ \xi_3 = E_{21},\ \xi_4 = E_{22},\ \xi_5 = x_1,\ \xi_6 = x_2,\ u_1 = n_1,\ u_2 = n_2\,.$$

The equations are then

$$\xi_1\xi_5 + \xi_2\xi_6 + u_1 - 1 = 0 \quad (5.5.11)$$

$$\xi_3\xi_5 + \xi_4\xi_6 + u_2 - 2 = 0\,. \quad (5.5.12)$$

The y_i are being treated as formally fixed, but the presence of u_1, u_2 actually represent their possible errors (the division into different elements of knowns and unknowns is not unique). Let there be an initial estimate,

$$\xi_1 = 2 \pm 1,\ \xi_2 = 2 \pm 2,\ \xi_3 = 0 \pm 0,\ \xi_4 = 3.5 \pm 1,\ \xi_5 = x_1 = 0 \pm 2,\ \xi_6 = 0 \pm 2,$$

with no imposed correlations so that $\mathbf{P}(0) = \mathrm{diag}[1 \;\; 4 \;\; 0 \;\; 1 \;\; 4 \;\; 4]$; the zero represents the requirement that E_{21} remain unchanged. Let $\mathbf{Q} = \mathrm{diag}[2 \;\; 2]$. Then a useful objective function is

$$J = (\xi_1\xi_5 + \xi_2\xi_6 - 1)^2/2 + (\xi_3\xi_5 + \xi_4\xi_6 - 2)^2/2 + (\xi_1 - 2)^2 + (\xi_2 - 2)^2/4 \\ + 10^6\xi_3^2 + (\xi_4 - 3.5)^2 + \xi_5^2/4 + \xi_6^2/4\,. \tag{5.5.13}$$

The 10^6 in front of the term in ξ_3^2 is a numerical approximation to the infinite value implied by a zero uncertainty in this term (an arbitrarily large value can cause numerical instability, characteristic of penalty and barrier methods; see Luenberger, 1984).

The finding of minima of expressions like (5.5.13) is usually known as *unconstrained optimization*, to distinguish it from *constrained optimization*, where the solution vector $\boldsymbol{\xi}$ may have to satisfy some relations exactly (hence, *constrained*) and can be imposed by Lagrange multipliers, just as we did in Chapter 3. Such objective functions define surfaces in spaces of the dimension of $\boldsymbol{\xi}$. Most procedures require the investigator to make a first guess at the solution, $\tilde{\xi}(0)$, and attempt to minimize J by going downhill from the guess. Various search algorithms have been developed over the years and are variants of steepest descent, conjugate gradient, Newton and quasi-Newton methods. The difficulties are numerous: The methods require computation or provision of the gradients of J with respect to $\boldsymbol{\xi}$, and the computational cost may become very great. The surfaces on which one is seeking to go downhill may become extremely tortuous, or very slowly changing. One can fall into local holes that are not the true minima. Nonetheless, the existing methods are very useful. The minimum of (5.5.13) corresponds to finding the solution of the nonlinear normal equations that would result from setting the partial derivatives to zero.

Let the true minimum be at $\boldsymbol{\xi}^*$. Assuming that the search procedure has succeeded, the objective function is locally

$$J = \text{constant} + (\boldsymbol{\xi} - \boldsymbol{\xi}^*)^T \mathcal{H}(\boldsymbol{\xi} - \boldsymbol{\xi}^*) + \Delta J \tag{5.5.14}$$

where $\mathcal{H}$ is the Hessian and ΔJ is a correction–assumed to be small. In the linear least-squares problem, (3.3.4), the Hessian is evidently $\mathbf{E}^T\mathbf{E}$, the second derivative of the objective function with respect to $\mathbf{x}$. The supposition is then that near the true optimum, the objective function is locally quadratic with a small correction. To the extent that this supposition is true, we can analyze the result in terms of the behavior of $\mathcal{H}$ as though it represented a locally defined version of $\mathbf{E}^T\mathbf{E}$. In particular, if $\mathcal{H}$ has a nullspace, or small eigenvalues, one can expect to see all the issues arising that we dealt with in Chapter 3, including ill-conditioning and solution variances that may become large in some elements. The machinery used in Chapter 3 (row and column scaling, nullspace suppression, etc.) thus becomes immediately relevant here and can be used to help conduct the search and to understand the solution.

It remains to find the minimum of (5.5.13). Tarantola and Valette (1982) suggested using a linearized search method, iterating from the initial estimate, which must be reasonably close to the correct answer. The method can be quite effective (e.g., Wunsch & Minster, 1982; Mercier, 1986). In a wider context, however, their method is readily recognizable as a special case of the many known methods for minimizing a general objective function. Most investigators are probably best advised to tackle problems such as finding the minimum of (5.5.13) by using one of the many general purpose numerical routines written by experts (Numerical Algorithms Group, 1988; Grace, 1990; Press et al., 1992; or equivalent) and discussed in detail in good textbooks (e.g., Luenberger, 1984; Gill, Murray, & Wright, 1981; Scales, 1985). Here, a quasi-Newton method (Grace, 1990) was employed to produce

$$E_{11} = 2.0001,\ E_{12} = 1.987,\ E_{21} = 0.0,$$
$$E_{22} = 3.5237,\ x_1 = -.0461,\ x_2 = 0.556$$

and the minimum of $J = 0.0802$. The inverse Hessian at the minimum is

$$\mathcal{H}^{-1} = \left\{\begin{matrix} 0.4990 & 0.0082 & -0.0000 & -0.0014 & 0.0061 & 0.0005 \\ 0.0082 & 1.9237 & 0.0000 & 0.0017 & -0.4611 & -0.0075 \\ -0.0000 & 0.0000 & 0.0000 & -0.0000 & -0.0000 & 0.0000 \\ -0.0014 & 0.0017 & -0.0000 & 0.4923 & 0.0623 & -0.0739 \\ 0.0061 & -0.4611 & -0.0000 & 0.0623 & 0.3582 & -0.0379 \\ 0.0005 & -0.0075 & 0.0000 & -0.0739 & -0.0379 & 0.0490 \end{matrix}\right\}.$$

The eigenvalues and eigenvectors of $\mathcal{H}$ are

$$\lambda_i = [2.075 \times 10^6 \quad 30.4899 \quad 4.5086 \quad 2.0033 \quad 1.9252 \quad 0.4859],$$

$$\mathbf{V} = \left\{\begin{matrix} 0.0000 & -0.0032 & 0.0288 & 0.9993 & 0.0213 & 0.0041 \\ -0.0000 & 0.0381 & -0.2504 & 0.0020 & 0.0683 & 0.9650 \\ -1.0000 & 0.0000 & 0.0000 & -0.0000 & -0.0000 & 0.0000 \\ 0.0000 & 0.1382 & 0.2459 & -0.0271 & 0.9590 & -0.0095 \\ -0.0000 & 0.1416 & -0.9295 & 0.0237 & 0.2160 & -0.2621 \\ 0.0000 & 0.9795 & 0.1095 & 0.0035 & -0.1691 & 0.0017 \end{matrix}\right\}.$$

The large jump from the first eigenvalue to the others is a reflection of the conditioning problem introduced by having one element, ξ_3, with almost zero uncertainty, characteristic of barrier methods. It is left to the reader to use this information about $\mathcal{H}$ to compute the uncertainty of the solution in the neighborhood of the optimal values–this would be the new uncertainty, $\mathbf{P}(1)$. A local resolution analysis follows from that of the SVD, employing knowledge of the $\mathbf{V}$, although Tarantola and Valette (1982) define *well-resolved parameters* as those for which there is a large reduction in the variance in $\mathbf{P}(0)$ to that in $\mathbf{P}(1)$, muddying the distinction between the variance and resolution. The particular system is too small for a proper statistical test of the result against the prior covariances, but the possibility should be clear. If $\mathbf{P}(0)$ etc., are simply regarded as nonstatistical weights, we are free to experiment with different values until a pleasing solution is found.

The geostrophic inverse problem in its many guises is thus a nonlinear regression or estimation problem for which a very large specialist literature exists (e.g., Seber & Wild, 1989). We have seen that linear estimation problems are ultimately reducible to the solution of an optimization problem. Nonlinear estimation problems are similarly reducible to nonlinear optimization problems, for whose solution a great deal is known. A discussion of such methods in any useful detail would make this book much longer than it already is, and we refer the reader to the references cited.

Wunsch (1994) applied these methods to a generalization of the β–spiral problem described in Chapter 4. As described there, the equations (4.7.2) were solved in simple overdetermined, least-squares form, leaving residuals in each layer of the water column. Suppose, however, that small disturbances to the observed density field could be found such that the residuals were reduced effectively to zero at all depths. The result would then be an estimate not only of the velocity field but of a density field that was completely consistent with it, in the sense that Equations (2.1.11) to (2.1.15) would be exactly satisfied.

To proceed, Wunsch (1994) wrote the β–spiral equation using the pressure field rather than isopycnal layer depths, and in spherical coordinates. Put

$$P \equiv \frac{p}{\rho_0} = -g \int_{z_0}^{z} \frac{\rho}{\rho_0} dz + P_0(\phi, \lambda) \tag{5.5.15}$$

where P_0 is an (unknown) reference pressure, and then equations (2.1.11)–(2.1.15) can be written as

$$\frac{\partial P}{\partial \phi}\left(\frac{\partial^3 P}{\partial z^3}\frac{\partial^2 P}{\partial \lambda \partial z} - \frac{\partial^2 P}{\partial z^2}\frac{\partial^3 P}{\partial \lambda \partial z^2}\right)$$
$$+ \frac{\partial P}{\partial \lambda}\left(\frac{\partial^2 P}{\partial z^2}\frac{\partial^3 P}{\partial \phi \partial z^2} - \frac{\partial^3 P}{\partial z^3}\frac{\partial^2 P}{\partial \phi \partial z} + \cot\phi\left(\frac{\partial^2 P}{\partial z^2}\right)^2\right)$$

$$= K2\Omega a^2 \sin\phi\cos\phi \left(\frac{\partial^2 P}{\partial z^2}\frac{\partial^4 P}{\partial z^4} - \left(\frac{\partial^3 P}{\partial z^3}\right)^2 \right) \tag{5.5.16}$$

where ϕ is latitude, λ is longitude, and K is a vertical-eddy coefficient, assumed to be constant in the vertical in deriving (5.5.16). Equation (5.5.16) was derived by Needler (1967) and is sometimes known as the *P-equation*; with $K = 0$, this was discussed in Chapter 4 as a variant of the β–spiral equation.

Substitution of an observed density field and a plausible initial reference pressure, $\tilde{P}_0(0)$, into (5.5.16) results in large values for K, which is here regarded primarily as a residual rather than as a manifestation of real mixing. Choosing P_0 is the same as having to choose an initial reference level in conventional dynamic computations. The problem was converted to an optimization one by asserting that K appears finite because the initial reference-level velocity of zero is not correct, and because there are also errors in the observed density field, called $\tilde{\rho}(0)$.

Both $\tilde{P}_0$ and $\tilde{\rho}$ were expanded in a three-dimensional set of Chebyshev polynomials, whose coefficients a_{nm}, α_{nml} respectively, become the parameters $\mathbf{x}$. The objective function, J, was formed by computing the sum of squares of K evaluated on a three-dimensional grid:

$$J = \sum_{ijk}(\tilde{K}_{ijk}\sin\phi_{ij}\cos\phi_{ij})^2 + R_r^{-2}\sum_{ijk}(\tilde{\rho}_{ijk}(1) - \tilde{\rho}_{ijk}(0))^2 . \tag{5.5.17}$$

The indices ijk refer to the horizontal and vertical grid positions where the expression (5.5.16) is evaluated after substitution of the Chebyshev expansion. The terms in $\hat{\rho}$ penalize deviations from the initial estimates with a weight given by R_r. The one in the argument denotes the modified estimate.

If the coefficients of $\tilde{\rho}$ are held fixed at their initial values as determined from the observations, one is optimizing only the reference pressure. The normal equations in that case are linear, and one is simply solving the ordinary β–spiral problem. The search for an improved density/pressure estimate was carried out using a so-called Levenburg-Marquardt method (e.g., Gill et al., 1981), and a minimum was found (shown in cross-section of two dimensions in Figure 5–4). The uncertainty of the final solution was then estimated using the Hessian evaluated in its vicinity. The final residuals are not zero but are sufficiently small that the solution was deemed acceptable. Figure 5–5 shows a before-and-after optimization estimate of the density field at one depth. Because slight changes in the prior density field are capable of producing a solution that implies conservation of potential vorticity, density, etc., within uncertainty limits, the result suggests that

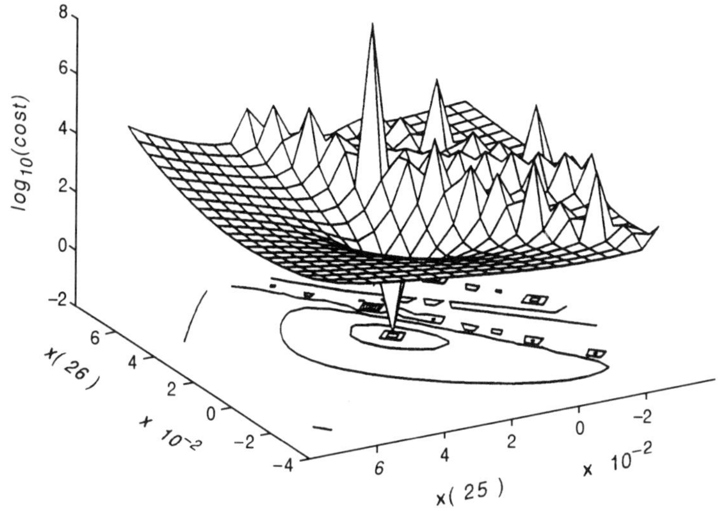

Figure 5–4. Shape of the objective function in a projection onto two dimensions of the nonlinear objective function of Wunsch (1994). Apparent minimum is found in the sharp hole. Note the complex surrounding terrain.

estimates of vertical mixing coefficients in the ocean differing from zero may well be nothing but artifacts of sampling errors. Davis (1994) draws a similar conclusion from entirely different reasoning.

5.5.3 Variant Nonlinear Methods, Including Combinatorial Ones

As with the linear least-squares problems discussed in Chapter 3, many possibilities exist for objective functions that are nonlinear in either data constraint terms or the model, and there are many variations on methods for searching for objective function minima.

The major variant on objective functions such as (5.5.7) is to introduce some of the constraints as hard ones using Lagrange multipliers. The partial derivatives of the objective function produce Equations (3.5.5)–(3.5.6) except that the model is nonlinear. But as with the linear methods, the use of residual terms in the hard constraints renders moot the distinction between hard and soft constraints. Barrier and penalty methods are techniques for converting such constrained minimization problems into unconstrained ones–for which the search methods tend to be somewhat simpler. But a few attempts have been made to use the Lagrange multiplier method directly for nonlinear, steady, ocean circulation inversions.

Schlitzer (1993) has produced an estimate of the circulation of the North and South Atlantic Oceans using a formal finite-difference model somewhat like the linear one of Martel and Wunsch (1993a) but invoking nonlinear advection of temperature and salt (analogous to the nonlinear constraints

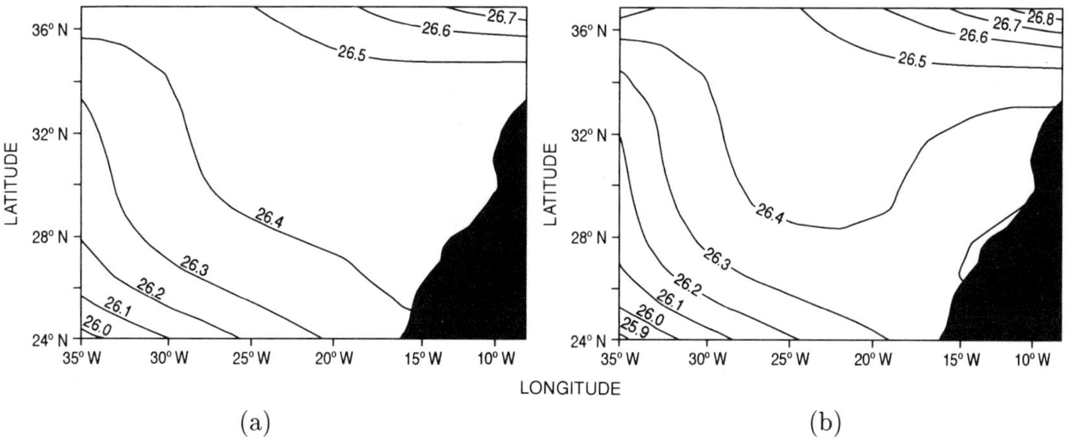

Figure 5–5. (a) Density field, σ_θ, at 100 meters prior to optimization described in the text, and (b) the density field following the optimization.

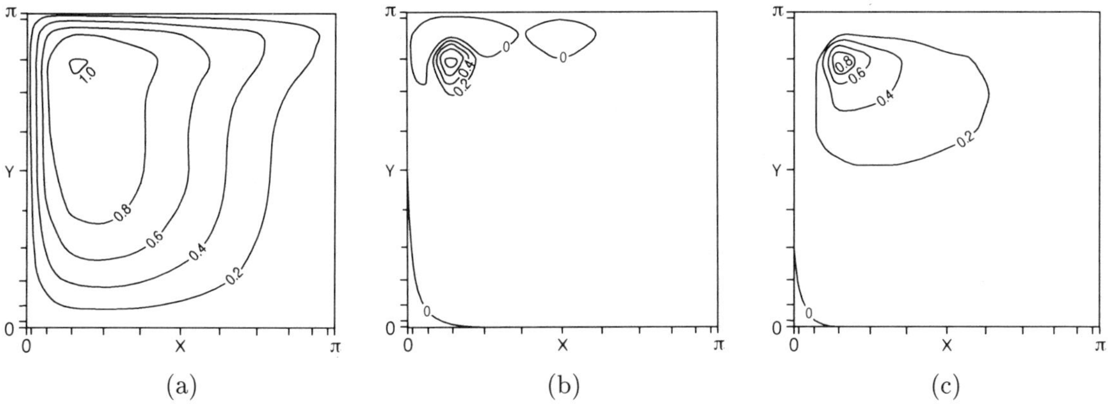

Figure 5–6. (a) Upper layer stream function computed by Schröter and Wunsch (1986) for a nonlinear baroclinic model. (b) The Lagrange multiplier solution (adjoint solution) corresponding to an objective function involving the model in (a), and the maximization of the western boundary current transport. (c) Lagrange multipliers for the model when the bottom friction coefficient was permitted to vary.

of Mercier et al., 1993, who used density). The model had a very coarse spatial resolution, which raises the usual issues concerning the ability to compute property fluxes realistically. The constraints involve mass, salt, oxygen, and nutrient balances. Schlitzer produces estimates of the flow field and boundary conditions resulting from his optimization, which is achieved by solving the normal equations through an iterative scheme. Schlitzer was able to force his model to have the high net meridional flux of heat roughly consistent with what is believed required, but his result is difficult to interpret, because his model does not require geostrophy to be satisfied accurately.

In another example, using another form of optimization, combining the nonlinear objective function with inequality constraints, Schröter and Wunsch (1986) discussed a steady, nonlinear, baroclinic GCM subject to imperfect observations of various elements of the model. They converted the constrained minimum problem into one of unconstrained minimization by adding the model to the cost function with a large weight and sought the minimum through a barrier method (e.g., see Luenberger, 1984). They did not use real data but demonstrated the possibility of constraining a nonlinear GCM with a variety of data types, using the Lagrange multipliers to understand the sensitivity of their objective function to the data uncertainties. The data uncertainties were posed by using inequality constraints on the wind-curl, model temperatures, etc. Figure 5–6a shows the upper-layer stream function in a two-layer version of their model, and Figure 5–6b is the Lagrange multipliers that result when the objective function sought to maximize the western boundary current transport in the presence of uncertainty in the wind-curl. The interpretation is that the transport is most sensitive to the wind-curl uncertainties in places where the Lagrange multipliers are largest. Schröter and Wunsch (1986) extended the method to include an uncertainty in a model parameter–for example, the bottom friction coefficient. Figure 5–6c shows the Lagrange multipliers indicating maximum sensitivity of the western boundary current transport to uncertainties in the coefficients. The generality of the methodologies is such that it is easy to extend the methods–the main complications being in the nonlinearity of the resulting equations.

By now it will be obvious that all of the inverse problems we have discussed are ultimately reduced to finding the minimum of an objective function, either in unconstrained or constrained form. Once the model is formulated, the objective function agreed on, and the data obtained in appropriate form (often the most difficult step), the formal solution is reduced to finding the constrained or unconstrained minimum.

When the problem involves strong nonlinearities, search methods seeking to go downhill can become lost in complicated high-dimensional spaces. In particular, they can be trapped in localized minima that do not represent either an acceptable solution or the best possible one. Figure 5–4 is an example from Wunsch (1994) that shows the complex "terrain" in the vicinity of what appeared to be a true minimum.

As with any mathematical subject dealing with nonlinearities, there are no fully general, guaranteed methods that prevent difficulties. But a very interesting and useful set of methods has been developed comparatively recently, called *combinatorial optimization*, which may well become oceanographically important. Combinatorial methods do not promise that the true minimum is found–merely that it is highly probable–because they search the space of solutions in clever ways which make it unlikely that one is very far from the true optimal solution. Two such methods, simulated annealing and genetic algorithms, have recently attracted considerable attention. For simulated annealing, the literature starts with Pincus (1970) and Kirkpatrick, Gelatt, and Vecchi (1983), and general discussions can be found in van Laarhoven and Aarts (1987), Ripley (1981), and Press et al. (1992). A simple oceanographic application to experiment design was discussed by Barth and Wunsch (1989). Simulated annealing searches randomly for solutions that reduce the objective function from a present best value. Its clever addition to purely random guessing is a willingness to accept the occasional uphill solution–one that actually raises the value of the objective function–as a way of avoiding being trapped in purely local minima. The probability of accepting an uphill value and the size of the tried random perturbations depend upon a parameter, a temperature defined in analogy to the real temperature of a slowly cooling (annealing) solid.

Genetic algorithms, as their name would suggest, are based upon searches generated in analogy to genetic drift in biological organisms. Goldberg (1989), Holland (1992), and Denning (1992) discuss the method, and Barth (1992) describes an oceanographic application and compares it to simulated annealing. Other comparisons of simulated annealing and genetic algorithms are provided by Scales, Smith, and Fisher (1992).

6

The Time-Dependent Inverse Problem

The discussion so far has treated the ocean circulation as though it were steady–apart from concerns about aliasing from components regarded mainly as noise. But one does not really expect a fluid system as complex as the coupled ocean and atmosphere, driven externally by the sun, to remain steady. Indeed, if there were certain elements, describable as a finite band of frequencies and/or wavenumbers, which were truly steady, it would be both a remarkable phenomenon and a very powerful theoretical tool. For this would be a statement that a spectral gap existed, permitting the separation of the flow field into different dynamical regimes, as sometimes assumed in turbulence theories (e.g., Monin & Yaglom, 1975).

Treatment of the ocean circulation as though it were time invariant carries the weight of history and tradition, and there is no doubt that a great deal has been learned about the ocean this way. But as oceanographers become more quantitative in their description of the circulation, it must be recognized that the steadiness assumption has been a consequence of necessity dictated by the paucity of data and intellectual inertia rather than a defensible deduction from fluid dynamics. One must question whether a steady, laminar flow field (as in Figures 2–7, 2–29) could conceivably depict the physics of a fluid that is turbulent in the sense suggested by Figure 1–1.[1]

Static treatment of the observations can be defended rigorously in two circumstances: (1) The data are truly synoptic, and one is literally constructing a snapshot of the circulation; or (2) it is known that the time-dependent elements are all noise with neither kinematical nor dynamical consequences

[1] One can describe the movement of fluid properties, such as a dye concentration in a turbulent pipe flow under an imposed pressure gradient, by adjusting a laminar flow law until the observed property distribution or flux is roughly captured. But the rule would have little skill in predicting what would happen should any parameter be changed. Thus, such pretty ideas as global conveyor belts are useful summary devices for unsophisticated audiences–fairy tales for adults. But they are grossly misleading as pictures of how the fluid moves and how it carries properties with it.

for any fluid property (such as a tracer or momentum distribution) on larger time and space scales. Unfortunately, the only way known to demonstrate the validity of (2) is by sampling densely enough to treat the data as in (1), permitting a demonstration that a large-scale mean, or other picture, is independent of the time-dependent elements.

Even the earliest observers (e.g., Maury, 1855; Helland-Hansen & Nansen, 1920) were quite aware of the time dependence of the system. But the working assumption of near-time independence was so central to progress in describing the circulation, and its removal so detrimental to any data interpretation, that the assumption tended to turn into an unexpressed axiom taken for granted. Thus, there is a large recent literature (e.g., Dickson, Meincke, Malmberg, & Lee, 1988) whose tone is one of surprise that some aspect of the circulation is seen to change (in their case, the salinity at high latitudes). But what would be truly surprising would be to find that some component did not change at all!

It is clear that some circulation elements are quasi-permanent over years to decades and longer, although it is not so easy to characterize exactly what this means. For example, the Gulf Stream seems to have existed as an entity for hundreds and probably for many thousands of years at least, although its mass and property transports, its positions, and many other of its characteristics have shifted (the last glacial cycle is an obvious example of a period of change).[2] The instrumental record is about 100 years long; it does show that the North Atlantic thermocline as observed by the Challenger in 1972 is qualitatively much like that of the 1990s (a pictorial comparison may be seen in Wunsch, 1981). How does one usefully characterize the effectively stationary elements over some fixed interval of time and distinguish them from the more variable elements? By useful is meant a quantitative measure and a dynamical model. In the work so far (Chapters 4 and 5), the methodologies might be characterized as addressing the question, "to what extent are the data quantitatively consistent with a steady dynamical model, and what are the implications of such a flow field?"

Known time dependences are of various types. Consider the so-called transient tracers. Here, even in the presence of a completely steady flow, the

[2] Part of the difficulty in describing steady components is that the usual vocabulary for depicting space/time variability, in terms of frequencies and wavenumbers, is inadequate for the description of the mean properties of a meandering jet. No really useful vocabulary or tool seems to exist. The difficulty can be traced to the inability to construct dynamical equations governing the behavior of a fluid in distinct frequency and/ or wavenumber bands. Even dynamically linear flows are scale-coupled kinematically through equations such as that for mass conservation.

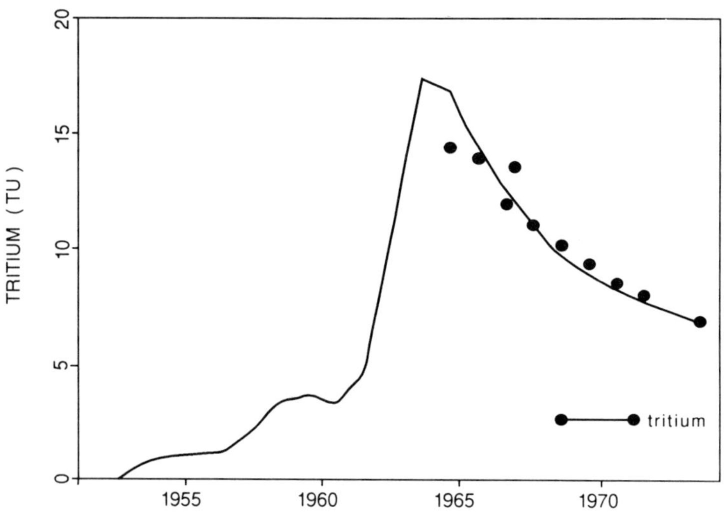

Figure 6–1. Estimate (after Dreisigacker & Roether, 1978) of the surface concentration of tritium in the North Atlantic between 20°and 60°N. These values are typical of time-dependent forcing (here for the tritium tracer concentration), which lead to time-dependent modeling, even should the flow field be steady. (Compare to the interior concentration in Figure 2–6b,c.) Points are from observations; the solid line is based upon inference from atmospheric values.

data distributions are complex functions of space and time. Figure 6–1, from Dreisigacker and Roether (1978), shows an estimate of the history of ocean surface concentrations of tritium. Tritium, an unstable isotope of hydrogen (written ^3H), decays with a half-life of 12.5 years into a stable isotope of helium (helium–3, written ^3He). Although tritium does not normally occur in measurable amounts in seawater, atmospheric testing of nuclear weapons beginning in 1945 showed a rise in atmospheric concentrations of ^3H, which reached a crescendo in the early 1960s, as seen in the figure. With the advent of the Partial Test Ban Treaty, the atmospheric concentrations diminished. The fallout products, including ^3H, entered the ocean by rainfall and gas exchanges (for details, see Weiss & Roether, 1980), and it was recognized that observations of the behavior of this decaying dye, along with its daughter product ^3He, might show in concrete fashion how the ocean circulation behaves.

Figure 2–6b,c depicts estimates of tritium from observations in the western North Atlantic in 1972 and 1981 (Östlund & Fine, 1979; and Östlund & Rooth, 1990) resulting from boundary conditions like those shown in Figure 6–1. Such data are obviously time dependent; indeed, it is the time dependence that renders them interesting. The need to learn how to use such data, beyond the mere drawing of pictures, has motivated many of the developments outlined below. The use of time-dependent distributions in the presence of a time-invariant flow field presents some very interesting problems; when the flow field is itself time varying, the problems of inferring the circulation represent some of the most challenging ones, mathematically and practically, that face oceanographers today. A central difficulty is that when there are few data directly constraining the flow field, one can often find a steady flow consistent with an observed tracer distribution, even though the real flow field is completely unsteady, perhaps even with zero-mean mass flux. The potential is very great for gravely distorted inferences.

Many of the methods outlined in this chapter have only recently found oceanographic application. Although only just coming into the mainstream of the science, I strongly believe that they are worth understanding–the future of oceanography surely lies in the quantitative combination of increasingly sophisticated, time-dependent, three-dimensional dynamical models with ever more massive observational data sets.

This chapter is in no sense exhaustive; many complete books are devoted to the material and its extensions, which are important for understanding and practical use. The intention is to lay out the fundamental ideas, which are generalizations of methods already described in Chapters 3 and 5, with the hope that they will permit the reader to penetrate the wider literature. Several very useful textbooks are available for readers who are not deterred by discussions in contexts differing from oceanography. I have found those by Liebelt (1967), Gelb (1974), Bryson and Ho (1975), Brown (1983), and Anderson and Moore (1979) to be especially helpful.

In meteorology, estimation for forecasting is called *assimilation*, and Daley (1991) has a good account. Oceanographic adaptation of meteorological methods is described in Section 6.7 as a special case of time-dependent inverse problems. For convenience, the methods of Chapters 3 and 5 sometimes will be referred to as *static* because time is not explicitly involved. As will be seen, the methods discussed below generally reduce to the earlier forms in various limits.

6.1 Some Basic Ideas and Notation

6.1.1 Models

Models are statements about the connections between flow motions in some place at some time, and those in all other places and times. The connections may be strong and immediate, or weak and distant. For the steady models considered in Chapters 4 and 5, information about boundary conditions could, for example, be used to infer the interior state of the fluid. We also saw how to find boundary conditions from information about the interior state, thus using the model, a statement of our knowledge, to estimate quantities we could not otherwise measure.

A time-dependent model is, completely analogously, a statement of information–what we know about oceanic physics (or chemistry or biology). If it is to be used quantitatively, formulation of a model should carry with it a statement of confidence (the expected accuracy) as well. The presence of time as a coordinate means that the model can be used to carry information not only from one spatial region to another but both backward and forward in time, using knowledge of the ocean at one time, to infer what it must have been doing earlier or later. In this context, the inverse problem is simply defined as that of estimating the state of the ocean and its governing physics, chemistry, etc., using all available data and applicable dynamics.

Consider any model, whether time dependent or steady. The state vector $\mathbf{x}$ is defined as those elements of the model employed to describe fully the physical state of the system at any time and all places. For the discrete Laplace/Poisson equation (1.2.4), $\boldsymbol{\phi}$ is the state vector. In a full oceanic general circulation model, the state vector might consist of three components of velocity and density at each of millions of grid points, and it will be a function of time, $\mathbf{x}(t)$, as well. Choice of the state vector is not unique. Consider the continuous formulation for linear, barotropic waves,

$$
\begin{aligned}
\frac{\partial u}{\partial t} - fv &= -\frac{\partial p}{\partial x} + \frac{\tau_x}{h} \\
\frac{\partial v}{\partial t} + fu &= -\frac{\partial p}{\partial y} + \frac{\tau_y}{h} \\
0 &= -\frac{\partial p}{\partial z} - g \\
\frac{\partial u}{\partial x} + \frac{\partial v}{\partial y} + \frac{\partial w}{\partial z} &= 0
\end{aligned}
\tag{6.1.1}
$$

where τ_x, τ_y are the windstress components and h is the constant depth. The remaining notation is all conventional, as used in Chapter 2. The coordinate, x, should not be confused with the state vector $\mathbf{x}$. One can reduce this system to a single partial differential equation in p, v, u, or w, the choice being at the investigator's convenience (on a β–plane, the equation in v is simplest but may be inconvenient, depending upon the boundary conditions; on an f–plane, all equations are the same). The arbitrariness in the choice of the dependent variables, be it all of $(u, v, p, \ldots)$ or a single partial differential equation in $u, v, \ldots$, corresponds to different choices for $\mathbf{x}$.

If we make the approximation of a rigid upper boundary on the ocean,

Equations (6.1.1) lead to an equation in the pressure,

$$\frac{\partial}{\partial t}(\nabla_h^2 p) + \beta \frac{\partial p}{\partial x} = \hat{\mathbf{k}} \cdot \nabla \times \boldsymbol{\tau}\,. \tag{6.1.2}$$

Here, p is the *state function.* But if it is solved by an expansion

$$p(x,\, y,\, t) = \sum_j a_{2j}(t) \cos(\mathbf{k}_j \cdot \mathbf{r}) + a_{2j-1}(t) \sin(\mathbf{k}_j \cdot \mathbf{r}) \tag{6.1.3}$$

$[\mathbf{k}_j = (k_x,\, k_y),\ \mathbf{r} = (x,\, y)]$, then $\mathbf{a}(t) = [a_1(t)\ a_2(t) \cdots a_{2j-1}(t) \ldots]^T$ becomes the state vector (for an example, see Gaspar & Wunsch, 1989).

In the most general terms, we can write any *discrete* model as a set of functional relations

$$\begin{aligned} &\mathcal{L}(\mathbf{x}(0), \ldots, \mathbf{x}(t-\Delta t),\, \mathbf{x}(t),\, \mathbf{x}(t+\Delta t), \ldots \mathbf{x}(t_f) \ldots,\, \mathbf{B}(t)\mathbf{q}(t)\,, \\ &\quad \mathbf{B}(t)\mathbf{q}(t+\Delta t), \ldots, t) = 0 \end{aligned} \tag{6.1.4}$$

where $\mathbf{B}(t)\mathbf{q}(t)$ represents a general form for boundary and initial conditions/sources/sinks.[3] We almost always choose the time units so that $\Delta t = 1$. The static system equation

$$\mathbf{A}\mathbf{x} = \mathbf{q} \tag{6.1.5}$$

is a special case. In practice, the collection of relationships (6.1.4) always can be rewritten as a time-stepping rule–for example

$$\mathbf{x}(t+1) = \mathbf{L}(\mathbf{x}(t),\, \mathbf{B}(t)\mathbf{q}(t),\, t) \tag{6.1.6}$$

or, if the model is linear,

$$\mathbf{x}(t+1) = \mathbf{A}(t)\, \mathbf{x}(t) + \mathbf{B}(t)\, \mathbf{q}(t)\,. \tag{6.1.7}$$

If the model is time invariant, $\mathbf{A}(t) = \mathbf{A}$. These relationships have complete analogues in the continuous-time case; for example, (6.1.7) would be

$$\frac{d\mathbf{x}(t)}{dt} = \mathbf{A}_1(t)\, \mathbf{x}(t) + \mathbf{B}(t)\, \mathbf{q}(t)\,. \tag{6.1.8}$$

A time-dependent model is a set of rules for computing the state vector at time $t+1$ from knowledge of its values at time t and the externally imposed forces and boundary conditions. It is generally true that any linear discretized model can be put into this canonical form, although it may take some work. By the same historical conventions described in Chapter 1, solution of systems like (6.1.6), subject to appropriate initial and boundary conditions, constitutes the forward or direct problem.

[3] Construction of computer codes for the fluid dynamical equations is a large and complex subject in its own right. Readers interested in the general problem of oceanic numerical models can consult textbooks such as O'Brien (1986).

Example: The Straight Line. The straight-line model, discussed in Chapter 3, can be put into the time-evolving framework. A straight line, $\xi(t)$, satisfies the rule

$$\frac{d^2\xi}{dt^2} = 0\,, \tag{6.1.9}$$

which can be discretized as

$$\xi(t + \Delta t) - 2\xi(t) + \xi(t - \Delta t) = 0\,. \tag{6.1.10}$$

Putting

$$x_1(t) = \xi(t)\,, \quad x_2(t) = \xi(t - \Delta t)\,,$$

one has

$$\mathbf{x}(t + \Delta t) = \mathbf{A}\mathbf{x}(t)$$

where

$$\mathbf{A} = \begin{Bmatrix} 2 & -1 \\ 1 & 0 \end{Bmatrix}\,,$$

which is of the standard form (6.1.7), if $\Delta t = 1$, $\mathbf{B} = \mathbf{0}$.

Example: The Mass-Spring Oscillator. The elementary mass-spring oscillator satisfies the differential equation

$$m\frac{d^2\xi(t)}{dt^2} + r\frac{d\xi(t)}{dt} + k\xi(t) = q(t)$$

where r is a damping constant. A simple one-sided time discretization produces

$$\begin{aligned} m(\xi(t + \Delta t) - 2\xi(t) + \xi(t - \Delta t)) + r\Delta t(\xi(t) - \xi(t - \Delta T)) \\ + k(\Delta t)^2\,\xi(t) = q(t)\,(\Delta t)^2 \end{aligned}$$

or

$$\begin{aligned} \xi(t + \Delta t) = {} & \left(2 - \frac{r\Delta t}{m} - \frac{k(\Delta t)^2}{m}\right)\xi(t) \\ & + \left(\frac{r\Delta t}{m} - 1\right)\xi(t - \Delta t) + (\Delta t)^2\,\frac{q(t)}{m}\,, \end{aligned}$$

which is

$$\begin{bmatrix} \xi(t + \Delta t) \\ \xi(t) \end{bmatrix} = \begin{Bmatrix} 2 - \frac{r}{m}\Delta t - \frac{k}{m}(\Delta t)^2 & \frac{r\Delta t}{m} - 1 \\ 1 & 0 \end{Bmatrix} \begin{bmatrix} \xi(t) \\ \xi(t - \Delta t) \end{bmatrix}$$

$$+\begin{bmatrix}(\Delta t)^2 \frac{q(t)}{m}(t) \\ 0\end{bmatrix} \tag{6.1.11}$$

and is the canonical form with $\mathbf{A}$ independent of time,

$$\mathbf{x}(t) = [\xi(t) \quad \xi(t-\Delta t)]^T, \quad \mathbf{B}(t)\mathbf{q}(t) = [(\Delta t)^2\, q(t)/m \quad 0]^T .$$

Example: The Barotropic Rossby Wave Equation. Discretizing the unforced Equation (6.1.2) and setting $\Delta t = 1$,

$$\nabla_d^2 \frac{\left(p_{i,j}^{t+1} - p_{ij}^{t-1}\right)}{2} + \frac{\beta}{\Delta x}\left(p_{i+1,j}^t - p_{i-1,j}^t\right) = 0 \tag{6.1.12}$$

where ∇_d^2 is the matrix forming numerical second derivatives,

$$\nabla_d^2 = \begin{Bmatrix} \cdot & \cdot & \cdot & \cdot & \cdot & \cdot & \cdot & \cdot & \cdot & \cdot & \cdot & \cdot & \cdot \\ 0 & 0 & \frac{1}{(\Delta x)^2} & -4\left(\frac{1}{(\Delta x)^2} + \frac{1}{(\Delta y)^2}\right) & \frac{1}{(\Delta x)^2} & 0 & 0 & \frac{1}{(\Delta y)^2} & 0 & \frac{1}{(\Delta y)^2} & \cdot & 0 \\ \cdot & \cdot & \cdot & \cdot & \cdot & \cdot & \cdot & \cdot & \cdot & \cdot & \cdot & \cdot & \cdot \end{Bmatrix} \tag{6.1.13}$$

and a centered time difference has been employed.

Then

$$p_{i,j}^{t+1} = p_{i,j}^{t-1} - \frac{2\beta}{\Delta x}\nabla_d^{-2}(p_{i+1,j}^t - p_{i-1,j}^t) \tag{6.1.14}$$

where ∇_d^{-2} is the inverse of the matrix in (6.1.13). Equation (6.1.14) can be written as

$$\begin{aligned} \mathbf{p}_1(t+1) &= \mathbf{A}_1\mathbf{p}_1(t) + \mathbf{p}(t-1), \\ \mathbf{p}_1(t) &= [\mathbf{p}(t) \quad \mathbf{p}(t-1)]^T, \end{aligned} \tag{6.1.15}$$

which is not yet quite the desired form. Put

$$\begin{aligned} \mathbf{x}(t) &= \begin{bmatrix}\mathbf{p}_1(t) \\ \mathbf{p}_1(t-1)\end{bmatrix} \\ \mathbf{A} &= \begin{Bmatrix}\mathbf{A}_1 & \mathbf{I} \\ \mathbf{I} & \mathbf{0}\end{Bmatrix}. \end{aligned} \tag{6.1.16}$$

Then (6.1.15) becomes

$$\mathbf{x}(t+1) = \mathbf{A}\mathbf{x}(t). \tag{6.1.17}$$

To obtain the canonical form, we have used a state vector that involves three time steps; in some circumstances, even more may be required.

Example: A Linear Difference Equation. A difference equation important in time-series analysis (e.g. Box & Jenkins, 1978) is

$$\xi(t) + a_1\xi(t-1) + a_2\xi(t-2) + \cdots + a_N\xi(t-N) = \eta(t) \qquad (6.1.18)$$

where $\eta(t)$ is a zero-mean, white-noise process [Equation (6.1.18) is an example of an autoregressive process (AR)]. To put this into the canonical form, write (e.g., Luenberger, 1979)

$$\begin{aligned} x_1(t) &= \xi(t-N) \\ x_2(t) &= \xi(t-N+1) \\ &\vdots \\ x_N(t) &= \xi(t-1) \\ x_N(t+1) &= -a_1\, x_{N-1}(t) - a_2\, x_{N-2}(t) \cdots - a_N\, x_1(t) + \eta(t)\,. \end{aligned} \qquad (6.1.19)$$

It follows that $x_1(t+1) = x_2(t)$, etc., or

$$\mathbf{x}(t+1) = \left\{ \begin{matrix} 0 & 1 & 0 & \cdots & 0 & 0 \\ 0 & 0 & 1 & \cdots & 0 & 0 \\ \cdot & \cdot & \cdot & \cdots & \cdot & \cdot \\ -a_N & -a_{N-1} & -a_{N-2} & \cdots & -a_2 & -a_1 \end{matrix} \right\} \mathbf{x}(t) + \begin{bmatrix} 0 \\ 0 \\ \cdot \\ 1 \end{bmatrix} \eta(t)\,. \qquad (6.1.20)$$

Coefficients of the form in (6.1.20) are known as *companion matrices*; Equation (6.1.20) connects this chapter with time-series analysis. Here, $\mathbf{B}(t) = [0 \;\; 0 \;\; \cdot \;\; 1]^T$, $\mathbf{q}(t) = \eta(t)$.

Given that most time-dependent models can be written as in (6.1.6) or (6.1.7), the forward model solution involves marching forward from known initial conditions at $t = 0$, subject to specified boundary values. So, for example, the linear model (6.1.7), with given initial conditions $\mathbf{x}(0) = \mathbf{x}_0$, involves the sequence

$$\begin{aligned} \mathbf{x}(1) &= \mathbf{A}(0)\,\mathbf{x}_0 + \mathbf{B}(0)\,\mathbf{q}(0) \\ \mathbf{x}(2) &= \mathbf{A}(1)\,\mathbf{x}(1) + \mathbf{B}(1)\,\mathbf{q}(1) \\ &= \mathbf{A}(1)\,\mathbf{A}(0)\,\mathbf{x}_0 + \mathbf{A}(1)\,\mathbf{B}(0)\,\mathbf{q}(0) + \mathbf{B}(1)\,\mathbf{q}(1) \\ &\vdots \\ \mathbf{x}(t_f) &= \mathbf{A}(t_f-1)\,\mathbf{x}(t_f-1) + \mathbf{B}(t_f-1)\,\mathbf{q}(t_f-1) \\ &= \mathbf{A}(t_f-1)\,\mathbf{A}(t_f-2)\ldots\mathbf{A}(0)\,\mathbf{x}_0 + \ldots. \end{aligned} \qquad (6.1.21)$$

Most of the basic ideas can be understood in the notationally simplest case of time-independent **A**, **B**, and that is usually the situation we will address with little loss of generality. Figure 6–3a depicts the time history for the harmonic oscillator, with the parameter choice $\Delta t = 1$, $k = 0.25$, $m = 1$, $r = 0$, so that

$$\mathbf{A} = \left\{ \begin{matrix} 1.75 & -1 \\ 1 & 0 \end{matrix} \right\}, \quad \mathbf{Bq}(t) = \begin{bmatrix} 1 \\ 0 \end{bmatrix} u(t)$$

where $< u(t)^2 > = 1$. The initial conditions were $\mathbf{x}(0) = [\xi(0)\, \xi(-1)]^T$. The general $t^{1/2}$ growth of the displacement of a randomly forced, undamped oscillator is apparent.

If the model is a "black box" in which $\mathbf{A}$ is unknown, it can be found in a number of ways. Suppose that $\mathbf{Bq}(t) = \mathbf{0}$, and Equation (6.1.7) is solved N times, starting at time t_0, subject to $\mathbf{x}(i)(t_0) =$ column i of $\mathbf{I}$–that is, the model is stepped forward for N different initial conditions corresponding to the N-different problems of unit initial condition at a single grid or boundary point, with zero initial conditions everywhere else. Let each column of $\mathbf{G}(t, t_0)$ correspond to the appropriate value of $\mathbf{x}(t)$–that is,

$$\begin{aligned}
\mathbf{G}(t_0, t_0) &= \mathbf{I} \\
\mathbf{G}(t_0 + 1, t_0) &= \mathbf{A}(t_0)\mathbf{G}(t_0, t_0) \\
\mathbf{G}(t_0 + 2, t_0) &= \mathbf{A}(t_0 + 1)\mathbf{G}(t_0 + 1, t_0) = \mathbf{A}(t_0 + 1)\mathbf{A}(t_0) \\
&\vdots \\
\mathbf{G}(t_0 + t, t_0) &= \mathbf{A}(t_0 + t - 1)\mathbf{A}(t_0 + t - 2) \cdots \mathbf{A}(t_0) .
\end{aligned} \tag{6.1.22}$$

We refer to $\mathbf{G}(t, t_0)$ as a *Green's function.* The solution for arbitrary initial conditions is then

$$\mathbf{x}(t) = \mathbf{G}(t, t_0)\mathbf{x}(t_0) , \tag{6.1.23}$$

and the modification for $\mathbf{Bq} \neq 0$ is straightforward.

We make the somewhat obvious point that $\mathbf{A}$ is necessarily square. It is also commonly true that $\mathbf{A}^{-1}$ exists. If $\mathbf{A}^{-1}$ can be computed, one can contemplate the possibility (important later) of running a model backward in time, for example, as

$$\mathbf{x}(t) = \mathbf{A}^{-1}\mathbf{x}(t + 1) - \mathbf{A}^{-1}\mathbf{B}(t)\, \mathbf{q}(t) .$$

Such a computation may be inaccurate if carried on for long times, but its possibility in principle must be noted.

Some attention must be paid to the structure of $\mathbf{B}(t)\, \mathbf{q}(t)$. The partitioning into these elements is not unique and can be done to suit one's convenience. The dimension of $\mathbf{B}$ is that of the size of the state vector by the dimension of $\mathbf{q}$, which typically would reflect the number of independent degrees of freedom in the forcing/boundary conditions. (*Forcing* is hereafter

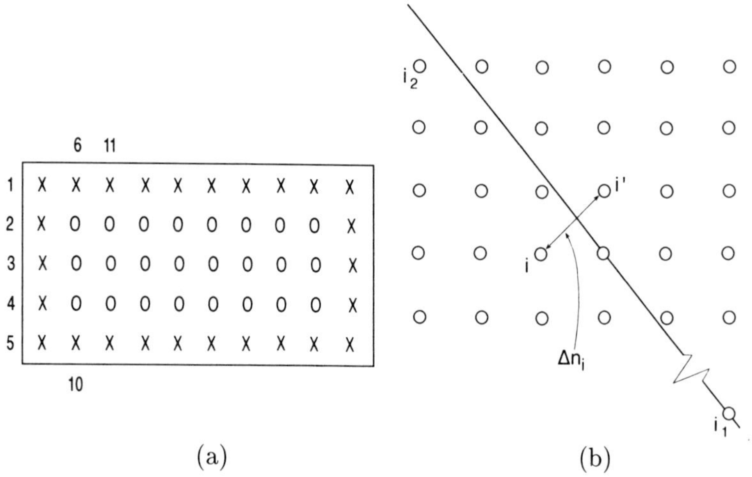

Figure 6–2. (a) Simple numerical grid for use of discrete form of model: x denote boundary grid point, and o are interior ones. Numbering is sequentially down the columns, as shown. (b) Tomographic velocity integral is assumed given between i_1 and i_2. Such a measurement is an average of the pressure gradient between grid points perpendicular to the line, approximated by the difference between pairs i and i'.

used to include boundary conditions, sources and sinks, and anything normally prescribed externally to the model.) Consider the model grid points displayed in Figure 6–2a. Suppose that the boundary grid points are numbered 1–5, 6, 10, 46–50, and all others are interior. If there are no interior forces, and all boundary values have a time history $q(t)$, then we could take

$$\mathbf{B} = [1 \quad 1 \quad 1 \quad 1 \quad 1 \quad 1 \quad 0 \quad 0 \quad 0 \cdots 1 \quad 1]^T . \tag{6.1.24}$$

Suppose, instead, that boundary grid point 2 has values $q_1(t)$, all other boundary conditions are zero, and interior point 7 has a forcing history $q_2(t)$ and all others are unforced; then

$$\mathbf{Bq}(t) = \begin{Bmatrix} 0 & 0 \\ 1 & 0 \\ 0 & 0 \\ 0 & 0 \\ 0 & 0 \\ 0 & 0 \\ 0 & 1 \\ \cdot & \cdot \\ 0 & 0 \end{Bmatrix} \begin{bmatrix} q_1(t) \\ q_2(t) \end{bmatrix} . \tag{6.1.25}$$

A time-dependent **B** would correspond to time-evolving positions at which forces were prescribed–a somewhat unusual situation for an oceanographic model, but not impossible, and even perhaps a convenient formulation, for example, if one were driving a model with a heat flux in the presence of a prescribed moving ice cover. One could also impose initial conditions using a time-dependent $\mathbf{B}(t)$, which would vanish after $t = 0$.

As with steady models, we need to be careful about understanding the propagation of errors in time and space. If we have some knowledge of the initial oceanic state, $\tilde{\mathbf{x}}(0)$, and are working at sea at a later time t, the prior information–the estimated initial conditions–carries information in addition to what we are currently measuring from shipboard. We seek to combine the two sets of information. How does information propagate forward in time? Formally, the rule (6.1.6) tells us exactly what to do. But because there are always errors in $\tilde{\mathbf{x}}(0)$, we need to be careful about assuming that a model computation of $\tilde{\mathbf{x}}(t)$ is useful. Depending upon the details of the model, one can qualitatively distinguish the behavior of the errors through time. (1) The model has decaying components. If the amplitudes of these components are partially erroneous, then for large enough t, these elements will have diminished, perhaps to the point where they are negligible. (2) The model has neutral components. At time t, the erroneous elements have amplitudes no larger than they were at $t = 0$. (3) The model has unstable components, at time t any erroneous parts may have grown to swamp everything else computed by the model.

A realistic oceanic GCM will contain all three types of behavior simultaneously. It thus becomes necessary to determine which elements of the forecast $\tilde{\mathbf{x}}(t)$ can be used to help estimate the oceanic state by combination with new data, and which should be suppressed as partially or completely erroneous. Simply assuming all components are equally accurate can be a disastrous recipe.

Before proceeding, it may be helpful to make the simple point that time need not be accorded a privileged position. Define the inclusive state vector, $\mathbf{x}$, to consist of the entirety of all the individual vectors $\mathbf{x}(t)$, so that

$$\mathbf{x} = \begin{bmatrix} \mathbf{x}(0) \\ \mathbf{x}(1) \\ \vdots \\ \mathbf{x}(t) \end{bmatrix} .$$

Then models of the form (6.1.7) can be written in the *whole-domain* form

$$\mathbf{A}_1\mathbf{x} = \mathbf{d}$$
$$\mathbf{A}_1 = \left\{ \begin{matrix} -\mathbf{A} & \mathbf{I} & \mathbf{0} & \cdot & \cdot & \mathbf{0} & \mathbf{0} \\ \mathbf{0} & -\mathbf{A} & \mathbf{I} & \mathbf{0} & \cdot & \mathbf{0} & \mathbf{0} \\ \cdot & \cdot & \cdot & \cdot & \cdot & -\mathbf{A} & \mathbf{I} \end{matrix} \right\}, \quad \mathbf{d} = \begin{bmatrix} \mathbf{B}\mathbf{q}(0) \\ \mathbf{B}\mathbf{q}(1) \\ \vdots \end{bmatrix}, \quad (6.1.26)$$

which is no different, except for size, from that of a static system.

6.1.2 Observations and Data

Here, observations are introduced into the modeling discussion so that they stand on an equal footing with the set of model equations (6.1.6) or (6.1.7) or (6.1.26). Observations will be represented as a set of linear simultaneous equations at time t,

$$\mathbf{E}(t)\,\mathbf{x}(t) + \mathbf{n}(t) = \mathbf{y}(t)\,, \tag{6.1.27}$$

a straightforward generalization of the previous static systems where t did not appear explicitly; here, $\mathbf{E}$ is sometimes called the *design* or *observation* matrix.

The requirement that the observations be linear combinations of the state-vector elements can be relaxed if necessary, but most common observations are of that form. [An obvious exception would be the situation in which the state vector included velocity components, $u(t)$, $v(t)$, but a current meter measured the speed, $\sqrt{(u(t)^2 + v(t)^2)}$, which would be a nonlinear relation between $y_i(t)$ and the state vector. Such systems are usually handled by some form of linearization; see, for example, Bryson and Ho (1975, p. 351).]

To be specific, the noise $\mathbf{n}(t)$ is supposed to have zero mean and known second-moment matrix

$$< \mathbf{n}(t) > = 0, \quad < \mathbf{n}(t)\,\mathbf{n}(t)^T > = \mathbf{R}(t)\,. \tag{6.1.28}$$

But

$$< \mathbf{n}(t)\,\mathbf{n}(t')^T > = \mathbf{0},\ t \neq t'\,. \tag{6.1.29}$$

That is, the observational noise should not be correlated from one measurement time to another; there is a considerable literature on how to proceed when this crucial assumption fails (called the *colored-noise problem*; see, for example, Stengel, 1986). Unless specifically stated otherwise, we will assume that (6.1.29) is valid.

The matrix $\mathbf{E}(t)$ can accommodate almost any form of linear measurement. If at some time there are no measurements, then $\mathbf{E}(t)$ vanishes, along with $\mathbf{R}(t)$. If a single element $x_i(t)$ is measured, then $\mathbf{E}(t)$ is a row vector that is zero everywhere except column i, where it is 1. It is particularly important to recognize that many measurements are weighted averages of the state-vector elements. Some measurements–for example, tomographic ones (Munk, Worcester, & Wunsch, 1995)–are explicitly spatial averages (integrals) obtained by measuring the speed with which an acoustic ray transits between two points. Any such data representing spatially filtered versions of the state vector can be written

$$y(t) = \sum \alpha_j x_j(t) \,. \tag{6.1.30}$$

Consider the tomographic observation of velocity, which for present purposes can be simply described as the measurement of the average velocity tangential to the line lying between two grid points i_0, i_1. Suppose that the state vector consists of the gridded elements, $p(i)$, of the pressure, and the model is that of Rossby waves. Because Rossby waves are quasi-geostrophic, a velocity measurement is very nearly the same as a measure of the horizontal derivative of p, normal to the line (see Figure 6–2b), so that a perfect tomographic measurement demands:

$$y(t) = v_{\text{tomog}}(t) = \frac{g}{f}\int_{r_{i_1}}^{r_{i_2}} \frac{\partial p}{\partial \mathbf{n}} ds \equiv \frac{g}{f} \sum_{i=\text{grid}\, i_1}^{\text{grid}\, i_2} \frac{p(i') - p(i)}{\Delta n_i} \Delta s_i \tag{6.1.31}$$

where i' is the grid point normal to the line lying along the path of integration. $\mathbf{E}$ is then composed of elements that form the numerical derivatives in (6.1.31).

Point observations often occur at positions not coincident with model grid positions (although many models, e.g., spectral ones, do not use grids). Then (6.1.27) is an interpolation rule, possibly either very simple or conceivably a full-objective mapping calculation, of the value of the state vector at the measurement point. Often the number of model grid points vastly exceeds the number of the data grid points; thus, it is convenient that the formulation (6.1.27) demands interpolation from the dense model grid to the sparse data positions. (In the oceanographically unusual situation where the data density is greater than the model grid density, one can readily restructure the problem so the interpolation goes the other way.)

More complex filtered measurements exist, of which satellite altimetry is a prime example (see the discussion in Wunsch, 1991). In particular, one may have measurements of a state vector only in specific wavenumber bands; but such *band-passed* observations are automatically in the form (6.1.27).

As with the model, the observations of the combined state vector can be concatenated into a single observational set

$$\mathbf{E}_1 \mathbf{x} + \mathbf{n}_1 = \mathbf{y}_1 \tag{6.1.32}$$

where

$$\mathbf{E}_1 = \begin{Bmatrix} 0 & 0 & 0 & \cdot & 0 \\ 0 & \mathbf{E}(1) & 0 & \cdot & 0 \\ 0 & 0 & \mathbf{E}(2) & \cdot & \cdot \\ \cdot & \cdot & \cdot & 0 & \mathbf{E}(t_f) \end{Bmatrix}, \quad \mathbf{n}_1 = \begin{bmatrix} 0 \\ \mathbf{n}(1) \\ \vdots \\ \mathbf{n}(t_f) \end{bmatrix}, \quad \mathbf{y}_1 = \begin{bmatrix} 0 \\ \mathbf{y}(1) \\ \vdots \\ \mathbf{y}(t_f) \end{bmatrix}.$$

If the size is no problem, the concatenated model and observations could be dealt with using any of the methods of Chapter 3. The rest of this chapter can be thought of as an attempt to produce from the model/data combination the same type of estimates as were found useful in Chapter 3, but exploiting the special structure of matrices $\mathbf{A}_1$ and $\mathbf{E}_1$ so as to avoid having to store them in the computer all at once, which may be physically impossible.

The Green's function formulation of (6.1.22) leads to a particularly simple reduced form if, for example, only the initial conditions $\mathbf{x}(0)$ are unknown, as one has immediately

$$\mathbf{y}(t) = \mathbf{E}(t)\mathbf{G}(t,0)\mathbf{x}(0) + \mathbf{n}(t), \qquad 1 \le t \le t_f,$$

which are readily solved in whole-domain form. If only a subset of the $\mathbf{x}(0)$ are thought to be nonzero, the columns of $\mathbf{G}$ need to be computed only for those elements (a method exploited by Stammer and Wunsch, 1996, to reduce computation).

6.2 Estimation

The central distinguishing feature of oceanography as a branch of fluid dynamics is the extraordinary difficulty of obtaining observations of the system adequate to describe what is going on (Chapter 2). Generally speaking, the oceanographer's strategy has to be to make inferences about the system by combining observations with whatever dynamics are believed to describe the system. The problems are of statistical inference with a host of specific subtypes. As we have seen in Chapter 3, such problems always confront one with a two-stage set of issues: stage 1 addresses the determination of consistency of the model with data; if consistency is found, stage 2 estimates various problem parameters through the data/model combination. The introduction of time into the system does not change this basic two-step process, although as with ordinary least squares, the steps may be algorithmically combined, since both stages of analysis are conducted after the computation is over.

One important type of problem might be thought of as a form of interpolation: Given a set of observations in space and time as described by Equation (6.1.27), use the dynamics as described by the model (6.1.6) or (6.1.7) to estimate various state-vector elements at various times of interest. Yet another, less familiar problem recognizes that some of the forcing terms $\mathbf{B}(t)\mathbf{q}(t)$ are partially or wholly unknown (e.g., no one believes that the windstress boundary conditions over the ocean are perfectly known), and one might seek to estimate them from whatever ocean observations are available and from the known model dynamics. Most real problems are actually some combination of these two major subtypes.

The forcing terms–representing boundary conditions as well as interior sources/sinks and forces–almost always need to be divided into two elements: the known and the unknown parts. The latter will often be perturbations about the known values. Rewrite (6.1.7) in the modified form

$$\mathbf{x}(t+1) = \mathbf{A}(t)\,\mathbf{x}(t) + \mathbf{B}(t)\,\mathbf{q}(t) + \boldsymbol{\Gamma}(t)\,\mathbf{u}(t) \tag{6.2.1}$$

where now $\mathbf{B}(t)\,\mathbf{q}(t)$ represent the known forcing terms and $\boldsymbol{\Gamma}(t)\,\mathbf{u}(t)$ the unknown ones, which we will generally refer to as the *controls*, or *control terms*. $\boldsymbol{\Gamma}(t)$ is known and plays the same role for $\mathbf{u}(t)$ as does $\mathbf{B}(t)$ for $\mathbf{q}(t)$. Usually $\mathbf{B}(t)$, $\boldsymbol{\Gamma}(t)$ will be treated as time independent, but this simplification is not necessary. Almost always, we can make some estimate of the size of the control terms, as for example,

$$< \mathbf{u}(t) >= 0, \; < \mathbf{u}(t)\,\mathbf{u}(t)^T >= \mathbf{Q}(t)\,. \tag{6.2.2}$$

The controls have a second, somewhat different role: They can also represent the model error. No fluid model is wholly accurate–approximations are always made to the equations describing any particular physical situation. One can expect that the person who constructed the model has some idea of the size and structure of the physics (or chemistry, ...) that has been omitted or distorted in the model construction (otherwise, why would one think to use the model?). In this context, $\mathbf{Q}$ represents the covariance of the model error, and the control terms represent the missing physics. The assumption $< \mathbf{u}(t) > = \mathbf{0}$ must be critically examined in this case, and in the event of failure, some modification of the model must be made or the variance artificially modified to attempt to accommodate what becomes a model bias error. But the most serious problem is that models are rarely produced with *any* quantitative description of their accuracy beyond one or two examples of comparison with known solutions, and one is left to make pure guesses at $\mathbf{Q}(t)$.

Collecting the standard equations of model and data:

$$\mathbf{x}(t+1) = \mathbf{A}(t)\mathbf{x}(t) + \mathbf{B}(t)\mathbf{q}(t) + \boldsymbol{\Gamma}(t)\mathbf{u}(t)\,, \quad 0 \le t \le t_f - 1 \tag{6.2.3}$$

$$\mathbf{E}(t)\mathbf{x}(t) + \mathbf{n}(t) = \mathbf{y}(t)\,, \quad t = 1 \le t \le t_f \tag{6.2.4}$$

$$< \mathbf{n}(t) > = 0\,, \quad < \mathbf{n}(t)\mathbf{n}(t')^T > = \mathbf{R}(t)\delta_{tt'} \tag{6.2.5}$$

$$< \mathbf{u}(t) > = 0\,, \quad < \mathbf{u}(t)\mathbf{u}(t)^T > = \mathbf{Q}(t) \tag{6.2.6}$$

$$\tilde{\mathbf{x}}(0) = \mathbf{x}_0\,, \ < (\tilde{\mathbf{x}}(0) - \mathbf{x}(0))(\tilde{\mathbf{x}}(0) - \mathbf{x}(0))^T > = \mathbf{P}(0) \tag{6.2.7}$$

where t_f defines the endpoint of the interval of interest.

The last equation, (6.2.7), treats the initial conditions of the model as a special case–the uncertain initialization problem, where $\mathbf{x}(0)$ is the true initial condition and $\tilde{\mathbf{x}}(0) = \mathbf{x}_0$ is the value actually used but with uncertainty $\mathbf{P}(0)$. An alternative would be to start the sequence of observations in (6.2.4) at $t = 0$; then Equation (6.2.7) would be redundant.

This general form permits one to grapple with reality: The quantities that oceanographers can most readily measure (traditionally noisy temperature and salinity in the oceanic interior; more recently seasurface elevation, float trajectories, etc.) are not at all the quantities that classical mathematical physics asserts are required to pose properly the governing partial differential equations (perfect initial and boundary conditions known everywhere at all times).

In the spirit of ordinary least squares and its intimate cousin, minimum-error variance estimation, consider the general problem of finding state vectors and controls that minimize an objective function

$$\begin{aligned} J = {} & (\mathbf{x}(0) - \mathbf{x}_0)^T \mathbf{P}(0)^{-1} (\mathbf{x}(0) - \mathbf{x}_0) \\ & + \sum_{t=1}^{t_f} (\mathbf{E}(t)\,\mathbf{x}(t) - \mathbf{y}(t))^T \mathbf{R}(t)^{-1} (\mathbf{E}(t)\,\mathbf{x}(t) - \mathbf{y}(t)) \\ & + \sum_{t=0}^{t_f - 1} \mathbf{u}(t)^T \mathbf{Q}(t)^{-1} \mathbf{u}(t)\,, \end{aligned} \tag{6.2.8}$$

subject to the model, (6.2.3) and (6.2.6). As written here, this choice of an objective function is somewhat arbitrary but perhaps reasonable as the direct analogue to (3.3.60). It seeks a state vector $\mathbf{x}(t)$, $0 \le t \le t_f$, and a set of controls, $\mathbf{u}(t)$, $0 \le t \le t_f - 1$, that satisfy the model and that agree with the observations to an extent determined by the weight matrices $\mathbf{R}(t)$ and $\mathbf{Q}(t)$, respectively. From the previous discussions of least squares and minimum-error variance estimation, the minimum-square requirement (6.2.8) will produce a solution identical to that produced with minimum variance estimation by the specific choice of the weight matrices as the corresponding prior uncertainties, $\mathbf{R}(t)$, $\mathbf{Q}(t)$, $\mathbf{P}(0)$. In a Gaussian system, it also proves to be the maximum likelihood estimate. The introduction of the controls, $\mathbf{u}(t)$, into the objective function, represents an acknowledgment that arbitrarily large controls (forces) would not be an acceptable solution.

Much of the rest of this chapter will be directed at solving the problem of finding the minimum of J subject to the model. Notice that J involves the state vector, the controls, and the observations over the entire time period under consideration, $0 \leq t \leq t_f$. This type of objective function is the one usually of most interest to oceanographers–in which data are stored and employed over a finite time. In some other types of problems, one of which is taken up immediately below, one has only the past measurements available, as would be the case in forecasting.

Although we will not keep repeating the warning each time an objective function such as (6.2.8) is encountered, the reader is reminded of the message of Chapter 3: The assumption that the model and observations are consistent and that the minimum of the objective function produces a meaningful and useful estimate must always be tested after the fact. That is, at the minimum, $\tilde{\mathbf{u}}(t)$ must prove consistent with $\mathbf{Q}(t)$, and $\tilde{\mathbf{x}}(t)$ must produce residuals consistent with $\mathbf{R}(t)$. Failure of these and other posterior tests should lead to rejection of the model. As always, one can thus reject a model [which includes $\mathbf{Q}(t)$, $\mathbf{R}(t)$] on the basis of a failed consistency with observations. One can never prove a model "correct," merely "consistent."[4]

6.2.1 The Kalman Filter

Let us begin by considering a special case. Suppose that by some means at time $t = 0$ we have an unbiased estimate, $\tilde{\mathbf{x}}(0)$, of the state vector with known uncertainty $\mathbf{P}(0)$. At time $t = 1$, observations from (6.2.4) are available. How would the information available best be used to estimate $\mathbf{x}(1)$?

The model permits a forecast of what $\mathbf{x}(1)$ should be, were $\tilde{\mathbf{x}}(0)$ known perfectly,

$$\tilde{\mathbf{x}}(1, -) = \mathbf{A}\tilde{\mathbf{x}}(0) + \mathbf{B}\mathbf{q}(0) \tag{6.2.9}$$

[4] Some modelers like to claim "validation" or "verification" of their models. This terminology is misleading and ought to be suppressed: What is usually meant is that the model has been tested for consistency with observations, but it is impossible ever to prove it "valid" in the sense of it being the truth. One can, however, readily prove a model invalid.

where the unknown control terms have been replaced by the best estimate we can make of them–their mean, which is zero, and $\mathbf{A}$ has been assumed to be time independent. A minus sign has been introduced into the argument of $\tilde{\mathbf{x}}(1,-)$ to show that no data have yet been used to make the estimate at $t=1$, in a notation we will generally use. How good is this forecast?

Suppose the erroneous components of $\tilde{\mathbf{x}}(0)$ are

$$\gamma(0) = \tilde{\mathbf{x}}(0) - \mathbf{x}(0)\,, \tag{6.2.10}$$

then the erroneous components of the forecast are

$$\begin{aligned}\gamma(1) \equiv \tilde{\mathbf{x}}(1,-) - \mathbf{x}(1) &= \mathbf{A}\tilde{\mathbf{x}}(0) + \mathbf{Bq}(0) - (\mathbf{Ax}(0) + \mathbf{Bq}(0) + \boldsymbol{\Gamma}\mathbf{u}(0)) \\ &= \mathbf{A}\gamma(0) - \boldsymbol{\Gamma}\mathbf{u}(0)\,,\end{aligned} \tag{6.2.11}$$

that is, composed of two distinct elements: the propagated erroneous portion of $\tilde{\mathbf{x}}(0)$ and the unknown control term. Their second moments are

$$\begin{aligned}<\gamma(1)\,\gamma(1)^T> &= <(\mathbf{A}\gamma(0) - \boldsymbol{\Gamma}\mathbf{u}(0))(\mathbf{A}\gamma(0) - \boldsymbol{\Gamma}\mathbf{u}(0))^T> \\ &= \mathbf{A}\ <\gamma(0)\,\gamma(0)^T>\ \mathbf{A}^T + \boldsymbol{\Gamma}\ <\mathbf{u}(0)\,\mathbf{u}(0)^T>\boldsymbol{\Gamma}^T \\ &= \mathbf{AP}(0)\mathbf{A}^T + \boldsymbol{\Gamma}\mathbf{Q}(0)\boldsymbol{\Gamma}^T \\ &\equiv \mathbf{P}(1,-)\end{aligned} \tag{6.2.12}$$

by the definitions of $\mathbf{P}(0)$, $\mathbf{Q}(0)$ and the assumption that the unknown controls are not correlated with the error in the state estimate at $t=0$. We now have an estimate of $\mathbf{x}(1)$ with uncertainty $\mathbf{P}(1,-)$ and a set of observations,

$$\mathbf{E}(1)\,\mathbf{x}(1) + \mathbf{n}(1) = \mathbf{y}(1)\,. \tag{6.2.13}$$

To combine the two sets of information, we use the recursive least-squares solution (3.7.8)–(3.7.11). By assumption, the uncertainty in $\mathbf{y}(1)$ is uncorrelated with that in $\tilde{\mathbf{x}}(1,-)$. Making the appropriate substitutions into those equations,

$$\begin{aligned}\tilde{\mathbf{x}}(1) &= \tilde{\mathbf{x}}(1,-) + \mathbf{K}(1)\,[\mathbf{y}(1) - \mathbf{E}(1)\,\tilde{\mathbf{x}}(1,-)]\,, \\ \mathbf{K}(1) &= \mathbf{P}(1,-)\,\mathbf{E}(1)^T[\mathbf{E}(1)\,\mathbf{P}(1,-)\,\mathbf{E}(1)^T + \mathbf{R}(1)]^{-1}\end{aligned} \tag{6.2.14}$$

with new uncertainty

$$\mathbf{P}(1) = \mathbf{P}(1,-) - \mathbf{K}(1)\,\mathbf{E}(1)\,\mathbf{P}(1,-)\,. \tag{6.2.15}$$

Thus there are four steps:

1. Make a forecast using the model (6.2.3) with the unknown control terms set to zero.

2. Calculate the uncertainty of this forecast, (6.2.12), which is made up of two separate terms.
3. Do a weighted average (6.2.14) of the forecast with the observations, the weighting as in Section 3.8 being chosen to reflect the relative uncertainties.
4. Compute the uncertainty of the final weighted average (6.2.15).

Such a computation is called a *Kalman filter*, after Kalman (1960). It is conventionally given a more formal derivation. $\mathbf{K}$ is called the *Kalman gain.* At the stage where the forecast (6.2.9) has already been made, the problem was reduced to finding the minimum of the objective function

$$\begin{aligned} J = &[\tilde{\mathbf{x}}(1,-) - \mathbf{x}(1)]^T \mathbf{P}(1,-)^{-1} [\tilde{\mathbf{x}}(1,-) - \mathbf{x}(1)] \\ &+ [\mathbf{y}(1) - \mathbf{E}(1)\,\mathbf{x}(1)]^T \mathbf{R}(1)^{-1} [\mathbf{y}(1) - \mathbf{E}(1)\,\mathbf{x}(1)]\,, \end{aligned} \tag{6.2.16}$$

which is a special case of the cost function in (3.7.6) used to define the recursive least-squares algorithm. In this final stage, the explicit model has disappeared, being present only implicitly through the uncertainty $\mathbf{P}(1,-)$. Notice that the model is being satisfied exactly; in the terminology introduced in Chapter 3, it is a hard or strong constraint. But again, as was true with the static models of Chapter 3, the hard constraint description is somewhat misleading, as the presence of the terms in $\mathbf{u}$ means that model errors are permitted.

A complete recursion can now be defined through the Equations (6.2.9)–(6.2.15), replacing all the $t = 0$ variables with $t = 1$ variables, the $t = 1$ variables becoming $t = 2$ variables, etc. In terms of arbitrary t, the recursion is

$$\tilde{\mathbf{x}}(t+1,-) = \mathbf{A}(t)\tilde{\mathbf{x}}(t) + \mathbf{B}(t)\mathbf{q}(t)\,, \tag{6.2.17}$$

$$\mathbf{P}(t+1,-) = \mathbf{A}(t)\,\mathbf{P}(t)\,\mathbf{A}(t)^T + \boldsymbol{\Gamma}(t)\,\mathbf{Q}(t)\,\boldsymbol{\Gamma}(t)^T\,, \tag{6.2.18}$$

$$\begin{aligned} &\tilde{\mathbf{x}}(t+1) = \tilde{\mathbf{x}}(t+1,-) + \mathbf{K}(t+1)\,[\mathbf{y}(t+1) - \mathbf{E}(t+1)\,\tilde{\mathbf{x}}(t+1,-)]\,, \\ &\mathbf{K}(t+1) = \mathbf{P}(t+1,-)\,\mathbf{E}(t+1)^T \times \\ &\qquad [\mathbf{E}(t+1)\mathbf{P}(t+1,-)\mathbf{E}(t+1)^T + \mathbf{R}(t+1)]^{-1}\,, \end{aligned} \tag{6.2.19}$$

$$\mathbf{P}(t+1) = \mathbf{P}(t+1,-) - \mathbf{K}(t+1)\,\mathbf{E}(t+1)\,\mathbf{P}(t+1,-)\,, \qquad 1 \le t \le t_f - 1\,. \tag{6.2.20}$$

Example: Straight Line. Let us reconsider the problem of fitting a straight line to data, as discussed in Chapter 3, but now in the context of a Kalman filter, using the canonical form derived from (6.1.10). "Data"

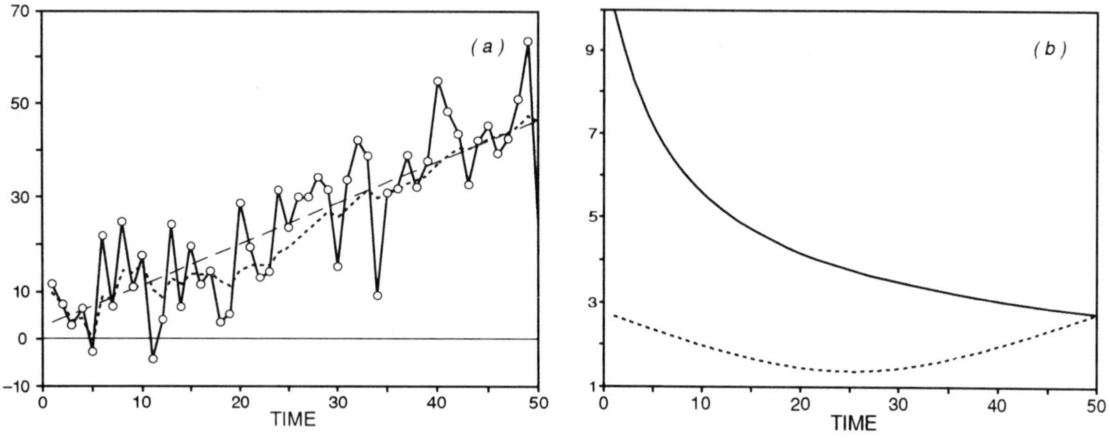

Figure 6–3. (a) For the model of a straight line in time, the "0" are the observations, $\mathbf{y}$, and the dotted line is the Kalman filter estimate for the first 50 time steps. There is an observation at each time-step. The filter does a reasonable job of tracking the straight line in the presence of noisy observations, converging to it at the end. The dashed line is the result of the RTS smoother applied to the results of the Kalman filter. Now the result is very near the correct straight-line state estimate. (b) Standard error of the Kalman filter estimate of x_1 (solid line), and the standard error of the smoothed estimate (dotted line). The correct values and the estimated values are generally within one standard error of each other, as they should be most of the time. Notice that the filter and smoother covariance begin with identical values at the terminal time with the smoother error reaching a minimum near the center of the time interval, increasing again toward $t = 0$ owing to the very large initial estimate of $\mathbf{P}(0)$. Again, the best estimates are, unsurprisingly, in the center of the observation period.

were generated as depicted in Figure 6–3a–a straight line plus noise. The observation equation is

$$y(t) = x_1(t) + n(t)\,,$$

that is, $\mathbf{E}(t) = \{1 \quad 0\}$, $R(t) = 100$. The model was assumed perfect, $Q(t) = 0$, but the initial state estimate was set erroneously as $\mathbf{x}(0) = [10 \quad 10]^T$ with an uncertainty

$$\mathbf{P}(0) = \begin{Bmatrix} 100 & 0 \\ 0 & 100 \end{Bmatrix}.$$

The result of the computation for the fit is shown in Figure 6–3a for 50 time steps. Figure 6–4b shows the standard error, $\sqrt{P_{11}(t')}$ of $x_1(t)$, and its decline as observations accumulate.

If the state vector is redefined to consist of the two model parameters a, b, then $\mathbf{x} = [a \quad b]^T$ and $\mathbf{A} = \mathbf{I}$. Now the observation matrix is $\mathbf{E} = [1 \quad t]$–that is, time-dependent. Notice the radical change in the state vector from a time-varying one to a constant. The same grossly incorrect estimates

$\tilde{\mathbf{x}}(0) = [10 \quad 10]^T$ were used, with the same $\mathbf{P}(0)$ (the correct values are $a = 1$, $b = 2$) and with the time histories of the estimates depicted in Figure 6–4a. At the end of 100 time steps, we have $\tilde{a} = 1.85 \pm 2.0$, $\tilde{b} = 2.0 \pm 0.03$, both of which are consistent with the correct values. For reasons the reader might wish to think about, the uncertainty of the intercept is much greater than for the slope.

Example: The Mass-Spring Oscillator. Consider the mass-spring oscillator described earlier with time history in Figure 6–5a. It was supposed that the initial conditions were perfectly known but that the forcing was completely unknown. Noisy observations of $x_1(t)$ were provided at every time step with a noise variance $R = 9$. The Kalman filter was computed by (6.2.17)–(6.2.20) and used to estimate the position at each time step. The result for part of the time history is in Figure 6–5b, showing the true value and the estimated value. The time history of the uncertainty of $x_1(t)$ is depicted in Figure 6–5c and is reaching an asymptote. Overall, the filter manages to track the position of the oscillator within two standard deviations.

It was then supposed that the same noisy observations were available but only at every 25th time step. In general, the presence of the model error, or control uncertainty, will accumulate over the 25 time steps as the model is run forward without observations $[\mathbf{P}(t+1) = \mathbf{P}(t+1, -)$ in that situation$]$, and one anticipates an error growth, until observations arrive. The expected error of such a system is shown for 150 time steps in Figure 6–5e. Notice (1) the growing envelope as uncertainty accumulates faster than the observation can reduce it; (2) the periodic nature of the error within the growing envelope; and (3) that the envelope appears to be asymptoting to a fixed upper bound for large t. The true and estimated time histories for a portion of the time history are shown in Figure 6–5d. With fewer available observations, as expected, the misfit of the estimated and true values is larger than with data at every time step. At every 25th point, the error norm drops as observations become available.

If the observation is that of the velocity $x_1(t+1) - x_2(t+1) = \xi(t+1) - \xi(t)$, then $\mathbf{E} = \{1 \quad -1\}$. A portion of the time history of the Kalman filtered estimate with a velocity observation available only at every 25th point may be seen in Figure 6–5f. Velocity observations are evidently useful for estimating position, owing to the connection between velocity and position provided by the model.

The Kalman filter does *not* produce the minimum of the objective function (6.2.8) because the data from times later than t are not being used to

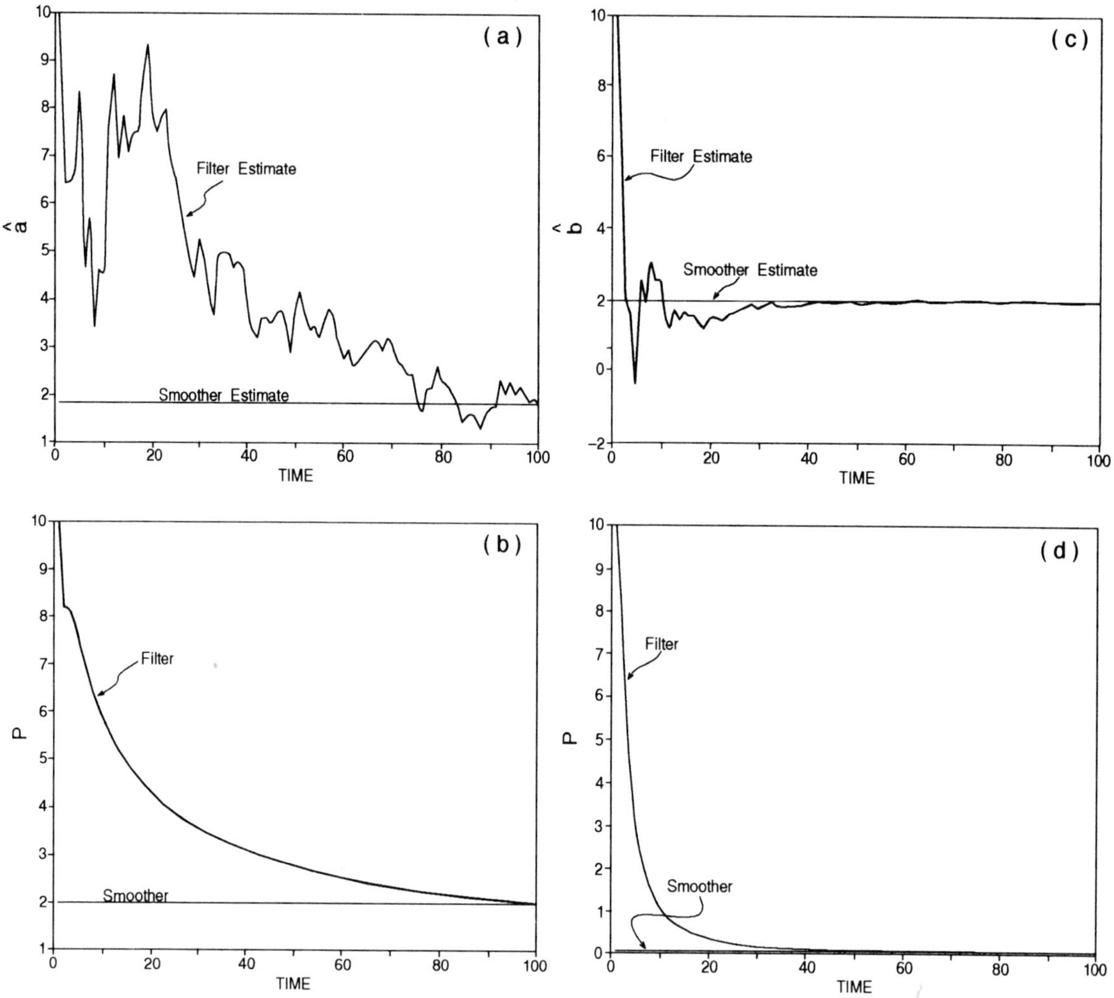

Figure 6–4. (a) Estimate of straight-line intercept, a, when the state vector was defined to be $\mathbf{x} = [a, b]^T$. The initial estimate was 10, when the correct value was 1. Notice that the smoothed estimate is a constant over the entire domain, consistent with the dynamical model for x. (b) Standard errors of the filter and smoother for estimates in (a). (c) Filter and smoother estimates of straight-line slope (correct value is $b = 2$) when slope was part of the state vector. Convergence is much more rapid than for the intercept. (d) The standard errors of the filtered and smoothed estimate in (c).

make estimates of the earlier values of the state vector or of $\mathbf{u}(t)$. At each step, the Kalman filter is instead minimizing an objective function of the form (6.2.16) where t corresponds to the previous $t = 0$, and $t + 1$ corresponds to $t = 1$. To obtain the needed minimum we have to consider what is called the *smoothing problem*, to which we will turn in a moment.

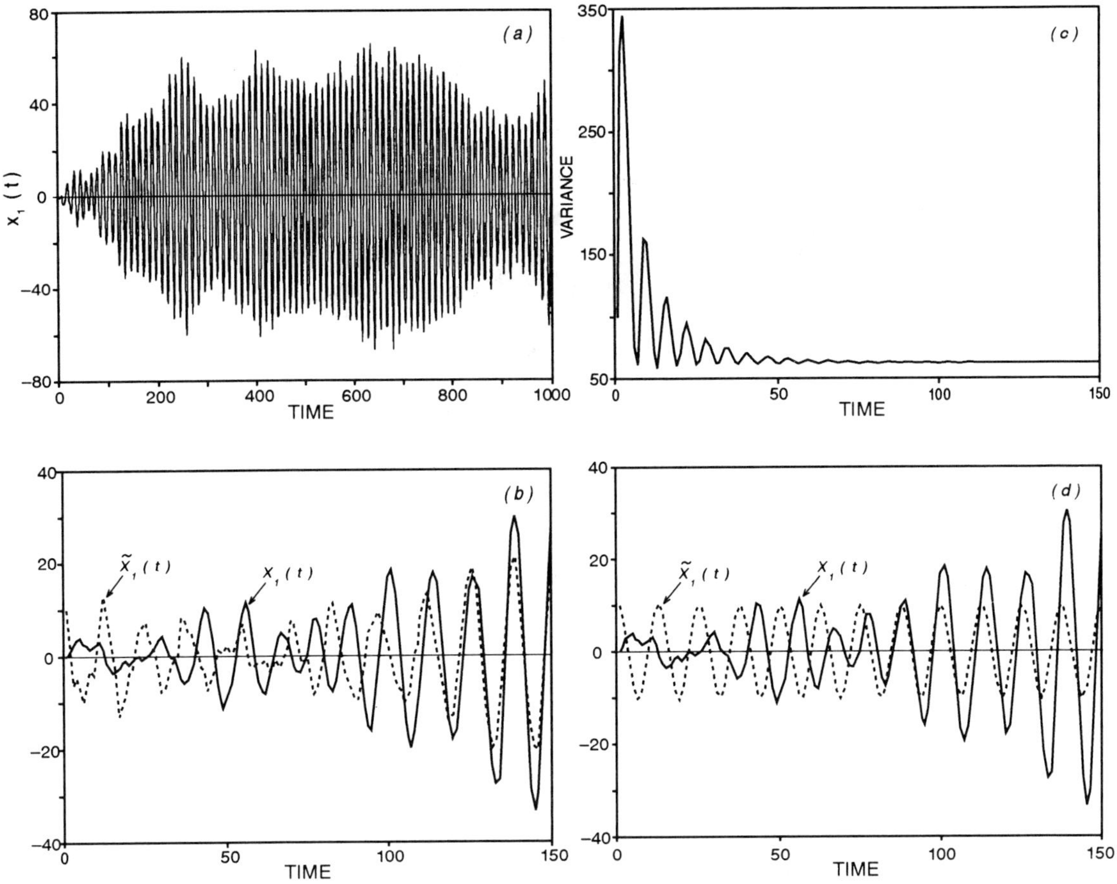

Figure 6–5a,b,c,d. (a) The time history of $x_1(t)$ from the state vector of a linearized discrete oscillator driven by white noise forcing (the control). Here, $\Delta t = 1$, $k = 0.25$, $r = 0$, $Q = 1$, $< u(t) > = 0$. The large excursions possible with a zero-mean small variance forcing are a simple illustration of the random-walk phenomenon. (b) Early portion of the time history $x_1(t)$ (solid line) displayed in (a), and the Kalman filter estimate of it (dashed). The filter was started with $\tilde{\mathbf{x}}(0) = [10, 10]^T$ (correct value is $\mathbf{x}(0) = [0, 0]^T$), $\mathbf{P}(0) = 100\mathbf{I}$, $\mathbf{E} = [1, 0]$, $\mathbf{Q} = 1$, and $\mathbf{R} = 1000$. The controls were treated as completely unknown. The system gradually converges to the true state and is consistent within one standard error of the truth. (c) The variance $P_{11}(t)$ of the Kalman filter estimate displayed in (b). Note the initial oscillatory behavior derived from the dynamics and then the asymptote as the incoming data stream just balances the uncertainties introduced by the unknown $u(t)$. (d) The same as (b) except that the data were available only every 25 time-steps ($t = 25, 50, \ldots$). Convergence still takes place but is much slower.

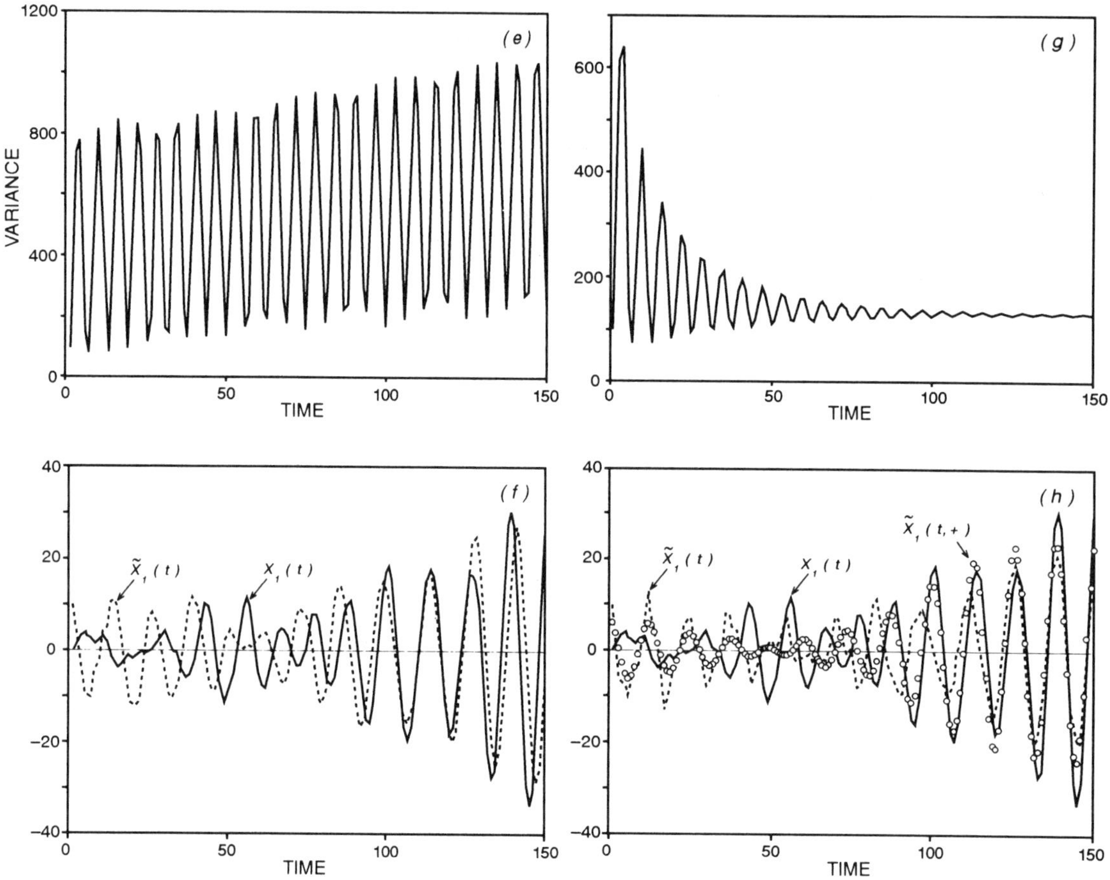

Fig. 6–5e,f,g,h. (e) The variance $P_{11}(t)$ for the Kalman filter estimate in Figure 6–5d, when data were available only every 25th time-step. An error asymptote is still achieved, but only after a longer time than shown and only in the envelope, with the values oscillating between data availability. Values are considerably higher than in Figure 6–5c. (f) The Kalman filter estimate for the first part of Figure 6–5a, with the same parameters, except here the observation is supposed to be of the "velocity": $y(t) = x_1(t) - x_2(t) + n(t)$, $(E = [1, -1])$, $\mathbf{R} = 1000$, at every time-step. Convergence to the correct state still takes place but is slow. (g) The variance, $P_{11}(t)$, for the filter estimate in (f), which can be compared to that in Figure 6–5c. The slower convergence to the correct solution in (f) is consistent with the slower asymptote and the higher ultimate value of the uncertainty that is achieved. (h) Smoothed estimate, $\tilde{x}_1(t,+)$ (shown as open circles, from the RTS smoother started from $t = t_f = 150$ from the Kalman filter estimate $\tilde{\mathbf{x}}(150)$ in Figure 6–5b. The correct value, $x_1(t)$, is the solid curve. As expected in a smoothing algorithm, the smoother significantly improves agreement with the true values in the center of the interval. Its value and uncertainty agree with that of the Kalman filter at $t = t_f$ and somewhat improves the values at the time origin.

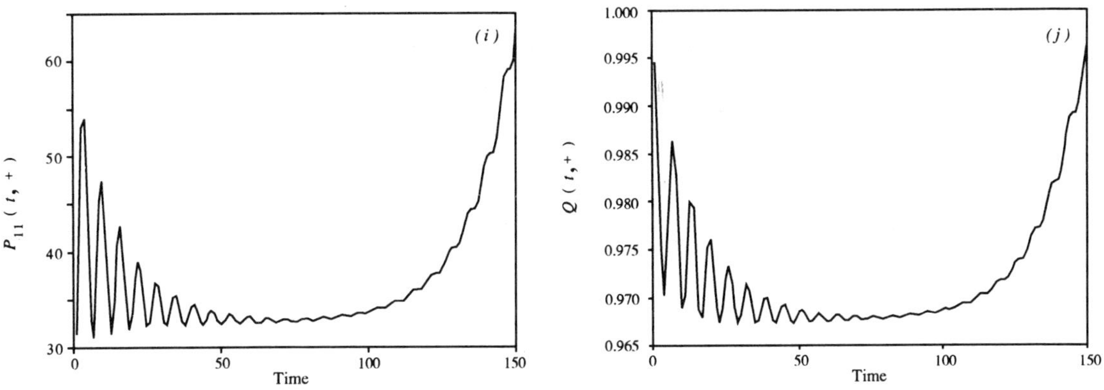

Figure 6–5i,j. (i) Uncertainty, $P_{11}(t,+)$, for the RTS smoother estimate in Figure 6–5h showing the minimum expected error in the center of the interval. This uncertainty should be compared to that of the filter in Figure 6–5c. (j) Uncertainty estimate $Q(t,+)$, for the control estimate, $\tilde{u}(t,+)$, from the RTS smoother. [$\tilde{u}(t,+)$ is not shown, being a set of random numbers, as is $u(t)$, and is uninteresting. The two do agree within the uncertainty shown. Note that the data error here was very large, and hence the final uncertainty is little reduced over the initial value of 1.]

But the Kalman filter is extremely important in practice for many problems. In particular, if one must literally make a forecast (e.g., such filters are used to help land airplanes or, in a primitive way, to forecast the weather), then the future data are simply unavailable, and the state estimate made at time $t+1$, using data up to and including time $t+1$, is the best one can do. Some history of the idea of the filter, its origins in the work of Wiener and Kolmogoroff and with a number of applications, can be found in Sorenson (1985). (Much of the discussion in the wider literature is in terms of continuous time. The continuous form of the filter is usually known as the *Kalman-Bucy filter*; see, for example, Bucy & Joseph, 1968). The filter is so important that entire books are devoted to its theory and practice, and many thousands of papers have been published on it and its extensions. Ghil, Cohn, Tavantzis, Bube, and Isaacson, (1981) discuss it in a meteorological context; Miller (1986), Wunsch (1988b), Miller and Cane (1989), Fu, Vazquez, and Perigaud (1991), and Gaspar and Wunsch (1989) discuss oceanographic applications; and Ghil and Malanotte-Rizzoli (1991) and Bennett (1992) provide reviews in the wider field of geophysical fluid dynamics.

From the oceanographer's point of view, the Kalman filter is only a first step owing to its failure to use data from the formal future. It also raises questions about computational feasibility. As with all recursive estimators, the uncertainties $\mathbf{P}(t,-)$, $\mathbf{P}(t)$ must be available so as to form the weighted averages. If the state vector contains N elements, then the model, (6.2.17), requires multiplying an N-dimensional vector by an $N \times N$ matrix at each time step. The covariance update (6.2.18) requires updating each of N columns of $\mathbf{P}(t)$ in the same way, and then doing it again [i.e., in practice, one forms $\mathbf{AP}(t)$, transposes it, and forms $\mathbf{A}(\mathbf{AP}(t))^T$, equivalent to running the model $2N$ times at each time step]. In many applications of oceanographic interest, this covariance update step dominates the calculation and renders it impractical.

The Kalman filter is nonetheless of crucial importance. Someday we will need to make forecasts (some limited forecasting of El Niño is already attempted; navies forecast acoustic features for antisubmarine warfare maneuvers; Heemink & Kloosterhuis, 1990, employ it operationally for storm surge forecasting, etc.). That the Kalman filter is the desired result is of great theoretical importance, and it becomes a guide in understanding potentially more practical, suboptimal methods. Furthermore, it is also a central element of most algorithms that do solve the problem of using all the data, as well as their suboptimal approximations.

The Kalman filter was derived heuristically as a simple generalization of the ideas used in Chapter 3. Unsurprisingly, the static inverse results are readily obtained from the filter in various limits. As one example, consider the nearly noise-free case in which both process and observation noise are very small, i.e. $\| \mathbf{Q} \|, \| \mathbf{R} \| \to 0$. Then if $\mathbf{P}(t+1,-)$ is nearly diagonal, $\mathbf{P}(t+1,-) \sim \delta^2 \mathbf{I}$, and

$$\mathbf{K}(t+1) \to \mathbf{E}^T \left(\mathbf{E}\mathbf{E}^T\right)^{-1},$$

assuming existence of the inverse, and

$$\begin{aligned} \tilde{\mathbf{x}}(t+1) \sim\ & \mathbf{A}\tilde{\mathbf{x}}(t) + \mathbf{B}\mathbf{q}(t) \\ & + \mathbf{E}^T(\mathbf{E}\mathbf{E}^T)^{-1}\left\{\mathbf{y}(t+1) - \mathbf{E}[\mathbf{A}\tilde{\mathbf{x}}(t) + \mathbf{B}\mathbf{q}(t)]\right\} \\ =\ & \mathbf{E}^T(\mathbf{E}\mathbf{E}^T)^{-1}\mathbf{y}(t+1) + \left[\mathbf{I} - \mathbf{E}^T(\mathbf{E}\mathbf{E}^T)^{-1}\mathbf{E}\right][\mathbf{A}\tilde{\mathbf{x}}(t) + \mathbf{B}\mathbf{q}(t)]\,. \end{aligned} \tag{6.2.21}$$

$\mathbf{E}^T\left(\mathbf{E}\mathbf{E}^T\right)^{-1}\mathbf{y}(t+1)$ is just the expression (3.3.53) for the direct estimate of $\mathbf{x}(t+1)$ from a set of full-rank, noise-free observations. It is the static estimate we would use at time $t+1$ if no dynamics were available to permit the use of prior observations. Equation (6.2.21) will be recognized as identical to (3.7.15) from the recursive least-squares solution. The only difference from (3.3.53) is that here the dynamical evolution equation transmits information about previous observations to help improve the static estimate. The columns of $\mathbf{I} - \mathbf{E}^T\left(\mathbf{E}\mathbf{E}^T\right)^{-1}\mathbf{E}$ are the nullspace of $\mathbf{E}$ (3.4.46), and (6.2.21) thus employs only those elements of the forecast lying in the nullspace of the observations–a sensible result given that the observations here produce perfect estimates of components of $\mathbf{x}(t+1)$ in the range of $\mathbf{E}$. Thus, in this particular limit, the Kalman filter computes from the noise-free observations those elements of $\mathbf{x}(t+1)$ that it can, and for those which cannot be so determined, it forecasts them from the dynamics. The reader ought to examine other limiting cases–retaining process and/or observational noise–including the behavior of the error covariance propagation.

The filter equations (6.2.17)–(6.2.21) can be rewritten, with the help of the matrix inversion lemma and other identities, in a variety of useful forms, including:

$$\mathbf{K}(t+1) = \mathbf{P}(t+1)\,\mathbf{E}(t+1)^T\mathbf{R}(t+1)^{-1} \tag{6.2.22}$$

(Stengel, 1986, Equation 4.3–22), and

$$\mathbf{P}(t+1) = \left[\mathbf{P}(t+1,-)^{-1} + \mathbf{E}^T(t+1)\mathbf{R}(t+1)^{-1}\mathbf{E}(t+1)\right]^{-1}, \tag{6.2.23}$$

$$\begin{aligned}\tilde{\mathbf{x}}(t+1) &= (\mathbf{A} - \mathbf{K}(t+1)\mathbf{E}(t+1))\,\tilde{\mathbf{x}}(t)\\ &+ \mathbf{B}(t)\mathbf{q}(t) + \mathbf{K}(t+1)\mathbf{y}(t+1)\,.\end{aligned} \tag{6.2.24}$$

A number of more general reformulations of the equations into algebraically equivalent forms are particularly important. In one form, one works not with the covariances, $\mathbf{P}(t+1,-),\ldots$, but with their inverses, the so-called information matrices, $\mathbf{P}(t+1,-)^{-1}$, etc. This *information filter* form may be more efficient if, for example, the information matrices are banded while the covariance matrices are not. In another formulation, one uses the square roots of the covariance matrices rather than the matrices themselves. The *square root filter* can be of great importance as there is a tendency for the computation of the updated values of $\mathbf{P}$ to become nonpositive-definite owing to round-off errors and other problems. The square root formulation guarantees positive definite covariances. Anderson and Moore (1979) discuss these and other variants of the Kalman filter equations.

Example: It is interesting to apply some of these expressions to the simple problem of finding the mean of a set of observations, considered before in §3.6.3. The model is of an unchanging scalar mean,

$$x(t+1) = x(t)$$

observed in the presence of noise,

$$y(t) = x(t) + n(t)$$

where $< n(t)^2 > = R$, so $E = 1$, $A = 1$. In contrast to the situation in §3.6.3, the machinery we have developed here requires that the noise be uncorrelated: $< n(t)n(t') > = 0$, $t \neq t'$, although as already mentioned, methods exist to overcome this restriction. Suppose that the initial estimate of the mean is 0–that is, $\tilde{x}(0) = 0$, with uncertainty $P(0)$. Then (6.2.18) is $P(t+1,-) = P(t)$, and the Kalman filter uncertainty, in the form (6.2.23), is

$$\frac{1}{P(t+1)} = \frac{1}{P(t)} + \frac{1}{R},$$

a difference equation, with known initial condition, whose solution by inspection is

$$\frac{1}{P(t)} = \frac{t}{R} + \frac{1}{P(0)}.$$

Starting (6.2.19) at $\mathbf{E} = 1$ and successively stepping forward (Bryson & Ho, 1975, p. 363, or Brown, 1983, p. 218) produces

$$\tilde{x}(t+1) = \frac{R}{R + (t+1)P(0)} \left\{ \frac{P(0)}{R} \sum_{j=1}^{t+1} y(j) \right\},$$

whose limit as $t \to \infty$ is

$$\tilde{x}(t+1) \to \frac{1}{t+1} \sum_{j=1}^{t+1} y(j),$$

the simple average, with uncertainty $P(t) \to 0$, as $t \to \infty$.

6.2.2 The Smoothing Problem

Minimization of J in (6.2.8) subject to the model is still the goal. Begin the discussion by again considering a one-step process (adapted here from Bryson & Ho, 1975, Chapter 13, whose notation is unfortunately somewhat difficult). Consider the problem (6.2.8) but where there are only two time steps involved, $t = 0, 1$. There is an initial estimate $\tilde{\mathbf{x}}(0)$, $\tilde{\mathbf{u}}(0) \equiv 0$ with uncertainties $\mathbf{P}(0)$, $\mathbf{Q}(0)$ for the initial state and control vectors, respectively, a set of measurements at time-step 1, and the model. The objective function is

$$\begin{aligned} J = {} & (\tilde{\mathbf{x}}(0,+) - \tilde{\mathbf{x}}(0))^T \mathbf{P}(0)^{-1} (\tilde{\mathbf{x}}(0,+) - \tilde{\mathbf{x}}(0)) \\ & + (\tilde{\mathbf{u}}(0,+) - \tilde{\mathbf{u}}(0))^T \mathbf{Q}(0)^{-1} (\tilde{\mathbf{u}}(0,+) - \tilde{\mathbf{u}}(0)) \\ & + (\mathbf{y}(1) - \mathbf{E}(1)\tilde{\mathbf{x}}(1))^T \mathbf{R}(1)^{-1} (\mathbf{y}(1) - \mathbf{E}(1)\tilde{\mathbf{x}}(1)), \end{aligned} \tag{6.2.25}$$

subject to the model

$$\tilde{\mathbf{x}}(1) = \mathbf{A}(0)\,\tilde{\mathbf{x}}(0,+) + \mathbf{B}(0)\mathbf{q}(0) + \boldsymbol{\Gamma}\tilde{\mathbf{u}}(0,+)\,, \tag{6.2.26}$$

with the weight matrices again chosen as the inverses of the prior covariances. A minimizing solution to this objective function would produce a new estimate of $\mathbf{x}(0)$, denoted $\tilde{\mathbf{x}}(0,+)$, with error covariance $\mathbf{P}(0,+)$; the $+$ denotes use of future observations, $\mathbf{y}(1)$, in the estimate. On the other hand, we would still denote the estimate at $t = 1$ as $\tilde{\mathbf{x}}(1)$, coinciding with the Kalman filter estimate, because only data prior to and at the same time would have been used. The estimate $\tilde{\mathbf{x}}(1)$ must be given by (6.2.14)–(6.2.15), but it remains to improve $\tilde{\mathbf{u}}(0)$, $\tilde{\mathbf{x}}(0)$.

The basic issue can be understood by observing that the initial estimates $\tilde{\mathbf{u}}(0)$, $\tilde{\mathbf{x}}(0)$ (the former is usually taken to be zero) lead to a forecast that disagrees with the final best estimate $\tilde{\mathbf{x}}(1)$. If either of $\tilde{\mathbf{u}}(0)$, $\tilde{\mathbf{x}}(0)$ were known perfectly, the forecast discrepancy could be ascribed to the other one, permitting ready computation of the required value. In practice, both are somewhat uncertain, and the modification must be partitioned between them. One would not be surprised to find that the partitioning proves to be proportional to their initial uncertainty.

To find the stationary point (we will not trouble to prove it a minimum rather than a maximum), set the differential of J with respect to $\tilde{\mathbf{x}}(0,+)$, $\tilde{\mathbf{x}}(1)$, $\tilde{\mathbf{u}}(0,+)$ to zero,

$$\begin{aligned}\frac{dJ}{2} &= d\tilde{\mathbf{x}}(0,+)^T\mathbf{P}(0)^{-1}[\tilde{\mathbf{x}}(0,+) - \tilde{\mathbf{x}}(0)] \\ &\quad + d\tilde{\mathbf{u}}(0,+)^T\mathbf{Q}(0)^{-1}[\tilde{\mathbf{u}}(0,+) - \tilde{\mathbf{u}}(0)] \\ &\quad - d\tilde{\mathbf{x}}(1)^T\mathbf{E}(1)^T\mathbf{R}(1)^{-1}[\mathbf{y}(1) - \mathbf{E}(1)\tilde{\mathbf{x}}(1)] = 0\,. \end{aligned} \tag{6.2.27}$$

The coefficients of the differentials cannot be set to zero separately because they are connected via the model, Equation (6.2.26), which provides a relationship

$$d\tilde{\mathbf{x}}(1) = \mathbf{A}(0)\,d\tilde{\mathbf{x}}(0,+) + \boldsymbol{\Gamma}(0)\,d\tilde{\mathbf{u}}(0,+)\,. \tag{6.2.28}$$

Eliminating $d\tilde{\mathbf{x}}(1)$,

$$\begin{aligned}\frac{dJ}{2} &= d\tilde{\mathbf{x}}(0,+)^T\Big[\mathbf{P}(0)^{-1}(\tilde{\mathbf{x}}(0,+) - \tilde{\mathbf{x}}(0)) - \\ &\qquad \mathbf{A}(0)^T\mathbf{E}(1)^T\mathbf{R}(1)^{-1}(\mathbf{y}(1) - \mathbf{E}(1)\tilde{\mathbf{x}}(1)\Big] \\ &\quad + d\tilde{\mathbf{u}}(0,+)^T\Big[\mathbf{Q}(0)^{-1}(\tilde{\mathbf{u}}(0,+) - \tilde{\mathbf{u}}(0)) + \\ &\qquad \boldsymbol{\Gamma}^T(0)\mathbf{E}(1)^T\mathbf{R}(1)^{-1}(\mathbf{y}(1) - \mathbf{E}(1)\tilde{\mathbf{x}}(1))\Big]\,. \end{aligned} \tag{6.2.29}$$

Now the differential vanishes, producing a stationary value of J only if the coefficients of $d\tilde{\mathbf{x}}(0,+)$, $d\tilde{\mathbf{u}}(0,+)$ separately vanish, yielding

$$\tilde{\mathbf{x}}(0,+) = \tilde{\mathbf{x}}(0) + \mathbf{P}(0)\mathbf{A}(0)^T\mathbf{E}(1)^T\mathbf{R}(1)^{-1}(\mathbf{y}(1) - \mathbf{E}(1)\tilde{\mathbf{x}}(1)) \quad (6.2.30)$$

$$\tilde{\mathbf{u}}(0,+) = \tilde{\mathbf{u}}(0) + \mathbf{Q}(0)\boldsymbol{\Gamma}(0)^T\mathbf{E}(1)^T\mathbf{R}(1)^{-1}(\mathbf{y}(1) - \mathbf{E}(1)\tilde{\mathbf{x}}(1)) \quad (6.2.31)$$

and

$$\tilde{\mathbf{x}}(1) = \tilde{\mathbf{x}}(1,-) + \mathbf{P}(1,-)\mathbf{E}(1)^T[\mathbf{E}(1)\mathbf{P}(1,-)\mathbf{E}(1)^T + \mathbf{R}(1)]^{-1}(\mathbf{y}(1) - \mathbf{E}(1)\tilde{\mathbf{x}}(1,-)) \quad (6.2.32)$$

using the previous definitions of $\tilde{\mathbf{x}}(1,-)$, $\mathbf{P}(1,-)$, which is recognized as the Kalman filter estimate. At this point we are essentially done: An estimate has been produced not only of $\mathbf{x}(1)$, but an improvement has been made in the prior estimate of $\mathbf{x}(0)$ using the future measurements, and we have estimated the control term. Notice the corrections to $\tilde{\mathbf{u}}(0)$, $\tilde{\mathbf{x}}(0)$ are proportional to $\mathbf{P}(0)$, $\mathbf{Q}(0)$, respectively, as anticipated. We still need to examine the uncertainties of these latter quantities.

First rewrite the estimates (6.2.30)–(6.2.31) as

$$\begin{aligned} \tilde{\mathbf{x}}(0,+) &= \tilde{\mathbf{x}}(0) + \mathbf{L}(1)\left(\tilde{\mathbf{x}}(1) - \tilde{\mathbf{x}}(1,-)\right), \\ \mathbf{L}(1) &= \mathbf{P}(0)\mathbf{A}(0)^T\mathbf{P}(1,-)^{-1}, \\ \tilde{\mathbf{u}}(0,+) &= \tilde{\mathbf{u}}(0) + \mathbf{M}(1)\left(\tilde{\mathbf{x}}(1) - \tilde{\mathbf{x}}(1,-)\right), \\ \mathbf{M}(1) &= \mathbf{Q}(0)\boldsymbol{\Gamma}(0)^T\mathbf{P}(1,-)^{-1}, \end{aligned} \quad (6.2.33)$$

which can be done by extended, but uninteresting, algebraic manipulation. The importance of these latter two expressions is that both $\tilde{\mathbf{x}}(0,+)$, $\tilde{\mathbf{u}}(0,+)$ are expressed in terms of their prior estimates in a weighted average with the difference between the prediction of the state at $t = 1$, $\tilde{\mathbf{x}}(1,-)$ and what was actually estimated there following the data use, $\tilde{\mathbf{x}}(1)$. [But the data do not appear explicitly in (6.2.33).] Then it is also possible to show that

$$\begin{aligned} \mathbf{P}(0,+) &= \mathbf{P}(0) + \mathbf{L}(1)(\mathbf{P}(1) - \mathbf{P}(1,-))\mathbf{L}(1)^T \\ \mathbf{Q}(0,+) &= \mathbf{Q}(0) + \mathbf{M}(1)(\mathbf{P}(1) - \mathbf{P}(1,-))\mathbf{M}(1)^T. \end{aligned} \quad (6.2.34)$$

Based upon this one-step derivation, a complete recursion for any time interval can be inferred. Suppose the Kalman filter has been run all the way to a terminal time t_f. The result is $\tilde{\mathbf{x}}(t_f)$ and its variance $\mathbf{P}(t_f)$. With no future data available, $\tilde{\mathbf{x}}(t_f)$ cannot be further improved. At time $t_f - 1$, we have an estimate $\tilde{\mathbf{x}}(t_f - 1)$ with uncertainty $\mathbf{P}(t_f - 1)$, which could be improved by knowledge of the future observations at t_f. But this situation is precisely the one addressed by the objective function (6.2.25) with t_f

being the equivalent of time $t = 1$, and $t_f - 1$ taking the place of $t = 0$. Now having improved the estimate at $t_f - 1$ and calling it $\tilde{\mathbf{x}}(t_f - 1, +)$ with uncertainty $\mathbf{P}(t_f - 1, +)$, this new estimate is used to improve the prior estimate $\tilde{\mathbf{x}}(t_f - 2)$, and we step all the way back to $t = 0$. The complete recursion is

$$\tilde{\mathbf{x}}(t,+) = \tilde{\mathbf{x}}(t) + \mathbf{L}(t+1)\left[\tilde{\mathbf{x}}(t+1,+) - \tilde{\mathbf{x}}(t+1,-)\right],$$
$$\mathbf{L}(t+1) = \mathbf{P}(t)\mathbf{A}(t)^T\,\mathbf{P}(t+1,-)^{-1} \tag{6.2.35}$$
$$\tilde{\mathbf{u}}(t,+) = \tilde{\mathbf{u}}(t) + \mathbf{M}(t+1)\left[\tilde{\mathbf{x}}(t+1,+) - \tilde{\mathbf{x}}(t+1,-)\right],$$
$$\mathbf{M}(t+1) = \mathbf{Q}(t)\boldsymbol{\Gamma}(t)^T\mathbf{P}(t+1,-)^{-1}, \tag{6.2.36}$$
$$\mathbf{P}(t,+) = \mathbf{P}(t) + \mathbf{L}(t+1)[\mathbf{P}(t+1,+) - \mathbf{P}(t+1,-)]\mathbf{L}(t+1)^T, \tag{6.2.37}$$
$$\mathbf{Q}(t,+) = \mathbf{Q}(t) + \mathbf{M}(t+1)[\mathbf{P}(t+1,+) - \mathbf{P}(t+1,-)]\mathbf{M}(t+1)^T, \tag{6.2.38}$$

with
$$\tilde{\mathbf{x}}(t_f,+) \equiv \tilde{\mathbf{x}}(t_f),\ \mathbf{P}(t_f,+) \equiv \mathbf{P}(t_f).$$

This recipe, which uses the Kalman filter on a first forward sweep to the end of the available data, and which then successively improves the prior estimates by sweeping backwards, is called the *RTS algorithm*, for Rauch, Tung, and Striebel (1965). The particular form has the advantage that the data are not involved in the backward sweep, since all the available information has been used in the filter calculation. It does have the potential disadvantage of requiring the storage at each time step of $\mathbf{P}(t)$, and the inversion of $\mathbf{P}(t,-)$. $\mathbf{P}(t,-)$ is readily recomputed from (6.2.18) and need not be stored. By direct analogy with the one-step objective function, the recursion (6.2.35)–(6.2.38) is seen to be the solution to the minimization of the objective function (6.2.26) subject to the model.

It is possible to examine limiting cases of the RTS smoother much as with the Kalman filter. For example, suppose again that $\mathbf{Q}$ vanishes, and $\mathbf{A}^{-1}$ exists. Then,

$$\mathbf{L}(t+1) \to \mathbf{P}(t)\mathbf{A}^T\left(\mathbf{A}\mathbf{P}(t)\mathbf{A}^T\right)^{-1} = \mathbf{A}^{-1} \tag{6.2.39}$$

for diagonal $\mathbf{P}(t)$, and Equation (6.2.35) becomes

$$\tilde{\mathbf{x}}(t,+) \to \mathbf{A}^{-1}\left[\tilde{\mathbf{x}}(t+1,+) - \mathbf{B}\mathbf{q}(t)\right], \tag{6.2.40}$$

a sensible backward estimate obtained by simply solving

$$\tilde{\mathbf{x}}(t+1) = \mathbf{A}\tilde{\mathbf{x}}(t) + \mathbf{B}\mathbf{q}(t) \tag{6.2.41}$$

for $\tilde{\mathbf{x}}(t)$. Other limits are also illuminating but are left to the reader.

Example: The Straight Line. The smoother result for the straight-line model (6.1.10) is shown in Figures 6–3 and 6–4 for both forms of state vector. The time-evolving estimate is now a nearly perfect straight line, whose uncertainty (e.g., Figure 6–3) has a terminal value at $t = 100$ equal to that for the Kalman filter estimate, as it must, and reaches a minimum near the middle of the estimation period, before growing again toward $t = 0$, where the initial uncertainty was very large. In the case where the state vector consisted of the constant intercept and slope of the line, both smoothed estimates are seen, in contrast to the filter estimate, to conform very well to the known true behavior. Again, the system has more information available to determine the slope value than it does the intercept. It should be apparent that the best-fitting, straight-line solution of Chapter 3 is also the solution to the smoothing problem, but with the data and model handled all at once, a whole-domain method, rather than sequentially.

Example: The Mass-Spring Oscillator. Figure 6–5h shows the state estimate for the mass-spring oscillator made from a smoothing computation run backward from $t = 150$, and its variance is shown in Figure 6–5i. On average (but not everywhere), the smoothed estimate is closer to the correct value than is the filtered estimate, as expected. The standard error is also smaller for the smoothed estimate. Figure 6–5j displays the variance, $\mathbf{Q}_{11}(t)$, of the estimate one can make of the scalar control variable $\mathbf{u}(\mathbf{t})$. $\tilde{\mathbf{u}}(\mathbf{t})$ is not shown because the actual value was white noise, and the plot is uninteresting [estimate and truth do agree consistent with $\mathbf{Q}_{11}(\mathbf{t})$].

Example: A Transient Tracer Problem. Consider a problem stimulated by the need to extract information from oceanic transient tracers, C, which are assumed to satisfy an equation like

$$\frac{\partial C}{\partial t} + \mathbf{v} \cdot \nabla C - \kappa \nabla^2 C = -\lambda C + q(\mathbf{r}, t) \tag{6.2.42}$$

where q represents sources/sinks and λ is a decay rate if the tracer is radioactive. To have a simple model that will capture the structure of this problem, the ocean is divided into a set of boxes as depicted in Figure 6–6a. The flow field, as depicted there, is represented by exchanges between boxes given by the $J_{ij} \geq 0$. That is, the J_{ij} are just a simplified representation of the effects of oceanic flows and mixing on a dye C (the relationship between such simple parameterizations and more formal and elaborate finite-difference schemes was discussed in Chapter 4). Here, it will only be remarked that

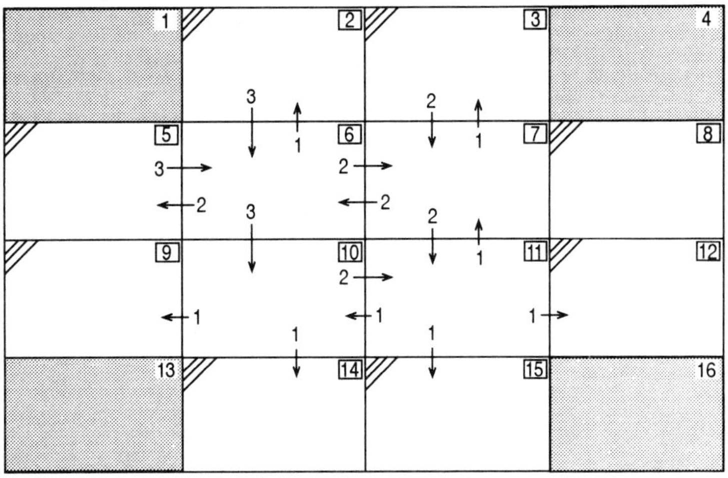

Figure 6–6a. Tracer box model (Wunsch, 1988b). J_{ij} represent fluxes between boxes and are chosen to conserve mass. Boxes with shaded corners are boundary boxes with concentrations externally prescribed. Boxes are referred to by number in the upper right-hand corner.

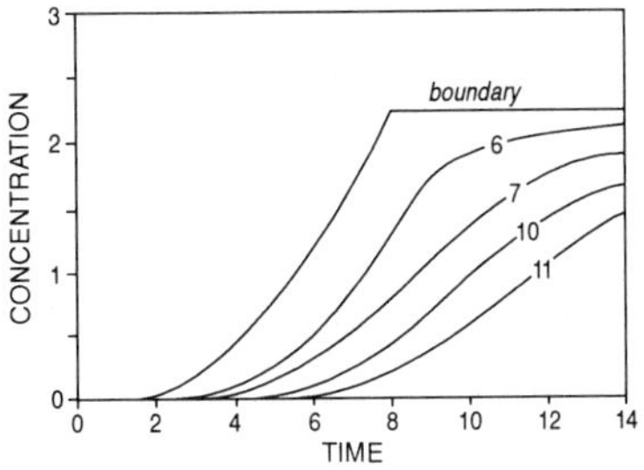

Figure 6–6b. Time histories of the forward computation in which boundary concentrations shown were prescribed, and values were computed for boxes 6, 7, 10, 11. These represent the "truth."

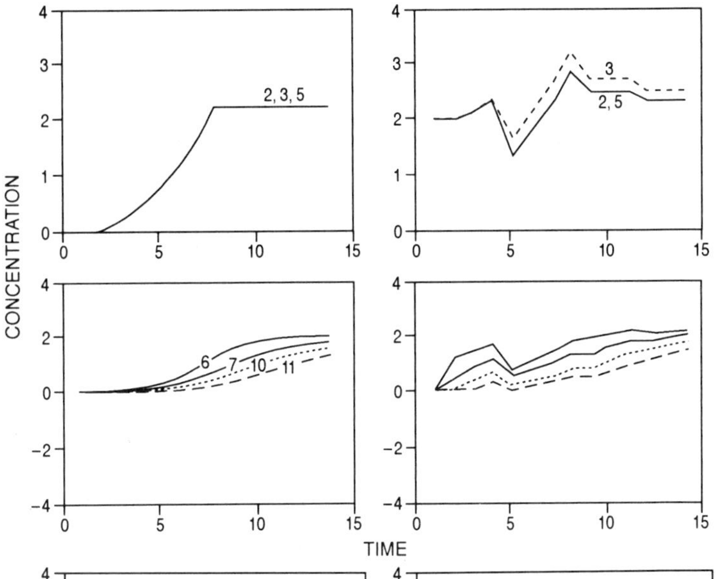

Figure 6–6c. The left two panels show the correct values for boundary boxes (upper panel) and interior ones (lower panel). The right panels show the results of the Kalman filter for boundary (upper panel) and interior boxes (lower panel) when noisy observations were provided at $t = 5$, 9, 12. At the observation times, estimates are pulled toward the observations but tend to diverge afterward. By the time the last observations are used, estimated and correct values are quite close. Although not displayed here, there is an uncertainty estimate at all times.

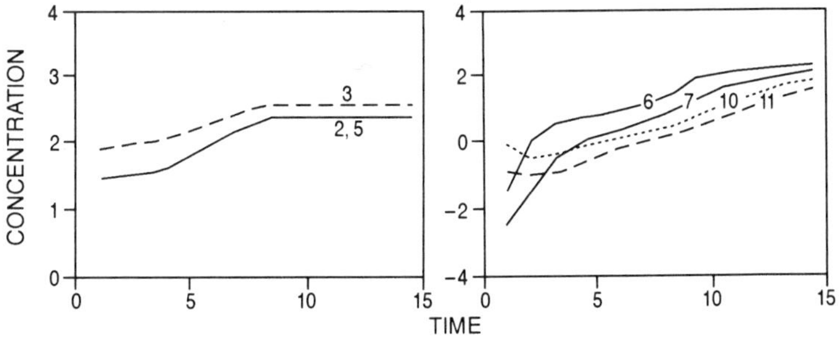

Figure 6–6d. Smoothed estimates of boundary (left panel) and interior (right panel) values of tracers.

J_{ij} are chosen to be mass conserving so that the sum over all J_{ij} entering and leaving a box vanishes. The discrete analogue of (6.2.42) is taken to be

$$C_i(t+1) = C_i(t) - \lambda \Delta t C_i(t) - \frac{\Delta t}{V} \sum_{j \in N(i)} C_i(t) J_{ij} + \frac{\Delta t}{V} \sum_{j \in N(i)} C_j(t) J_{ji} \tag{6.2.43}$$

where the notation $j \in N(i)$ denotes an index sum over the neighboring boxes to box i, V is a volume for the boxes, and Δt is the time step. This model easily can be put into the canonical form

$$\mathbf{x}(t+1) = \mathbf{A}\mathbf{x}(t) + \mathbf{B}\mathbf{q}(t) + \boldsymbol{\Gamma}\mathbf{u}(t)\,, \qquad \mathbf{Q} = \mathbf{0}\,, \tag{6.2.44}$$

with the state vector being the vector of box concentrations $C_i(t)$, $C_i(t-1)$.

The particular problem to be considered was stimulated by the need to understand how to use transient tracer data such as ^3H (Figures 2–6a,c) and chlorofluorocarbons (Freons) (Figures 2–2e). The problem is discussed in many places–for example, Broecker and Peng (1982) and Wunsch (1988a,b), from whom the present example is taken. A forward computation was run with initial concentrations everywhere of 0, using the boundary conditions depicted in Figure 6–6b, resulting in interior box values as shown. Based upon these correct values, noisy "observations" of the interior boxes only were constructed at times $t = 5, 9, 12$.

An initial estimate of tracer concentrations at $t = 0$ was taken (correctly) to be zero, but this estimate was given a large variance [diagonal $\mathbf{P}(0)$ with large norm]. The boundary box concentrations were set erroneously to $C = 2$ for all t and held at that value. A Kalman filter computation was run as shown in Figure 6–6c. Initially, the interior box concentration estimates rise erroneously (owing to the dye leaking in from the high nonzero concentrations in the boundary boxes). At $t = 5$, the first set of observations becomes available, and the combined estimate is driven much closer to the true values. By the time the last set of observations is used, the estimated and correct concentrations are quite close, although the time history of the interior is somewhat in error. The RTS algorithm was then applied to generate the smoothed histories shown in Figure 6–6d and to estimate the boundary concentrations (the controls). As expected, the smoothed estimates are closer to the true time history than are the filtered ones (the uncertainty estimates are not shown, but the results are consistent with the "truth" within statistical expectation). Unless further information is provided, no other estimation procedure could do better, given that the model is the correct one.

Example: Altimetry. In comparison to the Kalman filter, smoothing algorithms have not yet seen wide oceanographic use. Gaspar and Wunsch (1989) employed the RTS algorithm on altimetric data in the context of the barotropic Rossby wave Equation (6.1.2). They wrote the solution to the equation as in (6.1.3) so that the state vector consisted of the expansion coefficients,

$$\mathbf{x}(t) = [a_1(t) \quad a_2(t) \quad \ldots \quad a_{2N-1}(t) \quad a_{2N}(t)]^T .$$

As is usually true in real, as opposed to textbook, problems, most of the effort lay in specifying the covariances $\mathbf{P}(0)$, $\mathbf{R}$, $\mathbf{Q}$. Figure 6–7a shows an estimate of the initial values, $\tilde{\mathbf{x}}(0,+)$ from six months of data. Figure 6–7b displays the filtered and smoothed estimates of some of the wave amplitudes and phases over the observational duration. The smoothed estimates are indeed much smoother than those from the filter, as one would expect from what is basically an interpolation computation as opposed to the extrapolation of the pure filter. A similar application, but in a regime where a linear model was more suitable, can be seen in Fu, Fukumori, and Miller, (1993), as well as in Fukumori, Benveniste, Wunsch, and Haidvogel (1993), who apply a smoother to an ocean GCM.

There are many versions of smoother algorithms chosen for special purposes (e.g., successively improving an estimate at a fixed time as data continue to accumulate–the so-called fixed-point smoother–or to achieve various trade-offs of computations versus storage requirements). Consider one other approach to smoothing. Suppose the Kalman filter has been run forward to some time t_c, producing an estimate $\tilde{\mathbf{x}}(t_c)$ with uncertainty $\mathbf{P}(t_c)$. Now suppose, perhaps on the basis of some further observations, that at a *later* time t_f an independent estimate $\tilde{\mathbf{x}}(t_f)$ has been made, with uncertainty $\mathbf{P}(t_f)$. The independence is crucial–we suppose this latter estimate is made without using any observations at time t_c or earlier so that any errors in $\tilde{\mathbf{x}}(t_c)$ and $\tilde{\mathbf{x}}(t_f)$ are uncorrelated.

Let us run the model *backward* in time from t_f to $t_f - 1$:

$$\tilde{\mathbf{x}}_b(t_f - 1) = \mathbf{A}^{-1}\tilde{\mathbf{x}}(t_f) - \mathbf{A}^{-1}\mathbf{B}\mathbf{q}(t_f - 1) \qquad (6.2.45)$$

where the subscript b denotes a backward-in-time estimate. The reader may object that running a model backward in time will often be an unstable operation, and this objection needs to be addressed, but ignore it for the moment. The uncertainty of $\tilde{\mathbf{x}}(t_f - 1)$ is

$$\mathbf{P}_b(t_f - 1) = \mathbf{A}^{-1}\mathbf{P}(t_f)\mathbf{A}^{-T} + \mathbf{A}^{-1}\boldsymbol{\Gamma}\mathbf{Q}(t_f - 1)\boldsymbol{\Gamma}^T\mathbf{A}^{-T} \qquad (6.2.46)$$

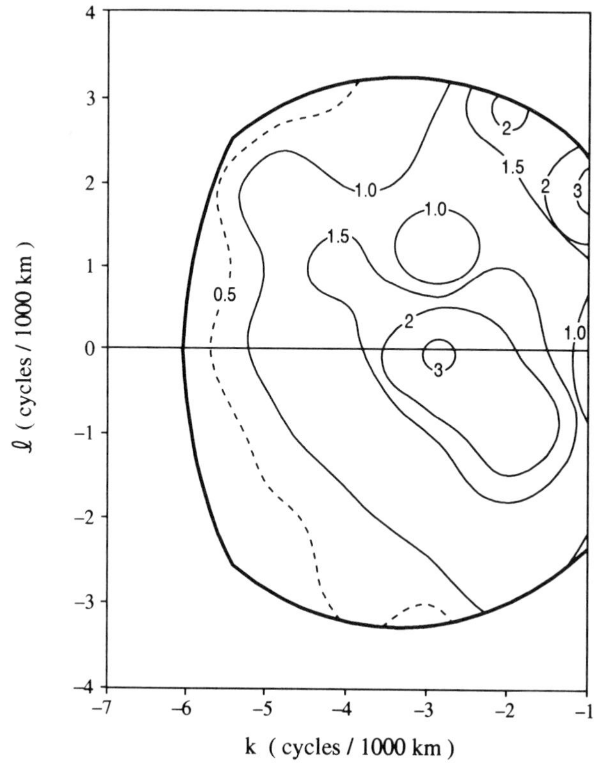

Figure 6–7a. Estimates of the initial values of the sea-surface elevation model amplitudes from the RTS smoother (Gaspar & Wunsch, 1989). The result is contoured in wavenumber space.

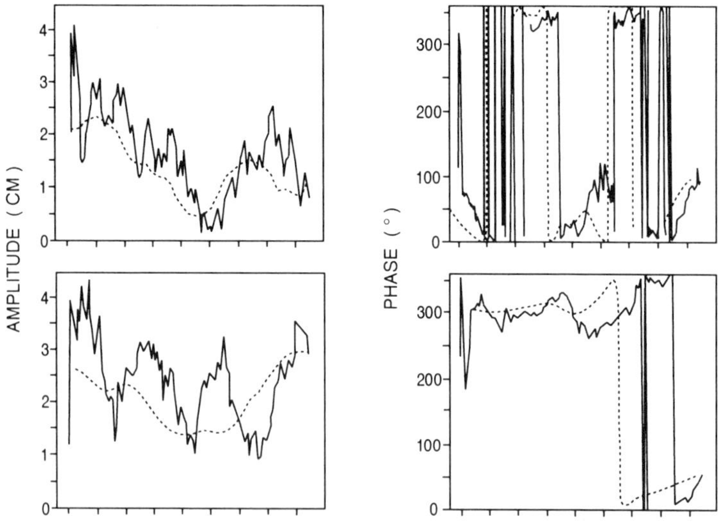

Figure 6–7b. Time histories from the Kalman filter and RTS smoother of two typical modal amplitudes (left panels) and phases (right panels) from six months of altimetric data (Gaspar & Wunsch, 1989).

in direct analogy to the forward model computation. This backward computation can be continued to time t_c, at which point we will have an estimate, $\tilde{\mathbf{x}}_b(t_c)$, with uncertainty $\mathbf{P}_b(t_c)$.

The two independent estimates of $\mathbf{x}(t_c)$ can be combined to make an improved estimate using the relations (3.7.23),

$$\tilde{\mathbf{x}}(t_c, +) = \tilde{\mathbf{x}}(t_c) + \mathbf{P}(t_c)(\mathbf{P}(t_c) + \mathbf{P}_b(t_c))^{-1}(\tilde{\mathbf{x}}_b(t_c) - \tilde{\mathbf{x}}(t_c)) \qquad (6.2.47)$$

and

$$\begin{aligned} \mathbf{P}(t_c) &= < (\tilde{\mathbf{x}}(t_c, +) - \mathbf{x}(t_c))\,(\tilde{\mathbf{x}}(t_c, +) - \mathbf{x}(t_c))^T > \\ &= (\mathbf{P}(t_c)^{-1} + \mathbf{P}_b(t_c)^{-1})^{-1}\,. \end{aligned} \qquad (6.2.48)$$

This estimate is the same as obtained from the RTS algorithm because the same objective function, model, and data have been employed.

Running the model backward may indeed be unstable if it contains any dissipative terms. A forward model may be unstable too, if there are unstable modes of motion, either real or numerical artifacts. But the expressions in (6.2.47)–(6.2.48) are stable, because the computation of $\mathbf{P}_b(t)$ and its use in the updating expression (6.2.48) automatically downweight unstable elements whose errors will be very large, which will not carry useful information about the earlier state. The same situation would occur if the forward model had unstable elements–these instabilities would amplify slight errors in the statement of their initial conditions, rendering them difficult to estimate from observations at later times (see Figure 6–8). Examination of the covariance propagation equation and the filter gain matrix shows that these elements are suppressed in the Kalman filter estimate, with correspondingly large uncertainties. Thus, the filter/smoother formalism properly accounts for unstable, and hence difficult to calculate, parameters by estimating their uncertainty as very large, thus handling very general ill-conditioning. In practice, one needs to be careful, for numerical reasons, of the pitfalls in computing and using matrices that may have norms growing exponentially in time. But the conceptual problem is solved.

As with the Kalman filter, it is possible to rewrite the RTS smoother expressions (6.2.35)–(6.2.38) in various ways for computational efficiency, storage reduction, and improved accuracy (see, for example, Gelb, 1974; Bryson & Ho, 1975; Anderson & Moore, 1979; Goodwin & Sin, 1984; Sorenson, 1985).

The dominant computational load in the smoother is again the calculation of the updated covariance matrices, whose size is square of the state-vector dimension, at every time step, leading to efforts to construct simplified

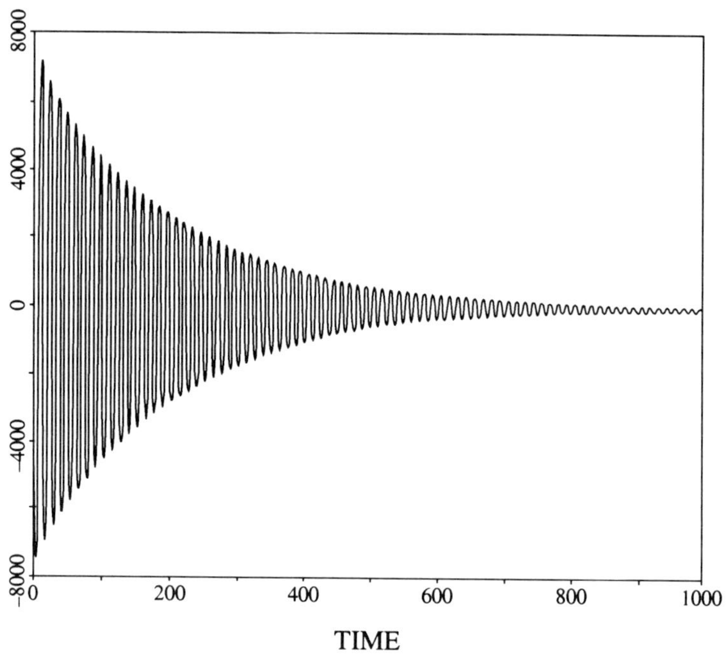

Figure 6–8. Mass-spring oscillator model with friction ($r = 0.01$) run backward in time from conditions specified at $t = 1000$. The system is unstable, and small uncertainties in the starting conditions would amplify. But the Kalman filter run backwards remains stable because its error estimate also grows, systematically downweighting the model forecast relative to any data that become available at earlier times. A model with unstable elements run in the forward direction would behave analogously, with unstable components being downweighted.

algorithms that retain most of the virtues of the filter/smoother combination but with reduced load. For example, it may have already occurred to the reader that in some of the examples displayed, the state vector uncertainties, $\mathbf{P}$, in both the filter and the smoother appear to approach rapidly a steady state. This asymptotic behavior in turn means that the gain matrices, $\mathbf{K}$, $\mathbf{L}$, $\mathbf{M}$ will also achieve a steady state, implying that one no longer needs to undertake the updating steps–fixed gains can be used. Such steady-state operators are known as *Wiener filters* and *smoothers* and if achieved, they represent a potentially very large computational savings. One needs to understand the circumstances under which such steady states can be expected to appear, and we will examine the problem in Section 6.5.

We turn now instead to another approach to reducing the computational load–the so-called adjoint methods. We will demonstrate how they work and their equivalence to smoothing.

6.3 Control Problems: The Pontryagin Principle and Adjoint Methods

6.3.1 Lagrange Multiplier Constraints

The results of the last section are recursive schemes for computing a filtered and then a smoothed estimate. As with recursive least squares, the necessity to combine two pieces of information to make an improved estimate demands knowledge of the uncertainty of the information. For static problems, the recursive methods of Section 3.8 may be required, either because all the data were not available initially or because one could not handle it all at once. But, in general, the computational load of the combined least-squares problem (3.7.3) is less than the recursive one, if one chooses not to compute any of the covariance matrices.

Because the covariance computation will usually dominate and potentially overwhelm the filter/smoother algorithms, it is very attractive, at least superficially, to find algorithms that do not require the covariances–that is, which employ the entire time domain of observations simultaneously, a whole-domain method. The algorithms that emerge are best known in the context of *control theory.* Essentially, there is a more specific focus upon determining the $\mathbf{u}(t)$: the control variables. In many nonoceanographic problems, one literally wishes to drive a system in desirable ways. Although oceanographers can drive their models in ways analogous to controlling a machine tool, they do not ever control the real ocean. It will help the reader who further explores these methods to recognize that we are still doing *estimation*, combining observations and models, but sometimes using algorithms best known under the control rubric.

To see the possibilities, consider again the two-point objective function (6.2.25) where $\mathbf{P}$, etc., are just weight matrices without necessarily having a statistical significance. We wish simply to find the minimum of the objective function subject to (6.2.26). To use a variant method, append the model equations as done in Chapter 3 [compare with (3.5.15)], with a vector of Lagrange multipliers $\boldsymbol{\mu}(1)$, for a new objective function

$$\begin{aligned} J = &(\tilde{\mathbf{x}}(0,+) - \tilde{\mathbf{x}}(0))\,\mathbf{P}(0)^{-1}(\tilde{\mathbf{x}}(0,+) - \tilde{\mathbf{x}}(0)) \\ &+ (\tilde{\mathbf{u}}(0,+) - \tilde{\mathbf{u}}(0))^T\mathbf{Q}(0)^{-1}(\tilde{\mathbf{u}}(0,+) - \tilde{\mathbf{u}}(0)) \\ &+ (\mathbf{y}(1) - \mathbf{E}(1)\tilde{\mathbf{x}}(1))^T\mathbf{R}(1)^{-1}(\mathbf{y}(1) - \mathbf{E}(1)\tilde{\mathbf{x}}(1)) \\ &- 2\boldsymbol{\mu}(1)^T(\tilde{\mathbf{x}}(1) - \mathbf{A}\tilde{\mathbf{x}}(0,+) - \mathbf{B}\mathbf{q}(0) - \boldsymbol{\Gamma}\tilde{\mathbf{u}}(0,+))\,. \end{aligned} \tag{6.3.1}$$

As with the filter and smoother, the model is being imposed as a hard constraint, but with the control term rendering the result indistinguishable from a soft one. In contrast to the approach in the last section, the presence of the Lagrange multiplier permits treating the differentials as independent; taking the derivatives of J with respect to $\tilde{\mathbf{x}}(0,+)$, $\tilde{\mathbf{x}}(1)$, $\tilde{\mathbf{u}}(0,+)$, $\boldsymbol{\mu}(1)$ and setting them to zero,

$$\mathbf{P}(0)^{-1}(\tilde{\mathbf{x}}(0,+) - \tilde{\mathbf{x}}(0)) + \mathbf{A}^T\boldsymbol{\mu}(1) = 0 \tag{6.3.2}$$

$$\mathbf{E}^T\mathbf{R}(1)^{-1}(\mathbf{y}(1) - \mathbf{E}\tilde{\mathbf{x}}(1)) + \boldsymbol{\mu}(1) = 0 \tag{6.3.3}$$

$$\mathbf{Q}(0)^{-1}(\tilde{\mathbf{u}}(0,+) - \tilde{\mathbf{u}}(0)) + \boldsymbol{\Gamma}^T\boldsymbol{\mu}(1) = 0 \tag{6.3.4}$$

$$\tilde{\mathbf{x}}(1) - \mathbf{A}\tilde{\mathbf{x}}(0,+) - \mathbf{B}\mathbf{q}(0) - \boldsymbol{\Gamma}\tilde{\mathbf{u}}(0,+) = 0\,. \tag{6.3.5}$$

Equation (6.3.2) is the adjoint model for $\boldsymbol{\mu}(1)$ involving the transpose of $\mathbf{A}$.

Because the objective function in (6.3.1) is identical with that used with the smoother and because the identical dynamical model has been imposed [in one case through the differentials in (6.2.28), in the other explicitly through the Lagrange multipliers], Equations (6.3.2)–(6.3.5) must produce the identical solution to that produced by the smoother. A demonstration that Equations (6.3.2)–(6.3.5) can be manipulated into the form (6.2.33) is an exercise in matrix identities. Some guidance is provided by Bryson and Ho (1975, pp. 390–5) or Liebelt (1967). In particular, Bryson and Ho (1975) introduce the Lagrange multipliers (their equations 13.2.7–13.2.8) simply as an intermediate numerical device for solving the smoother equations. As with smoothing algorithms, the finding of the solution of (6.3.2)–(6.3.5) can be arranged in a number of different ways, trading computation against storage, coding ease, convenience, etc.

Let us show explicitly the identity of smoother and adjoint solution for a restricted case–that for which the initial conditions are known exactly, so that $\tilde{\mathbf{x}}(0)$ is not modified by the later observations. For the one-term smoother, the result is obtained by dropping (6.3.2), as $\mathbf{x}(0)$ is no longer an adjustable parameter. Without further loss of generality, we may put $\tilde{\mathbf{u}}(0) = \mathbf{0}$, and set $\mathbf{R}(1) = \mathbf{R}$, reducing the system to

$$\tilde{\mathbf{x}}(1) = \mathbf{A}\tilde{\mathbf{x}}(0) + \mathbf{B}\mathbf{q}(0) + \boldsymbol{\Gamma}\tilde{\mathbf{u}}(0,+) \tag{6.3.6}$$

$$\begin{aligned}\tilde{\mathbf{u}}(0,+) &= -\mathbf{Q}(0)\boldsymbol{\Gamma}^T\boldsymbol{\mu}(1)\\ &= \mathbf{Q}(0)\boldsymbol{\Gamma}^T\mathbf{E}^T\mathbf{R}^{-1}[\mathbf{y}(1) - \mathbf{E}(1)\tilde{\mathbf{x}}(1)]\,.\end{aligned} \tag{6.3.7}$$

Eliminating $\tilde{\mathbf{u}}(0,+)$ from (6.3.6) produces

$$\tilde{\mathbf{x}}(1) = \mathbf{A}\tilde{\mathbf{x}}(0) + \mathbf{B}\mathbf{q}(0) + \boldsymbol{\Gamma}\mathbf{Q}(0)\boldsymbol{\Gamma}^T\mathbf{E}^T\mathbf{R}^{-1}[\mathbf{y}(1) - \mathbf{E}\tilde{\mathbf{x}}(1)]\,. \tag{6.3.8}$$

With no initial error in $\mathbf{x}(0)$, $\mathbf{P}(1,-) = \boldsymbol{\Gamma}\mathbf{Q}(0)\boldsymbol{\Gamma}^T$ and *defining*,

$$\tilde{\mathbf{x}}(1,-) \equiv \mathbf{A}\mathbf{x}(0) + \mathbf{B}\mathbf{q}(0) \tag{6.3.9}$$

so that (6.3.8) can be written

$$[\mathbf{I} + \mathbf{P}(1,-)\mathbf{E}^T\mathbf{R}^{-1}\mathbf{E}]\tilde{\mathbf{x}}(1) = \tilde{\mathbf{x}}(1,-) + \mathbf{P}(1,-)\mathbf{E}^T\mathbf{R}^{-1}\mathbf{y}(1) \tag{6.3.10}$$

or [factoring $\mathbf{P}(1,-)$]

$$\tilde{\mathbf{x}}(1) = \left\{\mathbf{P}(1,-)^{-1} + \mathbf{E}^T\mathbf{R}^{-1}\mathbf{E}\right\}^{-1} \mathbf{P}(1,-)^{-1}\tilde{\mathbf{x}}(1,-)$$
$$+ \left\{\mathbf{P}(1,-)^{-1} + \mathbf{E}^T\mathbf{R}^{-1}\mathbf{E}\right\}^{-1} \mathbf{E}^T\mathbf{R}^{-1}\mathbf{y}(1) . \qquad (6.3.11)$$

Applying the matrix inversion lemma in the form (3.1.24) to the first term on the right, and in the form (3.1.25) to the second term on the right,

$$\tilde{\mathbf{x}}(1) = \left\{\mathbf{P}(1,-) - \mathbf{P}(1,-)\mathbf{E}^T[\mathbf{E}\mathbf{P}(1,-)\mathbf{E}^T + \mathbf{R}]^{-1}\mathbf{E}\mathbf{P}(1,-)\right\} \cdot$$
$$\mathbf{P}(1,-)^{-1}\tilde{\mathbf{x}}(1,-) + \mathbf{P}\mathbf{E}^T[\mathbf{R} + \mathbf{E}\mathbf{P}(1,-)\mathbf{E}^T]^{-1}\mathbf{y}(1) \quad (6.3.12)$$

or

$$\tilde{\mathbf{x}}(1) = \tilde{\mathbf{x}}(1,-)+\mathbf{P}(1,-)\mathbf{E}^T[\mathbf{E}\mathbf{P}(1,-)\mathbf{E}^T+\mathbf{R}]^{-1}(\mathbf{y}(1)-\mathbf{E}\tilde{\mathbf{x}}(1,-)) . \quad (6.3.13)$$

This last result is the ordinary Kalman filter estimate, as it must be, but it results here from the adjoint formalism.

We need to consider the adjoint approach for the entire interval $0 \leq t \leq t_f$. Let us start with the objective function (6.2.8) and append the model consistency demand using Lagrange multipliers,

$$J = (\mathbf{x}(0) - \tilde{\mathbf{x}}(0))^T\mathbf{P}(0)^{-1}(\mathbf{x}(0) - \tilde{\mathbf{x}}(0))$$
$$+ \sum_{t=1}^{t_f}(\mathbf{E}(t)\mathbf{x}(t) - \mathbf{y}(t))^T\mathbf{R}(t)^{-1}(\mathbf{E}(t)\mathbf{x}(t) - \mathbf{y}(t))$$
$$+ \sum_{t=0}^{t_f-1} \mathbf{u}(t)^T\mathbf{Q}^{-1}\mathbf{u}(t)$$
$$- 2 \sum_{t=0}^{t_f-1} \boldsymbol{\mu}(t+1)^T[\mathbf{x}(t+1) - \mathbf{A}\mathbf{x}(t) - \mathbf{B}\mathbf{q}(t) - \boldsymbol{\Gamma}\mathbf{u}(t)] . \quad (6.3.14)$$

Note the differing lower limits of summation.

Setting all the derivatives to zero gives the normal equations,

$$\frac{1}{2}\frac{\partial J}{\partial \mathbf{u}(t)} = \mathbf{Q}^{-1}\mathbf{u}(t) + \boldsymbol{\Gamma}^T\boldsymbol{\mu}(t+1) = 0 , \quad 0 \leq t \leq t_f - 1 \qquad (6.3.15)$$

$$\frac{1}{2}\frac{\partial J}{\partial \boldsymbol{\mu}(t+1)} = \mathbf{x}(t+1) - \mathbf{A}\mathbf{x}(t) - \mathbf{B}\mathbf{q}(t) - \boldsymbol{\Gamma}\mathbf{u}(t) = 0 , \quad 0 \leq t \leq t_f - 1 \qquad (6.3.16)$$

$$\frac{1}{2}\frac{\partial J}{\partial \mathbf{x}(0)} = \mathbf{P}(0)^{-1}(\mathbf{x}(0) - \tilde{\mathbf{x}}(0)) + \mathbf{A}^T\boldsymbol{\mu}(1) = 0 , \qquad (6.3.17)$$

$$\frac{1}{2}\frac{\partial J}{\partial \mathbf{x}(t)} = \mathbf{E}(t)^T\mathbf{R}(t)^{-1}[\mathbf{E}(t)\mathbf{x}(t) - \mathbf{y}(t)] - \boldsymbol{\mu}(t) \qquad (6.3.18)$$

$$+ \mathbf{A}^T\boldsymbol{\mu}(t+1) = 0\,, \qquad 1 \le t \le t_f - 1$$

$$\frac{1}{2}\frac{\partial J}{\partial \mathbf{x}(t_f)} = \mathbf{E}(t_f)^T\mathbf{R}(t_f)^{-1}[\mathbf{E}(t_f)\mathbf{x}(t_f) - \mathbf{y}(t_f)] \qquad (6.3.19)$$

$$- \boldsymbol{\mu}(t_f) = 0$$

where the derivatives for $\mathbf{x}(t)$, at $t = 0,\ t = t_f$, have been computed separately for clarity. The so-called adjoint model is now given by (6.3.18). An equation count shows that the number of equations is exactly equal to the number of unknowns $[\mathbf{x}(t),\ \mathbf{u}(t),\ \boldsymbol{\mu}(t)]$. With a large enough computer, we could contemplate solving them by brute force. But for real oceanic models with large time spans and large state vectors, even the biggest supercomputers are easily swamped, and one needs to find other methods.

Before addressing the solution to these equations, consider a slight variation in their statement. The objective function, J, behaves like a Lagrangian in a mechanics problem–which is rendered stationary through the calculus of variations. In this discrete case, Equations (6.3.15)–(6.3.19) result. Define the function

$$H(\mathbf{x}(t),\, \mathbf{u}(t),\, \boldsymbol{\mu}(t)) = F(\mathbf{x}(t),\, \mathbf{u}(t)) + \boldsymbol{\mu}(t+1)^T(\mathbf{A}\mathbf{x}(t) + \mathbf{B}\mathbf{q}(t) + \boldsymbol{\Gamma}(t)\mathbf{u}(t))$$

$$F = \frac{(\mathbf{E}(t)\mathbf{x}(t) - \mathbf{y}(t))^T\mathbf{R}(t)^{-1}(\mathbf{E}(t)\mathbf{x}(t) - \mathbf{y}(t))}{2} + \frac{\mathbf{u}(t)^T\mathbf{Q}^{-1}\mathbf{u}(t)}{2}\,, \qquad (\mathbf{E}(0) = \mathbf{I},\ \mathbf{y}(0) \equiv \tilde{\mathbf{x}}(0))\,. \qquad (6.3.20)$$

Then it can be confirmed that Equations (6.3.15)–(6.3.19) are the same as

$$\boldsymbol{\mu}(t) = \frac{\partial H}{\partial \mathbf{x}(t)} \qquad (6.3.21)$$

$$\mathbf{x}(t) = \frac{\partial H}{\partial \boldsymbol{\mu}(t)} \qquad (6.3.22)$$

$$\text{min of } H \text{ w.r.t } \mathbf{u}(t): \quad \frac{\partial H}{\partial \mathbf{u}(t)} = 0\,, \qquad 0 \le t \le t_f - 1\,. \qquad (6.3.23)$$

Equations (6.3.21)–(6.3.23) are a discrete analogue to the usual Hamiltonian equations of motion, with $\boldsymbol{\mu}(t)$ playing the role of the canonical momentum. On the basis of this analogy, H is usually called the *Hamiltonian*, and Equations (6.3.21)–(6.3.23) represent a special case (linear dynamics) of the *Pontryagin minimum principle* (see, for example, Luenberger, 1979–who calls it the maximum principle, having changed a sign–or Stengel, 1986).

Equation (6.3.20) is similar to the usual relationship between a Lagrangian and a Hamiltonian. The Hamiltonian equations must be supplemented by the demand for minimization of H with respect to $\mathbf{u}$. There is an analogue of the Hamilton-Jacobi equation, usually called the *Hamilton-Jacobi-Bellman* (HJB) *equation*, which is solved using an interesting methodology called *dynamic programming* (e.g., see Brogan, 1985), not discussed here.

The Lagrange multipliers–that is, the adjoint model–have the same interpretation that they did for the steady models described in Chapter 3–that is, as a measure of the objective function sensitivity to the data,

$$\frac{1}{2}\frac{\partial J'}{\partial \mathbf{Bq}(t)} = \boldsymbol{\mu}(t+1)\,. \tag{6.3.24}$$

The physics of the adjoint model, as in Chapter 3, are again represented by the matrix $\mathbf{A}^T$. For a forward model that is both linear and self-adjoint ($\mathbf{A}^T = \mathbf{A}$), the adjoint solution would have the same physics as the state vector. If the model is not self-adjoint (the usual situation), the evolution of the adjoint may have a radically different interpretation than $\mathbf{x}(t)$. Insight into that physics is the road to understanding of information flow in the ocean. For example, if one employed a large general circulation model to compute the oceanic flux of heat, and wished to understand the extent to which the result was sensitive to the wind forcing at various places, or to a prescribed flux somewhere, the adjoint solution carries that information. In the future, one expects to see display and discussion of the results of the adjoint model on a nearly equal footing with that of the forward model.

Models and their adjoints are intimately connected in a number of ways. In the absence of observations and with known controls only, the forward and adjoint models are

$$\begin{aligned}\mathbf{x}(t+1) &= \mathbf{Ax}(t) + \mathbf{Bq}(t)\\ \boldsymbol{\mu}(t) &= \mathbf{A}^T\boldsymbol{\mu}(t+1)\,,\end{aligned}$$

and so

$$\begin{aligned}\boldsymbol{\mu}(t+1)^T\mathbf{x}(t+1) - \boldsymbol{\mu}(t)^T\mathbf{x}(t) &= \boldsymbol{\mu}(t+1)^T[\mathbf{Ax}(t) + \mathbf{Bq}(t)]\\ &\quad -\boldsymbol{\mu}(t+1)^T\mathbf{Ax}(t) = \boldsymbol{\mu}(t+1)^T\mathbf{Bq}(t)\,,\end{aligned} \tag{6.3.25}$$

which vanishes for an unforced model and is a form of "energy" conservation principle for the model and its adjoint (or model and antimodel). Identities such as (6.3.25) have uses including checks on numerical codes for forward and adjoint models and will be needed in Section 6.9.

The meteorological and oceanographic communities have taken to calling the Equation sets (6.3.15)–(6.3.19) or (6.3.21)–(6.3.23) *adjoint modeling*, when the much larger and older control theory literature refers to it as the *Pontryagin principle.* Here, one sees the origins of unnecessary cross-disciplinary semantic barriers. Some authors try to distinguish Pontryagin principle methods from sequential ones by labeling the former as *variational* techniques. But such usage suggests a misunderstanding: Both these (and other) methods represent variational principles if continuous independent variables are used, and both are algebraic if the independent variables are discrete. The classic continuous-time discussion of sequential methods is Bucy and Joseph (1968).

6.3.2 Terminal Constraint Problem: Open Loop Control

Consider first the adjoint approach in the context of the simple chemical box model already described and depicted in Figure 6–6. The following idealized situation was considered. At $t = 0$, the tracer concentrations in the boxes are known to vanish–that is, $\mathbf{x}(0) = \mathbf{x}_0 = \mathbf{0}$ (the initial conditions are supposedly known exactly). At $t = t_f$, a "ship" surveys the region, and the concentrations $\mathbf{y}(t_f) = \mathbf{E}(t_f)\mathbf{x}(t_f) + \mathbf{n}(t_f)$, $\mathbf{E}(t_f) \equiv \mathbf{I}$, $< \mathbf{n}(t) > = \mathbf{0}$, $< \mathbf{n}(t_f)\mathbf{n}(t_f)^T > = \mathbf{R}$ are known. No other observations are available. The question posed is: If the boundary conditions are all unknown a priori–that is $\mathbf{Bq} \equiv \mathbf{0}$–what boundary conditions would produce the observed values at t_f within the estimated error bars?

The problem is an example of a *terminal constraint control problem*–it seeks controls (forces, etc.) able to drive the system from an observed initial state to within a given tolerance of a required terminal state. Luenberger (1979) has a good discussion of this general class of problems. But in the present context, we interpret the result as an *estimate* of the actual boundary condition.

For this special case, take the objective function

$$J = (\mathbf{x}(t_f) - \mathbf{x}_d)^T \mathbf{R}(t_f)^{-1} (\mathbf{x}(t_f) - \mathbf{x}_d) + \sum_{t=0}^{t_f-1} \mathbf{u}^T(t) \mathbf{Q}^{-1} \mathbf{u}(t)$$
$$- 2 \sum_{0}^{t_f-1} \boldsymbol{\mu}(t+1)^T [\mathbf{x}(t+1) - \mathbf{A}\mathbf{x}(t) - \mathbf{Bq}(t) - \boldsymbol{\Gamma}\mathbf{u}(t)] \,. \quad (6.3.26)$$

We have written $\mathbf{x}_d = \mathbf{y}(t_f)$ to show that the observations at the terminal state, with an identity observation matrix, represents a desired terminal state. The Hamiltonian is

$$H(\mathbf{x}(t), \mathbf{u}(t)) = \frac{\mathbf{u}(\mathbf{t})^T \mathbf{Q}^{-1} \mathbf{u}(t)}{2} + \boldsymbol{\mu}(t+1)^T (\mathbf{A}\mathbf{x}(t) + \boldsymbol{\Gamma}(t)\mathbf{u}(t)) \,, \quad (6.3.27)$$

and the governing normal equations, either from (6.3.15)–(6.3.19) or (6.3.21)–(6.3.23), are

$$\boldsymbol{\mu}(t) = \mathbf{A}^T \boldsymbol{\mu}(t+1) \tag{6.3.28}$$

$$\boldsymbol{\mu}(t_f) = \mathbf{R}^{-1}(\mathbf{x}(t_f) - \mathbf{x}_d) \tag{6.3.29}$$

$$\mathbf{Q}^{-1}\mathbf{u}(t) = -\boldsymbol{\Gamma}^T \boldsymbol{\mu}(t+1) \tag{6.3.30}$$

plus the model.

Eliminating

$$\mathbf{u}(t) = -\mathbf{Q}\boldsymbol{\Gamma}^T \boldsymbol{\mu}(t+1) \tag{6.3.31}$$

and substituting into the model, the system to be solved is

$$\mathbf{x}(t+1) = \mathbf{A}\mathbf{x}(t) - \boldsymbol{\Gamma}\mathbf{Q}\Gamma^T \boldsymbol{\mu}(t+1)\,, \qquad \mathbf{x}(0) = \mathbf{x}_0 \equiv \mathbf{0}\,, \tag{6.3.32}$$

$$\boldsymbol{\mu}(t) = \mathbf{A}^T \boldsymbol{\mu}(t+1)\,, \qquad 1 \le t \le t_f - 1\,, \tag{6.3.33}$$

$$\boldsymbol{\mu}(t_f) = \mathbf{R}^{-1}(\mathbf{x}(t_f) - \mathbf{x}_d)\,. \tag{6.3.34}$$

The coupled problem must be solved for $\mathbf{x}(t)$, $\boldsymbol{\mu}(t)$ but with the initial conditions on the state vector provided at $t = 0$ and those for the Lagrange multipliers provided at the terminal time t_f, in terms of the still unknown $\mathbf{x}(t_f)$ [Equation (6.3.34)], recognizing that the estimated terminal state and the desired one will almost always differ.

This present problem can be solved in straightforward fashion without having to deal with the giant set of simultaneous equations by exploiting the special structure present in them. Using (6.3.34), step backward in time from t_f via (6.3.33) to produce

$$\begin{aligned}
\boldsymbol{\mu}(t_f - 1) &= \mathbf{A}^T\mathbf{R}^{-1}(\mathbf{x}(t_f) - \mathbf{x}_d) \\
&\vdots \\
\boldsymbol{\mu}(1) &= \mathbf{A}^{(t_f-1)T}\mathbf{R}^{-1}(\mathbf{x}(t_f) - \mathbf{x}_d)
\end{aligned} \tag{6.3.35}$$

so $\boldsymbol{\mu}(t)$ is given in terms of the known $\mathbf{x}_d$ and the still unknown $\mathbf{x}(t_f)$. Substituting into (6.3.32) generates

$$\begin{aligned}
\mathbf{x}(1) &= \mathbf{A}\mathbf{x}(0) - \boldsymbol{\Gamma}\mathbf{Q}\boldsymbol{\Gamma}^T\mathbf{A}^{(t_f-1)T}\mathbf{R}^{-1}(\mathbf{x}(t_f) - \mathbf{x}_d) \\
\mathbf{x}(2) &= \mathbf{A}\mathbf{x}(1) - \boldsymbol{\Gamma}\mathbf{Q}\boldsymbol{\Gamma}^T\mathbf{A}^{(t_f-2)T}\mathbf{R}^{-1}(\mathbf{x}(t_f) - \mathbf{x}_d) \\
&= \mathbf{A}^2\mathbf{x}(0) - \mathbf{A}\boldsymbol{\Gamma}\mathbf{Q}\boldsymbol{\Gamma}^T\mathbf{A}^{(t_f-1)T}\mathbf{R}^{-1}(\mathbf{x}(t_f) - \mathbf{x}_d) \\
&\quad - \boldsymbol{\Gamma}\mathbf{Q}\boldsymbol{\Gamma}^T\mathbf{A}^{(t_f-2)T}\mathbf{R}^{-1}(\mathbf{x}(t_f) - \mathbf{x}_d) \\
&\vdots \\
\mathbf{x}(t_f) &= \mathbf{A}^{t_f}\mathbf{x}(0) - \mathbf{A}^{(t_f-1)}\boldsymbol{\Gamma}\mathbf{Q}\boldsymbol{\Gamma}^T\mathbf{A}^{(t_f-1)T}\mathbf{R}^{-1}(\mathbf{x}(t_f) - \mathbf{x}_d)
\end{aligned}$$

$$- \mathbf{A}^{(t_f-2)}\boldsymbol{\Gamma}\mathbf{Q}\boldsymbol{\Gamma}^T\mathbf{A}^{(t_f-2)T}\mathbf{R}^{-1}(\mathbf{x}(t_f) - \mathbf{x}_d) - \cdots$$
$$- \boldsymbol{\Gamma}\mathbf{Q}\boldsymbol{\Gamma}^T\mathbf{R}^{-1}(\mathbf{x}(t_f) - \mathbf{x}_d)\,. \tag{6.3.36}$$

The last equation permits us to bring the terms in $\mathbf{x}(t_f)$ over to the left-hand side and solve for $\mathbf{x}(t_f)$ in terms of $\mathbf{x}_d$ and $\mathbf{x}(0)$:

$$\begin{aligned}
&\Big\{\mathbf{I} + \mathbf{A}^{(t_f-1)}\boldsymbol{\Gamma}\mathbf{Q}\boldsymbol{\Gamma}^T\mathbf{A}^{(t_f-1)T}\mathbf{R}^{-1} \\
&\quad + \mathbf{A}^{(t_f-2)}\boldsymbol{\Gamma}\mathbf{Q}\boldsymbol{\Gamma}^T\mathbf{A}^{(t_f-2)T}\mathbf{R}^{-1} + \cdots + \boldsymbol{\Gamma}\mathbf{Q}\boldsymbol{\Gamma}^T\mathbf{R}^{-1}\Big\}\,\mathbf{x}(t_f) \\
&= \mathbf{A}^{t_f}\mathbf{x}(0) + \Big\{\mathbf{A}^{(t_f-1)}\boldsymbol{\Gamma}\mathbf{Q}\boldsymbol{\Gamma}^T\mathbf{A}^{(t_f-1)T}\mathbf{R}^{-1} \\
&\quad + \mathbf{A}^{(t_f-2)}\boldsymbol{\Gamma}\mathbf{Q}\boldsymbol{\Gamma}^T\mathbf{A}^{(t_f-2)T}\mathbf{R}^{-1} + \cdots + \boldsymbol{\Gamma}\mathbf{Q}\boldsymbol{\Gamma}^T\mathbf{R}^{-1}\Big\}\,\mathbf{x}_d\,.
\end{aligned} \tag{6.3.37}$$

With $\mathbf{x}(t)$ now known, $\boldsymbol{\mu}(t)$ can be computed for all t from (6.3.35). Then the control $\mathbf{u}(t)$ is also known from (6.3.30) and the state vector can be found from (6.3.32). The resulting solution for $\tilde{\mathbf{u}}(t)$ is in terms of the externally prescribed $\mathbf{x}_0$, $\mathbf{x}_d$ and is usually known as *open-loop* control in contrast to a different form of solution discussed below.

Let $\tilde{\mathbf{x}}(t_f)$ be as shown in Figure 6–9a. The initial conditions were taken as zero. Then Figure 6–9b shows the controls able to produce these terminal conditions within a tolerance governed by $\mathbf{R}$. Wunsch (1988b) shows a variety of calculations as a function of variations in the terminal constraint accuracies. The adjoint solution is depicted in Figure 6–9c. A version of this methodology was used by Wunsch (1988a) and Memery and Wunsch (1990) to discuss the evolution of observed tritium in the North Atlantic.

The canonical form for a terminal constraint problem usually used in the literature differs slightly; it is specified in terms of a given, nonzero, initial condition $\mathbf{x}(0)$, and the controls are determined so as to come close to a desired terminal state, which is zero. The solution to this so-called *deadbeat* control (driving the system to rest) problem can be used to solve the problem for an arbitrary desired terminal state.

Suppose we have the solution to the deadbeat control problem, given by $\mathbf{u}(\mathbf{x}_0,\, \mathbf{x}(t_f) = \mathbf{0},\, t)$. Then from (6.3.35), $\boldsymbol{\mu}(t)$ is a function of $\mathbf{x}(t_f)$, and hence so is $\mathbf{u}(t)$ [from (6.3.30)]. It would therefore suffice to replace $\mathbf{x}(t_f)$ in the deadbeat controller with $\mathbf{u}(\mathbf{x}_0,\, \mathbf{x}(t_f) - \mathbf{x}_d,\, \mathbf{0},\, t)$, at least in the linear case.

The smoothing problem has been solved without computing the uncertainty and the advantage of adjoint/Pontryagin methods over the sequential estimators. Adjoint methods solve for the entire time domain at once; consequently, there is no weighted averaging of intermediate solutions and no need for the uncertainties. On the other hand, in view of everything that

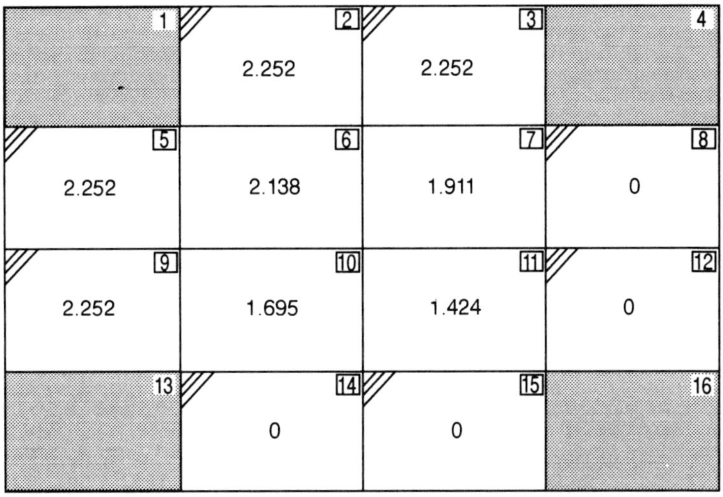

(a)

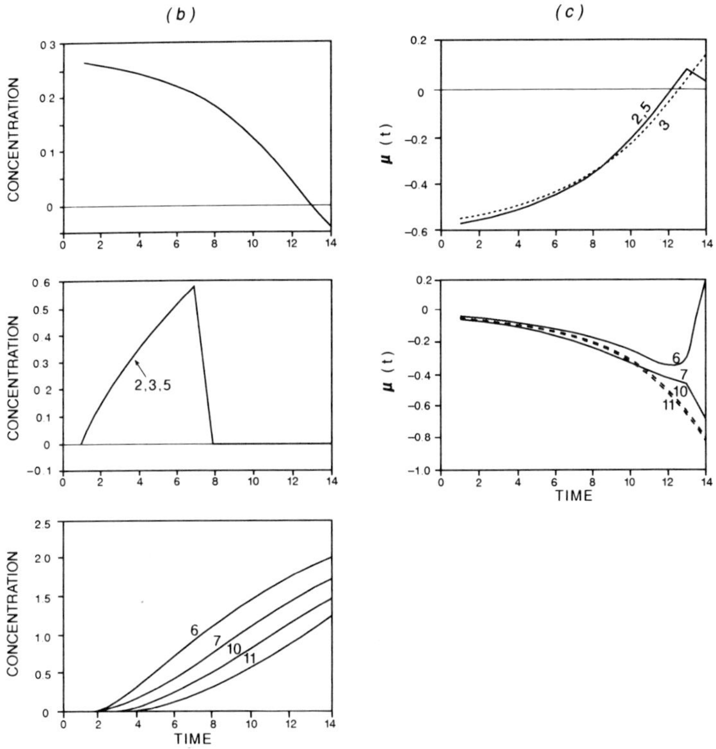

Figure 6–9a. Terminal time concentrations (corresponding to the time histories in Figure 6–6b), representing the terminal constraint (within error bars) of the control problem for transient tracers.
b,c. (b) The upper panel displays the concentration rate of change–that is, du/dt–of the boundary concentrations (controls) as estimated through the Pontryagin principle. The middle panel shows du/dt for the "truth," and the lower panel shows the time history of tracer concentrations in the interior boxes from the controls in the upper panel. These interior values can be compared to the "truth" in Figure 6–6b. The terminal constraint can be reached within the tolerances imposed from the initial conditions of zero tracer concentration by a set of boundary controls quite different from the truth, and the observations only poorly constrain the boundary conditions. (c) The upper panel shows the Lagrange multipliers (adjoint solution) for boundary boxes 2, 3, 5, and the lower panel for the interior boxes. As is typical of many control problems, the adjoint solution shows its greatest structure near the terminal time when comparatively large "forces" are exerted by the controls to push the system into the required terminal state.

has come before in this book, the utility of solutions without uncertainty estimates must be questioned. In the context of Chapter 1, problems of arbitrary posedness are being solved. The various methods using objective functions, prior statistics, etc., whether in time-evolving or static situations, permit stable, useful estimates to be made under almost any circumstances, using almost any sort of available information. But the reader will by now appreciate that the use of such methods can produce structures in the solution, pleasing or otherwise, that may be present because they are required by (1) the observations, (2) the model, (3) the prior statistics, (4) some norm or smoothness demand on elements of the solution, or (5) all of the preceding in concert. A solution produced in ignorance of these differing sources of structure can hardly be thought very useful, and it is the uncertainty matrices that are usually the key to understanding. Consequently, we will later spend some time examining the problem of obtaining the missing covariances. In the meantime, one should note that the covariances of the filter/smoother will also describe the uncertainty of the adjoint/Pontryagin solution, because they are the same solution to the same set of equations deriving from the same objective function.

There is one situation where a solution without uncertainty estimates is plainly useful–it is where one simply inquires, "Is there a solution at all?"–that is, when one wants to know if the observations actually contradict the model. In that situation, mere existence of an acceptable solution may be of greatest importance, suggesting, for example, that a model of adequate complexity is already available.

6.3.3 Method of Unit Solutions/Boundary Green's Functions

To gain further insight into the adjoint/Pontryagin equations, consider another approach to solving (6.3.32)–(6.3.34) with a so-called shooting method often used for two-point boundary values problems; see Roberts and Shipman (1972). We here follow Bryson and Ho (1975).

Rewrite (6.3.33) as

$$\boldsymbol{\mu}(t+1) = \mathbf{A}^{-T}\boldsymbol{\mu}(t), \tag{6.3.38}$$

thus reversing the sense of the recursion. Then substituting for $\boldsymbol{\mu}(t+1)$, (6.3.32)–(6.3.33) become [setting $\mathbf{x}_d = \mathbf{0}$ for the deadbeat case but with

$\tilde{\mathbf{x}}(0)$ not necessarily zero]

$$\begin{bmatrix} \mathbf{x}(t+1) \\ \boldsymbol{\mu}(t+1 \end{bmatrix} = \begin{Bmatrix} \mathbf{A} & -\boldsymbol{\Gamma}\mathbf{Q}\boldsymbol{\Gamma}^T\mathbf{A}^{-T} \\ \mathbf{0} & \mathbf{A}^{-T} \end{Bmatrix} \begin{bmatrix} \mathbf{x}(t) \\ \boldsymbol{\mu}(t) \end{bmatrix} \tag{6.3.39}$$

or

$$\mathbf{q}(t+1) = \mathbf{H}\mathbf{q}(t) \tag{6.3.40}$$

where $\mathbf{H}$ is known as a *Hamiltonian matrix* (but it is not the Hamiltonian of the Pontryagin principle).

Such Hamiltonian matrices have what is known as the symplectic property–that is, if

$$\mathbf{J} = \begin{Bmatrix} \mathbf{0} & \mathbf{I} \\ -\mathbf{I} & \mathbf{0} \end{Bmatrix}, \tag{6.3.41}$$

then

$$\mathbf{H}^T\mathbf{J}\mathbf{H} = \mathbf{J}, \tag{6.3.42}$$

and it follows by inspection (see Goldstein, 1980) that if we write

$$\mathbf{H} = \begin{Bmatrix} \mathbf{H}_{aa} & \mathbf{H}_{ab} \\ \mathbf{H}_{ba} & \mathbf{H}_{bb} \end{Bmatrix}, \tag{6.3.43}$$

then

$$\mathbf{H}^{-1} = \begin{Bmatrix} \mathbf{H}_{bb}^T & -\mathbf{H}_{ab}^T \\ -\mathbf{H}_{ba}^T & \mathbf{H}_{aa}^T \end{Bmatrix}. \tag{6.3.44}$$

Hence, this system could be solved backward in time from t_f except that the terminal values $\mathbf{x}(t_f)$ are unknown. Let us seek solutions to the system in which we successively set all the terminal values to zero, with the exception of one element of $\mathbf{x}(t_f)$. That is, let $\mathbf{x}^{(i)}(t)$, $\boldsymbol{\mu}^{(i)}(t)$ be the solution to the system under the assumption

$$\begin{aligned} x_j^{(i)}(t_f) &= 0, \qquad i \neq j \\ &= 1, \qquad i = j \\ \boldsymbol{\mu}_j^{(i)}(t_f) &= (\mathbf{R}^{-1})_{ij}. \end{aligned}$$

The last requirement on $\boldsymbol{\mu}(t_f)$ is simply the appropriate boundary condition from

$$\boldsymbol{\mu}(t_f) = \mathbf{R}^{-1}(\mathbf{x}(t_f) - \mathbf{x}_d) \tag{6.3.45}$$

with $\mathbf{x}_d = \mathbf{0}$.

With the terminal value of $\mathbf{x}^{(i)}$ supposedly known, we can solve the model backward in time as

$$\begin{aligned}\mathbf{q}^{(i)}(t-1) &= \mathbf{H}^{-1}\mathbf{q}^{(i)}(t) \\ \mathbf{q}^{(i)}(t) &= \begin{bmatrix}\mathbf{x}^{(i)}(t) \\ \boldsymbol{\mu}^{(i)}(t)\end{bmatrix}\end{aligned} \tag{6.3.46}$$

so that the implied values at $t = 0$, $\mathbf{x}^{(i)}(0)$, $\boldsymbol{\mu}^{(i)}(0)$ are known. Construct a matrix $\mathbf{X}(t)$ whose i-th column is $\mathbf{x}^{(i)}(t)$ and a matrix $\mathbf{M}(t)$ whose i-th column is $\boldsymbol{\mu}^{(i)}(t)$. Then it follows that

$$\mathbf{X}(t_f) = \mathbf{I},\ \mathbf{M}(t_f) = \mathbf{R}^{-1}. \tag{6.3.47}$$

If $\mathbf{x}(t_f)$ were known, the solution could be written

$$\mathbf{x}(t) = \mathbf{X}(t)\mathbf{x}(t_f)\,, \qquad \boldsymbol{\mu}(t) = \mathbf{M}(t)\mathbf{x}(t_f)\,. \tag{6.3.48}$$

But $\mathbf{x}(0)$ is known, and (6.3.48) produces

$$\mathbf{x}(t_f) = \mathbf{X}(0)^{-1}\mathbf{x}(0)\,, \tag{6.3.49}$$

and the solution to the problem is

$$\begin{aligned}\mathbf{x}(t) &= \mathbf{X}(t)\mathbf{X}(0)^{-1}\mathbf{x}_0 \\ \boldsymbol{\mu}(t) &= \mathbf{M}(t)\mathbf{X}(0)^{-1}\mathbf{x}_0\end{aligned} \tag{6.3.50}$$

with the control

$$\mathbf{u}(t) = -\mathbf{Q}\boldsymbol{\Gamma}^T\mathbf{M}(t+1)\mathbf{X}(0)^{-1}\mathbf{x}_0\,. \tag{6.3.51}$$

Because the control is given for the same objective function as in the last section, it must be the same solution. As with the previous solution, $t = 0$ can be replaced by an arbitrary time $t = t_0$, and the solution determines the control to be applied; $\mathbf{X}(t)\mathbf{X}(0)^{-1}$ is the Green's function for the problem.

This method can lead to numerical difficulties, as is to be anticipated any time a model with dissipation must be run backward in time. However, it is useful, both computationally and in theory. Nonetheless, one is led to consider other procedures.

6.3.4 The Control Riccati Equation

Let us solve the adjoint/Pontryagin principle equations in yet another way. (If the reader is wondering why such a fuss is being made about these equations, the answer, among others, is that it will turn out to be an important route to reducing the computational load required for the Kalman filter and various smoothing algorithms.) We look at the same special case of the objective function, (6.3.26), and the equations that follow from it [(6.3.28)–(6.3.30) plus the model]. Before proceeding, note that with the definition $\boldsymbol{\mu}(t_f+1) = \mathbf{0}$, the deadbeat terminal constraint requirement can be put into the adjoint equation in the form

$$\mu(t) = \mathbf{A}^T\boldsymbol{\mu}(t+1) + \mathbf{R}(t)^{-1}\mathbf{x}(t)\,, \qquad 1 \le t \le t_f\,, \tag{6.3.52}$$

stipulating that $\mathbf{R}(t)^{-1} = \mathbf{0}$, $t \neq t_f$.

Take a trial solution, an *ansatz*, in the form

$$\boldsymbol{\mu}(t+1) = \mathbf{S}(t+1)\mathbf{x}(t+1) \tag{6.3.53}$$

where $\mathbf{S}(t+1)$ is unknown. Then Equation (6.3.30) becomes

$$\mathbf{Q}^{-1}\mathbf{u}(t) + \boldsymbol{\Gamma}^T\mathbf{S}(t+1)\mathbf{x}(t+1) = \mathbf{0} \tag{6.3.54}$$

or, using the model,

$$\mathbf{Q}^{-1}\mathbf{u}(t) + \boldsymbol{\Gamma}^T\mathbf{S}(t+1)(\mathbf{A}\mathbf{x}(t) + \boldsymbol{\Gamma}\mathbf{u}(t)) = \mathbf{0}\,. \tag{6.3.55}$$

So

$$\begin{aligned}\mathbf{u}(t) &= -\left\{\boldsymbol{\Gamma}^T\mathbf{S}(t+1)\boldsymbol{\Gamma} + \mathbf{Q}^{-1}\right\}^{-1}\Gamma^T\mathbf{S}(t+1)\mathbf{A}\mathbf{x}(t)\\ &= -\mathbf{L}^{-1}\boldsymbol{\Gamma}^T\mathbf{S}(t+1)\mathbf{A}\mathbf{x}(t)\\ \mathbf{L} &= \boldsymbol{\Gamma}^T\mathbf{S}(t+1)\boldsymbol{\Gamma} + \mathbf{Q}^{-1}\,.\end{aligned} \tag{6.3.56}$$

Substituting this last expression, and (6.3.53) for $\boldsymbol{\mu}(t)$, into the adjoint model (6.3.52),

$$\begin{aligned}\{\mathbf{A}^T\mathbf{S}(t+1)\mathbf{A} - \mathbf{A}^T\mathbf{S}(t+1)\boldsymbol{\Gamma}\mathbf{L}(t+1)^{-1}\boldsymbol{\Gamma}^T\mathbf{S}(t+1)\mathbf{A} -\\ \mathbf{S}(t) + \mathbf{R}(t)^{-1}\}\,\mathbf{x}(t) = \mathbf{0}\,.\end{aligned} \tag{6.3.57}$$

Unless $\mathbf{x}(t)$ is to vanish identically,

$$\mathbf{S}(t) = \mathbf{A}^T\mathbf{S}(t+1)\mathbf{A} - \mathbf{A}^T\mathbf{S}(t+1)\boldsymbol{\Gamma}\mathbf{L}^{-1}\boldsymbol{\Gamma}^T\mathbf{S}(t+1)\mathbf{A} + \mathbf{R}(t)^{-1}\,, \tag{6.3.58}$$

a nonlinear difference equation, known as the *matrix Riccati equation*, which produces a backward recursion for $\mathbf{S}(t)$. Start the recursion with

$$\mathbf{S}(t_f)\mathbf{x}(t_f) = \mathbf{R}(t_f)^{-1}\mathbf{x}(t_f) \quad \text{or} \quad \mathbf{S}(t_f) = \mathbf{R}(t_f)^{-1}\,, \tag{6.3.59}$$

and step backward to $t = 0$. The problem has now been solved–by what is called the *sweep method* (Bryson & Ho, 1975). Notice that with $\mathbf{S}(t)$ known, the control is in the form

$$\boldsymbol{\Gamma}\mathbf{u}(t) = \mathbf{K}_c(t)\mathbf{x}(t)\,, \tag{6.3.60}$$

known as *feedback control* because the values to be applied are determined by the value of the state vector at that time. It contrasts with the open-loop control derived above but necessarily produces the identical answer.

With feedback control, the computation of the model update step would now be

$$\mathbf{x}(t+1) = (\mathbf{A} - \mathbf{K}_c)\mathbf{x}(t) + \mathbf{B}\mathbf{q}(t). \tag{6.3.61}$$

The structure of the matrix

$$\mathbf{A}' = \mathbf{A} - \mathbf{K}_c \tag{6.3.62}$$

is the center of a discussion of the stability of the scheme (see references).

There are thus several ways to solve the adjoint/Pontryagin condition equations. In particular, the very nonlinear Riccati equation (6.3.58) is equivalent to the linear system (6.3.39) involving the Hamiltonian matrix–that is, the nonlinear difference equation is algebraically identical to a set of linear equations but has twice the vector dimension.

Following Franklin, Powell, and Workman (1990), put

$$\begin{aligned}\mathbf{S}(t,-) &\equiv \mathbf{S}(t+1) - \mathbf{S}(t+1)\boldsymbol{\Gamma}[\boldsymbol{\Gamma}^T\mathbf{S}(t+1)\boldsymbol{\Gamma} + \mathbf{Q}^{-1}]^{-1}\boldsymbol{\Gamma}^T\mathbf{S}(t+1)\\ \mathbf{S}(t) &= \mathbf{A}^T\mathbf{S}(t,-)\mathbf{A} + \mathbf{R}(t)^{-1},\end{aligned} \tag{6.3.63}$$

a simple way of rewriting the Riccati equation–the significance of which is taken up in Section 6.4.

6.3.5 The Initialization Problem

Another special case of wide interest is determination of the initial conditions, $\tilde{\mathbf{x}}(0)$, from later observations, which has attracted wide meteorological notice (e.g., LeDimet & Talagrand, 1986). Let us attempt it using several different methods.

For notational simplicity and without loss of generality, assume that the known controls vanish so that the model is

$$\mathbf{x}(t+1) = \mathbf{A}\mathbf{x}(t) + \boldsymbol{\Gamma}\mathbf{u}(t), \tag{6.3.64}$$

that there is an existing estimate of the initial conditions, $\tilde{\mathbf{x}}_0(0)$, with estimated uncertainty $\mathbf{P}(0)$, and that there is a single terminal observation of the complete state,

$$\mathbf{y}(t_f) = \mathbf{E}\mathbf{x}(t_f) + \mathbf{n}(t_f), \quad \mathbf{E} = \mathbf{I} \tag{6.3.65}$$

where the observational noise covariance is again $\mathbf{R}(t_f)$.

We now can solve this problem in four seemingly different but nonetheless identical ways:

1. The terminal observations can be written explicitly in terms of the initial conditions as

$$\mathbf{y}(t_f) = \mathbf{A}^{t_f}\mathbf{x}(0) + \mathbf{A}^{t_f-1}\boldsymbol{\Gamma}\mathbf{u}(0) + \mathbf{A}^{t_f-2}\boldsymbol{\Gamma}\mathbf{u}(1) + \cdots + \boldsymbol{\Gamma}\mathbf{u}(t_f-1) + \mathbf{n}(t_f)\,, \qquad (6.3.66)$$

which is in canonical observation equation form,

$$\mathbf{y}(t_f) = \mathbf{E}_p\mathbf{x}(0) + \mathbf{n}_p(t_f)\,, \qquad \mathbf{E}_p = \mathbf{A}^{t_f}\,, \quad \mathbf{n}_p = \mathbf{A}^{t_f-1}\boldsymbol{\Gamma}\mathbf{u}(0) + \cdots + \boldsymbol{\Gamma}\mathbf{u}(t_f-1) + \mathbf{n}(t_f)\,, \qquad (6.3.67)$$

and where the covariance of this combined error is

$$\mathbf{R}_p \equiv <\mathbf{n}_p\mathbf{n}_p^T> = \mathbf{A}^{t_f-1}\boldsymbol{\Gamma}\mathbf{Q}\boldsymbol{\Gamma}^T\mathbf{A}^{(t_f-1)T} + \cdots + \boldsymbol{\Gamma}\mathbf{Q}\boldsymbol{\Gamma}^T + \mathbf{R}(t_f)\,. \qquad (6.3.68)$$

Then the least-squares recursive solution leads to

$$\tilde{\mathbf{x}}(0) = \tilde{\mathbf{x}}_0(0) + \mathbf{P}(0)\mathbf{E}_p^T\left(\mathbf{E}_p\mathbf{P}(0)\mathbf{E}_p^T + \mathbf{R}_p\right)^{-1}\left[\mathbf{y}(t_f) - \mathbf{E}_p\tilde{\mathbf{x}}_0(0)\right], \qquad (6.3.69)$$

and the uncertainty estimate follows immediately.
2. A second method (which the reader should confirm produces the same answer) is to run the Kalman filter forward to t_f and then run the smoother backward to $t = 0$. There is more computation here, but a byproduct is an estimate of the intermediate values of the state vectors, of the controls, and their uncertainty.
3. Write the model in backward form,

$$\mathbf{x}(t) = \mathbf{A}^{-1}\mathbf{x}(t+1) - \mathbf{A}^{-1}\boldsymbol{\Gamma}\mathbf{u}\,, \qquad (6.3.70)$$

and use the Kalman filter on this model, with time running backward. The observation equation (6.3.65) provides the initial estimate of $\mathbf{x}(t_f)$, and its error covariance becomes the initial estimate covariance $\mathbf{P}(t_f)$. At $t = 0$, the original estimate of $\tilde{\mathbf{x}}_0(0)$ is treated as an observation, with uncertainty $\mathbf{P}(0)$ taking the place of the usual $\mathbf{R}$. The reader should again confirm that the answer is the same as in (1).
4. The problem has already been solved using the adjoint/Pontryagin formalism.

6.3.6 Adjoints and Smoothing

The adjoint/Pontryagin principal approach can be used to solve the general smoothing problem–that is, finding the minimum of (6.2.26) subject to the model, which leads to the general Equations (6.3.31)–(6.3.34). In recent years, a significant oceanographic literature has grown up concerning the solution of these equations (Thacker & Long, 1988; Wunsch, 1988a,b; Tziperman & Thacker, 1989; Sheinbaum & Anderson, 1990; Tziperman et al., 1992a,b; Marotzke & Wunsch, 1993; and others). Generally, the large equation set is solved iteratively rather than by brute force or by use of any of the exact expressions discussed above. Iteration becomes essential if the dynamical model is nonlinear, as described below.

Tziperman et al. (1992b) carried out a computation differing somewhat from the smoothing calculation. Their model was a GCM describing the entire North Atlantic Ocean. The objective function was conventional in penalizing deviations of the model from observed temperature and salinity fields, as well as observed boundary conditions for windstress and thermodynamic fluxes. The equations were those used in Sections 2.1.1–2.1.5, including time dependence, but they used separate temperature and salinity conservation equations (with diffusion terms) instead of simple density conservation. They appended to the objective function a requirement minimizing the deviation of the results from a steady state. That is, they sought a model that would produce a steady flow, not one that followed the data through time. The data were thus treated as though representing a steady ocean. Figure 6–10 shows an estimate of the temperature fields along a fixed depth before and after the system is solved.

There is some similarity of the Tziperman et al. (1992a,b) calculation and the terminal constraint control problem, where the terminal constraint contains a requirement that the system must become essentially steady. But their objective function is somewhat different from a canonical one. Marotzke (1992) and Marotzke and Wunsch (1993) discuss some of the issues in solving this type of problem.

The issue of estimating the uncertainty of the solution remains, even if the system is linear. One approach is to calculate it from the covariance evolution equation of the filter/smoother. When one wishes to avoid that computational load, Thacker (1989) and Marotzke and Wunsch (1993) discuss the possibility of obtaining some limited information about it from the Hessian of the cost function at the solution point.

Understanding of the Hessian is central to quadratic norm optimization problems in general. Let $\boldsymbol{\xi}$ represent all of the variables being optimized,

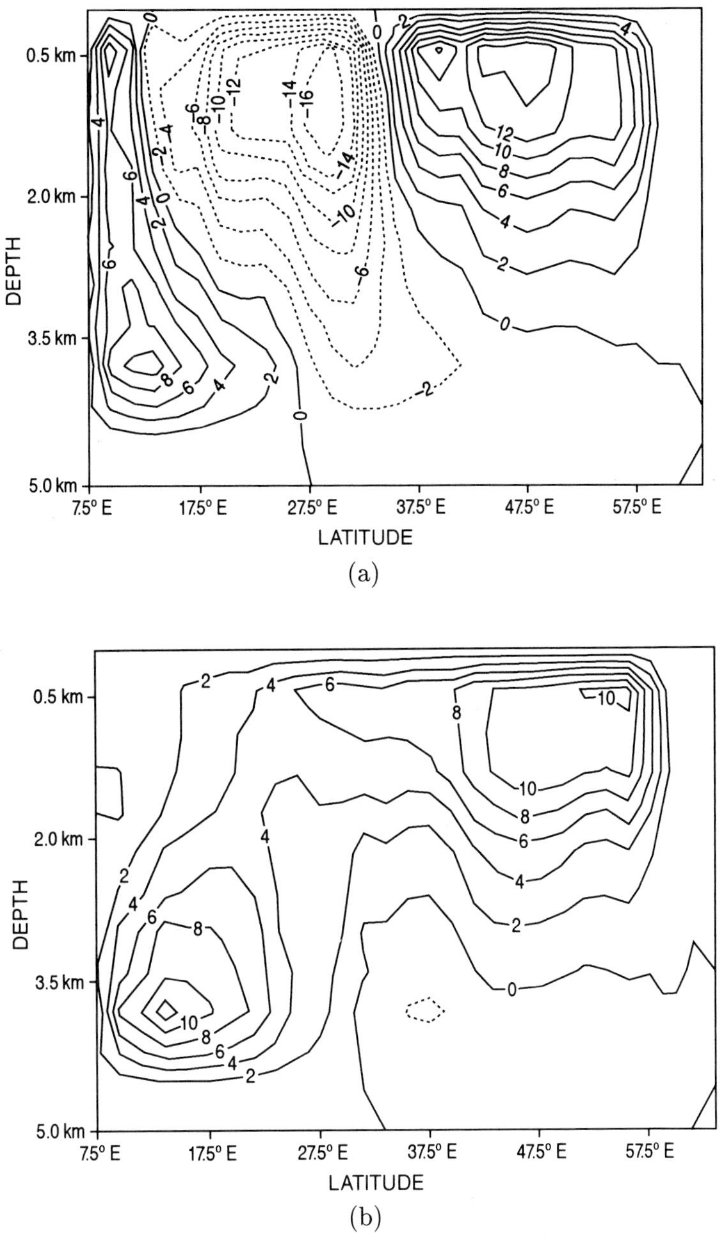

Figure 6–10. Computation by Tziperman et al. (1992b) of the zonally integrated circulation of the North Atlantic (a) before and (b) after optimization, using the adjoint method and a cost function that penalized deviations of the model from steady flow conditions.

including $\mathbf{x}(t)$, $\mathbf{u}(t)$ for all t. Let $\boldsymbol{\xi}^*$ be the optimal value that is sought. Then, as with the static problems of Chapter 3, if we are close enough to ξ^* in the search process, the objective function is locally

$$J = \text{constant} + (\boldsymbol{\xi} - \boldsymbol{\xi}^*)^T \, \mathcal{H}(\boldsymbol{\xi} - \boldsymbol{\xi}^*) + \Delta J$$

where $\mathcal{H}$ is the Hessian and ΔJ is a higher-order correction. The discussion of the behavior of the solution in the vicinity of the estimated optimal value proceeds then, exactly as before with row and column scaling being relevant, and issues of ill-conditioning, solution variances, etc., all depending upon the eigenvalues and eigenvectors of $\mathcal{H}$. For example, Tziperman et al. (1992b) grapple with ill-conditioning in their results; the ill-conditioning is interpretable as arising from a nullspace in the Hessian.

The only problem, albeit a difficult one, is that the dimensions of $\mathcal{H}$ are square of the dimensions of $\mathbf{x}(t)$ plus $\mathbf{u}(t)$ over the entire time history of the model and data. Finding ways to understand the solution structure and uncertainty with realistic oceanic circulation models and large-scale datasets remains as one of the most important immediate challenges in oceanography. (See the further discussion in Section 6.8.)

6.4 Duality and Simplification: The Steady-State Filter and Adjoint

For linear models, the adjoint/Pontryagin method and the filter/smoother algorithms produce identical solutions. In both cases, the computation of the uncertainty remains an issue–in the former case because it is not part of the solution, and in the latter because it may overwhelm the computation. However, if the uncertainty is computed for the sequential estimator solutions, it must also represent the uncertainty derived from the adjoint/ Pontryagin principle. In the interests of gaining insight into both methods, and of ultimately finding uncertainty estimates, consider the covariance propagation equation for the Kalman filter:

$$\mathbf{P}(t+1,-) = \mathbf{A}(t)\mathbf{P}(t)\mathbf{A}(t)^T + \boldsymbol{\Gamma}(t)\mathbf{Q}(t)\boldsymbol{\Gamma}^T \tag{6.4.1}$$

$$\begin{aligned}\mathbf{P}(t+1) = \mathbf{P}(t+1,-) \\ - \mathbf{P}(t+1,-)\mathbf{E}(t+1)^T[\mathbf{E}(t+1)\mathbf{P}(t+1,-)\mathbf{E}(t+1)^T \\ + \mathbf{R}(t+1)]^{-1}\mathbf{E}(t+1)\mathbf{P}(t+1,-)\end{aligned} \tag{6.4.2}$$

where we have substituted for $\mathbf{K}(t+1)$. If we make the substitutions shown in Table 6.1, the equations for evolution of the uncertainty of the Kalman filter are identical to those for the control matrix $\mathbf{S}(t)$, given in Equation (6.3.58); hence, the Kalman filter covariance *also satisfies a matrix Riccati equation*. Time runs backward in the control formulation and forward in the estimation problem, but this difference is not fundamental. The significance of this result is that simplifications and insights obtained from one problem can be employed on the other (some software literally

Table 6.1. *Correspondences between the variables of the control formulation and that of the Kalman filter, which lead to the Riccati equation. Note that time runs backward for control cases and forward for the filter.*

Adjoint/Control	Kalman Filter
$\mathbf{A}$	$\mathbf{A}^T$
$\mathbf{S}(t,-)$	$\mathbf{P}(t+1)$
$\mathbf{S}(t+1)$	$\mathbf{P}(t+1,-)$
$\mathbf{R}^{-1}$	$\boldsymbol{\Gamma}\mathbf{Q}\boldsymbol{\Gamma}^T$
$\boldsymbol{\Gamma}$	$\mathbf{E}^T$
$\mathbf{Q}^{-1}$	$\mathbf{R}$

makes the substitutions of Table 6.1 to compute the Kalman filter solution from the algorithm for solving the control Riccati equation).

This feature–that both problems produce a matrix Riccati equation–is referred to as the *duality* of estimation and control. It does *not* mean that they are the same problem; in particular, recall that the control problem is equivalent not to filtering but to smoothing.

As we have seen, the computation of the covariances often dominates the Kalman filter (and smoother) and sometimes leads to the conclusion that the procedures are impractical. But as with all linear least-square–like estimation problems, the state vector uncertainty does not depend upon the actual data values, only upon the prior error covariances. Thus, the filter and smoother uncertainties (and the filter and smoother gains) can be computed in advance of the actual application to data, and stored. The computation can be done either by stepping through the recursion in (6.4.1)–(6.4.2) starting from $t = 0$ or by converting to the equivalent linear equations involving the Hamiltonian through the relations in Table 6.1, or any other method producing the solution to the adjoint/Pontryagin equations.

Furthermore, it was pointed out that in Kalman filter problems, the covariances and Kalman gain approach a steady state, in which $\mathbf{P}(t)$, $\mathbf{P}(t,-)$, $\mathbf{K}(t)$ do not depend upon time after awhile. Physically, the growth in error from the propagation Equation (6.4.1) is just balanced by the reduction in uncertainty from the incoming data stream (6.4.2). [This simple description supposes the data come in at every time step; often the data appear only intermittently (but periodically), and the steady-state solution is periodic–errors displaying a characteristic saw-tooth structure between observation times. See the examples in Ghil et al. (1981) or Figure 6–5e.]

If one can find these steady-state values, then the necessity to update the covariances and gain matrix disappears, and the computational load is much reduced, potentially by many orders of magnitude. The equivalent steady state for the control problem is best interpreted in terms of the feedback gain control matrix $\mathbf{K}_c$, which can also become time independent, meaning that the value of the control to be applied depends only upon the state observed at time t and need not be recomputed at each time step.

The great importance of steady-state estimation and control has led to a large number of methods for obtaining the solution of the various steady-state Riccati equations requiring one of, ($\mathbf{S}(t+1) = \mathbf{S}(t)$, $\mathbf{S}(t+1,-) = \mathbf{S}(t,-)$, $\mathbf{P}(t) = \mathbf{P}(t+1)$, or $\mathbf{P}(t+1,-) = \mathbf{P}(t,-)$), many of which are discussed, for example, in Stengel (1986), Franklin *et al.* (1990), Anderson and Moore (1979), Bittanti, Laub, and Willems (1991a) and are used in an oceanographic context by Fukumori et al. (1993) and Fu et al. (1993). Some methods are based upon the Hamiltonian form, and some upon the Riccati equation itself. [The steady-state equation is often known as the *algebraic Riccati equation.* See Reid (1972) for a discussion of the history of the Riccati equation in general; it is intimately related to Bessel's equation and has been studied in scalar form since the 18th century. Bittanti et al. (1991a) discuss many different aspects of the matrix form.]

A steady-state solution to the Riccati equation would correspond not only to a determination of the steady-state filter and smoother covariances but also to the steady-state solution of the adjoint/Pontryagin equations–a so-called steady-state control. Generalizations to the steady-state problem exist; an important one is the possibility of a periodic steady state discussed by Bittanti et al. (1991b).

Before seeking a steady-state solution, one must determine whether one exists. That no such solution will exist in general is readily seen by considering, for example, a physical system in which certain components (elements of the flow) are not readily observed. If these components are initialized with partially erroneous values, then if they are unstable, they will grow without bound, and there will be no limiting steady-state value for the uncertainty, which will also have to grow without bound. Alternatively, suppose there are elements of the state vector whose values cannot be modified by the available control variables. Then no observations of the state vector produce information about the control variables; if the control vector uncertainty is described by $\mathbf{Q}$, then this uncertainty will accumulate from one time step to another, growing without bound with the number of time steps.

6.5 Controllability and Observability

In addition to determining whether there exists a steady-state solution either to the control or estimation Riccati equations, there are many reasons for examining in some detail the existence of many of the matrix operations that have been employed routinely throughout. Matrix inverses occur throughout the developments above, and the issue of whether they actually exist has been ignored. Ultimately, however, one must face up to questions of whether some of the computations (e.g., the determination of deadbeat controls, or the finding of initial conditions from later observations, or the steady-state solution to the Riccati equation) are actually possible. The questions are intimately connected to some very useful structural descriptions of models and data that we will now examine briefly. (The Fukumori et al., 1993, discussion of the steady-state Riccati equation solution relies heavily on the ideas that follow.)

Consider the question of whether controls can be found to drive a system from a given initial state $\mathbf{x}(0)$ to an arbitrary $\mathbf{x}(t_f)$. If the answer is "yes," the system is said to be *controllable*. To find an answer, consider for simplicity (following Franklin et al., 1990) a model with $\mathbf{B} = \mathbf{0}$ and with the control, u, a scalar. Then the model time steps can be written

$$\begin{aligned}
\mathbf{x}(1) &= \mathbf{A}\mathbf{x}(0) + \boldsymbol{\Gamma} u(0) \\
\mathbf{x}(2) &= \mathbf{A}\mathbf{x}(1) + \boldsymbol{\Gamma} u(1) \\
&= \mathbf{A}^2\mathbf{x}(0) + \mathbf{A}\boldsymbol{\Gamma} u(0) + \boldsymbol{\Gamma} u(1) \\
&\vdots \\
\mathbf{x}(t_f) &= \mathbf{A}^{t_f}\mathbf{x}(0) + \sum_{j=0}^{t_f-1} \mathbf{A}^{t_f-1-j}\boldsymbol{\Gamma} u(j) \\
&= \mathbf{A}^{t_f}\mathbf{x}(0) + [\boldsymbol{\Gamma}\ \mathbf{A}\boldsymbol{\Gamma} \cdots \mathbf{A}^{t_f-1}\boldsymbol{\Gamma}] \begin{bmatrix} u(t_f-1) \\ \vdots \\ u(0) \end{bmatrix} .
\end{aligned} \tag{6.5.1}$$

To determine $u(t)$, we must be able to solve the system

$$[\boldsymbol{\Gamma}\ \mathbf{A}\boldsymbol{\Gamma} \cdots \mathbf{A}^{t_f-1}\boldsymbol{\Gamma}] \begin{bmatrix} u(t_f-1) \\ \vdots \\ u(0) \end{bmatrix} = \mathbf{x}(t_f) - \mathbf{A}^{t_f}\mathbf{x}(0) ,$$

or

$$\mathbf{C}\mathbf{u} = \mathbf{x}(t_f) - \mathbf{A}^{t_f}\mathbf{x}(0) \tag{6.5.2}$$

for $u(t)$. The state vector dimension is N; therefore the dimension of $\mathbf{C}$

is N by the number of columns, t_f [a special case–$u(t)$ being scalar means that $\boldsymbol{\Gamma}$ is $N \times 1$]. Therefore, Equation (6.5.2) has no (ordinary) solution if t_f is less than N. If $t_f = N$ and $\mathbf{C}$ is nonsingular–that is, of rank N–there is a unique solution, and the system is controllable. If the dimensions of $\mathbf{C}$ are nonsquare, one could have a discussion, familiar from Chapter 3, of solutions for $u(t)$ with nullspaces present. If $t_f < N$, there is a nullspace of the desired output, and the system would not be controllable. If $t_f > N$, then there will still be a nullspace of the desired output, unless the rank is N, when $t_f = N$, and the system is controllable. The test can therefore be restricted to this last case.

This concept of controllability can be described in a number of interesting and useful ways (e.g., Franklin et al., 1990; Stengel, 1986) and generalized to vector controls and time-dependent models. To the extent that a model is found to be uncontrollable, it shows that some elements of the state vector are not connected to the controls, and one might ask why this is so and whether the model cannot then be usefully simplified.

The concept of *observability* is connected to the question of whether given N perfect observations, it is possible to infer all of the initial conditions. Suppose that the same model (6.5.1) is used, and that we have (for simplicity only) a scalar observation sequence

$$y(t) = \mathbf{E}(t)\mathbf{x}(t) + n(t)\,, \qquad 0 \le t \le t_f\,. \tag{6.5.3}$$

Can we find $\mathbf{x}(0)$? Again, following Franklin et al. (1990), the sequence of observations can be written, with $u(t) \equiv 0$, as

$$\begin{aligned} y(1) &= \mathbf{E}(1)\mathbf{x}(1) = \mathbf{E}(1)\mathbf{A}\mathbf{x}(0) \\ &\vdots \\ y(t_f) &= \mathbf{E}(t_f)\mathbf{A}^{t_f}\mathbf{x}(0)\,, \end{aligned} \tag{6.5.4}$$

which is,

$$\mathbf{O}\mathbf{x}(0) = \begin{bmatrix} y(1) \\ \vdots \\ y(t_f) \end{bmatrix}$$

$$\mathbf{O} = \begin{Bmatrix} \mathbf{E}(1)\mathbf{A} \\ \vdots \\ \mathbf{E}(t_f)\mathbf{A}^{t_f} \end{Bmatrix}. \tag{6.5.5}$$

If the *observability matrix* $\mathbf{O}$ is square–that is, $t_f = N$ and is full rank–there is a unique solution for $\mathbf{x}(0)$, and the system is said to be observable. Should it fail to be observable, it suggests that at least some of the initial conditions are not determinable by an observation sequence and are irrelevant. Determining why that should be would surely shed light on the model one was using. As with controllability, the test (6.5.5) can be rewritten in a number of ways, and the concept can be extended to more complicated systems. The concepts of *stabilizability*, *reachability*, *reconstructability*, and *detectability* are closely related; see Goodwin and Sin (1984) or Stengel (1986). These references and Fukumori et al. (1993) demonstrate both the connection between observability and controllability and the existence of a steady-state solution for the algebraic Riccati equations.

In the oceanic fluid flows that are our primary concern, one must distinguish between mathematical observability and controllability and practical limitations imposed by the realities of observational systems. It is characteristic of fluids that changes occurring in some region at a particular time are ultimately communicated to all locations, no matter how remote, at later times, although the delay may be considerable, and the magnitudes of the signal may be much reduced by dissipation and geometrical spreading. Nonetheless, one anticipates that there is almost no possible element of a fluid flow, no matter how distant from a particular observation, that is not in principle observable. So, for example, a fluctuation of the strength of the Deep Western Boundary Current in the North Atlantic might not be readily visible immediately to an altimeter satellite. But a shift in DWBC transport, persisting for a few days, must ultimately affect the mass and vorticity budgets of the upper ocean and therefore lead to a shift in the surface elevation. With an adequate model, accurate enough observations, and adequate sampling, one could deduce that the DWBC had shifted, perhaps long ago. Whether such a deduction can be made practically (as opposed to mathematically) is dependent upon the accuracy and distribution of the particular observation system and the model skill.

6.6 Nonlinear Models

Fluid flows are nonlinear by nature, and one must address the data/model combination problem where the model is nonlinear. (There are also, as noted above, instances in which the data are nonlinear combinations of the state vector elements.) Nonetheless, the focus here on linear models is hardly wasted effort. As with more conventional systems, there are not many general methods for solving nonlinear estimation or control problems; rather, as with forward modeling, each situation has to be analyzed as a special case. Much insight is derived from a thorough understanding of the linear case, and indeed it is difficult to imagine tackling any nonlinear situation without a thorough grasp of the linear one. Not unexpectedly, the most general approaches to nonlinear estimation/control are based upon linearizations.

A complicating factor in the use of nonlinear models is that the objective functions need no longer have unique minima. There can be many nearby, or distant, minima, and the one chosen by the usual algorithms may depend upon exactly where one starts in the parameter space and how the search for the minimum is conducted. Indeed, the structure of the cost function may come to resemble a chaotic function, filled with hills, plateaus, and valleys into which one may stumble, never to get out again. Miller, Ghil, and Gauthiez (1994) discuss some of the practical difficulties. The combinatorial methods described in Chapter 5 are a partial solution.

6.6.1 The Linearized and Extended Kalman Filter

If one employs a nonlinear model,

$$\mathbf{x}(t+1) = \mathbf{L}(\mathbf{x}(t),\, \mathbf{B}\mathbf{q}(t),\, \boldsymbol{\Gamma}(t)\mathbf{u}(t))\,, \tag{6.6.1}$$

then reference to the Kalman filter recursion shows that the forecast step can be taken as before,

$$\tilde{\mathbf{x}}(t+1,-) = \mathbf{L}(\tilde{\mathbf{x}}(t),\, \mathbf{B}\mathbf{q}(t),\, 0)\,, \tag{6.6.2}$$

but it is far from clear how to propagate the uncertainty from $\mathbf{P}(t)$ to $\mathbf{P}(t+1,-)$, the previous derivation being based upon the assumption that the error propagates linearly, independent of the true value of $\mathbf{x}(t)$ (or equivalently, that if the initial error is Gaussian in character, then so is the propagated error). A number of approaches exist to finding approximate solutions to this problem, but they can no longer be regarded as strictly optimal, representing different linearizations.

Suppose that we write

$$\mathbf{x}(t) = \mathbf{x}_o(t) + \Delta\mathbf{x}(t)\,, \quad \mathbf{q} = \mathbf{q}_0(t) + \Delta\mathbf{q}(t)\,, \tag{6.6.3}$$

$$\begin{aligned}\mathbf{L}(\mathbf{x}(t),\, \mathbf{B}\mathbf{q}(t),\, \boldsymbol{\Gamma}\mathbf{u}(t),\, t) &= \mathbf{L}_o(\mathbf{x}_o(t),\, \mathbf{q}_o(t), 0, t)\\ &+ \mathbf{L}_x(\mathbf{x}_o(t), 0)^T \Delta\mathbf{x}(t) + \mathbf{L}_q(\mathbf{x}_o(t),\, \mathbf{q}_o(t))^T \Delta\mathbf{q}(t)\\ &+ \mathbf{L}_u(\mathbf{x}_o(t),\, \mathbf{q}_o(t))^T \mathbf{u}(t)\end{aligned} \tag{6.6.4}$$

where

$$\mathbf{L}_x(\mathbf{x}_o(t),\, \mathbf{q}_o(t)) = \frac{\partial \mathbf{L}}{\partial \mathbf{x}(t)}\,,$$

$$\mathbf{L}_q(\mathbf{x}_o(t),\, \mathbf{q}_o(t)) = \frac{\partial \mathbf{L}}{\partial \mathbf{q}(t)}\,,$$
$$\mathbf{L}_u(\mathbf{x}_o(t),\, \mathbf{q}_o(t)) = \frac{\partial \mathbf{L}}{\partial \mathbf{u}(t)}\,. \tag{6.6.5}$$

Then

$$\mathbf{x}_o(t+1) = \mathbf{L}_o(\mathbf{x}_o(t),\, \mathbf{B}\mathbf{q}_o(t),\, 0,\, t) \tag{6.6.6}$$

defines a nominal solution, or trajectory, $\mathbf{x}_o(t)$. (The model is assumed to be differentiable in this manner.)

We have an equation for the solution perturbation:

$$\Delta\mathbf{x}(t+1) = \mathbf{L}_x(\mathbf{x}_o(t),\, \mathbf{q}_o(t))^T \Delta\mathbf{x}(t) + \mathbf{L}_q^T \Delta\mathbf{q}(t) + \mathbf{L}_u^T \mathbf{u}(t)\,, \tag{6.6.7}$$

which is linear and of the form already used for the Kalman filter, but with redefinitions of the governing matrices. The full solution would be the sum of the nominal solution $\mathbf{x}_o(t)$ and the perturbation $\Delta\mathbf{x}(t)$. This form of estimate is sometimes known as the *linearized Kalman filter*, or the *neighboring optimal estimator*. Its usage depends upon the existence of a nominal solution, differentiability of the model, and the presumption that the controls $\Delta\mathbf{q}$, $\mathbf{u}$ do not drive the system too far from the nominal trajectory.

The so-called *extended Kalman filter* is nearly identical except that the linearization is taken not about a nominal solution but about the most recent estimate $\tilde{\mathbf{x}}(t)$; that is, the partial derivatives in (6.6.4) are evaluated using not $\mathbf{x}_o(t)$ but $\tilde{\mathbf{x}}(t)$. This latter form is more prone to instabilities, but if the system drifts very far from the nominal trajectory, it could well be more accurate than the linearized filter. The references go into these questions in great detail. Problems owing to multiple minima in the cost function (e.g., Miller et al., 1994) can always be overcome by having enough observations to keep the estimates close to the true state. Linearized smoothing algorithms can be developed in analogous ways, and the inability to track strong model nonlinearities is much less serious with a smoother than with a filter. The usual posterior checks of model and data residuals are also a very powerful precaution against a model failing to track the true state adequately.

6.6.2 Parameter Estimation (Adaptive Estimation)

An important application of oceanic models with data is to the estimation of empirical parameters used to describe the circulation. An example would be an attempt to improve estimates of eddy coefficients in the model by successively correcting them as more data appeared. Even with a linear dynamical model, such estimation generates a nonlinear estimation problem.

Suppose that the system is indeed linear and that the model contains a vector of parameters $\mathbf{p}$ whose nominal values $\mathbf{p}_0$ we wish to improve upon while also estimating the state vector. Write the model as

$$\mathbf{x}(t+1) = \mathbf{A}(\mathbf{p}(t))\mathbf{x}(t) + \mathbf{B}\mathbf{q}(t) + \boldsymbol{\Gamma}\mathbf{u}(t) \tag{6.6.8}$$

where the time dependence in the parameters refers to the changing estimate of their value rather than a true physical time dependence. A general approach to solving this problem is to augment the state vector. That is,

$$\mathbf{x}_A(t) = \begin{bmatrix} \mathbf{x}(t) \\ \mathbf{p}(t) \end{bmatrix}. \tag{6.6.9}$$

Then write a model for this augmented state as

$$\mathbf{x}_A(t+1) = \mathbf{L}_A(\mathbf{x}_A(t),\, \mathbf{q}(t),\, \mathbf{u}(t)) \tag{6.6.10}$$

where

$$\mathbf{L}_A = \left\{ \begin{matrix} \mathbf{A}(\mathbf{p}(t)) & \mathbf{0} \\ \mathbf{0} & \mathbf{I} \end{matrix} \right\} \mathbf{x}_A + \mathbf{B}\mathbf{q}(t) + \boldsymbol{\Gamma}\mathbf{u}(t)\,. \tag{6.6.11}$$

The observation equation is augmented simply as

$$\begin{aligned} \mathbf{y}_A(t) &= \mathbf{E}_A(t)\mathbf{x}_A(t) + \mathbf{n}_A(t)\,, \\ \mathbf{E}_A(t) &= \{\mathbf{E}(t) \quad \mathbf{0}\}\,, \quad \mathbf{n}_A(t) = \mathbf{n}(t)\,, \end{aligned} \tag{6.6.12}$$

assuming that there are no direct measurements of the parameters. The evolution equation for the parameters can be made more complex than indicated here. A solution can be found by using the linearized Kalman filter, for example, linearizing about the nominal parameter values. Parameter estimation is a very large subject (e.g., Anderson & Moore, 1979; Goodwin & Sin, 1984; Haykin, 1986).

A major point of concern in estimation procedures based upon Gauss-Markov type methods lies in specification of the various covariance matrices, especially those describing the model error [here lumped into $\mathbf{Q}(t)$]. The reader will probably have concluded that there is, however, nothing precluding deduction of the covariance matrices from the model and observations, given that adequate numbers of observations are available. For example, it is straightforward to show that if a Kalman filter is operating properly, then the so-called innovation, $\mathbf{y}(t) - \mathbf{E}\tilde{\mathbf{x}}(t,-)$, should be uncorrelated with all previous measurements:

$$< \mathbf{y}(t')(\mathbf{y}(t) - \mathbf{E}\tilde{\mathbf{x}}(t,-)) > = 0\,, \qquad t' < t \tag{6.6.13}$$

[recall (3.7.16]. To the extent that (6.6.13) is not satisfied, the covariances need to be modified, and algorithms can be formulated for driving the system toward this condition. The possibilities for such procedures have been known for a long time and have an extended literature under the title *adaptive estimation.* Among textbooks that discuss this subject are those of Haykin (1986), Goodwin and Sin (1984), and Ljung (1987).

Where one really knows almost nothing of the covariance structure, the literature on *stochastic estimation*–for example, Robbins-Monro methods–can be used (see Albert & Gardner, 1967, or Goodwin & Sin, 1984). This approach basically starts the estimation procedure with simplified gain matrices and permits them to evolve as the database provides information about the true covariances. But we leave further discussion to the references.

6.6.3 Nonlinear Adjoint Equations; Searching for Solutions

Consider now a nonlinear model in the context of the adjoint/Pontryagin minimum approach. Let the model be nonlinear so that a typical objective function is

$$\begin{aligned} J = {} & (\mathbf{x}(0) - \tilde{\mathbf{x}}(0))^T \mathbf{P}(0)^{-1} (\mathbf{x}(0) - \tilde{\mathbf{x}}(0)) \\ & + \sum_{t=1}^{t_f} (\mathbf{E}(t)\mathbf{x}(t) - \mathbf{y}(t))^T \mathbf{R}(t)^{-1} (\mathbf{E}(t)\mathbf{x}(t) - \mathbf{y}(t)) \\ & + \sum_{t=0}^{t_f-1} \mathbf{u}(t)^T \mathbf{Q}(t)^{-1} \mathbf{u}(t) \\ & - 2 \sum_{t=0}^{t_f-1} \boldsymbol{\mu}(t+1)^T \Big(\mathbf{x}(t+1) \\ & - \mathbf{L}(\mathbf{x}(t), \mathbf{B}\mathbf{q}(t), \boldsymbol{\Gamma}\mathbf{u}(t)) \Big) . \end{aligned} \tag{6.6.14}$$

The normal equations are:

$$\frac{1}{2} \frac{\partial J}{\partial \mathbf{u}(t)} = \mathbf{Q}^{-1} \mathbf{u}(t) + \boldsymbol{\Gamma}^T \boldsymbol{\mu}(t+1) = 0 , \quad 0 \le t \le t_f - 1 \tag{6.6.15}$$

$$\frac{1}{2} \frac{\partial J}{\partial \boldsymbol{\mu}(t)} = \mathbf{x}(t) - \mathbf{L}(\mathbf{x}(t-1), \mathbf{B}\mathbf{q}(t-1), \boldsymbol{\Gamma}\mathbf{u}(t-1)) = 0 , \quad 1 \le t \le t_f \tag{6.6.16}$$

$$\frac{1}{2}\frac{\partial J}{\partial \mathbf{x}(0)} = \mathbf{P}(0)^{-1}(\mathbf{x}(0) - \tilde{\mathbf{x}}(0)) + \left(\frac{\partial \mathbf{L}}{\partial \mathbf{x}(0)}\right)^T \boldsymbol{\mu}(1) = 0\,, \tag{6.6.17}$$

$$\frac{1}{2}\frac{\partial J}{\partial \mathbf{x}(t)} = \mathbf{E}(t)^T\mathbf{R}(t)^{-1}(\mathbf{E}(t)\mathbf{x}(t) - \mathbf{y}(t)) - \boldsymbol{\mu}(t) \tag{6.6.18}$$

$$+ \left(\frac{\partial \mathbf{L}}{\partial \mathbf{x}(t)}\right)^T \boldsymbol{\mu}(t+1) = 0\,, \qquad 1 \leq t \leq t_f - 1$$

$$\frac{1}{2}\frac{\partial J}{\partial \mathbf{x}(t_f)} = \mathbf{E}(t_f)^T\mathbf{R}(t_f)^{-1}(\mathbf{E}(t_f)\mathbf{x}(t_f) - \mathbf{y}(t_f)) - \boldsymbol{\mu}(t_f) = 0\,. \tag{6.6.19}$$

These are nonlinear because of the nonlinear model (6.6.17)–although the adjoint model remains linear in $\boldsymbol{\mu}(t)$–and the linear methods of Section 6.3.1 are no longer useful. One might anticipate, however, that if the nonlinearity is not too great, perturbation methods might work. This notion leads to what is usually called *neighboring optimal control* (e.g., see Stengel, Chapter 5). Where the nonlinearity is too great, and this has been the case in the meteorological applications (e.g., LeDimet & Talagrand, 1986) and for oceanic GCMs (Tziperman et al. 1992a,b; Marotzke & Wunsch, 1993), the approach to solution has been essentially an iterative one.

Consider what one is trying to do. At the optimum, if we can find it, J will reach a stationary value in which the terms multiplying the $\boldsymbol{\mu}(t+1)$ will vanish identically. Essentially, one uses *search* methods that are able to find a solution (there may well be multiple such solutions, each corresponding to a local minimum of J).

There are many known ways to seek approximate solutions to a set of simultaneous equations, linear or nonlinear, using various search procedures. Many of them were listed in Chapter 5. Most such methods are based upon what are usually denoted as *Newton* or *quasi-Newton* methods, or variations on steepest descent. The most popular approach to tackling the set (6.6.15)–(6.6.19) has been a form of conjugate gradient or modified steepest descent algorithm. This method is discussed in all textbooks on optimization theory; Luenberger (1984) has a particularly clear account, and application to the adjoint/Pontryagin minimum problem has been made by Thacker and Long (1988), Tziperman and Thacker (1989), Smedstad and O'Brien (1991), Tziperman et al. (1992a,b), and Marotzke and Wunsch (1993), and others as well (see the papers in Anderson & Willebrand, 1989). The iteration cycles are commonly carried out by making a first estimate of the initial conditions and the boundary conditions–for example, setting $\mathbf{u} = \mathbf{0}$. One integrates (6.6.16) forward in time to produce a first guess for $\mathbf{x}(t)$. A first guess set of Lagrange multipliers is obtained by integrating (6.6.18) backward in time. Normally, (6.6.15) is not then satisfied, but because the values obtained provide information on the gradient of the objective function with respect to the controls, one knows the sign of the changes to make in the controls to reduce J. Perturbing the original guess for $\mathbf{u}(t)$ in this manner, one does another forward integration of the model and backward integration of the adjoint. In practice, the perturbations are determined by the conjugate gradient or other algorithm, continuing the iterations until convergence is obtained.

In this type of approximate solution, the adjoint solution, $\tilde{\boldsymbol{\mu}}(t)$, is really playing two distinct roles. On the one hand, it is a mathematical device to impose the model constraints; on the other, it is being used as a numerical convenience for determining the direction and step size to best reduce the objective function. The two roles are obviously intimately related, but as we have seen for the linear models, the first role is the primary one. The problem of possibly falling into the wrong minimum of the objective function remains here, too.

6.7 Assimilation

Assimilation is a term widely used in numerical weather prediction (NWP) to describe the process of combining a forecast with current observations in order to update a dynamical model in preparation for another forecast. The term has more recently come to describe loosely any data/model combination, including those we have looked at here. Much of NWP can be regarded as a form of engineering–in which the forecaster's major problem is to find something that works, and only secondarily to understand exactly why. There is now a large literature on meteorological assimilation (see the reviews by Lorenc, 1986; Daley, 1991; or Ghil & Malanotte-Rizzoli, 1991). Much of it involves simplified forms of objective mapping, in which the model dynamics are used in a primitive fashion to help choose covariances for interpolation. The formal uncertainties of the forecast are not usually computed–the forecaster learns empirically, and very quickly, whether and which aspects of his forecast are any good. If something works, then one keeps on doing it; if it doesn't work, one changes it. Because of the short time scale, feedback from the public, the military, farmers, the aviation industry, etc., is fast and vehement. Oceanographic examples are provided by Robinson and Leslie (1985) and Derber and Rosati (1989).

A lot of schemes can be understood by referring back to the Kalman filter averaging step,

$$\tilde{\mathbf{x}}(t+1) = \tilde{\mathbf{x}}(t+1,-) + \mathbf{K}(t+1)[\mathbf{y}(t+1) - \mathbf{E}\tilde{\mathbf{x}}(t+1,-)]\,. \qquad (6.7.1)$$

This equation has the form of a *predictor-corrector*–the dynamical forecast of $\tilde{\mathbf{x}}(t+1,-)$ is compared to the observations and corrected on the basis of the discrepancies. Some assimilation schemes represent guesses for $\mathbf{K}$ rather than the computation of the optimum choice, which we know is given by the Kalman gain, replacing (6.7.1) with

$$\tilde{\mathbf{x}}(t+1) = \tilde{\mathbf{x}}(t+1,-) + \mathbf{K}_m[\mathbf{y}(t+1) - \mathbf{E}\tilde{\mathbf{x}}(t+1,-)] \tag{6.7.2}$$

where $\mathbf{K}_m$ is a modified gain matrix. Thus, for example, Anthes (1974) proposed what is usually known as *nudging*, a scheme explored in an oceanographic context by Malanotte-Rizzoli and Holland (1986, 1988), who used simulated data. $\mathbf{K}_m$ is diagonal or nearly so with elements which are weights that the forecaster assigns to the individual observations. To the extent that the measurements have uncorrelated noise, as might be true of pointwise meteorological instruments like anemometers, and the forecast error owing to uncertain initial and boundary conditions is also nearly pointwise uncorrelated, nudging the model values pointwise to the data may be very effective. If in (6.7.2) the observations $\mathbf{y}(t+1)$ are direct pointwise measurements of state vector elements [e.g., if the state vector includes the density and $\mathbf{y}(t+1)$ represents observed densities], then $\mathbf{E}(t+1)$ is very simple–but only if the measurement point coincides with one of the model grid points. If, as is often true, the measurements occur between model grid points, $\mathbf{E}$ is an interpolation operator from the model grid to the data location. In the most usual situation, there are many more model grid points than data points, and this direction for the interpolation is the most reasonable and accurate. With more data points than model grid points, one might better interchange the direction of the interpolation. Formally, this interchange is readily accomplished by rewriting (6.7.1) as

$$\tilde{\mathbf{x}}(t+1) = \tilde{\mathbf{x}}(t+1,-) + \mathbf{K}_m\mathbf{E}(\mathbf{E}^+\mathbf{y}(t+1) - \tilde{\mathbf{x}}(t+1,-)) \tag{6.7.3}$$

where $\mathbf{E}^+$ is any right inverse of $\mathbf{E}$ in the sense of Chapter 3, for example, the Gauss-Markov optimal interpolator or some plausible approximation to it.

There are potential pitfalls of nudging, however. If the data have correlated errors as is true of many oceanographic observation systems, then the model is being driven toward spatial structures that are erroneous. More generally, the expected great variation in time and space of the relative errors of model forecast and observations cannot be accounted for with a fixed diagonal gain matrix. A great burden is placed upon the insights and skill of the investigator who must choose the weights. Finally, one can calculate the uncertainty of the weighted average (6.7.2), using this suboptimal gain, but it requires that one specify the true covariances. As noted, however, in NWP formal uncertainty estimates are not of much interest. But user feedback is rarely available to the oceanographer, who is still primarily in the business of trying to understand what he knows–the estimation problem–rather than forecasting the system for public consumption. When forecasts are made, the time scale is often so long as to preclude direct test of the result.

Sarmiento and Bryan (1982) introduced a form of oceanic GCM modeling in which the surface temperature and salinity fields were "restored to climatology" in a scheme often referred to as a *robust diagnostic* method. So, for example, they used a temperature evolution equation equivalent to the form,

$$\begin{aligned} \theta_i(t+1) &= \theta_i(t) + F(\boldsymbol{\theta}(t)) + \gamma(\theta_i^* - \theta_i(t)) \\ &\equiv \theta_i(t+1,-) + \gamma(\theta_i^* - \theta_i(t+1,-)) \qquad (6.7.4) \\ \theta_i(t+1,-) &= \theta_i(t) + 1/(1-\gamma)F(\boldsymbol{\theta}(t)) \qquad (6.7.5) \end{aligned}$$

for the temperature at grid point i, where F is a function of the full-state vector, $\boldsymbol{\theta}$. The coefficient γ was a simple constant, and $\boldsymbol{\theta}^*$ is an observed temperature climatology. This is a Kalman filterlike scheme, where the climatology acts as a set of observations, available at every time step at every grid point, with errors uncorrelated in both time and space. The relative errors of model and data are implied by the choice of γ.

Such schemes have much to recommend them, particularly when they work (usually meaning that the modeler likes the results), but without a clear understanding of the reliability of the climatology and the model error structure, the implication of the results is obscure. There is a real danger that a good model forecast is being driven toward erroneous components of the data and actually degraded.

Climatologies are always smoother than the ocean. The nudging in (6.7.4), which acts as a Kalman filter analogue, is in effect asserting that at every time step observations are available which show that no eddies are present. If the model is an eddy-resolving one and seeks to create eddies as part of the state vector, then the forcing to climatology is insisting, perpetually, that entire wavenumber bands are absent in the ocean and that the model should be suppressing them. What one really wants to do is to force the model, *at large scales*, to the climatology, leaving it to compute eddies and other high wavenumber phenomena as best it can. The complementary situation applies when observations are supposed to contain only the mesoscale, the larger space and time scales having been either unobserved or having been removed from the observations (this situation was common in handling of data from early altimetric satellite missions, when only the mesoscale variability was credible). A model can only be distorted if the observations continually forced on it are devoid of a general circulation; the model may respond in grotesque ways in an attempt to have an eddy field and no larger scales. Optimal filters and smoothers employ matrix operators to distinguish the wavenumber bands where one has confidence in the observations from those where one does not. The scalar weights used in most nudging schemes cannot make this distinction.

Another more flexible, approximate form of time-dependent estimation can also be understood in terms of the Kalman filter equations. In the Kalman filter update equation (6.2.19), at least in principle, all elements of the state vector are modified to some degree, given any difference between the measurements and the model-prediction of those measurements. The uncertainty of the statevector is *always* modified whenever data become available, even if the model should perfectly predict the observations. As time evolves, information from measurements in one part of the model domain is distributed by the model dynamics over the entire domain, leading to correlations in the uncertainties of all the elements. This distribution of information is intimately linked to the requirement that $\tilde{\mathbf{x}}(t)$ should satisfy the model equations.

One might suppose that some models propagate information in such a way that the uncertainty correlations diminish rapidly with increasing spatial and temporal separation. Supposing this to be true (and one must be aware that oceanic general circulation models are capable of propagating information, be it accurate or erroneous, over the entire globe), static approximations can be found in which the problem is reduced back to the objective mapping methods employed in Chapter 3. The model is used to make an estimate of the field at time t, $\tilde{\mathbf{x}}(t,-)$, and one then finds the prediction error $\Delta\mathbf{y}(t) = \mathbf{y}(t) - \mathbf{E}\tilde{\mathbf{x}}(t,-)$. A best estimate of $\Delta\mathbf{x}(t)$ is sought based upon the covariances of $\Delta\mathbf{y}(t)$, $\Delta\mathbf{x}(t)$, etc.–that is, objective mapping–and the improved estimate is

$$\tilde{\mathbf{x}}(t) = \tilde{\mathbf{x}}(t,-) + \Delta\tilde{\mathbf{x}}(t) = \tilde{\mathbf{x}}(t,-) + \mathbf{R}_{xx}\mathbf{E}^T(\mathbf{E}\mathbf{R}_{xx}\mathbf{E}^T + \mathbf{R}_{nn})^{-1}\Delta\mathbf{y}\,, \quad (6.7.6)$$

which has the form of a Kalman filter update, but in which the state uncertainty matrix, $\mathbf{P}$, is replaced in the gain matrix, $\mathbf{K}$, by $\mathbf{R}_{xx}$ representing the prior covariance of $\Delta\mathbf{x}$. $\mathbf{R}_{xx}$ is fixed, with no dynamical evolution of the gain matrix permitted. The operator in (6.7.6) is a Wiener filter (Wiener, 1949; Sorenson, 1985), which is a steady-state filter. The Kalman filter was originally developed to extend the Wiener theory to nonstationary situations, and steady-state limits of the Kalman filter reduce properly to the corresponding Wiener form (and similarly with smoothing operators). Viewed as a generalization of nudging, they permit one to specify spatial structure in the noise covariance through choice of a nondiagonal $\mathbf{R}_{nn}$. The weighting of the $\Delta\mathbf{y}$ and the modification for $\tilde{\mathbf{x}}$ is then much more complex than in pure nudging.

The major issues are the specification of $\mathbf{R}_{xx}$, $\mathbf{R}_{nn}$. Most attempts to use these methods have been simulations by modelers who were content to ignore the problem of determining $\mathbf{R}_{nn}$ or to assume that the noise was purely white. In principle, estimates of $\mathbf{R}_{xx}$ can be found either from observations or from the model itself. The advent of high-quality satellite altimetry (Fu et al., 1994; Stammer & Wunsch, 1994) has led to a focus on schemes capable of combining surface-pressure elevation data with general circulation models. For example, Haines (1991) and Ezer and Mellor (1994) use their models to generate the equivalent of the part of $\mathbf{R}_{xx}$ involving correlations between surface pressure and subsurface fields in the state vector, such as the temperature or potential vorticity. (But the latter authors found that observed climatological statistics appeared to produce better results.) The expression in (6.7.6) was used by Ezer and Mellor (1994) to account for local correlations in the horizontal, and they extended the expression so as to use data from a finite-time window through guessed covariances. Unlike a Kalman filter, or the true Wiener filter, which would potentially modify all elements of the state vector to account for model propagation of information from one region or depth to another, employing the information about the model contained in $\mathbf{P}$, methods like (6.7.6) are ad hoc local approximations to the Kalman filter, permitting information to influence the model only near the observation points. There is then no guarantee that the resulting new state vector is a solution to the model equations. Information is being discarded, but the results may prove adequate for particular purposes, and the methods are computationally much less demanding than optimal schemes. By way of example, Figure 6–11 shows a column of the approximate Kalman gain computed by I. Fukumori (personal communication, 1994) for a primitive equation model of the South Pacific Ocean. The datum is observed surface elevation at one point, and the contours show how the model streamfunction would be modified in the model/data averaging step for a particular set of choices of model and data errors. Significant changes are required at large distances from the observation point. The changes at remote locations in turn imply modifications at yet greater distances as time advances. Hypothetically, one could guess the structure of this information propagation through the model without formally comput-

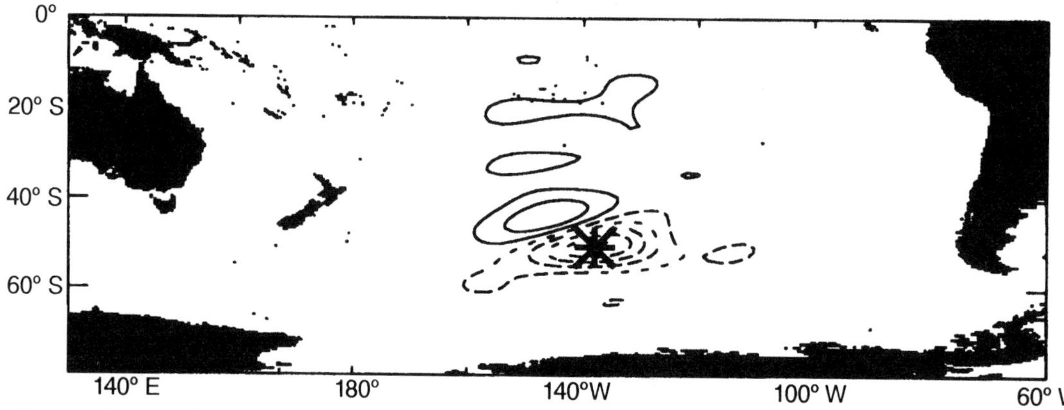

Figure 6–11. The approximate Kalman gain computed by I. Fukumori (personal communication, 1994) for a primitive equation model of the Pacific Ocean. It is supposed that the seasurface elevation is observed at the location of the asterisk with a given error budget. The contours (contour interval is $0.2 \times 10^{11}\text{cm}^3/\text{s}$) show the required modification of the model-forecast streamfunction at the time that the model state is averaged with the observations. Changes are required at large distances, implying, too, that the uncertainty will change at large distances from the observation point.

ing the Kalman gain–but actually doing so requires a deep understanding of the model physics and errors.

Methods that permit data to be employed from finite-time durations, weighting them inversely with their deviation from some nominal central time, are localized approximations to smoothing algorithms of the Wiener type. Many variations on these methods are possible, including (e.g., DeMey & Robinson, 1987) the replacement of $\mathbf{R}_{xx}$ by its eigenvectors (the EOFs), which again can be computed either from the model or from data. Improvements could be made by comparison of the covariance matrices used against the estimates emerging from the calculations of $\tilde{\mathbf{x}}(t)$, $\tilde{\mathbf{n}}(t)$.

All practical linearized assimilation methods are a weighted average of a model estimate of the oceanic state with one inferred from the observations. If the model and the observations are physically inconsistent, the forced combination will be impossible to interpret. Thus, the first step in any assimilation procedure has to be to demonstrate that model physics and data represent the same fluid–with disagreement being within the error bounds of both. Following this confirmation of physical consistency, one recognizes that the weighted average of model and data will be useful only if the weights make sense–chosen to at least well-approximate the relative uncertainties of these two. Otherwise, the result of the combination is an average of "apples and oranges."

A growing oceanographic literature now exists on time-dependent assimilation of real, or more often, simulated data using various plausible guesses about how the model and data should be combined. Few of these papers address the actual error structure of the observations or of the model forecasts. Some of the methods appear to work reasonably well. Some authors of these studies appear to believe that by not discussing the error structures of model and data that they have somehow avoided making statements about them. But all combinations of data and model estimates make implicit choices of these fields, and the results can be no better than the skill of the selection.

6.8 Other Minimization Methods and the Search for Practicality

All of the inverse problems we have discussed were reduced ultimately to finding the minimum of an objective function, either in unconstrained form [e.g., (3.5.2) or (6.2.25)] or constrained by exact relationships (e.g., models) (3.5.4) or (6.3.14). Once the model is formulated, the objective function agreed on, and the data obtained in appropriate form (often the most difficult step), the formal solution is reduced to finding the constrained or unconstrained minimum. *Optimization theory* is a very large, very sophisticated subject directed at finding such minima, and the methods we have described here–sequential estimation and adjoint/Pontryagin principles–are only two of a number of possibilities.

As we have seen, some of the methods stop at the point of finding a minimum and do not readily produce an estimate of the uncertainty of the solution. One can distinguish inverse methods from optimization methods by the requirement of the former for the requisite uncertainty estimates. Nonetheless, as noted before in some problems, mere knowledge that there is at least one solution may be of intense interest, irrespective of whether it is unique or whether its stability to perturbations in the data or model is well understood.

The reader interested in optimization methods generally is referred to the literature on that subject (e.g., Gill et al., 1981; Luenberger, 1984; Scales, 1985). The oceanographic general circulation problem falls into the category of extremely large, nonlinear optimization, a category which tends to preclude the general use of many methods that are attractive for problems of more modest size.

In the context of the more conventional methods, the continued exploration of ways to reduce the computational load without significantly degrading either the proximity to the true minimum or the information content (the uncertainties of the results) is a very high priority. Several approaches are known. We have already described the use of steady-state filters and smoothers where they exist. Textbooks (Albert & Gardner, 1967; Gelb, 1974; Goodwin & Sin, 1984; Anderson & Moore, 1979) discuss a variety of possibilities for simplifying various elements of the solutions. In addition to the steady-state assumption, methods include: (1) "state reduction" (e.g., Gelb, 1974)–attempting to remove from the model (and thus from the uncertainty calculation) elements of the state vector that are either of no interest or comparatively unchanging; (2) "reduced-order observers" (Luenberger, 1964; O'Reilly, 1983), in which some components of the model are so well observed that they do not need to be calculated; (3) proving or assuming that the uncertainty matrices (or the corresponding information matrices) are block diagonal or banded, permitting use of a variety of sparse algorithms. This list is not exhaustive. Todling and Cohn (1994) discuss some of these approaches, with an emphasis on meteorological experience.

Of particular, more recent, interest are the combinatorial methods, including genetic algorithms, briefly described in Chapter 5. To my knowledge, however, there is as yet no experience in their use with a full oceanic GCM, although Evensen (1994) does discuss a Monte Carlo approach to estimating the uncertainty in a sequential estimator.

6.9 Forward Models

The focus we have had on the solution of inverse problems has perhaps given the impression that there is some fundamental distinction between forward and inverse modeling. The point was made at the beginning of this book that inverse *methods* are important in solving forward as well as inverse *problems*. Almost all the inverse problems discussed here involved the use of an objective function, and such objective functions do not normally appear in forward modeling. The presence or absence of such functions thus might be considered a fundamental difference between the problem types.

But numerical models do not produce universal, uniformly accurate solutions to the fluid equations. Any modeler makes a series of decisions about which aspects of the flow are most important for accurate depiction–the energy flux, the large-scale velocities, the nonlinear cascades, etc.–and which cannot normally be achieved simultaneously with equal fidelity. It is rare that these goals are written explicitly, but they could be, and the modeler could choose the grid and differencing scheme, etc., to minimize a specific objective function. The use of such explicit objective functions would prove beneficial because it would quantify the purpose of the model.

One can also consider the solution of ill-posed forward problems. Indeed, Bennett and Chua (1994) point out that the conventional ocean circulation forward problem with open ocean boundaries is usually ill-posed. In view of the discussion throughout this book, the remedy is straightforward: One must introduce an explicit objective function of the now-familiar type, involving state vectors, observations, control, etc., and this approach is precisely that recommended by them. If a Lagrange multiplier method is adopted, then the relations (3.5.22)–(3.5.23) show that an over- or under-specified forward model produces a complementary under- or overspecified adjoint model, and it is difficult to sustain a claim that modeling in the forward direction is fundamentally distinct from that in the inverse sense.

Example: Consider the ordinary differential equation

$$\frac{d^2x(t)}{dt^2} - k^2x(t) = 0\,. \tag{6.9.1}$$

Formulated as an initial value problem, it is properly posed with Cauchy conditions $x(0) = x_0$, $x'(0) = x'_0$. The solution is

$$x(t) = A\exp(kt) + B\exp(-kt)\,, \tag{6.9.2}$$

with A, B determined by the initial conditions. If we add another condition–for example, at the end of the interval of interest, $x(t_f) = x_{t_f}$–the problem is ill-posed because it is now overspecified. To analyze and solve such a problem using the methods of this book, discretize it as

$$x(t+1) - (2+k^2)x(t) + x(t-1) = 0\,, \tag{6.9.3}$$

taking $\Delta t = 1$, with corresponding redefinition of k^2. A canonical form is

$$\begin{gathered}\mathbf{x}(t+1) = \mathbf{A}\mathbf{x}(t), \mathbf{x}(t) = [x(t), x(t-1)]^T\,,\\ \mathbf{A} = \left\{\begin{matrix} 2+k^2 & -1 \\ 1 & 0 \end{matrix}\right\}.\end{gathered} \tag{6.9.4}$$

Cauchy conditions correspond to specifying $\mathbf{x}(1) = [x(1), x(0)]^T$ [starting from $t = 1$ to avoid using $x(-1)$, which however, is a purely superficial change]. Like the continuous solution, there is an exponentially growing and an exponentially decaying component of $\mathbf{x}(t)$. One can choose initial conditions such that only the decaying component is excited (Figure 6–12a). But a very slight error in the appropriate $\mathbf{x}(1)$ excites the growing or *computational* mode for Figure 6–12a, and in that sense, the problem is numerically ill-conditioned to slight initial condition errors. We can make it further ill-conditioned by specifying $x(t_f)$ (i.e., the scalar terminal value),

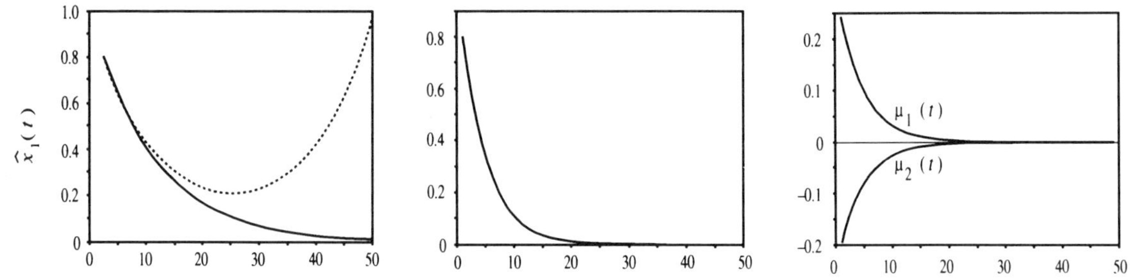

(a) (b) (c)

Figure 6–12. (a) Solutions $x_1(t)$ to the discretized version of Equation (6.9.1) for two slightly different initial conditions. The solid line shows the solution when the initial conditions are carefully chosen to avoid exciting the exponentially growing (computational) mode ($\mathbf{x}(1) = [0.800,\ 1.000]^T$). The dashed line shows the eventual emergence of the expondential mode out of the background after sufficient time has passed when the initial conditions are $\mathbf{x}(1) = [0.805,\ 1.000]^T$. (b) Solution to Equation (6.9.1) when the computational mode is excited by the initial conditions, but a terminal constraint is provided, which simultaneously overspecifies the system and stabilizes it through the objective function. (c) The two elements of the adjoint solution $\boldsymbol{\mu}(t)$ to the minimization of Equation (6.9.6).

rendering it also overspecified. Unless the initial and terminal values satisfy a solvability condition, the system is contradictory. To make the problem realistic, stipulate that the initial and final conditions are not perfect, instead treating them as *desired conditions*, $\mathbf{x}_d(1)$, $\mathbf{x}_d(t_f)$, respectively, with corresponding uncertainties $\mathbf{P}(t_0)$, $\mathbf{P}(t_f)$, respectively, and introduce an objective function

$$\begin{aligned} J = &(\mathbf{x}(1) - \mathbf{x}_d(1))\mathbf{P}(1)^{-1}(\mathbf{x}(1) - \mathbf{x}_d(1)) \\ &+ (\mathbf{x}(t_f) - \mathbf{x}_d(t_f))\mathbf{P}(t_f)^{-1}(\mathbf{x}(t_f) - \mathbf{x}_d(t_f)) \end{aligned} \tag{6.9.5}$$

subject to (6.9.3). We know several ways to minimize (6.9.5) constrained by the model–for example, the filter/smoother combination. Let us use the Pontryagin principle, modifying (6.9.5) to

$$\begin{aligned} J = &(\mathbf{x}(1) - \mathbf{x}_d(1))^T\mathbf{P}(1)^{-1}(\mathbf{x}(1) - \mathbf{x}_d(1)) \\ &+ (\mathbf{x}(t_f) - \mathbf{x}_d(t_f))^T\mathbf{P}(t_f)^{-1}(\mathbf{x}(t_f) - \mathbf{x}_d(t_f)) \\ &- 2\sum_{t=1}^{t_f-1} \boldsymbol{\mu}(t)^T[\mathbf{x}(t+1) - \mathbf{A}\mathbf{x}(t)], \end{aligned} \tag{6.9.6}$$

and the normal equations are

$$\frac{1}{2}\frac{\partial J}{\partial \mathbf{x}(t)} = -\boldsymbol{\mu}(t-1) + \mathbf{A}^T\boldsymbol{\mu}(t) = 0\,, \qquad 2 \le t \le t_f - 1\,, \quad (6.9.7)$$

$$\frac{1}{2}\frac{\partial J}{\partial \mathbf{x}(1)} = \mathbf{P}(1)^{-1}[\mathbf{x}(1) - \mathbf{x}_d(1)] + \mathbf{A}^T\boldsymbol{\mu}(1) = 0\,, \quad (6.9.8)$$

$$\frac{1}{2}\frac{\partial J}{\partial \mathbf{x}(t_f)} = \mathbf{P}(t_f)^{-1}[\mathbf{x}(t_f) - \mathbf{x}_d(t_f)] - \boldsymbol{\mu}(t_f - 1) = 0\,, \quad (6.9.9)$$

$$\frac{1}{2}\frac{\partial J}{\partial \boldsymbol{\mu}(t)} = -\mathbf{x}(t+1) + \mathbf{A}\mathbf{x}(t) = 0\,, \qquad 1 \le t \le t_f - 1\,. \quad (6.9.10)$$

These equations are easily solved by the same methodology used to obtain (6.3.36)–(6.3.37), involving a backward sweep of the adjoint model (6.9.7) to obtain $\boldsymbol{\mu}(1)$, which through (6.9.8) produces $\hat{\mathbf{x}}(1)$ in terms of $\mathbf{x}(t_f) - \mathbf{x}_d(t_f)$. A forward sweep of the model, to t_f, produces the numerical value of $\hat{\mathbf{x}}(t_f)$; the backward sweep of the adjoint model gives the corresponding numerical value of $\hat{\mathbf{x}}(1)$, and a final forward sweep of the model completes the solution. The subproblem forward and backward sweeps are always well posed. This recipe was run for

$$k^2 = 0.05,\ \Delta t = 1,\ \mathbf{x}_d(1) = [0.805,\ 1.0]^T,\ \mathbf{P}(1) = 10^{-2}\mathbf{I}\,,$$
$$x(t_f = 50) = 1.427 \times 10^{-5},\ \mathbf{P}(50) = \begin{Bmatrix} 10^{-4} & 0 \\ 0 & 10^4 \end{Bmatrix},$$

with results in Figure 6–12b,c. [The large subelement uncertainty in $\mathbf{P}(50)$, corresponding to scalar element $x(49)$, is present because we sought only to specify scalar element $x(50)$, in $\mathbf{x}(50)$.] If $\mathbf{x}_d(1) = [0.800,\ 1.0]^T$ instead of the value actually used, the unstable mode would not have been excited. Notice that the original ill-posedness in both overspecification and instability of the initial value problem have been dealt with. For a full GCM, the technical details are much more intricate, but the principle is not in doubt.

We are ending as we began–this result can be thought of as the solution to a forward problem, albeit ill-posed, or as the solution to a more or less conventional inverse one. The distinction between forward and inverse problems has nearly vanished.

Any forward model that is driven by observed conditions–for example, of the windstress or the buoyancy flux–is ill-posed in the sense that there can again be no unique solution, only a most probable one, smoothest one, etc. As with an inverse solution, forward calculations no more produce unique solutions in these circumstances than do inverse ones.

6.9.1 Empirical Linearization: POPs and Optimal Modes

A number of different threads can be brought together in the study of the results of large numerical models, whether or not data have been employed. As an example, consider a GCM that has been run to a statistical equilibrium and then disturbed sufficiently weakly that the evolution of the disturbance is approximately described by a constant coefficient model:

$$\Delta\mathbf{x}(t+1) = \mathbf{A}\Delta\mathbf{x}(t) + \mathbf{B}\Delta\mathbf{q}(t) \qquad (6.9.11)$$

where $\Delta\mathbf{q}(t)$ is assumed known. Equation (6.1.22) is a formal procedure for finding $\mathbf{A}$. As an alternative, form the product,

$$< \Delta\mathbf{x}(t+1)\Delta\mathbf{x}(t)^T > = \mathbf{A} < \Delta\mathbf{x}(t)\Delta\mathbf{x}(t)^T > \; + \mathbf{B} < \Delta\mathbf{q}(t)\Delta\mathbf{x}(t)^T > ;$$

assuming the last term vanishes, one has

$$\mathbf{A} = \mathbf{R}(1)\,\mathbf{R}(0)^{-1} \qquad (6.9.12)$$

where $\mathbf{R}(0) = < \Delta\mathbf{x}(t)\Delta\mathbf{x}(t)^T >$, $\mathbf{R}(1) = < \Delta\mathbf{x}(t+1)\Delta\mathbf{x}(t)^T >$–that is, $\mathbf{A}$ is proportional to the one time-step covariance matrix (see, for example, von Storch, Bruns, Fischer-Bruns, & Hasselmann, 1988). When $\mathbf{A}$ is a time-independent linear model, one is free to use multiple time-steps in this form–for example,

$$\Delta\mathbf{x}(t+r) = \mathbf{A}^r\Delta\mathbf{x}(t) + \text{forcing}\,, \qquad (6.9.13)$$

and $\mathbf{A}^r = \mathbf{R}(r)\,\mathbf{R}(0)^{-1}$. $\mathbf{R}$ would be estimated by averaging the model results.

Any linear model is automatically of the form (6.9.11). In general, $\mathbf{A}$, whether obtained analytically or through (6.9.11) or (6.9.12), will not be self-adjoint–that is, $\mathbf{A} \neq \mathbf{A}^T$. It nonetheless always has an SVD. Suppose we write $\mathbf{A} = \mathbf{U}\boldsymbol{\Lambda}\mathbf{V}^T$ and assume, as is normally true, that it is full rank. Then, because the singular vectors are a complete set, we can always write

$$\mathbf{x}(t) = \mathbf{V}\boldsymbol{\alpha}(t) \qquad (6.9.14)$$

where $\boldsymbol{\alpha}$ is a set of vector coefficients. We can also write the adjoint solution as

$$\boldsymbol{\mu}(t) = \mathbf{U}\boldsymbol{\beta}(t)\,, \qquad (6.9.15)$$

recalling (3.5.22)–(3.5.23). Then assuming, for simplicity only, that $\Delta\mathbf{q}(t) = 0$, (6.3.25) becomes

$$\boldsymbol{\beta}(t+1)^T\boldsymbol{\Lambda}\boldsymbol{\alpha}(t+1) - \boldsymbol{\beta}(t)\boldsymbol{\Lambda}\boldsymbol{\alpha}(t) = 0\,, \qquad (6.9.16)$$

which can be interpreted as an energy conservation principle, summed over modes, if the $\mathbf{u}_i$, $\mathbf{v}_i$ are regarded as eigenmodes of the model. They are also the eigenvectors of the one time-step covariance matrix and are thus ordinary EOFs. If $\mathbf{A}$ is replaced by $\mathbf{A}^r$, the entire analysis proceeds in exactly the same way, except the time difference is taken over r time-steps, and the singular vectors are the eigenvectors of $\mathbf{R}(r)\,\mathbf{R}(0)^{-1}$.

Farrell (1989) and Farrell and Moore (1993) have suggested using the r–step singular vectors of $\mathbf{A}$ as optimal modes to describe the dominant patterns of oscillation in time-evolving models. That is, having chosen r, the Eckart-Young-Mirsky theorem shows that the change in the field over r time-steps is given most efficiently by the singular vectors corresponding to the largest singular values, just as is true of a static field. They exploit (6.9.16) as an energy principle.

Hasselmann (1988) (also see von Storch et al., 1988, 1993) suggested using the eigenvectors of $\mathbf{A}$ directly. He called these *principal oscillation patterns*, or POPs. Because $\mathbf{A}$ is usually not symmetric (not self-adjoint), the eigenvalues are usually complex, and there is no guarantee that the eigenvectors are a complete set. But assuming that they provide an adequate expansion basis, the eigenvectors are used in pairs corresponding to complex conjugate pair eigenvalues. The expansion coefficients of the time-evolving field are readily shown to be the eigenvectors of $\mathbf{A}^T$–that is, the eigenvectors of the adjoint model. Assuming that the eigenvectors are not grossly deficient as a basis, and/or one is interested in only a few dominant modes of motion, the POP approach gives a reasonably efficient representation of the field.

6.10 A Last Word

The introduction of time evolution makes the dimension of the inclusive state vector, $\mathbf{x}$, jump from N with a steady model to $(t_f+1)\times N$, potentially approaching infinity. Time-evolving turbulent models thus contain degrees of freedom whose numbers have always swamped any reasonable estimate of the number M of ocean data available. Progress will lie with understanding the solution structures that are determinable, and those which remain uncertain, to guide future observational strategies and to render forecasts (if that is the goal) usable by provision of an understanding of what is most trustworthy, and what is least so.

A general solution to the observational difficulties that were described at the outset of this book cannot be found by dealing in isolation with either the technologies or pure modeling. Everything in the history of fluid dynamics tells us that understanding emerges only through a very tight coupling of theory and observational tests and trials. For the oceanographer, a general strategy is apparent: Modelers must be aware of and learn to employ those observations which technologies make most readily available (hydrography, altimetry, etc.), and to exploit them to estimate the quantities which observations may never readily produce directly or with adequate coverage (property fluxes, boundary conditions, etc.). Similarly, proponents of observation methods must become aware of the information content of their data relative to the elements of state estimates produced by models and be prepared to exploit the prior information that models contain. A true model-data synthesis is a necessity. The methods outlined in this book, for which no claim to completeness is made, suggest that the road to be traveled is clear, but many interesting challenges lie ahead. The reward for traveling that road will be greatly increased understanding of the ocean circulation and its consequences.

References

Abel, N. H. (1826). Auflösung einer mechanischen Aufgabe. *Crelle Journal*, 1, 153–157.

Aki, K., & P. G. Richards. (1980). *Quantitative Seismology.* W. H. Freeman, San Francisco, CA, 2 vols., 932pp.

Albert, A. E., & L. A. Gardner, Jr. (1967). *Stochastic Approximation and Nonlinear Regression.* The MIT Press, Cambridge, MA, 204pp.

Amos, A. F., A. L. Gordon, & E. D. Schneider. (1971). Water masses and circulation patterns in the region of the Blake-Bahama Outer Ridge. *Deep-Sea Res.*, 18, 145–165.

Anderson, B. D. O., & J. B. Moore. (1979). *Optimal Filtering.* Prentice-Hall, Englewood Cliffs, NJ, 357pp.

Anderson, D. L. T., & J. Willebrand, eds. (1989). *Oceanic Circulation Models: Combining Data and Dynamics.* Klüwer, Dordrecht, 605pp.

Anderson, T. W. (1984). *An Introduction to Multivariate Statistical Analysis*, 2nd ed., Wiley, New York, 675pp.

Anthes, R. A. (1974). Data assimilation and initialization of hurricane prediction models. *J. Atm. Scis.*, 39, 701–719.

Armi, L., & H. Stommel. (1983). Four views of a portion of the North Atlantic subtropical gyre. *J. Phys. Oc.*, 13, 828–857.

Armstrong, M. (ed.). (1989). *Geostatistics*, 2 Vols. Klüwer, Dordrecht, The Netherlands, 1038pp.

Arons, A. B., & H. Stommel. (1967). On the abyssal circulation of the world ocean–III. An advection-lateral mixing model of the distribution of a tracer property in an ocean basin. *Deep-Sea Res.*, 14, 441–457.

Arthnari, T. S., & Y. Dodge. (1981). *Mathematical Programming in Statistics*, Wiley, New York, 413pp.

Backus, G. E. (1970a). Inference from inadequate and inaccurate data, I. *Proc. Nat. Acad. Scis. U.S.A.*, 65, 1–7.

Backus, G. E. (1970b). Inference from inadequate and inaccurate data, II. *Proc. Nat. Acad. Scis. U.S.A.*, 65, 281–287.

Backus, G. E. (1988a). Bayesian inference in geomagnetism. *Geophys. J.*, 92, 125–142.

Backus, G. E. (1988b). Comparing hard and soft bounds in geophysical inverse problems. *Geophys. J.*, 94, 249–261.

Backus, G. E., & J. F. Gilbert. (1967). Numerical applications of a formalism for geophysical inverse theory. *Geophys. J. Roy. Astron. Soc.*, 13, 247–276.

Backus, G. E., & J. F. Gilbert. (1968). The resolving power of gross earth data. *Geophys. J. Roy. Astron. Soc.*, 16, 169–205.

Backus, G. E., & J. F. Gilbert. (1970). Uniqueness in the inversion of inaccurate gross earth data. *Phil. Trans. Roy. Soc. London* A, 266, 123–192.

Bacon, S. (1994). Skill in an inversion solution: Convex–91 hydrographic results compared with ADCP measurements. *J. Atm. Oc. Tech.*, 11, 1569–1591.

Baker, D. J., Jr. (1981a). A note on Sverdrup balance in the Southern Ocean. *J. Mar. Res.*, 40 (supplement), 21–26.

Baker, D. J, Jr. (1981b). Ocean instruments and experiment design. In *Evolution of Physical Oceanography. Scientific Surveys in Honor of Henry Stommel*, pp. 396–433, B. A. Warren & C. Wunsch (eds.). The MIT Press, Cambridge.

Barth, N. H., & C. Wunsch. (1989). Oceanographic experiment design by simulated annealing. *J. Phys. Oc.*, 20, 1249–1263.

Barth, N. H. (1992). Oceanographic experiment design II: Genetic algorithms. *J. of Atmospheric and Oceanic Tech.*, 9, 434–443.

Batchelor, G. K. (1967). *Introduction to Fluid Dynamics.* Cambridge Univ. Press, Great Britain, 615pp

Beardsley, R. C., & J. F. Festa. (1972). A numerical model of convection driven by a surface stress and a non-uniform horizontal heating. *J. Phys. Oc.*, 2, 444–455.

Bennett, A. F. (1978). Poleward heat fluxes in Southern Hemisphere Oceans. *J. Phys. Oc.*, 8, 785–798.

Bennett, A. F. (1992). *Inverse Methods in Physical Oceanography.* Cambridge Univ. Press, Cambridge, 346pp.

Bennett, A. F., & B. S. Chua. (1994). Open-ocean modeling as an inverse problem: the primitive equations. *Mon. Wea. Rev.*, 122, 1326–1336.

Berkooz, G., P. Holmes, & J. L. Lumley. (1993). The proper orthogonal decomposition in the analysis of turbulent flows. *Ann. Revs. Fluid Mech.*, 25, 539–575.

Bigg, G. R. (1986). Sensitivity studies of a simple inverse method applied to the Cox and Bryan model. *J. Geophys. Res.*, 91, 9639–9654.

Bindoff, N., & C. Wunsch. (1992). Comparison of synoptic and climatologically mapped sections in the South Pacific Ocean. *J. of Climate*, 5, 631–645.

Bindoff, N. L., & J. A. Church. (1992). Warming of the water column in the southwest Pacific Ocean. *Nature*, 357, 59–62.

Bingham, F. M., & L. D. Talley. (1991). Estimates of Kuroshio transports using an inverse technique. *Deep-Sea Res.*, 38, Supplement, J. Reid Volume, S21–S43.

Bittanti, S., A. J. Laub, & J. C. Willems (eds.). (1991a). *The Riccati Equation.* Springer, New York, 352pp.

Bittanti, S., P. Colaneri, & G. De Nicolao. (1991b). The periodic Riccati equation. In *The Riccati Equation*, pp. 127–162, S. Bittanti, A. J. Laub, & J. C. Willems (eds.). Springer-Verlag, Berlin, 127–162.

Bogden, P. S., R. E. Davis, & R. Salmon. (1993). The North Atlantic circulation: Combining simplified dynamics with hydrographic data. *J. Mar. Res.*, 51, 1–52.

Bolin, B., A. Björkström, K. Holmen, & B. Moore. (1983). The simultaneous use of tracers for ocean circulation studies. *Tellus*, 35B, 206–236.

Bolin, B., A. Björkström, & B. Moore. (1989). Uptake by the Atlantic Ocean of excess atmospheric carbon dioxide and radiocarbon. In *Understanding Climate Change*, pp. 57–78, A. Berger, R. E. Dickinson, & J. W. Kidson (eds.). American Geophysical Union, Geophysical Monograph, 52.

Böning, C. W., R. Döscher, & H.-J. Isemer. (1991). Monthly mean wind stress and Sverdrup transports in the North Atlantic: A comparison of the Hellerman-Rosenstein and Isemer-Hasse climatologies. *J. Phys. Oc.*, 21, 221–235.

Box, G. E. P., & G. M. Jenkins. (1978). *Time Series Analysis, Forecasting and Control.* Holden-Day, San Francisco, CA, 575pp.

Bracewell, R. N. (1978). *The Fourier Transform and Its Applications.* McGraw-Hill, New York, 444pp.

Bradley, S. P., A. C. Hax, & T. L. Magnanti. (1977). *Applied Mathematical Programming.* Addison-Wesley, Reading, MA, 761pp.

Bretherton, F. P., R. E. Davis, & C. Fandry. (1976). A technique for objective analysis and design of oceanographic instruments applied to MODE–73. *Deep-Sea Res.*, 23, 559–582.

Brewer, P. G., C. Goyet, & D. Dyrssen. (1989). Carbon dioxide transport by ocean currents at 25°N latitude in the Atlantic Ocean. *Science*, 246, 477–479.

Broecker, W. W. (1974). "NO" a conservative water mass tracer. *Earth & Plan. Sci. Letters*, 23, 100–107.

Broecker, W. S., & T. H. Peng. (1982). *Tracers in the Sea.* Eldigio Press, Palisades, NY, 690pp.

Broecker, W. S., T. Takahashi, & Y.-H. Li. (1976). Hydrography of the central Atlantic–I. The two-degree discontinuity. *Deep-Sea Res.*, 23, 1083–1104.

Brogan, W. L. (1985). *Modern Control Theory*, 2nd ed. Prentice-Hall, Englewood Cliffs, NJ, 509pp.

Brown, E. D., W. B. Owens, & H. L. Bryden. (1986). Eddy potential vorticity fluxes in the Gulf Stream recirculation. *J. Phys. Oc.*, 16, 523–531.

Brown, R. G. (1983). *Introduction to Random Signal Analysis and Kalman Filtering.* Wiley, New York, 347pp.

Bryan, K. (1962). Measurements of meridional heat transports by ocean currents. *J. Geophys. Res.*, 67, 3403–3414.

Bryan, F. O., C. W. Böning, & W. R. Holland. (1995). On the mid-latitude circulation in a high-resolution model of the North Atlantic. *J. Phys. Oc.*, 25, 289–305.

Bryden, H. L. (1977). Geostrophic comparisons from moored measurements of current and temperature during the Mid-Ocean Dynamics Experiment. *Deep-Sea Res.*, 24, 667–681.

Bryden, H. L. (1979). Poleward heat flux and conversion of potential energy in Drake Passage. *J. Mar. Res.*, 37, 1–22.

Bryden, H. L. (1980). Geostrophic vorticity balance in midocean. *J. Geophys. Res.*, 85, 2825–2828.

Bryden, H., & M. Hall. (1980). Heat transport by ocean currents across 25°N in the Atlantic Ocean. *Science*, 207, 884–886.

Bryden, H. L., & R. A. Heath. (1985). Energetic eddies at the northern edge of the Antarctic Circumpolar Current in the southwest Pacific. *Prog. in Oceanog.*, 14, 65–87.

Bryden, H., & N. P. Fofonoff. (1977). Horizontal divergence and vorticity estimates from velocity and temperature measurments in the MODE region. *J. Phys. Oc.*, 7, 329–337.

Bryden, H. L., D. H. Roemmich, & J. A. Church. (1991). Oceanic heat transport across 24°N in the Pacific. *Deep-Sea Res..*, 37, 297–324.

Bryson, A. E. Jr., & Y.-C. Ho. (1975). *Applied Optimal Control*, revised printing. Hemisphere, New York, 481 pp.

Bucy, R. S., & P. D. Joseph. (1968). *Filtering for Stochastic Processes with Applications to Guidance.* Wiley, New York, 195pp.

Bunker, A. F. (1976). Computation of surface energy flux and annual air-sea interaction cycles of the North Atlantic Ocean. *Monthly Wea. Rev.*, 104, 1122–1140.

Businger, P. A., & G. H. Golub. (1969). Singular value decomposition of a complex matrix. *Comm. ACM*, 12, 10,564–10,565.

Butzer, P. L., & R. L. Stens. (1992). Sampling theory for not necessarily band-limited functions: A historical overview. *SIAM Review*, 34, 40–53.

Cacuci, D. G. (1981). Sensitivity theory for nonlinear systems. I. Nonlinear functional analysis approach. *J. Math. Phys.*, 22, 2794–2802.

Cane, M. A. (1989). A mathematical note on Kawase's study of the deep-ocean circulation. *J. Phys. Oc.*, 19, 548–550.

Carissimo, B. C., A. H. Oort, & T. H. Vonder Haar. (1985). Estimating the meridional energy transports in the atmosphere and ocean. *J. Phys. Oc.*, 15, 82–91.

Cessi, P., & J. Pedlosky. (1986). On the role of topography in the ocean circulation. *J. Mar. Res.*, 44, 445–471.

Chandrasekhar, S. (1961). *Hydrodynamic and Hydromagnetic Stability.* Oxford Un. Press, London, 652pp.

Charney, J. G. (1955). The Gulf Stream as an inertial boundary layer. *Proc. Nat. Acad. Scis. U. S. A.*, 41, 731–740.

Charney, J. G., & G. R. Flierl. (1981). Oceanic analogues of large-scale atmospheric motions. In *Evolution of Physical Oceanography. Scientific Surveys in Honor of Henry Stommel*, pp. 504–548, B. A. Warren & C. Wunsch (eds.). The MIT Press, Cambridge.

Chereskin, T. K., & D. Roemmich. (1991). A comparison of measured and wind-derived Ekman transport at 11°N in the Atlantic Ocean. *J. Phys. Oc.*, 21, 869–878.

Coachman, L. K., & K. Aagaard. (1988). Transports through Bering Strait: Annual and interannual variability. *J. Geophys. Res.*, 93, 15,535–15,539.

Craig, H. (1969). Abyssal carbon and radiocarbon in the Pacific. *J. Geophys. Res.*, 74, 5491–5506.

Craik, A. D. D. (1985). *Wave Interaction and Fluid Flows.* Cambridge Univ. Press, 322pp.

Cramér, H. (1946). *Mathematical Methods of Statistics.* Princeton Univ. Press, Princeton, NJ, 574pp.

Daley, R. (1991). *Atmospheric Data Analysis.* Cambridge Univ. Press, Cambridge, 457pp.

Danabagoglu, G., J. C. McWilliams, & P. R. Gent. (1994). The role of mesoscale tracer transport in the general circulation of the oceans. *Science*, 264(5162), 1123–1126.

Dantzig, G. B. (1963). *Linear Programming and Extensions.* Princeton Univ. Press, Princeton, NJ, 625pp.

Davenport, W. B. Jr., & W. L. Root. (1958). *An Introduction to the Theory of Random Signals and Noise.* McGraw-Hill, New York, 393pp.

David, M. (1988). *Handbook of Applied Advanced Geostatistical Ore Reserve Estimation.* Elsevier, Amsterdam, 216pp.

Davis, R. E. (1978a). Estimating velocity from hydrographic data. *J. Geophys. Res.*, 83, 5507–5509.

Davis, R. E. (1978b). Predictability of sea level pressure anomalies over the North Pacific Ocean. *J. Phys. Oc.*, 8, 233–246.

Davis, R. E. (1985). Objective mapping by least-squares fitting. *J. Geophys. Res.*, 90, 4773–4777.

Davis, R. E. (1994). Diapycnal mixing in the ocean: Equations for large-scale budgets. *J. Phys. Oc.*, 24, 777–800.

Davis, P. J., & I. Polonsky. (1965). Numerical interpolation, differentiation and integration. In *Handbook of Mathematical Functions*, pp. 875–924, M. Abramowitz & I. A. Stegun (eds.). Dover, New York.

Defant, A. (1941). *Quantitative Untersuchungen zur Statik und Dynamik des Atlantischen Ozeans. Die absolute Topographie des physikalischen Meeresniveaus und der Druckflächen sowie die Wasserbewegungen im Raum des Atlantischen Ozeans.* In *Wissenschaftliche Ergebnisse der Deutschen Atlantischen Expedition auf dem Forschungs-und Vermessungsschiff "Meteor" 1925–27*, 6:2nd part, 1, pp. 191–260.

Defant, A. (1961). *Physical Oceanography*, vol. 1. Pergamon, New York, 598pp.

DeMey, P., & A. R. Robinson. (1987). Assimilation of altimetric fields in a limited-area quasi-geostrophic model. *J. Phys. Oc.*, 17, 2280–2293.

Denning, P. J. (1992). Genetic algorithms. *Amer. Sci.*, 80, 12–14.

Derber, J., & A. Rosati. (1989). A global oceanic data assimilation system. *J. Phys. Oc.*, 19, 1333–1347.

deSzoeke, R. A. (1985). Wind-driven mid-ocean baroclinic gyres over topography; extension of the Sverdrup relation, *J. Mar. Res.*, 43, 793–824.

deSzoeke, R. A., & M. D. Levine. (1981). The advective flux of heat by mean geostrophic motions in the Southern Ocean. *Deep-Sea Res.*, 28, 1057–1085.

Dickson, R. R., E. M. Gmitrowicz, & A. J. Watson. (1990). Deep-water renewal in the northern North Atlantic. *Nature*, 344(6269), 848–850.

Dickson, R. R., J. Meincke, S.-A. Malmberg, & A. J. Lee. (1988). The great salinity anomaly. *Prog. in Oceanog.*, 20, 103–151.

Draper, N. R., & H. Smith. (1982). *Applied Regression Analysis.* Wiley, New York, 709pp.

Dreisigacker, E., & W. Roether. (1978). Tritium and 90-Sr in North Atlantic surface water. *Earth and Planet. Sci. Lett.*, 38, 301–312.

Eckart, C., & G. Young. (1939). A principal axis transformation for non-Hermitian matrices. *Bull. Amer. Math. Soc.*, 45, 118–121.

Ekman, V. W. (1905). On the influence of the earth's rotation on ocean-currents. *Arkiv for Matematik, Astronomi och Fysik*, 2(11), 52pp.

Esbensen, K., & V. Kushnir. (1981). The heat budget of the global oceans: An atlas based on estimates from marine surface observations. *Climate Res. Inst.*, Rep. No. 29, 27pp. & 188 figs.

Evensen, G. (1994). Sequential data assimilation with a nonlinear quasi-geostrophic model using Monte Carlo methods to forecast error statistics. *J. Geosphy. Res.*, 99, 10,143–10,162.

Ezer, T., & G. L. Mellor. (1994). Continuous assimilation of Geostat altimeter data into a three-dimensional primitive equation Gulf Stream model. *J. Phys. Oc.*, 24, 832–847.

Faller, A. J. (1981). The origin and development of laboratory models and analogues of the ocean circulation. In *Evolution of Physical Oceanography. Scientific Surveys in Honor of Henry Stommel*, pp. 462–482, B. A. Warren & C. Wunsch (eds.). The MIT Press, Cambridge.

Farrell, B. F. (1989). Optimal excitation of baroclinic waves. *J. Atmos. Sci.*, 46, 1193–1206.

Farrell, B. F., & A. M. Moore. (1993). An adjoint method for obtaining the most rapidly growing perturbations to oceanic flows. *J. Phys. Oc.*, 22, 338–349.

Feller, W. (1957). *An Introduction to Probability Theory and Its Applications*, 2nd ed. Wiley, New York, 461pp.

Festa, J. F., & R. L. Molinari. (1992). An evaluation of the WOCE volunteer observing ship-xbt network in the Atlantic. *J. Atm. Oc. Tech.*, 9, 305–317.

Fiacco, A. V., & G. P. McCormick. (1968). *Nonlinear Programming: Sequential Unconstrained Minimization Techniques.* John Wiley, New York (reprinted by SIAM, 1992), 210pp.

Fiadeiro, M. E., & G. Veronis. (1982). On the determination of absolute velocities in the ocean. *J. Mar. Res.*, 40(supplement), 159–182.

Fofonoff, N. P. (1962). Dynamics of ocean currents. In *The Sea: Ideas and Observations of Progress in the Study of the Seas*, 1: *Physical Oceanography*, pp. 323–395, M. N. Hill (ed.). Wiley-Interscience, New York, 323–395.

Franklin, G. F., J. D. Powell, & M. L. Workman. (1990). *Digital Control of Dynamic Systems*, 2nd ed. Addison-Wesley, Reading, MA, 837pp.

Freeman, H. (1965). *Discrete-Time Systems. An Introduction to the Theory.* John Wiley, New York, 241pp.

Fu, L. L. (1981). The general circulation and meridional heat transport of the subtropical South Atlantic determined by inverse methods. *J. Phys. Oc.*, 11, 1171–1193.

Fu, L. L. (1984). Comments on "On the determination of absolute velocities in the ocean," by M. E. Fiadeiro & G. Veronis. *J. Mar. Res.*, 42, 259–262.

Fu, L. (1986). Mass, heat and freshwater fluxes in the South Indian Ocean. *J. Phys. Oc.*, 16, 1683–1693.

Fu, L.-L., E. J. Christensen, C. A. Yamarone, Jr., M. Lefebvre, Y. Menard, M. Dorrer, & P. Escudier. (1994). TOPEX/POSEIDON mission overview. *J. Geophys. Res.*, 99, 24,369–24,382.

Fu, L.-L., I. Fukumori, & R. N. Miller. (1993). Fitting dynamical models to the Geosat sea level observations in the tropical Pacific Ocean. Part II: A linear wind-driven model. *J. Phys. Oc.*, 23, 2162–2189.

Fu, L.-L., T. Keffer, P. Niiler, & C. Wunsch. (1982). Observations of mesoscale variability in the western North Atlantic: A comparative study. *J. Mar. Res.*, 40, 809–848.

Fu, L.-L., J. Vazquez, & C. Perigaud. (1991). Fitting dynamic models to the Geosat sea level observations in the tropical Pacific Ocean. Part I: A free wave model. *J. Phys. Oc.*, 21, 798–809.

Fuglister, F. C. (1960). *Atlantic Ocean Atlas of Temperature and Salinity Profiles and Data from the International Geophysical Year of 1957–1958.* Woods Hole Oceanographic Institution Atlas Series: I, 209pp.

Fukumori, I. (1991). Circulation about the Mediterranean Tongue: An analysis of an EOF-based ocean. *Prog. in Oceanog.*, 27, 197–224.

Fukumori, I., & C. Wunsch. (1991). Efficient representation of the North Atlantic hydrographic and chemical distributions. *Prog. in Oceanog.*, 27, 111–195.

Fukumori, I., F. Martel, & C. Wunsch. (1991). The hydrography of the North Atlantic in the early 1980's. An atlas. *Prog. in Oceanog.*, 27, 1–110.

Fukumori, I., J. Benveniste, C. Wunsch, & D. Haidvogel. (1993). Assimilation of seasurface topography into an ocean circulation model using a steady-state smoother. *J. Phys. Oc.*, 23, 1831–1855.

Gandin, L. S. (1965). *Objective Analysis of Meteorological Fields.* Israel Program for Scientific Translations, Jerusalem, 242pp.

Garçon, V. C., & J.-F. Minster. (1988). Heat, carbon and water fluxes in a 12–box model of the world ocean. *Tellus*, 40B, 161–177.

Gargett, A. E. (1989). Ocean turbulence. *Ann. Rev. Fl. Mech.*, 21, 419–451.

Garrett, C., P. MacCready, & P. Rhines. (1993a). Boundary mixing and arrested Ekman layers: Rotating stratified flow near a sloping boundary. *Ann. Rev. Fl. Mech.*, 25, 292–323.

Garrett, C., R. Outerbridge, & K. Thompson. (1993b). Interannual variability in Mediterranean heat and buoyancy fluxes. *J. Climate*, 6, 900–910.

Gaspar, P., & C. Wunsch. (1989). Estimates from altimeter data of barotropic Rossby waves in the northwestern Atlantic Ocean. *J. Phys. Oc.*, 19, 1821–1844.

Gelb, A. (Ed.). (1974). *Applied Optimal Estimation.* The MIT Press, Cambridge, MA, 374pp.

Gent, P. R., & J. C. McWilliams. (1990). Isopycnal mixing in ocean circulation models. *J. Phys. Oc.*, 20, 150–155.

Ghil, M., & P. Malanotte-Rizzoli. (1991). Data assimilation in meteorology and oceanography. *Adv. in Geophys.*, 33, 141–266.

Ghil, M., S. Cohn, J. Tavantzis, K. Bube, & E. Isaacson. (1981). Applications of estimation theory to numerical weather prediction. In *Dynamic Meteorology. Data Assimilation Methods*, pp. 139–224, L. Bengtsson, M. Ghil, & E. Källén (eds.). Springer-Verlag, New York.

Gill, A. E. (1982). *Atmosphere-Ocean Dynamics.* Academic Press, New York, 662pp.

Gill, A. E., & A. J. Clarke. (1974). Wind-induced upwelling, coastal currents and sea-level changes. *Deep-Sea Res.*, 21, 325–345.

Gill, P. E., W. Murray, & M. H. Wright. (1981). *Practical Optimization.* Academic Press, London, 401pp.

Goldberg, D. E. (1989). *Genetic Algorithms in Search, Optimization and Machine Learning.* Addison-Wesley, Reading, MA, 412pp.

Goldstein, H. (1980). *Classical Mechanics*, 2nd ed. Addison-Wesley, Reading, MA, 672pp.

Golub, G. H., & C. F. Van Loan. (1980). An analysis of the total least squares problem. *SIAM J. Num. Anal.*, 17, 883–893.

Golub, G. H., & C. F. Van Loan. (1989). *Matrix Computation*, 2nd ed. Johns Hopkins Univ. Press, Baltimore, 642pp.

Goodwin, G. C., & K. S. Sin. (1984). *Adaptive Filtering Prediction and Control.* Prentice-Hall, Englewood Cliffs, NJ, 540pp.

Grace, A. (1990). *Optimization Toolbox, For Use With MATLAB.* The MathWorks, Natick, MA, 56pp.

Gregg, M. C. (1987). Diapycnal mixing in the thermocline: A review. *J. Geophys. Res.*, 92, 5249–5286.

Grose, T. S., G. R. Bigg, & J. A. Johnson. (1994). The Bernoulli inverse method: Theory and practice. *Deep-Sea Res.*, 41, 767–786.

Haines, K. (1991). A direct method for assimilating sea surface height data into ocean models with adjustments to the deep circulation. *J. Phys. Oc.*, 21, 843–868.

Hall, M. C. G., & D. G. Cacuci. (1984). Systematic analysis of climatic model sensitivity to parameters and processes. In *Climate Processes and Climate Sensitivity*, pp. 171–179, J. E. Hansen & T. Takahashi (eds.). American Geophysical Union, Washington, D.C.

Hall, M. M., & H. L. Bryden. (1982). Direct estimates and mechanisms of ocean heat transport. *Deep-Sea Res.*, 29, 339–359.

Halpern, D. E., A. Hollingsworth, & F. Wentz. (1994). ECMWF and SSM/I surface wind speeds. *J. Atm. and Oc. Tech.*, 11, 779–788.

Hamann, I. M., & J. H. Swift. (1991). A consistent inventory of water mass factors in the intermediate and deep Pacific Ocean derived from conservative tracers. *Deep-Sea Res.*, 38, Supplement, J. Reid vol., S129–S169.

Hamming, R. W. (1973). *Numerical Methods for Scientists and Engineers.* Dover, New York, 721pp.

Han, Y.-J., & S. W. Lee. (1981). *A new analysis of monthly mean wind stress over the global ocean*, Rep. 26, Climate Res. Inst. Oregon State Univ., 148pp.

Hansen, P. C. (1992). Analysis of discrete ill-posed problems by means of the L-curve. *SIAM Rev.*, 34, 561–580.

Harrison, D. E. (1989). On climatological mean stress and wind stress curl fields over the world ocean. *J. Climate*, 2, 57–70.

Hasselmann, K. (1988). PIPs and POPs: The reduction of complex dynamical systems using principal interactions and oscillation patterns. *J. Geophys. Res.*, 93, 11,015–11,021.

Hastenrath, S. (1984). On meridional heat transports in the world ocean. *J. Phys. Oc.*, 12, 922–927.

Hautala, S. L., D. H. Roemmich, & W. J. Schmitz, Jr. (1994). Is the North Pacific in Sverdrup balance along 24°N? *J. Geophys. Res.*, 99, 16,041–16,052.

Haykin, S. (1986). *Adaptive Filter Theory.* Prentice-Hall, Englewood Cliffs, NJ, 590pp.

Heemink, A. W., & H. Kloosterhuis. (1990). Data assimilation for non-linear tidal models. *Int. J. Num. Methods in Fluids*, 11, 1097–1112.

Helland-Hansen, B., & F. Nansen. (1920). Temperature variations in the North Atlantic Ocean and in the atmosphere. *Smithsonian Misc. Collect.*, 70:4, 408pp.

Hellerman, S., & M. Rosenstein. (1983). Mean monthly wind stress over the world ocean with error estimates. *J. Phys. Oc.*, 13, 1093–1104.

Hidaka, K. (1940a). Absolute evaluation of ocean currents in dynamic calculations. *Proc. Imp. Acad. Tokyo*, 16, 391–393.

Hidaka, K. (1940b). Practical evaluation of ocean currents. *Proc. Imp. Acad. Tokyo*, 16, 394–397.

Hoerl, A. E., & R. W. Kennard. (1970a). Ridge regression. Biased estimation for non-orthogonal problems. *Technometrics*, 12, 55–67.

Hoerl, A. E., & R. W. Kennard. (1970b). Ridge regression. Applications to non-orthogonal problems. *Technometrics*, 12, 69–82.

Hogg, N. G. (1981). Topographic waves along 70°W on the continental rise. *J. Mar. Res.*, 39, 627–649.

Hogg, N. G. (1987). A least-squares fit of the advective-diffusive equations to Levitus Atlas data. *J. Mar. Res.*, 45, 347–375.

Holland, J. H. (1992). *Adaptation in Natural and Artificial Systems: An Introductory Analysis with Applications to Biology, Control and Artificial Intelligence.* The MIT Press, Cambridge, MA, 211pp.

Horton, C., & W. Sturges. (1979). A geostrophic comparison during MODE. *Deep-Sea Res.*, 26A, 521–533.

Hsiung, J. (1985). Estimates of global oceanic meridional heat transport. *J. Phys. Oc.*, 15, 1405–1413.

Hunkins, K. (1966). Ekman drift currents in the Arctic Ocean. *Deep-Sea Res.*, 13, 607–620.

Isemer, H.-J., & L. Hasse. (1985). *The Bunker Climate Atlas of the North Atlantic Ocean*; vol. 1: *Observations*; vol. 2: *Air-Sea Interactions.* Springer-Verlag, Berlin, 218pp & 252pp.

Isemer, H.-J., J. Willebrand, & L. Hasse. (1989). Fine adjustment of large scale air-sea energy flux parameterization by direct estimates of ocean heat transport. *J. of Climate*, 2, 1173–1184.

Jackson, J. D. (1975). *Classical Electrodynamics*, 2nd ed. John Wiley, New York, 848pp.

Jackson, J. E. (1991). *A User's Guide to Principal Components.* Wiley, New York, 569pp.

Jeffreys, H. W. (1925). On fluid motions produced by differences of temperature and humidity. *Q. J. Roy. Met. Soc.*, 51, 347–356.

Jeffreys, H. W. (1961). *Theory of Probability*, 3rd ed. Oxford, 447pp.

Jenkins, W. J., & P. B. Rhines. (1980). Tritium in the deep North Atlantic Ocean. *Nature*, 286, 877–881.

Jerri, A. J. (1977). The Shannon sampling theorem–Its various extensions and applications: A tutorial review. *Proc. IEEE*, 65, November, 1565–1596.

Jolliffe, I. T. (1986). *Principal Component Analysis.* Springer Verlag, New York, 271pp.

Joyce, T. M., D. S. Bitterman, Jr., & K. Prada. (1982). Shipboard acoustic profiling of upper ocean currents. *Deep-Sea Res.*, 29, 903–913.

Joyce, T. M., C. Wunsch, & S. D. Pierce. (1986). Synoptic Gulf Stream velocity profiles through simultaneous inversion of hydrographic and acoustic doppler data. *J. Geophys. Res.*, 91, 7573–7585.

Jung, G. H. (1952). Note on the meridional transport of energy by the oceans. *J. Mar. Res.*, 11, 139–146.

Kalman, R. E. (1960). A new approach to linear filtering and prediction problems. *J. Basic. Eng.*, 82D, 35–45.

Kamenkovich, V. M. (1977). *Fundamentals of Ocean Dynamics.* Elsevier, Amsterdam, 249pp.

Kawase, M. (1987). Establishment of deep ocean circulation driven by deep water production. *J. Phys. Oc.*, 17, 2294–2317.

Kawase, M., & J. L. Sarmiento. (1985). Nutrients in the Atlantic thermocline. *J. Geophys. Res.*, 90, 8961–8979.

Keeling, C. D., & B. Bolin. (1967). The simultaneous use of chemical tracers in oceanic studies. I. General theory of reservoir models. *Tellus*, 19, 566–581.

Kendall, M. G., & A. Stuart. (1976). *The Advanced Theory of Statistics*, 3 vols. Hafner, New York.

Killworth, P. D. (1979). A note on velocity determination from hydrographic data. *J. Geophys. Res.*, 84, 5093–5094.

Killworth, P. D. (1983a). Absolute velocity calculations from single hydrographic sections. *Deep-Sea Res.*, 30, 513–542.
Killworth, P. D. (1983b). Deep convection in the world ocean. *Revs. Geophys. & Space Phys.*, 21, 1–26.
Kirkpatrick, S., C. D. Gelatt, & M. P. Vecchi. (1983). Optimization by simulated annealing. *Science*, 220, 671–680.
Klema, V. C., & A. J. Laub. (1980). The singular value decomposition: Its computation and some applications. *IEEE Trans. Automatic Control*, AC–25, 164–176.
Körner, T. W. (1988). *Fourier Analysis.* Cambridge Univ. Press, Cambridge, 591pp.
Krauss, W. (1986). The North Atlantic Current. *J. Geophys. Res.*, 91, 5061–5074.
Lanczos, C. (1961). *Linear Differential Operators.* Van Nostrand, Princeton, 564pp.
Lanczos, C. (1970). *The Variational Principles of Dynamics*, 4th ed. Dover reprint, New York, 418pp.
Landau, H. J., & H. O. Pollak. (1962). Prolate spheroidal wave functions, Fourier analysis and uncertainty–III: The dimensions of the space of essentially time and bandlimited signals. *Bell System Tech. J.*, 41, 1295–1336.
Landau, L. D., & E. M. Lifschitz. (1987). *Fluid Mechanics*, 2nd ed., J. B. Sykes & W. H. Reid (trans.). Pergamon, Oxford, 539pp.
Large, W. G., & S. Pond. (1981). Open ocean momentum flux measurements in moderate to strong winds. *J. Phys. Oc.*, 11, 324–336.
Large, W. G., J. C. McWilliams, & S. C. Doney. (1994). Oceanic vertical mixing: A review and a model with nonlocal boundary layer parameters. *Revs. Geophys.*, 32, 363–403.
Lawson, C. L., & R. J. Hanson. (1974). *Solving Least-Squares Problems.* Prentice-Hall, Englewood Cliffs, NJ, 340pp.
Lazier, J. R. N. (1973). The renewal of Labrador Sea Water. *Deep-Sea Res.*, 20, 341–353.
LeDimet, F.-X., & O. Talagrand. (1986). Variational algorithms for analysis and assimilation of meteorological observations. *Tellus*, 38A, 97–110.
Ledwell, J. R., A. J. Watson, & C. S. Law. (1993). Evidence for slow mixing across the pycnocline from an open ocean tracer-release experiment. *Nature*, 364, 701–703.
Leetmaa, A., & A. F. Bunker. (1978). Updated charts of the mean annual wind stress, convergences in the Ekman layers and Sverdrup transports in the North Atlantic. *J. Mar. Res.*, 36, 311–322.
Leetmaa, A., J. P. McCreary, Jr., & D. W. Moore. (1981). Equatorial currents: Observations and theory. In *Evolution of Physical Oceanography. Scientific Surveys in Honor of Henry Stommel*, B. A. Warren & C. Wunsch (eds.). The MIT Press, Cambridge, 184–197.
Leetmaa, A., P. Niiler, & H. Stommel. (1977). Does the Sverdrup relation account for the Mid-Atlantic circulation? *J. Mar. Res.*, 35, 1–10.
Levitus, S. (1982). *Climatological Atlas of the World Ocean.* NOAA Professional Paper 13, 173pp.
Levitus, S. (1988). Ekman volume fluxes for the world ocean and individual ocean basins. *J. Phys. Oc.*, 18, 271–279.
Levitus, S. (1989). Interpentadal variability of temperature and salinity in the deep North Atlantic, 1970–74 versus 1955–59. *J. Geophys. Res.*, 94,

16,126–16,131 (see correction, *J. Geophys. Res.*, 94, 14,599).

Lewis, F. L., & R. G. Perkin. (1978). Salinity: Its definition and calculation. *J. Geophys. Res.*, 83, 466–478.

Liebelt, P. B. (1967). *An Introduction to Optimal Estimation.* Addison-Wesley, Reading, MA, 273pp.

Ljung, L. (1987). *System Identification. Theory for the User.* Prentice-Hall, Englewood Cliffs, NJ, 519pp.

Lorenc, A. C. (1986). Analysis methods for numerical weather prediction. *Q. J. Royal Met. Soc.*, 112, 1177–1194.

Luenberger, D. G. (1964). Observing the state of a linear system. *IEEE Trans. Military Electr.*, Mil–8.

Luenberger, D. G. (1969). *Optimization by Vector Space Methods.* John Wiley & Sons, New York, 326pp.

Luenberger, D. G. (1979). *Introduction to Dynamic Systems. Theory, Models and Applications.* John Wiley, New York, 446pp.

Luenberger, D. G. (1984). *Linear and Non-Linear Programming*, 2nd ed. Addison-Wesley, Reading, MA, 491pp.

Luyten, J. R., J. Pedlosky, & H. Stommel. (1983). The ventilated thermocline. *J. Phys. Oc.*, 13, 292–309.

Luyten, J., H. Stommel, & C. Wunsch. (1985). A diagnostic study of the North Atlantic subpolar gyre. *J. Phys. Oc.*, 15, 1344–1348.

Macdonald, A. (1993). Property fluxes at 30°S and their implications for the Pacific-Indian throughflow and the global heat budget. *J. Geophys. Res.*, 98, 6851–6868.

Macdonald, A. M. (1995). *Oceanic fluxes of mass, heat and freshwater: A global estimate and perspective.* Unpublished Ph.D. thesis, MIT/WHOI, 326pp.

Mackas, D. L., K. L. Denman, & A. F. Bennett. (1987). Least squares multiple tracer analysis of water mass composition. *J. Geophys. Res.*, 92, 2907–2918.

Magnus, J. R., & H. Neudecker. (1988). *Matrix Differential Calculus with Applications in Statistics and Econometrics.* John Wiley, Chichester, 393pp.

Malanotte-Rizzoli, P., & W. R. Holland. (1986). Data constraints applied to models of the ocean general circulation, Part I: The steady case. *J. Phys. Oc., 16*, 1665–1687.

Malanotte-Rizzoli, P., & W. R. Holland. (1988). Data constraints applied to models of the ocean general circulation, Part II: The transient, eddy resolving case. *J. Phys. Oc.*, 18, 1093–1107.

Marotzke, J. (1992). The role of integration time in determining a steady state through data assimilation. *J. Phys. Oc.*, 22, 1556–1567.

Marotzke, J., & C. Wunsch. (1993). Finding the steady state of a general circulation model through data assimilation: Application to the North Atlantic Ocean. *J. Geophys. Res.*, 98, 20,149–20,167.

Martel, F., & C. Wunsch. (1993a). The North Atlantic circulation in the early 1980's–an estimate from inversion of a finite difference model. *J. Phys. Oc.*, 23, 898–924.

Martel, F., & C. Wunsch. (1993b). Combined inversion of hydrography, current meter data and altimetric elevations for the North Atlantic circulation. *Manus. Geodaetica*, 18, 219–226.

Matear, R. J. (1993). Circulation in the Ocean Storms area located in the Northeast Pacific Ocean determined by inverse methods. *J. Phys. Oc.*, 23, 648–658.

Maury, M. F. (1855). *The Physical Geography of the Sea and Its Meteorology.* Harper & Bros., New York (reprinted by Harvard University Press, J. Leighly, ed., 1963), 432pp.
McDougall, T. G. (1987a). Neutral surfaces. *J. Phys. Oc.*, 17, 1950–1964.
McDougall, T. J. (1987b). Thermobaricity, cabelling and water-mass conversion. *J. Geophys. Res.*, 92, 5448–5464.
McIntosh, P. C., & G. Veronis. (1993). Solving underdetermined tracer inverse problems by spatial smoothing and cross validation. *J. Phys. Oc.*, 23, 716–730.
Mellor, G. L., C. R. Mechoso, & E. Keto. (1982). A diagnostic calculation of the general circulation of the Atlantic Ocean. *Deep-Sea Res.*, 29, 1171-1192.
Memery, L., & C. Wunsch. (1990). Constraining the North Atlantic circulation with tritium data. *J. Geophys. Res.*, 95, 5229–5256.
Menke, W. (1989). *Geophysical Data Analysis: Discrete Inverse Theory,* 2nd ed. Academic Press, New York, 289pp.
Mercier, H. (1986). Determining the general circulation of the ocean: A non-linear inverse problem. *J. Geophys. Res.*, 91, 5103–5109.
Mercier, H., M. Ollitrault, & P. Y. Le Traon. (1993). An inverse model of the North Atlantic general circulation using Lagrangian float data. *J. Phys. Oc.*, 23, 689–715.
Miller, R. N. (1986). Toward the application of the Kalman filter to regional open ocean modelling. *J. Phys. Oc.*, 16, 72–86.
Miller, R. N., & M. A. Cane. (1989). A Kalman filter analysis of sea level height in the tropical Pacific. *J. Phys. Oc.*, 19, 773–790.
Miller, R. N., M. Ghil, & F. Gauthiez. (1994). Advanced data assimilation in strongly nonlinear dynamical models. *J. Atmos. Scis.*, 51, 1037–1056.
Millero, F. J., C.-T. Tung, A. Bradshaw, & K. Schleicher. (1980). A new high pressure equation of state for seawater. *Deep-Sea Res.*, 27A, 255–264.
Millero, F. J., G. Perron, & J. E. Desnoyers. (1973). The heat capacity of sea-water solutions from 5 to 35°C and 0.5 to 22o/oo chlorinity. *J. Geophys. Res.*, 78, 4499–4507.
Milliff, R. F., & J. C. McWilliams. (1994). The evolution of boundary pressure in ocean basins. *J. Phys. Oc.*, 24, 1317–1338.
Model, F. (1950). Warmwasserheizung Europas. *Ber. Deut. Wetterdienstes*, 12, 51–60.
Molinari, R. L., W. D. Wilson, & K. Leaman. (1985). Volume and heat transport of the Florida Current: April 1982 through August 1983. *Science*, 227, 295–297.
Monin, A.S., & A. M. Yaglom. (1975). *Statistical Fluid Mechanics: Mechanics of Turbulence*, 2 vols, trans. from Russian edition, 1965. The MIT Press, 874pp.
Montgomery, R. B. (1974). Comment on "Seasonal Variability of the Florida Current," by Niiler and Richardson. *J. Mar. Res.*, 32, 533–535.
Moore, D. W. (1968). *Planetary-gravity waves in an equatorial ocean.* Unpublished Ph.D. thesis, Harvard U., Cambridge, MA, 207pp.
Moore, D. W., & S. G. H. Philander. (1977). Modeling of the tropical oceanic circulation. In *The Sea: Ideas and Observations on Progress in the Study of the Seas*, 6: *Marine Modeling*, E. D. Goldberg, I. N. McCave, J. J. O'Brien, & J. H. Steele (eds.). Wiley-Interscience, New York, 319–361.
Moritz, H. (1978). Least-squares collocation. *Revs. Geophs. Space Phys.*, 16, 421–430.

Morse, P. M., & H. Feshbach. (1953). *Methods of Theoretical Physics*, 2 vols. McGraw-Hill, New York, 1978pp.

Müller, T. J., & G. Siedler. (1992). Multi-year current time series in the eastern North Atlantic Ocean. *J. Mar. Res.*, 50, 63–98.

Munk, W. (1950). On the wind-driven ocean circulation. *J. of Meteor.*, 7, 79–93.

Munk, W. H. (1966). Abyssal recipes. *Deep-Sea Res.*, 13, 707–730.

Munk, W. (1981). Internal waves and small-scale processes. In *Evolution of Physical Oceanography. Scientific Surveys in Honor of Henry Stommel*, B. A. Warren & C. Wunsch (eds.). The MIT Press, Cambridge, 264–291.

Munk, W. (1984). Affairs of the sea. In *It's The Water that Makes You Drunk, a Celebration in Geophysics and Oceanography–1982. In Honor of Walter Munk on his 65th Birthday, October 19, 1982.* Scripps Institution of Oceanography Ref. Ser. 84–85, 3–23.

Munk, W., & C. Wunsch. (1982). Up/down resolution in ocean acoustic tomography. *Deep-Sea Res.*, 29, 1415–1436.

Munk, W., P. Worcester, & C. Wunsch. (1995). *Ocean Acoustic Tomography.* Cambridge Univ. Press, 433pp.

Needler, G. (1967). A model for the thermohaline circulation in an ocean of finite depth. *J. Mar. Res.*, 25, 329–342.

Needler, G. (1972). Thermocline models with arbitrary barotropic flow. *Deep-Sea Res.*, 18, 895–903.

Needler, G. T. (1985). The absolute velocity as a function of conserved measurable quantities. *Prog. Oceanog.*, 14, 421–429.

Needler, G. T., & R. A. Heath. (1975). Diffusion coefficients calculated from the Mediterranean salt anomaly in the North Atlantic ocean. *J. Phys. Oc.*, 5, 173–182.

Nelkin, H. (1987). *Thermally driven circulation.* Unpublished Ph.D. thesis, MIT/WHOI, Cambridge, MA, 186pp.

Noble, B., & J. W. Daniel. (1977). *Applied Linear Algebra*, 2nd ed. Prentice-Hall, Englewood Cliffs, NJ, 477pp.

Numerical Algorithms Group. (1988). *The NAG Fortran Manual–Mark–13*, 7 Vols. Downers Grove, IL.

O'Brien, J. J. (Ed.). (1986). *Advanced Physical Oceanography Numerical Modelling.* D. Reidel, Dordrecht, 608pp.

O'Reilly, J. (1983). *Observers for Linear Systems.* Academic, London, 246pp.

Olbers, D., & M. Wenzel. (1989). Determining diffusivities from hydrographic data by inverse methods with application to the Circumpolar Current. In *Oceanic Circulation Models: Combining Data and Dynamics*, pp. 95–140, D. L. T. Anderson & J. Willebrand (eds.). Klüwer, Dordrecht.

Olbers, D., J. M. Wenzel, & J. Willebrand. (1985). The inference of North Atlantic circulation patterns from climatological hydrographic data. *Revs. Geophys.*, 23, 313–356.

Olbers, D. J., & J. Willebrand. (1984). The level of no motion in an ideal fluid. *J. Phys. Oc.*, 14, 203–212.

Oort, A. H., & T. H. Vonder Haar. (1976). On the observed annual cycle in the ocean-atmosphere heat balance over the northern hemisphere. *J. Phys. Oc.*, 6, 781–800.

Osborne, T. L., & C. S. Cox. (1972). Oceanic fine structure. *Geophys. Fl. Dyn.*, 3, 321–345.

Östlund, H. G., & R. A. Fine. (1979). Oceanic distribution and transport of tritium. In *Behavior of Tritium in the Environment*, pp. 303–314. International Atomic Energy Agency, Vienna.

Östlund, H. G., & C. G. H. Rooth. (1990). The North Atlantic tritium and radiocarbon transients 1972–1983. *J. Geophys. Res.*, 95, 20,147–20,165.

Paige, C. C., & M. A. Saunders. (1982). LSQR: An algorithm for sparse linear equations and sparse least-squares. *ACM Trans. Math. Software*, 8, 195–209.

Parker, R. L. (1977). Understanding inverse theory. *Ann. Revs. Earth and Planetary Sci.*, 5, 35–64.

Parker, R. L. (1994). *Geophysical Inverse Theory.* Princeton U. Press, Princeton, NJ, 377pp.

Parrilla, G., A. Lavin, H. Bryden, M. Garcia, & R. Millard. (1994). Rising temperatures in the subtropical North Atlantic over the past 35 years. *Nature*, 369, 48–51.

Pedlosky, J. (1987a). *Geophysical Fluid Dynamics*, 2nd ed. Springer-Verlag, 710pp.

Pedlosky, J. (1987b). Thermocline theories. In *General Circulation of the Oceans*, pp. 55–101, H. D. I. Abarbanel & W. R. Young (eds.). Springer-Verlag, New York.

Peixoto, J. P., & A. H. Oort. (1983). The atmospheric branch of the hydrological cycle and climate. In *Variations in the Global Water Budget*, pp. 5–65, A. Street-Perrott (ed.). D. Reidel, New York.

Petersen, D. P., & D. Middleton. (1962). Sampling and reconstruction of wave-number-limited functions in N-dimensional Euclidean space. *Inform. and Control*, 5, 279–323.

Phillips, N. A. (1963). Geostrophic motion. *Revs. Geophys.*, 1, 123–176.

Phillips, N. A. (1966). The equations of motion for a shallow rotating atmosphere and the "traditional approximation," *J. Atm. Sci.*, 23, 626–628.

Pickard, G. L., & W. J. Emery. (1982). *Descriptive Physical Oceanography. An Introduction*, 5th ed. Pergamon, Oxford, 249pp.

Pickart, R. S. (1992). Water mass components of the North Atlantic deep western boundary current. *Deep-Sea Res.*, 39, 1557–1572.

Pickart, R. S., & D. R. Watts. (1990). Deep western boundary current variability at Cape Hatteras. *J. Mar. Res.*, 48, 765–791.

Pierce, S. D., & T. M. Joyce. (1988). Gulf Stream velocity structure through inversion of hydrographic and acoustic data. *J. Geophys. Res.*, 93, 2227–2336.

Pillsbury J. E. (1891). The Gulf Stream–A description of the methods employed in the investigation, and the results of the research. In *Report of the Superintendent of the U. S. Coast and Geodetic Survey Showing the Progress of the Work during the Fiscal Year Ending with June, 1890*, pp. 459–620. Appendix No. 10, Washington, D.C.

Pincus, M. (1970). A Monte-Carlo method for the approximate solution of certain types of constrained optimization problems. *Oper. Res.*, 18, 1225–1228.

Pollard, R. T. (1983). Mesoscale (50–100 km) circulations revealed by inverse and classical analysis of the JASIN hydrographic data. *J. Phys. Oc.*, 13, 377–394.

Pond, S., & G. L. Pickard. (1983). *Introductory Dynamical Oceanography*, 2nd ed. Pergamon, Oxford, 329pp.

Preisendorfer, R. W. (1988). *Principal Component Analysis in Meteorology and Oceanography*, posthumously compiled and edited by C. D. Mobley. Elsevier, Amsterdam, The Netherlands, 425pp.

Press, W. H., B. P. Flannery, S. A. Teukolsky, & W. T. Vetterling. (1992). *Numerical Recipes*, 2nd ed. Cambridge Univ. Press, Cambridge, 963pp.

Price, J. F., R. A. Weller, & R. R. Schudlich. (1987). Wind-driven ocean currents and Ekman transport. *Science*, 238, 1534–1538.

Priestley, M. B. (1981). *Spectral Analysis and Time Series*. Vol. 1: *Univariate Series*; Vol. 2: *Multivariate Series, Prediction and Control*. Academic, London, 890pp. plus appendices (combined edition).

Provost, C., & R. Salmon. (1986). A variational method for inverting hydrographic data. *J. Mar. Res.*, 44, 1–34.

Rauch, H. E., F. Tung, & C. T. Striebel. (1965). Maximum likelihood estimates of linear dynamic systems. *AIAA J.*, 3, 1445–1450 (reprinted in Sorenson, 1985).

Redi, M. H. (1982). Oceanic isopycnal mixing by coordinate rotation. *J. Phys. Oc.*, 12, 1154–1158.

Reid, J. L. (1965). *Intermediate Waters of the Pacific Ocean*, Johns Hopkins Oceanographic Studies No. 2. Johns Hopkins Univ. Press, Baltimore, 85pp.

Reid, J. L. (1981). On the mid-depth circulation of the world ocean. In *Evolution of Physical Oceanography. Scientific Surveys in Honor of Henry Stommel*, B. A. Warren & C. Wunsch (eds.), The MIT Press, 70–111.

Reid, J. L. (1986). On the total geostrophic circulation of the South Pacific Ocean: Flow patters, tracers and transports. *Prog. in Oceanog.*, 16, 1–61.

Reid, J. L. (1989). On the total geostrophic circulation of the South Atlantic Ocean: Flow patterns, tracers, and transports. *Prog. in Oceanog.*, 23, 149–244.

Reid, W. T. (1972). *Riccati Differential Equations*. Academic, New York, 216pp.

Richardson, P. L., & K. Mooney. (1975). The Mediterranean outflow–A simple advection diffusion model. *J. Phys. Oc.*, 5, 476–482.

Riley, G. A. (1951). Oxygen, phosphate and nitrate in the Atlantic Ocean. *Bulletin of the Bingham Oceanographic Collection*, 13:1, 126pp.

Rintoul, S. (1991). South Atlantic interbasin exchange. *J. Geophys. Res.*, 96, 2675–2692.

Rintoul, S. R., & C. Wunsch. (1991). Mass, heat, oxygen and nutrient fluxes in the North Atlantic Ocean. *Deep-Sea Res.*, 38, Supplement, J. Reid Volume, S355–S377.

Ripley, B. D. (1981). *Spatial Statistics*. Wiley, New York, 252pp.

Roache, P. J. (1976). *Computational Fluid Dynamics*. Hermosa, Albuquerque, NM, 446pp.

Roberts, S. M., & J. S. Shipman. (1972). *Two-Point Boundary Value Problems: Shooting Methods*. American Elsevier, New York, 269pp + index.

Robinson, A. R. (Ed.). (1983). *Eddies in Marine Science*. Springer-Verlag, Berlin, 609pp.

Robinson, A. R., & H. Stommel. (1959). The oceanic thermocline and the associated thermohaline circulation. *Tellus*, 3, 295–308.

Robinson, A. R., & W. Leslie. (1985). Estimation and prediction of oceanic eddy fields. *Prog. Oceanog.*, 14, 485–510.

Rockafellar, R. T. (1993). Lagrange multipliers and optimality. *SIAM Rev.*, 35, 183–238.

Roemmich, D. (1980). Estimation of the meridional heat flux in the North Atlantic by inverse methods. *J. Phys. Oc.*, 10, 1972–1983.

Roemmich, D. (1981). Circulation of the Caribbean Sea: A well-resolved inverse problem. *J. Geophys. Res.*, 86, 7993–8005.

Roemmich, D. (1983). Optimal estimation of hydrographic station data and derived fields. *J. Phys. Oc.*, 13, 1544–1549.

Roemmich, D., & C. Wunsch. (1984). Apparent change in the climatic state of the deep North Atlantic Ocean. *Nature*, 307, 447–450.

Roemmich, D., & C. Wunsch. (1985). Two transatlantic sections: Meridional circulation and heat flux in the subtropical North Atlantic Ocean. *Deep-Sea Res.*, 32, 619–664.

Roemmich, D., & T. McCallister. (1989). Large scale circulation of the North Pacific Ocean. *Prog. in Oceanog.*, 22, 171–204.

Rogers, G. S. (1980). *Matrix Derivatives.* Marcel Dekker, New York, 209pp.

Rossby, C. G. and Collaborators. (1939). Relation between variations in the intensity of the zonal circulation of the atmosphere and the displacements of the semi-permanent centers of action. *J. Mar. Res.*, 2, 38–55.

Rossby, H. T. (1965). On thermal convection driven by non-uniform heating from below: An experimental study. *Deep-Sea Res.*, 12, 9–16.

Salmon, R. (1986). A simplified linear ocean circulation theory. *J. Mar. Res.*, 44, 695–711.

Saltzman, B. (1962). *Selected Papers on the Theory of Thermal Convection, with Special Application to the Earth's Planetary Atmosphere.* Dover, NY, 461pp.

Sandström, J. W. (1908). Dynamische Versuche mit Meerwasser. *Ann. Hydr. Mar. Met.*, 6.

Sarkisiyan, A. S., & A. F. Pastukhov. (1970). The density field as the main indicator of steady sea current. *Isvestiya, Atmospheric and Oceanic Physics* (English trans.), 6, 34–40.

Sarmiento, J. L., & K. Bryan. (1982). An ocean transport model for the North Atlantic. *J. Geophys. Res.*, 87, 394–408.

Sasaki, Y. (1970). Some basic formalisms in numerical variational analysis. *Mon. Wea. Rev.*, 98, 875–883.

Saunders, P. M. (1981). Practical conversion of pressure to depth. *J. Phys. Oc.*, 11, 573–574.

Scales, J. A., M. L. Smith, & T. L. Fischer. (1992). Global optimization methods for multimodal inverse problems. *J. of Comp. Phys.*, 103, 258–268.

Scales, L. E. (1985). *Introduction to Non-Linear Optimization.* Springer-Verlag, New York, 243pp.

Schlitzer, R. (1988). Modeling the nutrient and carbon cycles of the North Atlantic, 1. Circulation, mixing coefficients, and heat fluxes. *J. Geophys. Res.*, 93, 10699–10723.

Schlitzer, R. (1989). Modeling the nutrient and carbon cycles of the North Atlantic, 2. New production, particle fluxes, CO_2, gas exchange, and the role of organic nutrients. *J. Geophys. Res.*, 94, 12781–12794.

Schlitzer, R. (1993). Determining the mean, large-scale circulation of the Atlantic with the adjoint method. *J. Phys. Oc.*, 23, 1935–1952.

Schmitt, R., P. Bogden, & C. E. Dorman. (1989). Evaporation minus precipitation and density fluxcs for the North Atlantic. *J. Phys. Oc.*, 19, 1208–1221.

Schmitz, W. J., Jr., & W. S. Richardson. (1968). On the transport of the Florida Current. *Deep-Sea Res.*, 15, 679–694.

Schott, F., & H. Stommel. (1978). Beta spirals and absolute velocities in different oceans. *Deep-Sea Res.*, 25, 961–1010.

Schott, F., & R. Zantopp. (1979). Calculation of absolute velocities from different parameters in the western North Atlantic. *J. Geophys. Res.*, 84, 6990–6994.

Schott, F. A., T. N. Lee, & R. Zantopp. (1988). Variability of structure and transport of the Florida Current in the period range of days to seasonal. *J.*

Phys. Oc., 18, 1209–1230.
Schröter, J., & C. Wunsch. (1986). Solution of non-linear finite difference ocean models by optimization methods with sensitivity and observational strategy analysis. *J. Phys. Oc.*, 16, 1855–1874.
Seber, G. A. F. (1977). *Linear Regression Analysis.* John Wiley & Sons, New York, 465pp.
Seber, G. A. F., & C. J. Wild. (1989). *Nonlinear Regression.* John Wiley, New York, 768pp.
Semtner, A. J., Jr., & R. M. Chervin. (1992). Ocean general circulation from a global eddy-resolving model. *J. Geophys. Res.*, 97, 5493–5550.
Sewell, M. J. (1987). *Maximum and Minimum Principles. A Unified Approach with Applications.* Cambridge Univ. Press, Cambridge, 468pp.
Shepard, J. (1980). Cautious non-linear optimisation: A new technique for allocation problems. *J. Operational Res. Soc.*, 31, 993–1000.
Sheinbaum, J., & D. L. T. Anderson. (1990). Variational assimilation of XBT data, Part I. *J. Phys. Oc.*, 20, 672–688.
Shoji, D. (1972). Time variations of the Kuroshio south of Japan. In *Kuroshio: Its Physical Aspects*, pp. 217–234, H. Stommel & K. Yoshida (eds.). U. of Tokyo Press.
Smedstad, O. M., & J. J. O'Brien. (1991). Variational data assimilation and parameter estimation in an equatorial ocean model. *Prog. Ocean.*, 26, 179–241.
Sorenson, H. W. (ed.). (1985). *Kalman Filtering: Theory and Application.* IEEE Press, New York, 457pp.
Speer, K., & E. Tziperman. (1992). Rates of water mass formation in the North Atlantic Ocean. *J. Phys. Oc.*, 22, 93–104.
Stammer, D., & C. Wunsch. (1994). Preliminary assessment of the accuracy and precision of TOPEX/POSEIDON altimeter data with respect to the large scale ocean circulation. *J. Geophys. Res.*, 99, 22,584–24,604.
Stammer, D., & C. Wunsch. (1996). The determination of the large-scale circulation of the Pacific Ocean from satellite altimetry using model Green's functions. (Submitted for publication.)
Starr, V. P. (1968). *Physics of Negative Viscosity Phenomena.* Mc-Graw Hill, New York, 256pp.
Stengel, R. F. (1986). *Stochastic Optimal Control.* Wiley-Interscience, New York, 638pp.
Stewart, G. W. (1993). On the early history of the singular value decomposition. *SIAM Rev.*, 35, 551–556.
Stommel, H. (1948). The westward intensification of wind-driven ocean currents. *Trans. Am. Geophys. Un.*, 29, 202–206.
Stommel, H. (1956). On the determination of the depth of no meridional motion. *Deep-Sea Res.*, 3, 273–278.
Stommel, H. (1957). A survey of ocean current theory. *Deep-Sea Res.*, 4, 149–184.
Stommel, H. (1958). The abyssal circulation. *Deep-Sea Res.*, 5, 80–82.
Stommel, H. (1962). On the smallness of sinking regions in the ocean. *Proc. Natl. Acad. Scis., U.S.A.*, 48, 766–772.
Stommel, H. (1965). *The Gulf Stream: A Physical and Dynamical Description*, 2nd ed. Univ. Calif. Press, Berkeley, CA, 248pp.
Stommel, H. (1980). Asymmetry of interoceanic fresh-water fluxes. *Proc. Nat. Acad. Scis.*, 77, 2377–2381.

Stommel, H. (1982). Response to award of Bowie Medal. *EOS, Trans. Am. Geophys. Un.*, 63, 572.

Stommel, H. (1984). The delicate interplay between wind-stress and buoyancy input in ocean circulation: the Goldsbrough variations. *Tellus*, 36A, 111–119.

Stommel, H., & A. B. Arons. (1960a). On the abyssal circulation of the world ocean–I. Stationary planetary flow patterns on a sphere. *Deep-Sea Res.*, 6, 140–154.

Stommel, H., & A. B. Arons. (1960b). On the abyssal circulation of the world ocean–II. An idealized model of the circulation pattern and amplitude in oceanic basins. *Deep-Sea Res.*, 6, 217–233.

Stommel, H., & F. Schott. (1977). The beta spiral and the determination of the absolute velocity field from hydrographic station data. *Deep-Sea Res.*, 24, 325–329.

Stommel, H., A. B. Arons, & A. J. Faller. (1958). Some examples of stationary planetary flow patterns in bounded basins. *Tellus*, 10, 179–187.

Stommel, H., E. D. Stroup, J. L. Reid, & B. A. Warren. (1973). Transpacific hydrographic sections at Lats. 43°S and 28°S: The SCORPIO Expedition–I. Preface. *Deep-Sea Res.*, 20, 1–7.

Stommel, H., P. P. Niiler, & D. Anati. (1978). Dynamic topography and recirculation of the North Atlantic Ocean. *J. Mar. Res.*, 36, 449–468.

Stone, R. E., & C. A. Tovey. (1991). The simplex and projective scaling algorithms as iteratively reweighted least squares methods. *SIAM Rev.*, 33, 220–237.

Strang, G. (1986). *Introduction to Applied Mathematics.* Wellesley-Cambridge Press, Wellesley, MA, 758pp.

Strang, G. (1988). *Linear Algebra and Its Applications*, 3rd ed. Harcourt, Brace Jovanovich, San Diego, 505pp.

Sverdrup, H. U. (1947). Wind-driven currents in a baroclinic ocean; with application to the equatorial currents of the eastern Pacific. *Proc. Nat. Acad. Scis., U.S.A.*, 33, 318–326.

Sverdrup, H. U., M. W. Johnson, & R. H. Fleming. (1942). *The Oceans.* Prentice-Hall, Englewood Cliffs, NJ, 1087pp.

Swallow, J. C., & L. V. Worthington. (1961). An observation of a deep countercurrent in the western North Atlantic. *Deep-Sea Res.*, 8, 1–19.

Swallow, J. C. (1977). An attempt to test the geostrophic balance using the Minimode current measurements. In *A Voyage of Discovery: George Deacon 70th Anniversary Volume* (Supplement to *Deep-Sea Res.*, pp. 165–176, M. Angel (ed.). Pergamon, Oxford.

Talley, L. (1984). Meridional heat transport in the Pacific Ocean. *J. Phys. Oc.*, 14, 231–241.

Tarantola, A. (1987). *Inverse Problem Theory. Methods for Data Fitting and Model Parameter Estimation.* Elsevier, Amsterdam, 613pp.

Tarantola, A., & B. Valette. (1982). Generalized nonlinear inverse problems solved using the least squares criterion. *Revs. Geophys. & Space Phys.*, 20, 219–232.

Tatro, P. R., & E. L. Mollo-Christensen. (1967). Experiments on Ekman layer instability. *J. Fluid Mech.*, 28, 531–543.

Thacker, W. C. (1989). The role of the Hessian matrix in fitting models to measurements. *J. Geophys. Res.*, 94, 6177–6196.

Thacker, W. C., & R. B. Long. (1988). Fitting dynamics to data. *J. Geophys. Res.*, 93, 1227–1240.

Thièbaux, H. J., & M. A. Pedder. (1987). *Spatial Objective Analysis: With*

Applications in Atmospheric Science. Academic, London, 299pp.
Thompson, R. O. R. Y., & G. Veronis. (1980). Transport calculations in the Tasman and Coral Seas. *Deep-Sea Res.*, 27, 303–323.
Thorpe, S. A. (1987). Transitional phenomena and the development of turbulence in stratified fluids. *J. Geophys. Res.*, 92, 5231–5248.
Todling, R., & S. E. Cohn. (1994). Suboptimal schemes for atmospheric data assimilation based on the Kalman filter. *Mon. Wea. Rev.*, 122, 2530–2557.
Tomczak, M., & D. G. B. Large. (1989). Optimum multiparameter analysis of mixing in the thermocline of the East Indian Ocean. *J. Geophys. Res.*, 94, 16141–16150.
Toole, J. M, & M. R. Raymer. (1985). Heat and fresh water budget of the Indian Ocean revisited. *Deep-Sea Res.*, 32, 917–928.
Toole, J. M., K. L. Polzin, & R. W. Schmitt. (1994). Estimates of diapycnal mixing in the abyssal ocean. *Science*, 264(5164), 1120–1123.
Trenberth, K., W. G. Large, & J. G. Olson. (1989a). The effective drag coefficient for evaluating wind stress over the oceans. *J. of Climate*, 2, 1507–1516.
Trenberth, K., J. G. Olson, & W. G. Large. (1989b). *A Global Ocean Wind Stress Climatology Based on ECMWF Analysis.* NCAR/TN-338TSTR, Aug., 93pp.
Tziperman, E. (1987). The Mediterranean outflow as an example of a deep buoyancy-driven flow. *J. Geophys. Res.*, 92, 14,510–14,520.
Tziperman, E. (1988). Calculating the time-mean oceanic general circulation and mixing coefficients from hydrographic data. *J. Phys. Oc.*, 18, 519–525.
Tziperman, E., & A. Hecht. (1987). A note on the circulation in the eastern Levantine basin by inverse methods. *J. Phys. Oc.*, 18, 506–518.
Tziperman, E., & P. Malanotte-Rizzoli. (1991). The climatological seasonal circulation of the Mediterranean Sea. *J. Mar. Res.*, 49, 411–434.
Tziperman, E., & W. C. Thacker. (1989). An optimal control/adjoint equations approach to studying the oceanic general circulation. *J. Phys. Oc.*, 19, 1471–1485.
Tziperman, E., W. C. Thacker, R. B. Long, & S.-M. Hwang. (1992a). Oceanic data analysis using a general circulation model. Part I: Simulations. *J. Phys. Oc.*, 22, 1434–1457.
Tziperman, E., W. C. Thacker, R. B. Long, S.-M. Hwang, & S. R. Rintoul. (1992b). Oceanic data analysis using a general circulation model. Part II: A North Atlantic model. *J. Phys. Oc.*, 22, 1458–1485.
Van Huffel, S., & J. Vandewalle. (1991). *The Total Least Squares Problem. Computational Aspects and Analysis.* SIAM, Philadelphia, 300pp.
van Laarhoven, P. J. M., & E. H. L. Aarts. (1987). *Simulated Annealing: Theory and Practice.* Klüwer, Dordrecht, The Netherlands, 186pp.
Van Trees, H. L. (1968). *Detection, Estimation and Modulation Theory.* Part I. *Detection, Estimation and Linear Modulation Theory.* J. Wiley, New York, 697pp.
Vaníček, P., & E. J. Krakiwsky. (1986). *Geodesy: The Concepts.* North-Holland, Amsterdam, 697pp.
Veronis, G. (1969). On theoretical models of the thermocline circulation. *Deep-Sea Res.*, 16 (supplement, Frederick C. Fuglister Sixtieth Anniversary Volume), 301–323.
Veronis, G. (1981). Dynamics of large-scale ocean circulation. In *Evolution of Physical Oceanography. Scientific Surveys in Honor of Henry Stommel*, pp. 140–183, B. A. Warren & C. Wunsch (eds.). The MIT Press, Cambridge.

von Storch, H., G. Bürger, R. Schnur, & J.-S. von Storch. (1993). *Principal Oscillation Patterns.* Report No. 113, Max-Planck Institut fuër Meteorologie, Hamburg, 46pp.

von Storch, H., T. Bruns, I. Fischer-Bruns, & K. Hasselmann. (1988). Principal oscillation pattern analysis of the 30–to–60 day oscillation in a general circulation model equatorial troposphere. *J. Geophys. Res.*, 93, 11,022–11,036.

Vonder Haar, T. H., & A. H. Oort. (1973). New estimate of annual poleward energy transport by Northern Hemisphere oceans. *J. Phys. Oc.*, 2, 169–172.

Wagner, H. M. (1969). *Principles of Operations Research. With Applications to Managerial Decisions.* Prentice-Hall, Englewood Cliffs, NJ, 937pp.

Wahba, G. (1990). *Spline Models for Observational Data.* Soc. for Indust. Appl. Maths., Philadelphia, 169pp.

Wahba, G., & J. Wendelberger. (1980). New mathematical methods for variational objective analysis using splines and cross validation. *Mon. Wea. Rev.*, 108, 1122–1143.

Wallace, J. M. (1972). Empirical-orthogonal representation of time series in the frequency domain. Part II: Application to the study of tropical wave disturbances. *J. Appl. Met.*, 11, 893–900.

Wallace, J. M., & R. E. Dickinson. (1972). Empirical-orthogonal representation of time series in the frequency domain. Part I: Theoretical considerations. *J. Appl. Met.*, 11, 887–892.

Warren, B. A. (1972). Insensitivity of subtropical mode water characteristics to meteorological fluctuations. *Deep-Sea Res.*, 19, 1–20.

Warren, B. A. (1981). Deep circulation of the world ocean. In *Evolution of Physical Oceanography. Scientific Surveys in Honor of Henry Stommel,* pp. 6–41, B. A. Warren & C. Wunsch (eds.). The MIT Press, Cambridge.

Warren, B. A., & W. B. Owens. (1988). Deep currents of the central subarctic Pacific Ocean. *J. Phys. Oc.*, 18, 529–551.

Welander, P. (1971). The thermocline problem. *Phil. Trans. Roy. Soc., A*, 270, 69-73.

Weiss, R. E., J. C. Bullister, R. H. Gammon, & M. J. Warner. (1985). Atmospheric chlorofluoromethanes in the deep equatorial Atlantic. *Nature*, 314, 608–610.

Weiss, R. E., M. J. Warner, P. K. Salameh, F. A. Van Woy, & K. G. Harrison. (1993). *South Atlantic Ventilation Experiment: SIO Chlorofluorocarban Measurements.* SIO Reference 93–49 (December).

Weiss, W., & W. Roether. (1980). The rates of tritium input to the world ocean. *Earth, Planet. Sc. Lett.*, 49, 435–446.

Weller, R. A. (1981). Observations of the velocity response to wind forcing in the upper ocean. *J. Geophys. Res.*, 86, 1969–1977.

Wenzel, M. (1986). *Die mittlere Zirkulation des Nordatlantik auf der Grundlage klimatologischer hydrographischer Daten.* Berichte Institut für Meereskunde, Kiel 153, 109pp.

Whittaker, E., & G. Robinson. (1944). *The Calculus of Observations.* Blackie & Sons, Glasgow, 391pp.

Wiener, N. (1949). *The Extrapolation, Interpolation and Smoothing of Stationary Time Series.* The Technology Press of MIT and J. Wiley, New York, 163pp.

Wiggins, R. A. (1972). The general linear inverse problem: Implication of surface waves and free oscillations for earth structure. *Revs. Geophys. and Space Phys.*, 10, 251–285.

Wijffels, S. E., R. W. Schmitt, H. L. Bryden, & A Stigebrandt. (1992). On the transport of freshwater by the oceans. *J. Phys. Oc.*, 94, 249–261.

Worthington, L. V. (1976). *On the North Atlantic Circulation.* Johns Hopkins Univ. Press, Baltimore, MD, 110pp.

Worthington, L. V. (1981). The water masses of the world ocean: some results of a fine-scale census. In *Evolution of Physical Oceanography. Scientific Surveys in Honor of Henry Stommel*, pp. 42–69. B. A. Warren and C. Wunsch (eds.). The MIT Press, Cambridge.

Wunsch, C. (1977). Determining the general circulation of the oceans: A preliminary discussion. *Science*, 196, 871–875.

Wunsch, C. (1978). The North Atlantic general circulation west of 50°W determined by inverse methods. *Revs. Geophys. and Space Phys.*, 16, 583–620.

Wunsch, C. (1980). Meridional heat flux of the North Atlantic Ocean. *Proc. Nat. Acad. Scis, USA*, 77, 5043–5047.

Wunsch, C. (1981). Low frequency variability of the sea. In *Evolution of Physical Oceanography: Scientific Surveys in Honor of Henry Stommel*, pp. 342–374, B. A. Warren & C. Wunsch, eds. The MIT Press, Cambridge, MA.

Wunsch, C. (1984). An eclectic Atlantic Ocean circulation model. Part I: The meridional flux of heat. *J. of Phys. Oc.*, 14, 1712–1733.

Wunsch, C. (1985). Can a tracer field be inverted for velocity? *J. Phys. Oc.*, 15, 1521–1531.

Wunsch, C. (1986). Using transient tracers: The regularization problem. *Tellus*, 39B, 477–492.

Wunsch, C. (1988a). Eclectic modelling of the North Atlantic. II: Transient tracers and the ventilation of the eastern basin thermocline. *Phil. Trans. Roy. Soc., A*, 201–236.

Wunsch, C. (1988b). Transient tracers as a problem in control theory. *J. Geophys. Res.*, 93, 8099–8110.

Wunsch, C. (1989). Sampling characteristics of satellite orbits. *J. Atmo. and Oceanic Tech.*, 6, 891–907.

Wunsch, C. (1991). Global-scale sea surface variability from combined altimetric and tide gauge measurements. *J. Geophys. Res.*, 96, 15053–15082.

Wunsch, C. (1992). Decade-to-century changes in the ocean circulation. *Oceanography*, 5, 99–106.

Wunsch, C. (1994). Dynamically consistent hydrography and absolute velocity in the eastern North Atlantic Ocean. *J. Geophys. Res.*, 99, 14,071–14,090.

Wunsch, C., & B. Grant. (1982). Towards the general circulation of the North Atlantic Ocean. *Prog. in Oceanog.*, 11, 1–59.

Wunsch, C., D.-X. Hu, & B. Grant. (1983). Mass, heat, salt and nutrient fluxes in the South Pacific Ocean. *J. Phys. Oc.*, 13, 725–753.

Wunsch, C., & J.-F. Minster. (1982). Methods for box models and ocean circulation tracers: Mathematical programming and non-linear inverse theory. *J. Geophys. Res.*, 87, 5647–5662.

Wunsch, C., & D. Roemmich. (1985). Is the North Atlantic in Sverdrup balance? *J. Phys. Oc.*, 15, 1876–1880.

Wüst, G. (1924). *Florida- und Antillenstrom.* Berlin U., Institut f. Meereskunde, Veröff., N. F., A. Geogr.-naturwiss. Reihe, Heft 12, 48pp.

Wüst, G. (1935). *Schichtung und Zirkulation des Atlantischen Ozeans. Die Stratosphare.* In *Wissenschaftliche Ergebnisse der Deutschen Atlantischen Expedition auf dem Forschungs-und Vermessungsschiff "Meteor" 1925-1927,*

6:1st Part, 2, 180pp. (Reprinted as *The Stratosphere of the Atlantic Ocean*, W. J. Emery (ed.), 1978, Amerind, New Delhi, 112pp.)

Yih, C. S. (1965). *Dynamics of Non-homogeneous Fluids.* Macmillan, New York, 301pp.

Author Index

Aagard, K., 223
Aarts, E. H. L., 323
Abel, N. H., 10
Aki, K. 10., 14
Albert, A. E., 389, 398
Amos, A. F., 64
Anati, D., 59
Anderson, B. D. O., 327, 351, 361, 382, 388
Anderson, D. L. T., 378, 390
Anderson, T. W., 246
Armi, L., 291
Armstrong, M., 311
Arons, A. B., 42, 44, 73
Arthnari, T. S., 305–306

Backus, G. E., 14, 211, 306
Bacon, S., 256
Baker, D. J. Jr., 22, 56
Barth, N. H., 323
Batchelor, G. K., 18
Beardsley, R. C., 42
Bennett, A. F., xi, 14, 31, 247, 263, 267, 307, 311–312, 349, 399
Benveniste, J., 359
Berkooz, G. P., 307
Bigg, G. R., 295
Bindoff, N., 274
Bingham, F. M., 253, 256
Bittanti, S., 382
Bitterman, D. S., 256
Björkström, A., 285
Bogden, P. S., 68, 295
Bolin, B., 284, 285
Boning, C. W., 52–54, 58, 61–62
Box, G. E. P., 117–118, 332
Bracewell, R. N., 110, 191, 193
Bradley, S. P., 302
Bradshaw, A., 237
Bretherton, F. P., 189, 191, 199
Brewer, P. G., 271
Broecker, W. S., 34, 250, 358
Brogan, W. L., 141, 204, 367
Brown, E. D., 268
Brown, R. G., 327, 352
Bruns, T., 402
Bryan, F. O., 62, 262, 264
Bryan, K., 260, 263, 393
Bryden, H. L., 49–50, 57, 62, 247, 263–267, 268, 274, 290
Bryson, A. E. Jr., 327, 336, 352, 361, 364, 372, 375
Bube, K., 349
Bucy, R. S., 349, 368
Bunker, A. F., 45, 54, 260
Businger, P. A., 309
Butzer, P. L., 191

Cacuci, D. G., 303
Cane, M. A., 45, 349
Carissimo, B. C., 260
Cessi, P., 62

Chandrasekhar, S., 41
Charney, J. G., 5, 58
Chereskin, T. K., 46
Chervin, R. M., 2
Christensen, E. J., 76
Chua, B. S., 399
Church, J. A., 247, 274
Clarke, A. J., 47
Coachman, L. K., 223
Cohn, S., 349, 398
Cox, C. S., 282
Craig, H., 282
Craik, A. D. D., 45
Cramér, H., 101, 105, 109, 110, 122

Daley, R., 199, 327, 391
Danabagoglu, G., 268
Daniel, J. W., 92, 150
Dantzig, G. B., 302
Davenport, W. B. Jr., 307
David, M., 310
Davis, P. J., 195, 280
Davis, R. E., 189, 291, 295, 308, 312, 320
De Szoeke, R. A., 62, 265
Defant, A., 41, 80, 83, 84, 87, 291
DeMey, P., 396
Denman, K. L., 31
Denning, P. J., 323
Derber, J., 391
Desnoyers, J. E., 261
Dickinson, R., 308
Dickson, R. R., 67, 72, 325
Dodge, Y., 305–306
Doney, S., 261
Dorman, C. E., 68
Dorrer, M., 76
Döscher, R., 52
Draper, N. R., 105, 186
Dreisigacker, E., 326
Dyrssen, D., 271

Eckart, C., 150
Ekman, V. W., 34
Emery, W. J., 24
Escudier, P., 76
Evensen, G., 398
Ezer, T., 395

Faller, A. J., 44
Fandry, C., 189
Farrell, B. F., 403
Feller, W., 101
Feshbach, H., 129–130, 175, 177
Festa, J. F., 42, 312
Fiacco, A. V., 299
Fiadeiro, M. E., 242
Fine, R. A., 326
Fischer, T. L., 323
Fischer-Bruns, I., 402
Flannery, B. P., 98
Fleming, M. W., 82
Flierl, G. R., 5
Fofonoff, N. P., 17, 38, 50
Franklin, G. F., 376, 383–384
Freeman, H., 191, 194
Fu, L.-L., 76, 241, 247, 253, 263, 268, 300, 349, 359, 382, 395
Fuglister, F. C., 29, 306
Fukumori, I., 28, 31, 59, 200, 246, 279, 294, 307, 309, 359, 382–383, 385, 395–396

Gandin, L. S., 199
Garcia, M., 262
Garçon, V. C., 285
Gardner, L. A. Jr., 389, 398
Gargett, A. E., 279, 282
Garrett, C., 261, 279
Gaspar, P., 329, 349, 359, 360
Gauthiez, F., 386
Gelatt, C. D., 323
Gelb, A., 327, 361, 398
Gent, P. R., 268
Ghil, M., 349, 381, 386, 391
Gilbert, J. F., 14
Gill, A. E., 17–18, 47, 70
Gill, P. E., 317, 319, 397
Gleick, J., 303n
Gmitrowicz, E. M., 67

Goldberg, D. E., 323
Goldstein, H., 373
Golub, G. H., 96, 98, 109, 309, 313
Goodwin, G. C. 361, 385, 388–389, 398
Gordon, A. R., 64
Goyet, C., 271
Grace, A., 317
Grant, B., 64, 69, 219, 253, 257–258
Gregg, M. C., 279
Grose, T. S., 295

Haidvogel, D., 359
Haines, K., 395
Hall, M. C. G., 303
Hall, M. M., 57, 263–267, 274
Halpern, D. E., 54
Hamann, I., 31, 307
Hamming, R. W., 193
Han, Y.-J., 221
Hansen, P. C., 168
Hanson, R. J., 98, 101, 109, 168, 299–300
Harrison, D. E., 54
Hasse, L., 52, 61, 67–68, 260
Hasselmann, K., 402
Hastenrath, S., 261
Hautala, S. L., 63
Hax, A. C., 302
Haykin, S., 99, 150, 388, 389
Heath, R. A., 268, 279, 319
Hecht, A., 253, 275, 283, 300
Heemink, A. W., 350
Helland-Hansen, B., 325
Hellerman, S., 47, 52, 61
Hidaka, K., 80, 82, 85, 87
Ho, Y.-C., 327, 336, 352, 361, 364, 372, 375
Hoerl, A. E., 168
Hogg, N. G., 279–280, 310
Holland, J. H., 323
Holland, W., 62, 392
Hollingsworth, A., 54
Holmes, P., 307
Horton, C., 49
Hsiung, J., 261
Hu, D.-X., 64, 69, 253
Hunkins, K., 45
Huxley, T., 13

Illari, L., 54
Isaacson, E., 349
Isemer, H.-J., 52, 61, 67–68, 260

Jackson, J. D., 11
Jackson, J. E., 307
Jeffreys, H. W., 41, 101
Jenkins, G. M., 117–118, 332
Jenkins, W. J., 65
Jerri, A. J., 191, 194
Johnson, M. W., 82, 295
Jolliffe, I. T., 307
Joseph, P. D., 349, 368
Joyce, T. M., 253, 256
Jung, G. H., 260, 263

Kalman, R. E., 343
Kamenkovich, V. M., 18
Kawase, M., 32, 45, 70, 72
Keeling, C. D., 284
Keffer, T., 247
Kendall, M. G., 246
Kennard, R. W., 168
Keto, E., 295
Killworth, P. D., 88, 291
Kipling, R., 5
Kirkpatrick, S., 323
Klema, V. C., 150
Kloosterhuis, H., 350
Korner, T. W., 13
Krakiwsky, E. J., 312
Krauss, W., 84

Lanczos, C., 9, 130, 150, 175–178
Landau, H. J., 194
Landau, L. D., 19
Large, D. G. B., 307
Large, W. G., 54, 260
Laub, A. J., 150, 382
Lavin, A., 262
Law, C. S., 282
Lawson, C. L., 98, 101, 109, 168, 299–300
Lazier, J. R. N., 69

Le Traon, P.-Y., 275, 309
Leaman, K., 55
LeDimet, F.-X., 376, 390
Ledwell, J. R., 282
Lee, A. J., 325
Lee, S. W., 221
Lee, T. N., 55
Leetmaa, A., 45, 47, 50, 52, 54, 57
Lefebvre, M., 76
Leslie, W., 391
Levine, M. D., 265
Levitus, S., 47, 239, 274, 286
Lewis, F. L., 24, 234
Li, Y.-H., 34
Liebelt, P. B., 99, 182, 327, 364
Lifschitz, E. M., 19
Ljung, L., 389
Long, R. B., 378, 390
Lorenc, A. C., 391
Luenberger, D. G., 202, 302–303, 316–317, 322, 332, 366–368, 397–398
Lumley, J. L., 307
Luyten, J. R., 47, 62

Macdonald, A., 253, 263
Mackas, D. L., 31, 307, 310–311
Macready, P., 279
Magnanti, T. L., 302
Magnus, J. R., 99
Malanotte-Rizzoli, P., 253, 349, 391–392
Malmberg, S.-A., 325
Marotzke, J., 277, 378, 390
Martel, F., 199, 256, 271, 275–277, 284–285, 320
Matear, R. J., 256, 275
Maury, M. F., 325
McCallister, T., 27, 253
McCormick, G. P., 299
McCreary, J. P. Jr., 51, 255
McDougall, T. G., 20, 250
McIntosh, P. C., 295
McWilliams, J. C., 72, 261, 268
Mechoso, C. R., 295
Meincke, J., 325
Mellor, G. L., 295, 395
Memery, L., 286, 370
Menard, Y., 76
Menke, W., 14
Mercier, H. (See Le Traon), 275, 277, 308–309, 317, 321
Middleton, D., 194
Millard, R., 262
Miller, R. N., 349, 359, 386–387
Millero, F. J., 237, 261
Millif, R. F., 72
Minster, J.-F., 128, 285, 305, 317
Model, F., 260
Molinari, R. L., 55, 221, 312
Mollo-Christensen, E. L., 45
Monin, A. S., 324
Montgomery, R. B., 261
Mooney, K., 279
Moore, A. M., 403
Moore, B., 285
Moore, D. W., 51, 70
Moore, J. B., 327, 351, 361, 382, 388
Moritz, H., 312
Morse, P. M., 129, 130, 175, 177
Müller, T. J., 64, 66
Munk, W., 73, 128, 171, 279, 281–282, 336
Murray, W., 317

Nansen, F., 325
Neduecker, H., 99
Needler, G. T., 44, 47, 89–91, 279, 290
Nelkin, H., 42
Niiler, P. P., 52, 59, 247
Noble, B., 92, 150
Numerical Algorithms Group, 317

O'Brien, J. J., 329, 390
O'Reilly, J., 398
Olbers, D., 70, 71, 88, 282, 289, 291, 293
Ollitrault, M., 275, 309
Olson, J. G., 54
Oort, A. H., 260, 274
Osborne, T. L., 282
Östlund, H. G., 42, 326

Outerbridge, R., 261
Owens, W. B., 67, 268

Paige, C. C., 210, 277
Parker, R. L., 14
Parrilla, G., 262, 274
Pastukhov, A. F., 295
Pedder, M. A., 191, 199
Pedlosky, J., 17–18, 36, 38, 40, 45, 47, 58, 62, 70, 89
Peixoto, J. P., 274
Peng, T.-H., 358
Perigaud, C., 349
Perkin, R. G., 24, 237
Perron, G., 261
Petersen, D. P., 194
Philander, S. G. H., 70
Phillips, N. A., 17
Pickard, G. L., 14, 24, 38, 237
Pickart, R. S., 66–67
Pierce, S. D., 253
Pillsbury, J. E., 47, 49, 76
Pincus, M., 323
Pollak, H. O., 194
Pollard, R. T., 256
Polonksy, I., 194
Polzin, K., 282
Pond, S., 14, 38, 237, 260
Powell, J. D., 376
Prada, K., 256
Preisendorfer, R. W., 307
Press, W. H., 98, 317, 323
Price, J. F., 45–46
Priestley, M. B., 193, 309
Provost, C., 242, 295

Rauch, H. E., 355
Raymer, M. R., 267
Redi, M. H., 250n
Reid, J. L., 27–28, 59, 69, 78, 83
Reid, W. T., 382
Rhines, P. B., 65, 279
Richards, P. G., 10, 14
Richardson, P. L., 279
Richardson, W. S., 55
Riley, G. A., 82, 87, 263
Rintoul, S. R., 249–254, 258, 262–263, 267–268, 271
Ripley, B. D., 198, 310, 323
Roache, P. J., 284
Roberts, S. M., 372
Robinson, A. R., 44, 63, 391, 396
Robinson, G., xi
Rockafellor, R. T., 303
Roemmich, D., 27, 29, 45, 52, 59, 63, 219–220, 247, 253, 255, 258–259, 262–263, 267, 274
Roether, W., 326
Rogers, G. S., 99
Root, W. L., 307
Rooth, C. G. H., 42, 326
Rosati, A., 391
Rosenstein, M., 47, 52, 61
Rossby, C. G., 20
Rossby, H. T., 42

Salmon, R., 242, 279, 295
Saltzman, B., 41
Sandström, J. W., 41
Sarkisiyan, A. S., 295
Sarmiento, J. L., 312, 393
Sasaki, Y., 132
Saunders, M. A., 210, 275
Saunders, P. M., 24
Scales, J. A., 323
Scales, L. E., 317, 397
Schleicher, K., 237
Schlitzer, R., 275, 286, 304–305, 320
Schmitt, R., 68, 274, 282
Schmitz, W. J. Jr., 55, 63
Schneider, E. D., 64
Schott, F. A., 55, 87, 288–292
Schröter, J., 174–175, 321–322
Schuldich, R. R., 45
Seber, G. A. F., 101, 105, 117, 173, 284, 312, 318
Semtner, A. J. Jr., 2
Sewell, M. J., 131
Sheinbaum, J., 378
Shepard, J., 305
Shipman, J. S., 372
Shoji, D., 287

Siedler, G., 64, 66
Sin, K. S., 361, 385, 388–389, 398
Smedstad, O. M., 390
Smith, H., 105, 186
Smith, M. L., 323
Sorenson, H. W., 361, 394
Speer, K., 30
Stammer, D., 3, 338, 395
Starr, V. P., 35, 297n
Stengel, R. F., 204, 336, 366, 382, 384–385, 390
Stens, R. L., 191
Stewart, G. W., 150
Stigebrandt, A., 274
Stommel, H., 3, 37, 40–42, 44, 47, 52, 57, 59, 62, 69, 73, 88, 271, 274, 288–292
Stone, R. E., 303
Strang, G., 92, 150, 299, 303
Striebel, C. T., 355
Stroup, E. D., 69
Stuart, A., 246
Sturges, W., 49
Sverdrup, H. U., 38, 49–51, 82–83
Swallow, J. C., 49, 67
Swift, J. H., 31, 307

Takahashi, T., 34
Talagrand, O., 376, 390
Talley, L., 253, 256, 261
Tarantola, A., 14, 211, 314–315, 317
Tatro, P. R., 45
Tavantzis, J., 349
Teukoslky, S. A., 98
Thacker, W. C., 378, 390
Thièbaux, H. J., 191, 199
Thompson, K., 261
Thompson, R. O. R. Y., 253
Thomson, W. (Lord Kelvin), 5
Thorpe, S. A., 279
Todling, R., 398
Tokmakian, R., 2
Tomczak, M., 31, 307
Toole, J. M., 267, 282
Tovey, C. A., 303
Trenberth, K., 54
Tung, F., 237, 355
Tziperman, E., 30, 253, 275, 279, 283, 300, 378–379, 390

Valette, B., 314–315, 317–318
Van Huffel, S., 308, 313–314
van Laarhoven, P. J. M., 323
Van Loan, F., 96, 98, 109, 313
Van Trees, H. L., 113, 122
Vandewalle, J., 308, 313–314
Vaniček, P., 312
Vazquez, J., 349
Vecchi, M. P., 323
Veronis, G., 17–18, 44, 242, 253, 295
Vetterling, W. T., 98
Viterbo, P., 54
Von Der Haar, T. H., 260
von Storch, H., 402–403

Wagner, H. M., 305
Wahba, G., 295, 307, 310
Wallace, J. M., 308, 310
Warren, B. A., 41, 67, 69
Watson, A. J., 67
Watts, D. R., 66–67
Weiss, R. E., 32
Weiss, W., 326
Welander, P., 44–45, 89
Weller, R. A., 45
Wendelberger, J., 312
Wentz, F., 54,
Wenzel, M., 69–70, 282, 293–294
Whittaker, E., xi
Wiener, N., 394
Wiggins, R. A., 135, 150
Wijffels, S. E., 274
Wild, C. J., 284, 318
Willebrand, J., 67–69, 88, 291, 293, 390
Willems, J. C., 382
Wilson, W. D., 55
Worcester, P., 336
Workman, M. L., 376
Worthington, L. V., 24, 33, 67, 76, 83–84
Wright, M. H., 317
Wunsch, C., 3, 29, 31, 52, 59, 62, 64,

Wunsch, C. (*cont.*), 69, 80, 88, 89, 128, 171, 174, 194, 199, 219, 239, 244, 246, 247, 249–259, 262–263, 265, 267–268, 271, 274–277, 279–281, 284, 286, 290, 303–307, 309, 317–318, 320, 322–323, 325, 329, 336–338, 349, 358–360, 370, 378, 390, 395
Wüst, G., 25, 26, 47, 49, 67, 78, 217, 306
Yaglom, A. M., 323
Yamarone, C. A. Jr., 76
Yih, C. S., 19
Young, G., 150

Zantopp, R., 55, 291

Subject Index

Abel's problem, 10–11
absolute velocity; *see* reference level velocity
abyssal circulation; *see* Stommel-Arons theory
acoustic Doppler technique, 256
adaptive estimation, 387–389
adjoint models and solutions (Pontryagin Principle), 130; *see also* Lagrange multipliers; Riccati equation
  energy principle for, 366, 402
  examples of, 173–175, 372
  Hamilton-Jacobi-Bellman equation, 367
  in ill-posed forward models, 399–401
  interpretation of, 130–131
  method of unit solutions, 372–374
  nonlinear problems in, 389–391
  and steady models, 170–177
  and singular value decomposition, 176
  smoothing, relation to, 378+
  solution uncertainty, 378
  Stommel model example, 174–175
  in time-dependent estimation, 363+
ageostrophic flow
  in Ekman layer, 35
aliasing, 29, 192; *see also* sampling theorems
  and eddies, 64
  in inverse models, 238
altimetry, 3
  assimilation of, 395
  in filtering and smoothing, 359–360
  and surface pressure, 36
  and surface slope, 76
assimilation, 207, 327, 391+; *see also* inverse methods; adjoint models and solutions (Pontryagin Principle); least-squares; Kalman filter; smoothers
  in meteorology, 6
  nudging, 392
  robust diagnostic method, 393
autocovariance and autocorrelation, 118
  and time-dependent models, 339+
autoregressive processes (AR), 332

badly-posed problems; *see* ill-posed problems
band-limited signals, 194
Bayesian methods, 14
Bernoulli function, 89
beta-spiral, 87, 288+
  and "P-equation," 290
  variants of, 291–294
bias; *see also* least-squares; singular value decomposition (SVD)
  of equation solutions, 143
  in singular value decomposition, 157
bottom wedge, problem of, 220–221
boundary conditions, inference of, 8, 9
  as control variables, 363

Boussinesq approximation, 18
box inverse method, 85–88
  examples, 221
  finite-difference versions, 274–278
  large box examples, 249+
  nonlinear method, 315
box models (geochemical), 285

carbon flux, 271–272
characteristic function, 110
chi-square distribution, 111
  in least-squares, 122–123
Cholesky decomposition, 109, 126
circulation diagrams, 59–61, 81, 84, 258
classical problem (of oceanography); *see* reference level velocity
climatologies, 284+
  influence on solutions, 288
collocation; *see* Gauss-Markov estimation
colored-noise problem, 336
companion matrix, 332
conservation equations, 19, 23, 215
conservation laws (dynamic), 88–91
constraints: strong, weak, hard, soft, 132
control theory; *see also* adjoint models and solutions (Pontryagin Principle); smoothers
  feed-back control, 376
  open-loop control, 370
controllability, 383–384
  and Riccati equation, 385
controls, 339, 362+
convection, 41, 69
convolution, 110
correlation and covariation, 102; *see also* statistics, concepts
  coefficient of, 104
  and predictable power, 105
  and prediction, 105
  in property fluxes, 246–248
  in residuals, *see* least-squares
current meter records, 66
current meters
  in inverse models, 256

deep western boundary currents (DWBC), 44, 64–66
Defant's method, *see* reference level velocity
diagnostic methods, 295
direct models/problems; *see* forward problems
Dirichlet problem, 7–9, 11
dual, duality
  as adjoint, 172
  of control and Kalman filter, 381
  in linear programming, 303
dynamic method, 6, 74–76

eastern boundary currents, theory of, 73
Eckart-Young-Mirsky theorem, 308, 403
eclectic methods/models, 255–257, 303–305
eddy coefficients, 34–35
  and inequality constraints, 297
  in inverse models, 216
eddy field
  discovery of, 63
  effect on heat flux, 268–270
  measurements of, 63–65
eigenvectors and eigenvalues, 136; *see also* singular value decomposition (SVD)
  in simultaneous equations, 140–141
  vanishing eigenvalues, 137
Ekman depth, 35
Ekman layers, 31, 34–38
  divergence of (pumping), 36
  errors in property fluxes, 243–244
  and Goldsbrough circulation, 37
  inverse models use in, 215, 239
  mass fluxes, 221
  observations of, 45–48
  pumping from, 46–49
    and Sverdrup relation, 52
  theory of 31, 34-38
  time variation of, 48
  transport, 36
empirical orthogonal functions (EOFs), 306+

equation of state; *see* seawater, equation of state
equatorial properties, 32, 33
errors in variables; *see* total least-squares
errors, in circulation models, 236–239
estimation; *see also* Gauss-Markov estimation; least-squares; weighted and tapered least-squares; singular value decomposition (SVD); Kalman filter, smoothers, suboptimal estimation methods
  summary discussion of, 209–211
evaporation and precipitation, 68
extended Kalman filter; *see* Kalman filter

finite-differences; *see* models
Florida Straits; *see* Gulf Stream
fluid dynamics, equations of, 17–23; *see also* Ekman layer, Sverdrup relation; Stommel-Arons theory
  Boussinesq approximation, 18
  conservation equations, 19
  on sphere, 17–20
fluorocarbons (CFCs), 32
  sections 32–33
forward problems, 7, 13, 329
Freon; *see* fluorocarbons
freshwater, flux of, 272–274
functional analysis, 14

gauge transformations, 11
Gauss-Markov estimation,
  basic theorem, 180–183
  basis functions in, 186
  bias in, 189
  collocation, 312
  covariance specification for, 185
  estimates of circulation, 224–226
  least-squares, relation to, 185
  and mapmaking, 190–200
  mapmaking in higher dimensions, 199–200
  mean determination, 188–189;
  and simultaneous equations, 183–186
  singular matrices in, 198
Gaussian, *see* normal distribution (Gaussian)
general solution (to simultaneous equations), 139
generalized inverse, 95
geophysical fluid dynamics, 3, 5–6
geostrophic inverse method; *see* box inverse method
geostrophy, 6, 20, 36; *see also* thermal wind
  in inverse models, 215+
  observational tests of, 47, 49–50
  supposed failure in Worthington's results, 78, 83–87
Goldsbrough circulation, 37
Gram-Schmidt process, 101–101
Green's functions, 177, 333
  in adjoint/Pontryagin principle, 374
  and initial condition estimation, 338
  and over-specified problems, 178–179
Gulf Stream, 30, 256
  Florida Straits (Current), observations, 49, 55, 76
  fluxes in, 250
  in inverse models, 218–219, 221+
  theories of, 58–59
gyres, subpolar and subtropical, 46

Hamiltonian function, 366, 369
Hamiltonian matrix, 373
heat flux
  in atmospheric boundary layer, 261
  budget for, 233–236
  bulk formulas, 260
  calculations of, 260+;
  decomposition of, 267
  in ocean, 261+;
  relation to Sverdrup balance, 57–58
  transfer to atmosphere, 67–68
Hessian, 98, 176, 316
  and adjoint/Pontryagin Principle, 379–380
  and singular value decomposition, 176
Hidaka's method, *see* reference level velocity

Huxley, Thomas, 13

idempotent property, 143, 151
ill-posed problems, 10–11
  forward models, 400–403
inequality constraints, 297+
information filter; *see* Kalman filter
initial value estimation, 376+; *see also* smoothers, adjoint models and solutions (Pontryagin Principle)
interpolation, 195–197; *see also* Gauss-Markov theorem; objective mapping
  Aitken-Lagrange formula, 195, 197
  from model grid, 337
inverse (matrix), 15, 95; *see also* generalized inverse
inverse methods, 7, 9
  alleged failure of, 295–296
  nonlinear; *see* total least-squares; total inversion
inverse problems, 7, 9, 13; *see also* total inversion
  Abel's problem, 10–11
  in geophysics, 13–14
  mathematical, 12
iterative solution methods, 389–391

Kalman filter, 341+
  alternate forms, 351
  in assimilation, 393–396
  in control problem, 365
  extended and linearized, 386–387;
  information filter, 351
  square-root filter, 351
  in RTS smoother, 354–355
  reduction to static case, 350–351
Kalman-Bucy filter, 349
Karhunen-Loève theorem; *see* empirical orthogonal functions (EOFs)
Karmackar algorithm, 303
Kelvin waves, 70, 72
Kelvin, Lord (Sir William Thomson), 5, 13n
Kipling, Rudyard, 5n
Krige, David; *see* kriging
kriging, 311–312; *see also* least-squares
Kuhn-Tucker-Karush theorem, 299–300
Kuroshio, 287

laboratory experiments, 42, 44
Lagrange multipliers, 129–132; *see also* adjoint models and solutions (Pontryagin Principle)
  in control problems, 362+
Lagrangian, 175–176, 367
least-squares, 114–119; *see also* weighted and tapered least-squares; singular value decomposition; total least-squares basic formulation
  correlated residuals, 118
  and kriging, 311–312;
  nonnegative, 300, 311
  residuals in, 115, 117
  standard error in solutions, 116
  uncertainty of solutions, 116, 117
level-of-no-motion, 50; *see also* thermal wind
linear programming, 301+
linearization of forward models, 402

mapmaking; *see* Gauss-Markov estimation; objective mapping; interpolation
matrix
  basic operations and definitions, 92–97
  companion, 332;
  completing the square, 99–100
  eigenvectors and eigenvalues, 98
  Hessian, definition, 98
  identities, 97–100
  Jacobian, 98
  matrix inversion lemma, 99
  norm, 100
  observation, 336
matrix inverse; *see* inverse (matrix)
matrix inversion lemma, 99
maximum likelihood, 112–113, 122n
mean, determination of; *see* statistics, concepts; Gauss-Markov estimation
Mediterranean salt-tongue, 298

mesoscale eddies; *see* eddy field
meteorology, 5, 6
microstructure, 4
mixing in ocean, 215
  canonical values (Munk), 282
  determination in inverse models, 279+
  tensor, 250n
models,
  errors in, 116, 236–239
  finite-difference inverse, 274+
  ill-posed forward, 398–401
  of ocean circulation, 215
  time-dependent, 329+
  "validation" and "verification" of, 341n
Moore-Penrose inverse, 131

Needler's formula; *see* reference level velocity
Neumann problem, 9, 140
  solution by singular value decomposition, 154–156
neutral surfaces, 250n
nitrate,
  section, 32
nonlinear inverse problems, 312+; 320+; *see also* total least-squares; total inversion; optimization, nonlinear
  combinatorial methods, 323
  and Stommel model, 320–323
  time-dependent, 385+
nonnegative least-squares, 300, 311
norm,
  Frobenius, 100
  matrix, 100
  1-norm, 305
  2-norms, 95, 295
  semi-norm, 312
  vector, 95–96
normal distribution (Gaussian), 105
  multivariable normal distribution, 106–109
normal equations; *see* least-squares; weighted and tapered least-squares; objective functions
nudging; *see* assimilation
nullspaces; *see* ranges and nullspaces
nutrients,
  flux of, 253–254, 270–272

objective functions, 115; *see also* Lagrange multipliers
  and adjoint/Pontryagin Principle, 363, 365–366
  expected value of, 122
  in forward models, 400
  and Kalman filter, 343
  nonlinear time-dependent, 389–390
  and smoother, 353
  for terminal constraint problem, 368
  for time-dependent models, 340
  in underdetermined system, 128–129, 131
objective mapping, 200–207; *see also* Gauss-Markov theorem
  of derivatives, 200, 202
observability, 384–385
  and Riccati equation, 385
observers; *see* suboptimal methods
oceanography, general, 4
  technology influence on, 22
optimal modes; *see* Eckart-Young-Mirsky theorem
optimization, nonlinear, 317, 397–398
  and adjoint/Pontryagin Principle, 390–391
  and P-equation, 318–320
overdetermined problems
  example, 152–153
  and SVD, 156
overflows (North Atlantic), 67, 72
oxygen, 25, 34, 250
  apparent oxygen utilization rate (AOUR), determination of, 128
  flux of, 253

parallel fields, 290
parameter estimation; *see* adaptive estimation
particular solution (simultaneous equations), 138
  in inverse model, 240

phosphate, 34
"PO," 250
Pontryagin Principle; *see* adjoint models and solutions (Pontryagin Principle)
potential density, 20
precipitation; *see* evaporation and precipitation
predictor-corrector, 205
pressure-depth relation; *see* seawater
principal components; *see* empirical orthogonal functions (EOFs)
principal oscillation patterns (POPs), 403
property fluxes, 242; *see also* heat flux
  Ekman component of, 243
  spatial resolution requirement for, 267
property-property diagrams; *see* water masses

radiocarbon
  inverse calculation, 281–282
  modeled, 128
ranges and nullspaces
  and inequality constraints, 300–301
  in property flux computations, 265–267
  in simultaneous equations, 138, 140
  in singular value decomposition; *see* singular value decomposition (SVD)
rank
  deficiency, 156
  effective, 158
recursive methods, 202–209
  estimation by, 208–209;
  and Kalman filter, 341–343
  for least-squares, 203–204
  limiting cases of, 205
  objective mapping as a recursive method, 206
  in smoothing, 361
reference depth, 22
reference level,
  changing in model, 240
  initial choice of, 217
reference level velocity, 22, 76–82; *see also* box inverse method; beta-spiral
  and the classical problem of oceanography, 73, 212
  Defant's method, 80–81
  estimates from inverse models, 222+
  Hidaka's method, 80–82
  Needler's formula, 88–91
regression, 105; *see also* correlation and covariation; statistics, concepts
  and kriging, 311–312
  ridge, 168
relative velocity, 22
residuals; *see* least-squares; simultaneous equations
resolution (in inversions)
  in box model inversions, 257–260
  compact resolution, 135
  matrix, 134
  in singular value decomposition, 165–166
resolution (spatial), 8, 267–269
Riccati equation
  algebraic, 382
  control Riccati equation, 374–376
  history of, 382
  and observability/controllability, 385
  solution by sweep method, 375
  uncertainty equation, 381
ridge regression; *see* regression
robust diagnostic method; *see* assimilation
Rossby waves, 328, 331
  and tomography, 337
row-scaling; *see* weighted and tapered least-squares
RTS smoother; *see* smoothers

salinity, 26; *see also* seawater
  fluxes from inverse models, 229+
  Practical scale, 24n
  sections, 26
salt; *see* salinity
sampling theorems, 191–195; *see also* aliasing

scaling, row and column; *see* least-squares; singular value decomposition
seawater; *see also* oxygen, salinity; temperature, time changes in
  density of, 23, 36
  equation of state, 21
  pressure-depth relation, 24
  salinity scale, 24n
sequential estimation; *see* Kalman filter; smoothers
Shannon-Whittaker theorem; *see* sampling theorems
silicate, 30, 31
  sections, 30
Simplex method, 302
  and eclectic modeling, 303–304
simultaneous equations, 12, 96–97; *see also* underdetemined systems
  formally just-determined, 139
  and least-squares, 115–117
  residuals in, 139
  and singular vector expansion, 135
singular value decomposition (SVD); *see also* range and nullspaces; rank
  and adjoint, 176
  complex SVD, 309–310
  and constraints, 299–300
  derivation, 145–148
  and empirical orthogonal functions, 307–308
  generalized, 168
  history of, 150
  and inequality constraints, 300
  inverse methods, use in, 150
  matrix decomposition, 148–150
  ocean circulation estimates from, 226–229
  particular solution, 241
  relation to least-squares, 156–159
  row and column scaling in, 159–165;
  resolution of solutions, 165–166
  relation to tapered and weighted least-squares, 166–169
  symmetric matrices, relation to, 144–145
  and total least-squares, 313
  uniquely determined components, 241
singular vector expansion, 133–144; *see also* singular value decomposition (SVD)
  resolution matrix for, 134
smoothers, 344, 346+, 352+
solvability conditions, 138, 139
  in SVD, 147
spanning sets, 93
spectra, 193–194
square-root filter; *see* Kalman filter
state vector, 323+
state-reduction; *see* suboptimal methods
stationarity, 112
statistics, concepts; *see also* correlation and covariation, normal distribution (Gaussian), chi-square distribution; stationarity; maximum likelihood
  bias, definition, 103;
  bias in sample variance, 104
  Central Limit Theorem, 105
  conditional probability, 102
  degrees of freedom, 111–112
  dispersion, 103
  mean, 101
  moments, defined, 108
  probability density, 102
  random variables, functions of, 109–110
  sample mean, 101
  sums of random variables, 110–111
  variance, 103
stochastic estimation, 389
Stommel-Arons theory, 41–44
  observations of, 65, 67–72
suboptimal estimation methods, 397–398
SVD; *see* singular value decomposition
Sverdrup balance
  and bottom topography, 62
  and Ekman layer, 40
  and geostrophic interior, 40

Sverdrup balance (*cont.*)
observations, 50–52; tests of, 52–63
theory of, 38–40
sweep method; *see* Riccati equation

tapered least-squares; *see* weighted and tapered least-squares
Taylor-Proudman theorem, 21
temperature sections 27–30, 214
temperature, time changes in, 29, 262, 274
terminal constraint control problem, 368+
thermal wind, 6, 22, 49, 219
in beta-spiral, 289–290
errors in, 244
and tomography, 171
thermocline theory, 44, 47
tomography, 334, 337
total inversion, 314; *see also* optimization, nonlinear
total least-squares, 313–314
tracers, 79–80; *see also* tritium; carbon flux
in eclectic modeling, 305
and inequality constraints, 297–299
in smoother, 356–359
transient, 325–326, 373
transient tracers; *see* tracers
tritium, 41–43, 65, 326; *see also* tracers
and filtering, smoothing, 358
turbulent flow, 324

underdetermined systems, 127–133, 153–154
Gauss-Markov solutions with basis functions, 187
underparameterized models, 127–128
upwelling, 41, 44, 47
and Stommel-Arons theory, 69–71

variogram; *see* kriging
vectors, basic definitions; *see also* matrix
Gram-Schmidt process, 100–101
norm, 95–96
velocity sections, 49, 64, 214, 227, 241, 252
vertical velocity; *see* upwelling
vorticity balance, 22

water masses, 30, 33, 34
definition, 30;
and empirical orthogonal functions, 307–311
names of, 34
relation to reference levels, 217–219
weighted and tapered least-squares, 118–127; *see also* singular value decomposition (SVD)
normal equations in, 124
resolution in, 169–170
row-scaling in, 121
trade-off parameter, 124
white noise, definition, 116
whole-domain form (estimation), 335
Wiener filters and smoothers, 362, 394–396
wind-stress
and drag coefficients, 52
numerical values of, 36
observations, 46–47, 53–54